Applied Measurement Systems

Applied Measurement Systems

Editor:

Harinirina Randrianarisoa, Ph.D.

ArclerPress

www.arclerpress.com

Applied Measurement Systems

Editor: Harinirina Randrianarisoa, Ph.D.

© 2017 Arcler Press LLC
708 3rd Avenue, 6th Floor
New York
NY 10017
United States of America
www.arclerpress.com

ISBN: 978-1-68094-368-9

Library of Congress Control Number: 2016951045

About the Editor

Harinirina Randrianarisoa, Ph.D.

Hari Randrianarisoa obtained is Ph.D from Institute of Electronics, Microelectronics and Nanotechnology (IEMN) of the University of Science and Technology – Lille (France). His thesis focused for designing and manufacturing of thermal micro sensors. His interests span from on renewable energy especially solar energy. He has a strong knowledge in technology transfer field. He is currently working as a Business advisor for economic development organization in Québec City and as a consultant Scientist advisor.

Contents

List of Contributors

Rodrigo Munguía
University of Guadalajara, Guadalajara, Jalisco, Mexico

Emmanuel Nuño
University of Guadalajara, Guadalajara, Jalisco, Mexico

Carlos I. Aldana
University of Guadalajara, Guadalajara, Jalisco, Mexico

Sarquis Urzua
University of Guadalajara, Guadalajara, Jalisco, Mexico

Runxia Guo
College of Electronics Information and Automation, Civil Aviation University of China, Tianjin 300300, China

Kai Guo
College of Electronics Information and Automation, Civil Aviation University of China, Tianjin 300300, China

Quan Gan
College of Electronics Information and Automation, Civil Aviation University of China, Tianjin 300300, China

Junwei Zhang
College of Electronics Information and Automation, Civil Aviation University of China, Tianjin 300300, China

Jiankang Dong
College of Electronics Information and Automation, Civil Aviation University of China, Tianjin 300300, China

Lanping Bai
Process Engineering Department, Freescale Semiconductor (China) Limited, Tianjin 300300, China

Chuan Li
School of Mechanical Engineering, Dongguan University of Technology, Dongguan 523808, China

René-Vinicio Sánchez
Department of Mechanical Engineering, Universidad Politécnica Salesiana, Cuenca 010105, Ecuador

Grover Zurita
Department of Mechanical Engineering, Universidad Politécnica Salesiana, Cuenca 010105, Ecuador

Mariela Cerrada
Department of Mechanical Engineering, Universidad Politécnica Salesiana, Cuenca 010105, Ecuador

Diego Cabrera
Department of Mechanical Engineering, Universidad Politécnica Salesiana, Cuenca 010105, Ecuador

Joseph M. Sherwood
Department of Bioengineering, Faculty of Engineering, Imperial College London, London, United Kingdom,

Ester Reina-Torres
Department of Bioengineering, Faculty of Engineering, Imperial College London, London, United Kingdom,

Jacques A. Bertrand
Department of Bioengineering, Faculty of Engineering, Imperial College London, London, United Kingdom,

Barnaby Rowe
Department of Physics and Astronomy, Faculty of Mathematical and Physical Sciences, University College London, London, United Kingdom

Darryl R. Overby
Department of Bioengineering, Faculty of Engineering, Imperial College London, London, United Kingdom,

Johan Nohlert
Department of Signals and Systems, Chalmers University of Technology, SE-41296 Göteborg, Sweden

Livia Cerullo
Department of Signals and Systems, Chalmers University of Technology, SE-41296 Göteborg, Sweden

Johan Winges
Department of Signals and Systems, Chalmers University of Technology, SE-41296 Göteborg, Sweden

Thomas Rylander
Department of Signals and Systems, Chalmers University of Technology, SE-41296 Göteborg, Sweden

Tomas McKelvey
Department of Signals and Systems, Chalmers University of Technology, SE-41296 Göteborg, Sweden

Anders Holmgren
AstraZeneca R&D, Pharmaceutical Development, SE-431 83 Mölndal, Sweden

Lubomir Gradinarsky
AstraZeneca R&D, Pharmaceutical Development, SE-431 83 Mölndal, Sweden

Staffan Folestad
AstraZeneca R&D, Pharmaceutical Development, SE-431 83 Mölndal, Sweden

Mats Viberg
Department of Signals and Systems, Chalmers University of Technology, SE-41296 Göteborg, Sweden

Anders Rasmuson
Department of Chemical and Biological Engineering, Chalmers University of Technology, SE-41296 Göteborg, Sweden

L. Cordova
Heat and Vacuum Laboratory of the Physikalisch-Technische Bundesanstalt, Germany

T. Lederer
Heat and Vacuum Laboratory of the Physikalisch-Technische Bundesanstalt, Germany

M. Liukkonen
Department of Environmental Science, University of Eastern Finland, P.O. Box 1627, Kuopio FIN-70211, Finland.

T. Hiltunen
Foster Wheeler Power Group, P.O. Box 201, Varkaus FIN-78201, Finland

Juan Hea
School of Physics & Electronics Science, Fuyang Normal College, Fuyang, 236037, People's Republic of China

Shuai Xu
Institute of Plasma Physics, Chinese Academy of Sciences, Hefei, 230031, People's Republic of China

Liu Ye
School of Physics & Material Science, Anhui University, Hefei, 230601, People's Republic of China

Christine Charles
Space Plasma, Power and Propulsion Laboratory, Research School of Physics and Engineering, The Australian National University, Canberra, ACT, Australia

RoderickW.Boswell
Space Plasma, Power and Propulsion Laboratory, Research School of Physics and Engineering, The Australian National University, Canberra, ACT, Australia

AndrewBish
Space Plasma, Power and Propulsion Laboratory, Research School of Physics and Engineering, The Australian National University, Canberra, ACT, Australia

VadimKhayms
Lockheed Martin Space Systems Company, Sunnyvale, CA, USA

Edwin F.Scholz
Lockheed Martin Space Systems Company, Littleton, CO, USA

VyacheslavI.Yukalov
Department of Management, Technology and Economics, ETH Zürich, Zürich, Switzerland
Bogolubov Laboratory of Theoretical Physics, Joint Institute for Nuclear Research, Dubna, Russia

DidierSornette
Department of Management, Technology and Economics, ETH Zürich, Zürich, Switzerland
Swiss Finance Institute, University of Geneva, Geneva, Switzerland

Rudolf Thalmann
Federal Institute of Metrology METAS, 3003 Bern-Wabern, Switzerland

Felix Meli
Federal Institute of Metrology METAS, 3003 Bern-Wabern, Switzerland

Alain Küng
Federal Institute of Metrology METAS, 3003 Bern-Wabern, Switzerland

Adam Gąska
Laboratory of Coordinate Metrology, Cracow University of Technology, Kraków 31-155, Poland

Piotr Gąska
Laboratory of Coordinate Metrology, Cracow University of Technology, Kraków 31-155, Poland

Maciej Gruza
Laboratory of Coordinate Metrology, Cracow University of Technology, Kraków 31-155, Poland

Heng-Yi Su
Department of Electrical Engineering, Feng Chia University (FCU), No.100, Wenhwa Road, Seatwen, Taichung 40724, Taiwan

Yi-Chung Chen
Department of Information Engineering and Computer Science, Feng Chia University (FCU), No.100, Wenhwa Road, Seatwen, Taichung 40724, Taiwan

Yu-Liang Hsu
Department of Automatic Control Engineering, Feng Chia University (FCU), No.100, Wenhwa Road, Seatwen, Taichung 40724, Taiwan

So Ito
Department of Finemechanics, Tohoku University, Sendai 980-8579, Japan

Hirotaka Kikuchi
Department of Finemechanics, Tohoku University, Sendai 980-8579, Japan

Yuanliu Chen
Department of Finemechanics, Tohoku University, Sendai 980-8579, Japan

Yuki Shimizu
Department of Finemechanics, Tohoku University, Sendai 980-8579, Japan

Wei Gao
Department of Finemechanics, Tohoku University, Sendai 980-8579, Japan

Kazuhiko Takahashi
Engineering Department, MMC RYOTEC Corporation, Gifu 503-2301, Japan

Toshihiko Kanayama
Engineering Department, MMC RYOTEC Corporation, Gifu 503-2301, Japan

Kunmei Arakawa
Engineering Department, MMC RYOTEC Corporation, Gifu 503-2301, Japan

Atsushi Hayashi
Engineering Department, MMC RYOTEC Corporation, Gifu 503-2301, Japan

Nathalie Godin
INSA de Lyon, MATEIS (UMR CNRS 5510), 7 avenue Jean Capelle, 69621 Villeurbanne Cedex, France

Pascal Reynaud
INSA de Lyon, MATEIS (UMR CNRS 5510), 7 avenue Jean Capelle, 69621 Villeurbanne Cedex, France

Mohamed R'Mili
INSA de Lyon, MATEIS (UMR CNRS 5510), 7 avenue Jean Capelle, 69621 Villeurbanne Cedex, France

Gilbert Fantozzi
INSA de Lyon, MATEIS (UMR CNRS 5510), 7 avenue Jean Capelle, 69621 Villeurbanne Cedex, France

Yoshinari Takao
Department of Applied Chemistry, Graduate School of Engineering, Kyushu University, 744, Motooka, Nishi-ku, Fukuoka 819-0395, Japan

Tomoko Imasaka
Laboratory of Chemistry, Graduate School of Design, Kyushu University, 4-9-1, Shiobaru, Minami-ku, Fukuoka 815-8540, Japan

Yuichiro Kida
Department of Applied Chemistry, Graduate School of Engineering, Kyushu University, 744, Motooka, Nishi-ku, Fukuoka 819-0395, Japan

Totaro Imasaka
Department of Applied Chemistry, Graduate School of Engineering, Kyushu University, 744, Motooka, Nishi-ku, Fukuoka 819-0395, Japan
Division of Optoelectronics and Photonics, Center for Future Chemistry, Kyushu University, 744, Motooka, Nishi-ku, Fukuoka 819-0395, Japan

Sergio Aguado
Design and Manufacturing Engineering Department, Universidad de Zaragoza, Zaragoza 50018, Spain

Jorge Santolaria
Design and Manufacturing Engineering Department, Universidad de Zaragoza, Zaragoza 50018, Spain

David Samper
Design and Manufacturing Engineering Department, Universidad de Zaragoza, Zaragoza 50018, Spain

Jesús Velázquez
Design and Manufacturing Engineering Department, Universidad de Zaragoza, Zaragoza 50018, Spain

Carlos Javierre
Mechanical Engineering Department, Universidad de Zaragoza, Zaragoza 50018, Spain

Ángel Fernández
Mechanical Engineering Department, Universidad de Zaragoza, Zaragoza 50018, Spain

Rosa Weigand
Departamento de Óptica, Facultad de Ciencias Físicas, Universidad Complutense de Madrid, Avda. Complutense s/n, Madrid 28040, Spain

Helder M. Crespo
Departamento de Física e Astronomia e IFIMUP-IN, Faculdade de Ciências, Universidade do Porto, R. do Campo Alegre 687, Porto 4169-007, Portugal

Preface

The construction of an appropriate performance measurement system is an important objective that every company is facing today. Measurement is a multidisciplinary experimental science. This book discussed the synergy between science and engineering. With regards of Applied Measurement Systems, this book illustrated the diversity of measurement systems, and provided in-depth guidance for specific practical problems and applications. Also discusses new methods to provide fundamental data for research, design and development, control of processes and operations, and facilitate safe and economic performance of systems. This study permits to understand which are the decisive factors in a measurement system. This literature could help too engineers and technical managers responsible for timely delivery of clearly valid test data in testing laboratories or whose product quality organizations depend on these tests. That summarized various applications in different field. This book is suitable for scientists and engineers and for practicing measurement.

Editor
Harinirina Randrianarisoa, Ph.D.

CHAPTER 1

A VISUAL-AIDED INERTIAL NAVIGATION AND MAPPING SYSTEM

Rodrigo Munguía, Emmanuel Nuño, Carlos I. Aldana and Sarquis Urzua

University of Guadalajara, Guadalajara, Jalisco, Mexico

ABSTRACT

State estimation is a fundamental necessity for any application involving autonomous robots. This paper describes a visual-aided inertial navigation and mapping system for application to autonomous robots. The system, which relies on Kalman filtering, is designed to fuse the measurements obtained from a monocular camera, an inertial measurement unit (IMU) and a position sensor (GPS). The estimated state consists of the full state of the vehicle: the position, orientation, their first derivatives and the parameter errors of the inertial sensors (i.e., the bias of gyroscopes and accelerometers). The system also provides the spatial locations of the visual features observed by the camera. The proposed scheme was designed by considering the limited resources commonly available in small mobile robots, while it is intended to be applied to cluttered environments in order to perform fully vision-based navigation in periods where the position sensor is not available. Moreover, the estimated map of visual features would be suitable for multiple tasks: i) terrain analysis; ii) three-dimensional (3D) scene reconstruction; iii) localization, detection or perception of obstacles and generating trajectories to navigate around these obstacles; and iv) autonomous exploration. In this work, simulations and experiments with real data are presented in order to validate and demonstrate the performance of the proposal.

INTRODUCTION

In recent years, many researchers have addressed the issue of making vehicles more and more autonomous. In this context, the state estimation of the six degrees of freedom (DOFs) of the vehicle (i.e., its attitude and position) is a fundamental necessity for any application involving autonomy.

Even though, this problem is seemingly solved for open outdoor domains by an on-board Global Positioning System (GPS) unit and inertial measurements units (IMU), as well an integrated version: an inertial navigation system (INS). Contrary, unknown, cluttered and GPS-denied environments still pose a considerable challenge. While attitude estimation is well handled by available systems, GPS-based position estimation can have some drawbacks. GPS is not a reliable service as its availability can be limited by urban canyons and it is completely unavailable in indoor environments. The aforementioned issues have prompted recent studies to move towards the use of cameras for performing visual-based

Citation: Rodrigo Munguía, Emmanuel Nuño, Carlos I. Aldana and Sarquis Urzua. A Visual-aided Inertial Navigation and Mapping System. Int J Adv Robot Syst, 2016, 13:94. doi: 10.5772/64011

navigation in periods or circumstances where the position sensor (i.e., GPS) is not available. Cameras are well adapted for embedded systems: they are light, cheap and power-saving.

Related Work

Several strategies related to visual-based navigation can be found in the literature. Some systems that use vision for navigation require previous knowledge of the whole environment. In this case, some kind of guided or automated pretraining is needed in order to learn visual patterns in the environment, which are recognized and used later as visual clues at the navigation stage. Examples of guided pretraining methods are [1] and [2]. In [3, 4], during the navigation, the position is determined matching the online image with previously recorded images. Usually, the application of this kind of method is limited to environments where the training phase takes place.

Another kind of system has been designed to perceive the environment as it is navigated through. Behaviour-based navigation systems can rely on qualitative image information (examples are [5, 6] and [7]). In this kind of system, it is avoided as much as possible using, computing or generating accurate numerical data, such as distances, position coordinates, velocity projections from the image plane onto the real world plane, or contact time to the obstacles. Due to its critical dependence on unprocessed sensorial data, this kind of method is highly dependent on the change in the imaging conditions.

Approaches, such as [8] or [9], rely on optical flow information, as do [10] and [11], although the optical flow is combined with stereo information in their case. A related approach to the optical flow scheme is the tracking of visual features. In this case, however, instead of estimating the visual flow of the whole image (i.e., every pixel), a set of strong visual features, such as corners, lines and objects outlines, are detected and tracked over time for as much as it may be possible. Examples of methods based on the former approach are [12] and [13], in which the computation of the camera motion is obtained from a homography. In [14], inertial data obtained from an IMU is combined with visual tracking information.

Some systems can build a map of the environment while exploring it. The maps can have a topological representation, as in [15] and [16], or a metric representation. Occupancy grid-based strategies represent one of the early approaches to representing metric maps; some examples are [17] and [18] However, occupancy grid-based strategies can be computationally inefficient for path planning and localization, especially in complex and great indoor environments.

Another family of approaches relies on simultaneous localization and mapping (SLAM) strategies. In this case, the system must operate in an a priori unknown environment using only on-board sensors to simultaneously build a map of its surroundings and localize itself [19]. Usually, this kind of system also relies on the tracking of a set of prominent visual features. As these visual features typically have a spatial meaning, the information obtained from the visual measurements can be used to replace range sensors: the lasers are often expensive and heavy, while the operation range of ultrasonics is limited. In this sense, the appearance and spatial information provided by cameras can be useful for multiple tasks, such as: i) terrain analysis; ii) 3D scene reconstruction; iii) localization, detection or perception of obstacles and generating trajectories to navigate around these obstacles; and for iv) autonomous exploration.

Currently, there are two main approaches to implementing vision-based SLAM systems: i) filtering-based methods (see [20-22] and [23]), and ii) optimization-based methods (see [24] and [25]). While the former approach is shown to give accurate results when the availability of computational power is enough, filtering-based SLAM methods might still be beneficial if limited processing power is available [26]. While the benefit obtained from constructing a map of the robot's environment is self-evident, it comes together with a considerable increase in computational requirements. The former issue is due to the fact that, in visual SLAM, the visual features are included in the state vector. When the number of map features in the system state increases, computational cost grows rapidly, such that it consequently becomes difficult to maintain real-time performance. The above becomes more evident when limited hardware is available. On the other hand, in a context of visual odometry, this problem can be partially mitigated by removing old features from the state in order to stabilize the computational cost per frame [27].

In applications involving mobile robots, it is very common to have a set of sensors available for use. Incorporating complementary or redundant data from multiple sensors can improve the robustness and accuracy of the system. In this sense, sensor fusion that uses, for instance, IMU, GPS and cameras is not novel. Some examples of multi-sensor navigation algorithms are [28-31] and [32].

Objective

This work presents a novel method for implementing a visual odometry and mapping system. The method is mainly intended to be applied to autonomous mobile robots with limited computational resources. The proposed scheme is related, on the one hand, to visual odometry methods, which make use of the optical flow obtained from prominent visual features in order to compute the camera movement. On the other hand, it is related to visual SLAM methods in the sense that a map of the environment is built and used for refining the estimations. Some contributions and differences from previous work are:

- In contrast with visual SLAM approaches, the system state in the proposed scheme is never augmented with the state of map features. Instead, each map feature is managed independently from the state of the vehicle, as well as from other map features. The above technique, which perhaps represents the main contribution of this work, allows a reduction in the computational requirements.
- A novel technique based on two-step architecture is proposed in order to initialize and measure the visual features in a robust and efficient manner. The proposed scheme makes use of the information obtained from monocular measurements and allows advantage to be taken of the information provided by visual features, whose depth estimation is not well conditioned.
- The use of an extended Kalman filter (EKF) in direct configuration allows a straightforward integration with additional sensors, such as inertial sensors (i.e.,

gyroscopes and accelerometers), and GPS, when it is available.

It is important to note that the proposed system should not be considered as a true SLAM system; rather, and just in case, it should be considered as an approximation of it. This is because the cross-correlation information of the feature uncertainty, which is contained in the system covariance matrix, is discarded on behalf of computational cost reduction. Fortunately, the experimental results would suggest that the proposed scheme is useful when applied to a context of visual-based navigation.

Paper Outline

The paper is organized as follows. Section 2 presents the system parameterization and assumptions. Section 3 describes the proposed system in detail. In Section 4, experimental results, carried out using simulations and real data, are presented in order to validate and demonstrate the performance of the proposal. Finally, Section 5 presents conclusions and final remarks.

SYSTEM PARAMETRIZATION AND ASSUMPTIONS

The main goal of the proposed method is to estimate the following system state x

$$x = \left[q^{NR}, \omega^{R}, r^{N}, v^{N}, a^{N}, \chi_{g}, \chi_{a}\right]^{T} \quad (1)$$

where $q^{NR} = [q_1, q_2, q_3, q_4]$ represents the orientation of the vehicle with respect to the world (navigation) frame by a unit quaternion. $r^N = [x_v, y_v, z_v]$ Represents the position of the vehicle (robot) expressed in the navigation frame. $\omega^R = [\omega_x, \omega_y, \omega_z]$ is the bias-compensated velocity rotation of the robot expressed in the same frame of reference. $v^N = [v_x, v_y, v_z]$ denotes the linear velocity of the robot expressed in the navigation frame. $a^N = [a_x, a_y, a_z]$ represents the bias-compensated acceleration of the robot expressed in the navigation frame. $\chi_g = [\chi_{g_x}, \chi_{g_y}, \chi_{g_z}]$ is the bias of the gyroscopes. $\chi_a = [\chi_{a_x}, \chi_{a_y}, \chi_{a_z}]$ is the bias of the accelerometers.

The proposed method is intended for local autonomous vehicle navigation. In this case, the local tangent frame is used as the navigation reference frame and, in turn, the initial position of the vehicle defines its origin. The navigation coordinate frame follows the NED (north-east-down) convention. In this work, the magnitudes expressed in the i) navigation frame, ii) vehicle (robot) frame and iii) camera frame are respectively denoted by the superscripts ^{N}, ^{R} and C (see Figure 1). All the coordinate systems are right-handed. In this work, it is assumed that the coordinate frame of the IMU is aligned with the coordinate frame of the vehicle ^{R}.

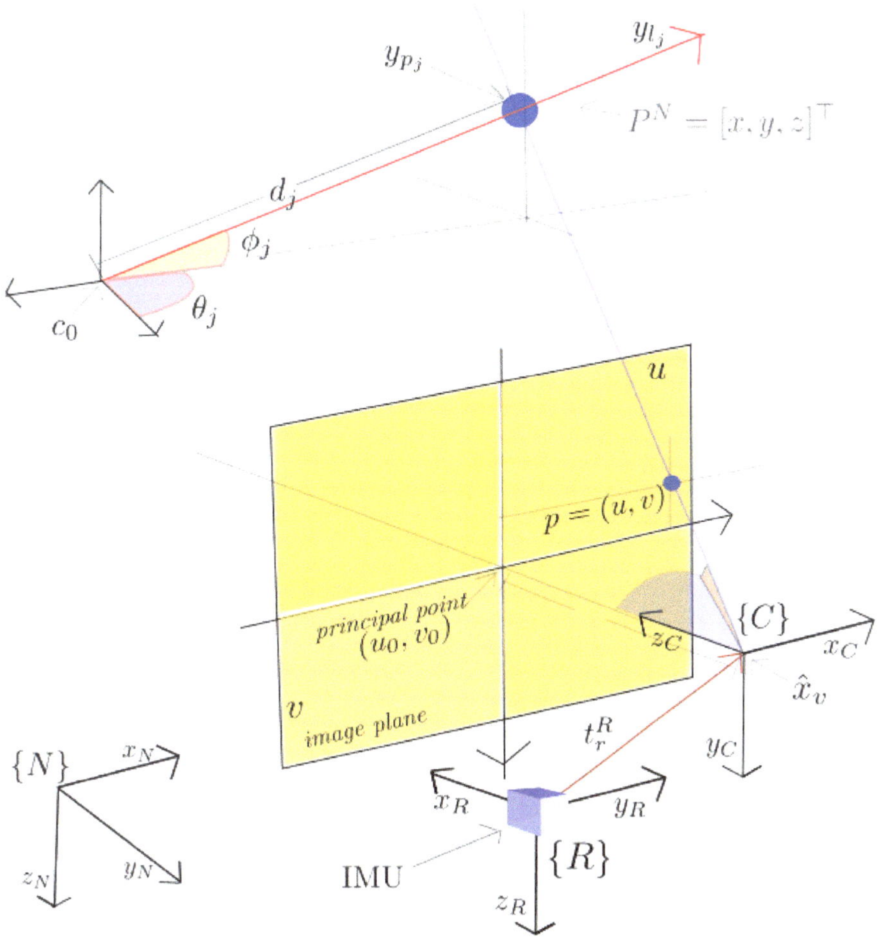

Figure 1: System parametrization. The local tangent frame is used as the navigation reference frame ^{N}. Furthermore, it is assumed that the coordinate frame of the IMU is aligned with the origin of the coordinate frame of the robot ^{R}.

The position and orientation of the camera, in respect of the vehicle coordinate frame ^{R}, is assumed to be known and fixed: R^{CR} is the camera to robot rotation matrix, while t_c^R is the position of the optical center of the camera, expressed in the vehicle frame. The superscript AB denotes a reference frame ^{B} expressed in respect of the reference ^{A}.

Visual Measurements

A standard monocular camera is considered (see Figure 1). In this case, the availability, at each k frame, of a list of $m_{(k)}=[z_{uv(1)},z_{uv(2)},...,z_{uv(n)}]$ visual measurements of n static landmarks of the environment is assumed. Visual measurements z_{uv} are modelled by:

$$z_{uv(j)} = y_{uv(j)} + v_{uv} \tag{2}$$

where $y_{uv(j)}=[u_{d_j},v_{d_j}]$ represents the true position (in the image plane) of the projection of a j landmark. u_{dj} and v_{dj} are distorted pixel coordinates. The term v_{uv} represents the uncertainty associated with visual measurements and is modelled by a Gaussian white noise with power spectral density (PSD) σ_{uv}^2. In this work, the list of visual measurements $m_{(k)}$ is obtained using the Kanade-Lucas-Tomasi (KLT) tracker [33], but any other small baseline tracker could be used.

A central projection camera model is assumed. In this case, the image plane is located in front of the camera's origin, upon which a non-inverted image is formed (Figure 1.) The camera frame C is right-handed with the z-axis pointing to the field of view.

The $\mathbb{R}^3 \Rightarrow \mathbb{R}^2$ projection of a 3D point, located at $p^N=(x,y,z)^T$ to the image plane p =(u,v), is defined by:

$$u = \frac{x'}{z'} \quad v = \frac{y'}{z'} \tag{3}$$

where u and v are the coordinates of the image point p expressed in pixels units and:

$$\begin{bmatrix} x' \\ y' \\ z' \end{bmatrix} = \begin{bmatrix} f & 0 & u_0 \\ 0 & f & v_0 \\ 0 & 0 & 1 \end{bmatrix} p^C \tag{4}$$

being p C the same 3D point p^N, but expressed in the camera frame ^{C} by $p^C=R^{NC}(p^N-t_c^N)$, and where $R^{NC}=(R^{RN}R^{CR})^T$ is the rotation matrix transforming from navigation frame ^{N} to the camera frame C. R^{CR} is known and R RN is computed from the current robot quaternion q^{NR}. $t_c^N=r^N+R^{RN}t_c^R$ Is the position of the camera's optical center as expressed in the navigation frame?

Inversely, a directional vector $h^C=[h_x^C,h_y^C,h_z^C]^T$ can be computed from the image point coordinates u and v.

$$h^C(u,v) = \left[\frac{u_0-u}{f}, \frac{v_0-v}{f}, 1\right]^T \tag{5}$$

The vector h^C points from the camera's optical center position to the 3D point location. h^C can be expressed in the navigation frame by $h^N=R^{CN}h^C$, where $R^{CN}=R^{RN}R^{CR}$ is the camera to navigation rotation matrix. Note that for the $\mathbb{R}^2 \Rightarrow \mathbb{R}^3$ mapping case, as defined in Equation 5, depth information is lost.

The distortion caused by the camera lens is considered in relation to the model described in [34]. Using the former model (and its inverse form), undistorted pixel coordinates (u,v) can be obtained from (u_d, v_d), as well as conversely. In this case, it is assumed that the intrinsic parameters of the camera are already known: (i) focal length f, (ii) principal point u_0, v_0, and (iii) radial lens distortion $k_1,...,k_n$.

The angular rate ω^R of the vehicle, measured by the gyroscopes (in the same frame ^{R}) and indicated as y_g, can be modelled by:

$$y_g = \omega^R + x_g + v_g \tag{6}$$

where x_g is the gyroscope bias and v_g is a Gaussian white noise with a PSD σ_g^2.

The acceleration of the vehicle a^R, measured by the accelerometers (in the same frame ^{R}) as y_a, can be modelled by:

$$y_a = a^R + g^R + x_a + v_a \tag{7}$$

where g^R is the gravity vector expressed in the vehicle frame R, x_a is the accelerometer bias and v_a is a Gaussian white noise with PSD σ_a^2.

GPS Measurements

Position and course measurements (y_r and y_θ, respectively) are obtained from the GPS, whenever possible, and are modelled by:

$$y_r = r^N + v_r$$
$$y_\theta = \theta^N + v_\theta \tag{8}$$

where v_r is a Gaussian white noise with PSD σ_r^2. θ^N is the course angle of the vehicle, measured by the GPS unit with respect to the geographic north, and v_θ is a Gaussian white noise with PSD σ_θ^2. Commonly, position measurements are obtained from GPS devices in geodetic coordinates (latitude, longitude and height). Therefore, in Equation 8, it is assumed that GPS position measurements have been previously transformed into their corresponding local tangent frame coordinates. It is also assumed that the offset between the GPS antenna and the vehicle frame was taken into account in the previous transformation.

METHOD DESCRIPTION

The architecture of the system is defined by the typical loop of prediction-update steps in the EKF in direct configuration, in which the EKF propagates the vehicle state. Whenever possible, the filter is updated by the filter update equations, in which visual information is obtained from the monocular camera as well as the data obtained from the additional sensors. The estimation of the features' map location is carried out, in an independent manner, by the main EKF process.

An initial period of time is used for system initialization tasks. During this period, the vehicle is assumed to be nonaccelerating. The initial system state x_{ini} is determined as follows:

$$\hat{x}_{ini} = \left[q_{ini}^{NR}, 0_{1\times3}, 0_{1\times3}, 0_{1\times3}, 0_{1\times3}, x_{g(ini)}, 0_{1\times3} \right]^T \quad (9)$$

where q_{ini}^{NR} and $x_{g(ini)}$ are estimated in same manner as in [35]. In this case, magnetometers are only used for obtaining an initial estimation of the heading (yaw). The average of the initial GPS lectures is subsequently used as a constant for transforming from geodetic to local tangent coordinates.

System Prediction

At every step k, when gyroscope and accelerometer measurements are available, the predicted system state $\hat{x}$ is taken a step forward by the following (discrete) non-linear model:

$$\begin{cases} q_{k|k-1}^{NR} = \left(\cos\|w_k\| I_{4\times4} + \dfrac{\sin\|w_k\|}{\|w_k\|} W_k \right) q_{k-1|k-1}^{NR} \\ \omega_{k|k-1}^{R} = y_{g_k} - x_{g_{k-1|k-1}} \\ r_{k|k-1}^{N} = r_{k-1|k-1}^{N} + v_{k-1|k-1}^{N}\Delta t + a_{k-1|k-1}^{N}\dfrac{\Delta t^2}{2} \\ v_{k|k-1}^{N} = v_{k-1|k-1}^{N} + a_{k-1|k-1}^{N}\Delta t \\ a_{k|k-1}^{N} = R_k^{RN}\left(y_{a_k} - x_{a_{k-1|k-1}} \right) - g^N \\ x_{g_{k|k-1}} = (I - \lambda_{xg}\Delta t) x_{g_{k-1|k-1}} \\ x_{a_{k|k-1}} = (I - \lambda_{xa}\Delta t) x_{a_{k-1|k-1}} \end{cases} \quad (10)$$

where R_k^{RN} is computed from the quaternion $q_{k-1|k-1}^{NR}$. Parameters λ_{xg} and λ_{xa} are correlation time factors, which respectively model how fast the bias of the gyroscopes and accelerometers varies. g^N is the gravity vector expressed in the navigation frame. Δt is the sample time of the system. In the model defined in Equation 10, a closed form solution of $\dot{q} = 1/2(W)q$ is used for integrating the velocity rotation ω R over the quaternion q^{NR}. In this case, $w_k = [\omega_{k|k-1}^{R}\Delta t/2]^T$ and:

$$W_k = \begin{bmatrix} 0 & -w_1 & -w_2 & -w_3 \\ w_1 & 0 & -w_3 & w_2 \\ w_2 & w_3 & 0 & -w_1 \\ w_3 & -w_2 & w_1 & 0 \end{bmatrix} \quad (11)$$

The state covariance matrix P is taken a step forward by:

$$P_{k|k-1} = \nabla F_x P_{k-1|k-1} \nabla F_x^T + \nabla F_u U \nabla F_u^T \quad (12)$$

The Jacobian ∇F_x is formed by the partial derivatives of the non-linear prediction model (Equation 10) with respect to the system state $\hat{x}$. The Jacobian ∇F_u is formed by the partial derivatives of the non-linear prediction model with respect to the system inputs y_g and y_a

The measurement noise of the gyroscopes and accelerometers (respectively, v_g and v_a) are incorporated into the system by means of the process noise covariance matrix U through parameters σ_g^2 and σ_a^2:

$$U = \mathrm{diag}\left[\sigma_g^2 I_{3\times3}, \sigma_a^2 I_{3\times3} \right] \quad (13)$$

Attitude and Position Updates

In order to correct and limit drift in the estimated system state $\hat{x}$, referenced information, along with assumptions about the system dynamics, are incorporated into the filter, whenever possible, through the EKF update equations:

$$\begin{cases} \hat{x}_{k|k} = \hat{x}_{k|k-1} + K_k(z_i - h_i) \\ P_{k|k} = P_{k|k-1} - K_k S_k K_k^T \\ K_k = P_{k|k-1}\nabla H_i^T S_k^{-1} \\ S_k = \nabla H_i P_{k|k-1} \nabla H_i^T + R_i \end{cases} \quad (14)$$

where z_i is the current measurement and $h_i = h(\hat{x})$ is the predicted measurement. K is the Kalman gain. S is the innovation covariance matrix ∇H_i is the Jacobian formed by the partial derivatives of the measurement prediction model h ($\hat{x}$) with respect to the state $\hat{x}$. R_i is the measurement noise covariance matrix.

Each filter update is carried out using Equation 14, along with z_i, h_i, ∇H_i and R_i, which are determined according to each kind of i measurement. The following filter updates are considered:

Roll and pitch Measurement Model

If the vehicle is not accelerating (i.e., $a^R \approx 0$), then Equation 7 can be approximated as $y_a \approx g^R + v_a$ (x_a is neglected). In this situation, accelerometer measurements y_a provide noisy observations of the gravity vector (in the robot frame). The gravity vector is used as an external reference for correcting roll and pitch estimations. In order to detect the time (corresponding to k instants) when the vehicle is in a nonaccelerating mode, the detector described in [36] is used.

The gravity vector g is predicted to be measured by the accelerometers as h_g:

$$h_g = R^{NR}\left[0, 0, g_c\right]^T \quad (15)$$

where g_c is the gravity constant and R^{NR} is computed using the current quaternion q^{NR}. For updating the filter: $z_i = y_a$,

$$h_i = h_g,\ R_i = I_{3\times3}\sigma_a^2 \text{ and } \nabla H_i = \frac{\partial h_g}{\partial x}.$$

Position and heading measurement model

In this work, a loosely coupled approach is used for incorporating data provided by the GPS unit, whenever it is possible. In a loosely coupled approach, the high level outputs provided by the GPS unit are directly incorporated into the system state through their corresponding measurement prediction model.

The model h_r, which is used for predicting GPS measurements about the vehicle position, is simply:

$$h_r = r^N \quad (16)$$

The course (heading) of the vehicle is predicted by the following measurement model h_γ:

$$h_\gamma = \mathrm{atan2}\left(2(q_2q_3 - q_1q_4), 1 - 2(q_3^2 + q_4^2)\right) \quad (17)$$

where $q^{NR}=[q_1,q_2,q_3,q_4]$ is the current quaternion. atan2 is a two-argument function that computes the arc tangent of a / b, given as a and b, within the range $[-\pi,\pi]$. Alternatively, magnetometers could also be used in order to incorporate information into the system about the heading, as ex- plained in [35]. For updating the filter: $z_i=[y_r,y_\theta]^T$, $h_i=[h_r,h_y]^T$, $R_i=\text{diag}[I_{3\times3}\sigma_r^2,\sigma_\theta^2]$ and $\nabla H_i=[\frac{\partial h_r}{\partial_r x},\frac{\partial h_y}{\partial_r x}]^T$

Dynamical constraints

Prior knowledge about the vehicle dynamics can be incorporated into the system through filter updates. For instance, in the case of a car-like vehicle, nonholonomic constraints are incorporated in the form of zero velocity updates. For this purpose, it can be assumed that the velocity of the vehicle, in its own reference frame, should be measured as $v^R=[v_x^R,0+v_v,0+v_v]$, where v_v is a Gaussian white noise with a PSD σ_v^2 . The measurement prediction model h_{v_0} is simply

$$h_{v_0}=\left[v_y^R,v_z^R\right]^T \tag{18}$$

where $[v_x^R,v_y^R,v_z^R]^T=R^{NR}v^N$. For updating the filter: $z_i=[0,0]^T$, $h_i=h_{v_0}$, $R_i=I_{2\times2}\sigma_v^2$ and $\nabla H_i=\frac{\partial h_{v_0}}{\partial x}$.

Visual Aid

Initialization of Visual Features

At the first instant k, when a new visual feature z_{uv} is detected, the following entry f_l is stored:

$$f_i=\left[(t_{c_0}^N)^T,\theta_0,\phi_0,d\right] \tag{19}$$

where $y_{l_j}=[t_c^N,\theta,\phi]$ models a 3D semi-line, defined on one side by the vertex t_c^N, corresponding to the current optical centre coordinates of the camera, as expressed in the navigation frame, while pointing to infinity on the other side with azimuth and elevation, θ and φ, respectively d indicates the depth of the feature, which is initialized with a NaN (not a number) value, and:

$$\theta=\text{atan2}(h_y^N,h_x^N)$$

$$\phi=\text{acos}\left(\frac{h_z^N}{\sqrt{(h_x^N)^2+(h_y^N)^2+(h_z^N)^2}}\right) \tag{20}$$

where $h^N=[h_x^N,h_y^N,h_z^N]^T$ is computed as indicated in Section 2.1.

Applying Epipolar Constraints

Depth information cannot be obtained in a single measurement when bearing sensors (i.e., monocular camera) are used. To infer the depth of a feature, the sensor must observe it repeatedly as it moves through its environment, estimating the angle from the feature to the sensor centre. The difference between those angle measurements is the parallax, which is the key that allows estimating feature depth. Nevertheless, features with unknown depth or with huge uncertainty in depth can still provide useful information, such as for very far away landmarks, which will never produce a parallax, but can improve camera pose estimates. The following technique, which is based on epipolar constraints, is used for incorporating information from visual features with unknown or ill-conditioned depth. Epipolar constraints have been successfully used in the past for improving estimates in visual-based navigation systems (e.g., [37]). In this work, a version of the constraint is derived by considering the particularities of the proposed system.

If the camera moves from the location at which a feature is initialized, then this feature, which is first modelled as a 3D semi line y_{l_j}, can be projected onto the current image plane. The above projection, l_r, is the two-dimensional line, expressed in the image plane, while defined by the epipole e_r and the point, x_r (see Figure 2).

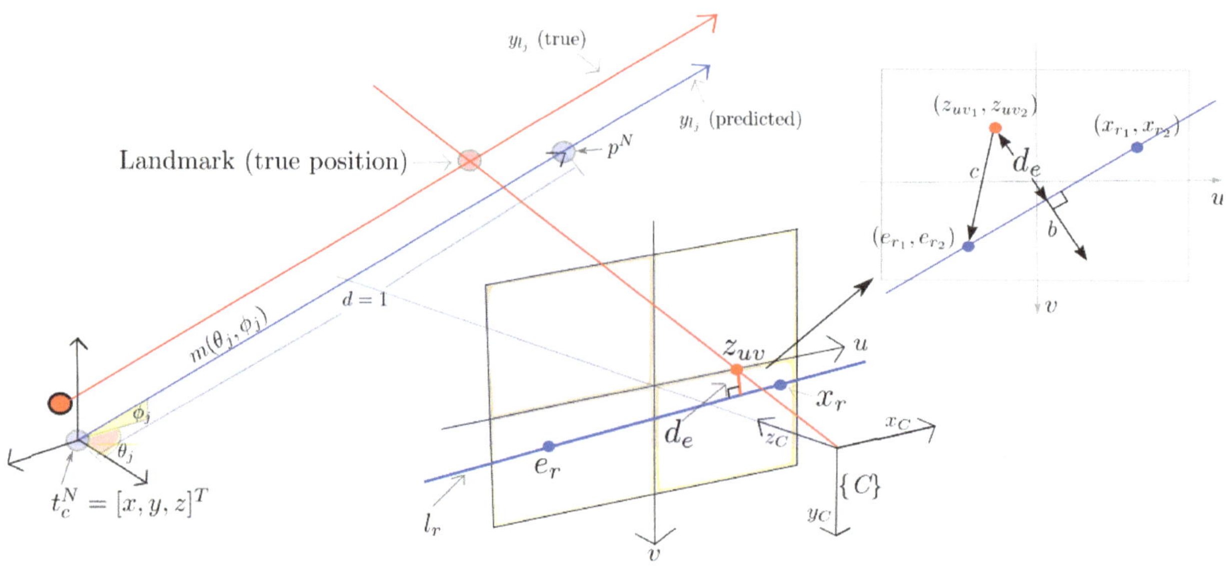

Figure 2: Model measurement for visual features with ill-conditioned depth

The epipole e_r is computed by projecting $t_{c_0}^N$, which is stored in f_l, onto the current image plane by Equations 3 and 4. The point x_r is computed by projecting the 3D point p N, as defined by the data stored in f l, through Equations 3 and 4, while assuming a depth equal to one (d =1). In this case, p^N models a 3D point located at:

$$p^N = t_c^N + d\left(m\left(\theta_j, \phi_j\right)\right) \tag{21}$$

where $m(\theta_j, \varphi_j)$ is a directional unitary vector defined by: $m(\theta_j, \varphi_j) = (\cos\theta\sin\varphi, \sin\theta\sin\varphi, \cos\varphi)^T$. The epipolar constraint implies that new undistorted measurements of the visual feature z_{uv} should lie over the line l_r. Therefore, d_e, which is the distance from the current measurement z_{uv} to the line l_r, is used as innovation error (measurementprediction) in order to update the filter. The distance de can be computed as follows (see Figure 2, upper right): If the line l_r is specified by two points, $e_r = (e_{r_1}, e_{r_2})$ and $x_r = (x_{r_1}, x_{r_2})$, then a vector perpendicular to the line is given by $b = [(x_{r_2} - e_{r_2}), -(x_{r_1} - e_{r_1})]^T$. Let c be a vector from the point $z_{uv} = (z_{uv_1}, z_{uv_2})$ to the first point on the line. $c = [(e_{r_1} - z_{uv_1}), (e_{r_2} - z_{uv_2})]^T$. The distance from z_{uv} to the line l_r is then given by projecting c onto b, resulting in:

$$d_e = \left\|b \cdot c\right\| = \frac{\left|(x_{r_2} - e_{r_2})(e_{r_1} - z_{uv_1}) - (x_{r_1} - e_{r_1})(e_{r_2} - z_{uv_2})\right|}{\sqrt{\left(x_{r_1} - e_{r_1}\right)^2 + \left(x_{r_2} - e_{r_2}\right)^2}} \tag{22}$$

Assuming that, for the current frame, n visual measurements are available for features with ill-conditioned depth, the filter is updated as follows: $z_i = [0_1, 0_2, \ldots, 0_n]^T$, $R_i = (I_{2n \times 2n})\sigma_{uv}^2$ and $\nabla H_i = \left[\frac{\partial h_{uv_1}}{\partial \hat{x}}, \frac{\partial h_{uv_2}}{\partial \hat{x}}, \ldots, \frac{\partial h_{uv_n}}{\partial \hat{x}}\right]^T$. For simplicity, the measurement noise is modelled as a Gaussian white noise with PSD σ_e^2.

Estimating Feature Depth

For minimizing computational cost, and unlike typical filtered-based SLAM methods, where the vector state is augmented in order to incorporate the map features, the map in the proposed method is maintained in an independent manner of the state of the vehicle; hence, the state of the features cannot be updated directly through the filter update equations when new measurements are available. Instead, when a new measurement z_{uv} is available (and along the entire life of the feature), the following procedure is applied for estimating the feature depth:

First, a hypothesis of depth d_j is computed by:

$$d_j = \frac{\left\|e_l\right\| \sin\gamma}{\sin\alpha} \tag{23}$$

where $\alpha_j = \pi - (\beta + \gamma)$ is the parallax. $e_l = t_{c_0}^N - t_c^N$ Indicates the displacement of the camera, since it was first observed to its current position, and:

$$\beta = \cos^{-1}\left(\frac{h_1 \cdot e_l}{\left\|h_1\right\|\left\|e_l\right\|}\right) \quad \gamma = \cos^{-1}\left(\frac{h_2 \cdot -e_l}{\left\|h_2\right\|\left\|e_l\right\|}\right) \tag{24}$$

where β is the angle defined by h_1 and e_l. h_1 is the normalized directional vector $m(\theta_j, \varphi_j)$, which is computed by taking θ_j, φ_j from f_l.

γ is the angle defined by h_2 and $-e_l$. $h_2 = h^N$, which is the directional vector pointing from the current camera optical centre to the feature location, is computed as indicated in Section 2.1. The current estimation of depth d is computed by passing the hypothesis d_j through a low-pass filter. This is due to the fact that the depth estimated by triangulation varies considerably, especially for a low parallax. Finally, f_l is updated with the new value of d.

3D map management

As previously stated before, the parallax produced by sensor movements permits the estimation of the depth of the features. In the case of an indoor sequence, a displacement of centimeters is enough to produce a parallax; on the other hand, the more distant the feature, the more the sensor has to travel to produce a parallax.

In [38], it is shown that only a few degrees of parallax are enough to reduce the uncertainty related to depth estimation. Then, when the parallax of a j feature is greater than a threshold ($\alpha_j > \alpha_{min}$), it is assumed that its depth is sufficiently well-conditioned in order to consider the feature as component of the 3D map. From this moment, it is assumed that the data, stored in f_l, model a 3D point $y_{p_j} = [t_c^N, \theta, \phi, d]$, with the location determined by Equation 21. Therefore, instead of updating the filter with epipolar constraints, the following prediction model $h_{uv} = h(\hat{x})$ is used.

First, the Euclidean representation p^N of y_{pj} is computed using Equation 21. Then the undistorted pixel coordinates of the feature (u,v) are found using the central projection camera model defined by Equations 3 and 4. Finally, the distorted pixel coordinates $h_{uv} = (u_d, v_d)$ are obtained from (u,v) by applying the corresponding distortion model.

Assuming that, for the current frame, n visual measurements are available for 3D features, the filter is updated as follows: $z_i = [z_{uv_1}, z_{uv_2}, \ldots, z_{uv_n}]^T$, $h_i = [h_{uv_1}, h_{uv_2}, \ldots, h_{uv_n}]^T$, $R_i = (I_{2n \times 2n})\sigma_{uv}^2$ and $\nabla H_i = \left[\frac{\partial h_{uv_1}}{\partial \hat{x}}, \frac{\partial h_{uv_2}}{\partial \hat{x}}, \ldots, \frac{\partial h_{uv_n}}{\partial \hat{x}}\right]^T$.

EXPERIMENTAL RESULTS

Experiments, using both synthetic data obtained by simulations and real data, were performed in order to validate the performance of the proposed method. A MATLAB implementation was used for this purpose

Experiments with Simulations

In experiments, a mobile robot was simulated moving through environments composed by 3D points, which can be used by the system as visual landmarks. The simulated vehicle is equipped with: i) inertial sensors, ii) a position sensor, and iii) a monocular camera. The position sensor is used only at the beginning of the trajectory for recovering the metric scale of the environment, after which it is turned off for the purpose of validating the performance of the method in which visual-based navigation is used.

The following parameters values were used in simulations: i) for emulating inertial sensors, σ_a =.002m/s^2, σ_g =.05/s and initial bias x_a , x_g , randomly determined with σ_{x_a} = 4d-4m/s^2 and σ_{x_g} =.25/s, respectively (these values were taken from the data sheet of a commercial degree IMU); ii) for emulating GPS measurements, σ_r =.5; and iii) for emulating the precision of the visual features tracking method, σ_{uv} = 1 pixel.

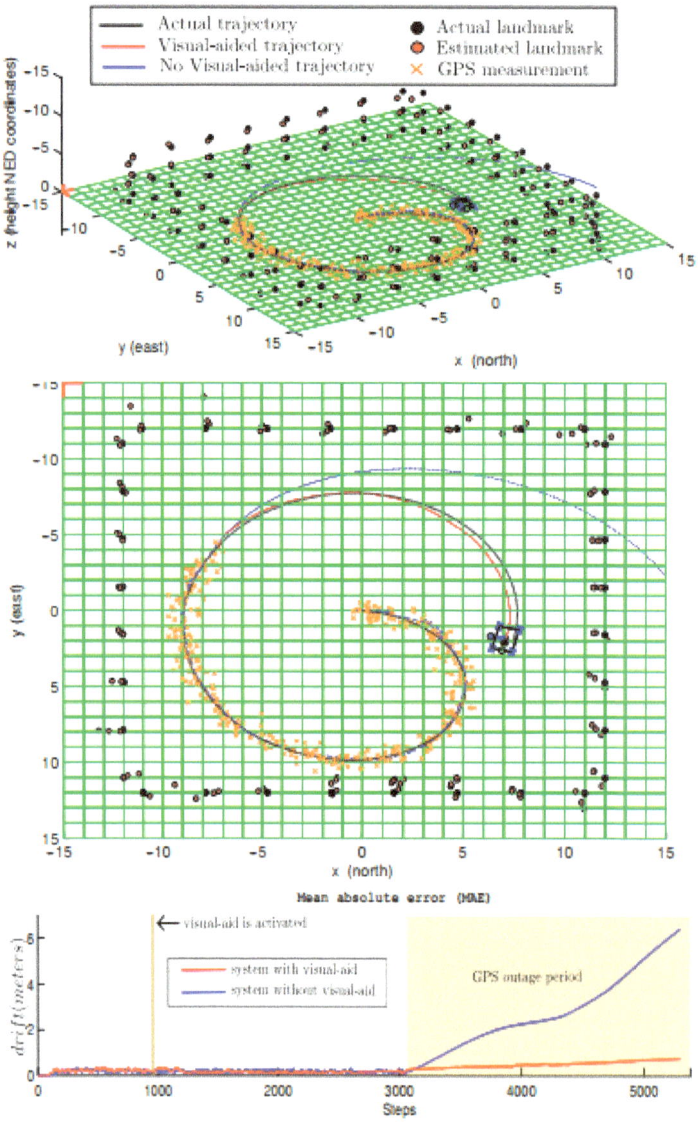

Figure 3: A Monte Carlo simulation was carried out in order to validate the improvement obtained when visual information is incorporated into the system through the proposed scheme. The upper and middle plots give an example of the trajectories obtained with and without visual information. The lower plot shows the average mean absolute error (MAE) in the position obtained after 20 runs. The experimental results clearly show that position's drift is considerable lower when the system makes use of the visual information.

Visual aiding vs. no visual aiding

Figure 3 (upper plot) shows the simulated 3D environment, which was used to perform this experiment. In this case, the land vehicle was moved in order to follow a spiral-like trajectory (see middle plot). The area, in which the vehicle is moving, is surrounded by a wall of landmarks that can be used by the system as visual features. As expected, the drift in estimations is considerable lower when visual information is incorporated into the system. Figure 3 (lower plot) shows the average MAE in the position obtained after 20 Monte Carlo runs of the simulation.

Proposed method vs. standard EKF-SLAM

Figure 4 shows two of the simulated environments used in this experiment. In the first series of experiments (a), the robot was moved in order to follow a spiral-like trajectory. The area, where the vehicle is moving is surrounded by a wall of landmarks. In these experiments, the camera was fixed, pointing toward to the axis of movement of the robot. In the second series of experiments (b), the robot was moved, following a wave-like trajectory. The environment, where the vehicle is moving, is composed by clouds of landmarks located in random positions. In these experiments, the camera was fixed, pointing perpendicular to the axis of movement of the robot.

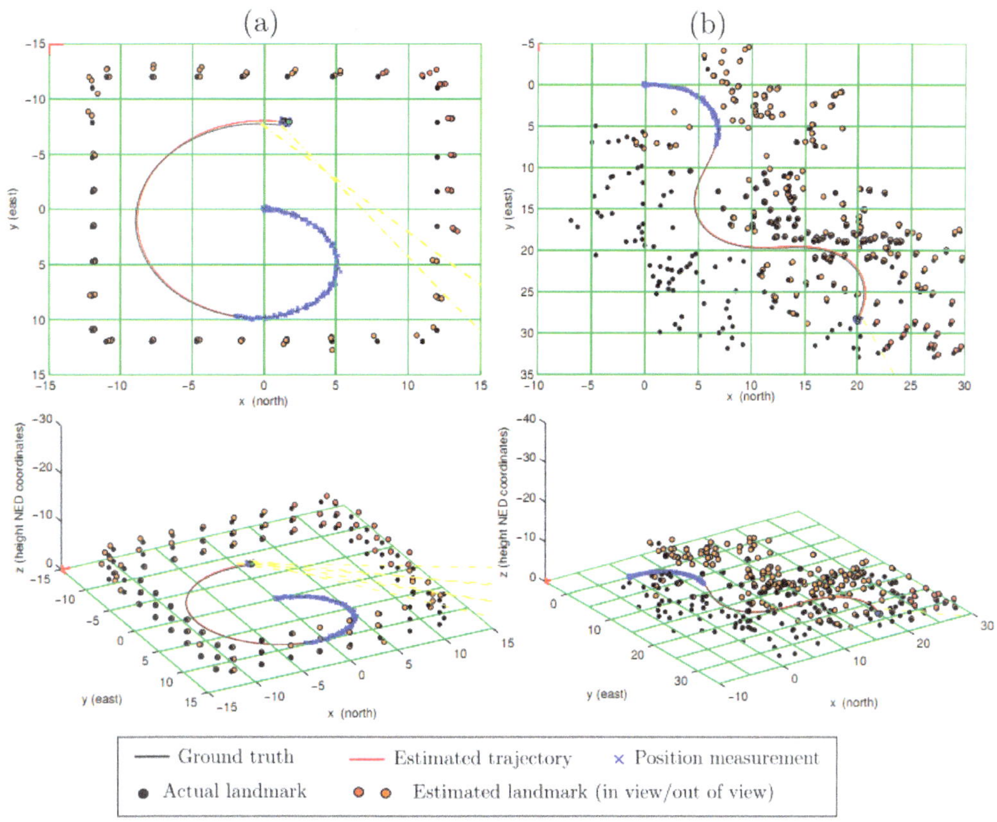

Figure 4: In simulations, a land vehicle was moved through different environments, which were composed by 3D points, emulating potential visual landmarks. The position sensors were only used at the beginning of the trajectory. The rest of the trajectory was performed under visual navigation conditions

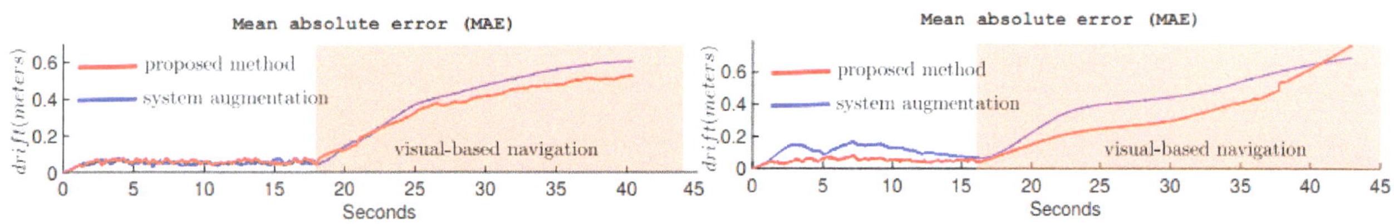

Figure 5: MAE: Proposed method vs. system state augmentation. Experiment (a) - left plot. Experiment (b) - right plot

In order to gain more insight about the performance of the proposed scheme, a variant of the proposal was also implemented, in this case by using system state augmentation (standard EKF-SLAM methodology). Figure 5 shows the average MAE in the position obtained after 20 Monte Carlo runs of the simulation, using the proposed method and system state augmentation. It can be observed that the progression of the MAE over time, which was obtained in scenarios (a) and (b), is very similar. On the other hand, Table 1 clearly shows the benefits obtained with the proposed method in terms of computational cost. In this case, the performance of the variant, which uses system state augmentation, quickly degrades as the number of visual features being tracked increases. In experiments (a), where the average number of visual features being tracked per frame is low, the execution time of the variant is slightly higher than the proposed method. However, in experiments (b), where the average number of features being tracked is higher, the execution time of the variant is almost three times the execution time of the proposed method.

Table 1: Execution time obtained using the proposed method and system augmentation. The third column indicates the average number of visual features being tracked per frame.

Experiment	Proposed method	System augmentation	Visual features per frame
(a)	19.0 s	25.0 s	23.9 ±3.4σ
(b)	44.2 s	128.1 s	63.1 ±38.5σ

Experiments with real data

In order to test the performance of the proposed method with real data, the MATLAB implementation was executed offline using the Malaga data set [39] as input signals. This data set is freely available online and contains several sources of sensor signals along with accurate ground truth. In the experiments, the following signals were used: i) inertial (accelerometer and gyroscope) measurements, available at 100; ii) GPS measurements, available at 1; iii) visual information, obtained from a frontal camera at seven frames per second (fps), and with a resolution of 512×384 pixels. The sensors were mounted over an electric buggytype vehicle, (see [39] for a complete description). The experiments presented here correspond to a vehicle moving across a car park.

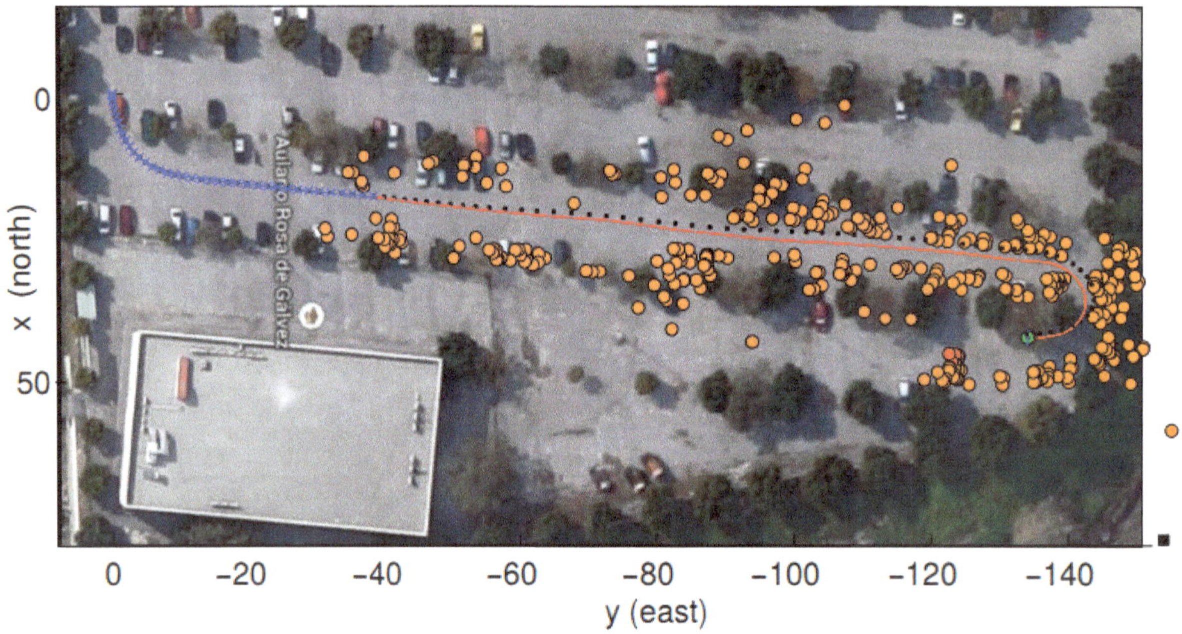

Figure 6: Aerial view of the estimated map and trajectory, using real data from sensors mounted in an electric buggy-type vehicle, which is moving in a car park. The GPS signal (blue) was only used at the beginning of the trajectory for recovering the metric scale. The visual-based estimated trajectory is indicated in red.

In the same manner as the simulations, the proposed method was compared against the standard EKF-SLAM methodology (system augmentation), except that real data were used in this case. Figure 6 shows the map and trajectory obtained by the proposed method for a route with a duration of 110. Note that GPS was only used at the beginning of the trajectory for recovering the metric scale of the world. Table 2 shows the experimental results obtained with both the proposed method and the standard EKF-SLAM methodology

Table 2: Experimental results with real data

Method	Execution time	Visual features per frame	Average Euclidean error
Proposed method	30.89 s	49.9 ±17.6σ	1.51m±1.25σ
System augmentation	88.37.0s	49.9 ±17.6σ	2.32m±2.64σ

The experiment was executed on a desktop PC with a i7-3.4 processor. For the presented case, the non-optimized MATLAB implementation of the proposed method ran almost three times faster than the standard EKF-SLAM methodology. The above execution time does not include the KTL execution time. For this experiment, the proposal also showed a lower average Euclidean error

Figure 7 shows the map and trajectory obtained for the entire data set. This route has a longitude of 524 and an approximate duration of 320. In the original data set, there is GPS availability almost over the whole trajectory. Nevertheless, GPS outages were artificially introduced in order to test the performance of the method under visualbased navigation conditions. In this case, the system has been able to determine the actual trajectory of the vehicle fairly well, even in periods where the GPS signal is not available. Figure 8 shows two different instances corresponding to periods when the system is working without a GPS signal (i.e., visual-based navigation mode). The spatial significance of the map can be appreciated. It is important to note that the video sequence, which is used in this experiment, contains a few dynamic elements, such as a few pedestrians or a moving car. There is also a lot of variation in lighting conditions, which are typical in outdoor environments.

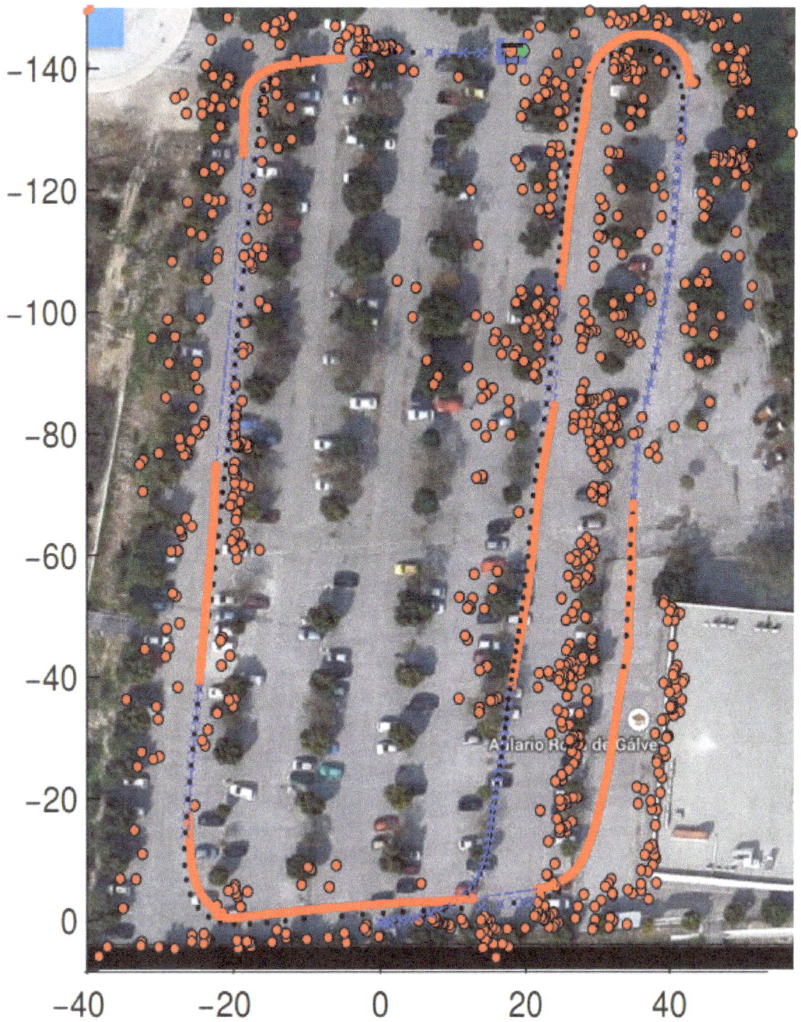

Figure 7: Map and trajectory estimated for the entire data set. Segments, in which the GPS is available, are indicated in blue; however, most of the trajectory was determined in visual-based navigation mode.

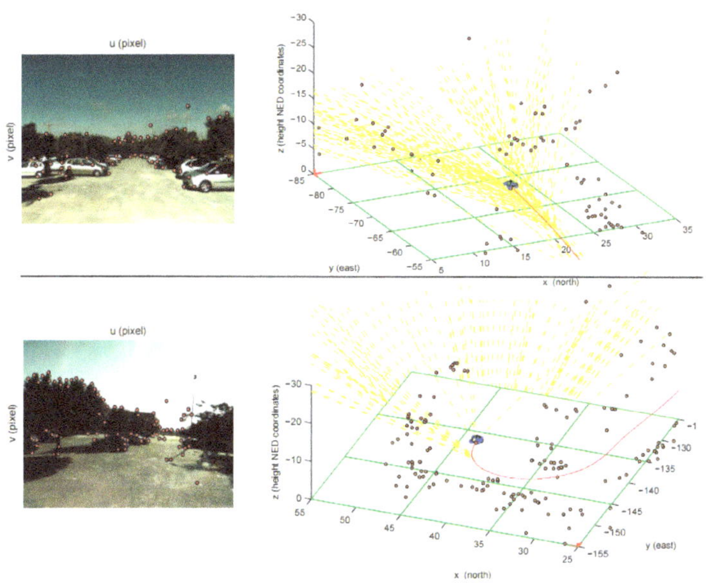

Figure 8: Estimated trajectory and map, corresponding to two different instants of time during periods of visual-based navigation. Visual features y_{l_i}', whose depth is ill-conditioned, are indicated by yellow lines. Visual features y_{p_j}' whose depth is considered as known, are indicated by small spheres. The corresponding video frames are shown to the left, while the visual features being tracked are indicated by small red circles. When comparing visual features, which correspond to elements in the environment (like trees or cars), with the estimated map, it can be appreciated that the physical structure of the environment is partially recovered.

CONCLUSION

In this work, a practical method for implementing a visual-aided inertial navigation and mapping system for autonomous robots has been presented. The testing of the method has been performed over real data collected from a land robot. However, it should be straightforward to extend the application of the method to other kinds of autonomous robot (e.g., flying robots).

A novel scheme has been introduced for taking advantage of visual information provided by a monocular camera. The proposed technique allows for handling, in an efficient and robust manner, the measurements corresponding to visual features. Experimental results, obtained through simulations and also with real data, show that the proposed method is able to estimate the trajectory of the vehicle fairly well, even in periods where the GPS signal is not available. In this case, the drift in estimation is considerably minimized, especially when it is compared with the results obtained using inertial sensors only. It is important to highlight that, in experiments, a video sequence with a low resolution and a low frame rate was used. In addition, the proposed method can operate with a moderate number of visual features being tracked.

The proposed method, different from visual odometry approaches, makes it possible to estimate the spatial location of visual features, whenever possible. Both the appearance and spatial information provided by the system could be potentially used for multiple tasks in future versions of the system.

Filter-based methods are convenient when limited processing power is a concern. However, the computational cost of Kalman filters scale poorly in relation to the incremental size of the state vector. For this reason, the proposed method treats each map feature independently, instead of augmenting the system state vector, as it is common in monocular SLAM approaches. The above characteristics should mean that the method is suitable for application to robots with limited resources.

REFERENCES

1. P. Sadeghi-Tehran, S. Behera, P. Angelov, and J. Andreu. Autonomous visual self-localization in completely unknown environment. In Evolving and Adaptive Intelligent Systems (EAIS), 2012 IEEE Conference on, pages 90–95, May 2012.
2. Ricardo C. B. Rodrigues, Sergio Pellegrino, and Hemerson Pistori. Artificial Intelligence and Soft Computing: 11th International Conference, ICAISC 2012, Zakopane, Poland, April 29-May 3, 2012, Proceedings, Part I, chapter Combining Color and Haar Wavelet Responses for Aerial Image Classification, pages 583–591. Springer Berlin Heidelberg, Berlin, Heidelberg, 2012.
3. Sinisa Segvic, Anthony Remazeilles, Albert Diosi, and FranÃ§ois Chaumette. A mapping and localization framework for scalable appearance-based navigation. Computer Vision and Image Understanding, 113(2):172 – 187, 2009.
4. Z. Chen and S. T. Birchfield. Qualitative vision-based path following. IEEE Transactions on Robotics, 25(3):749–754, June 2009.
5. Pablo De Cristóforis, Matías A. Nitsche, Tomáš Krajník, and Marta Mejail. Real-time monocular image-based path detection. Journal of Real-Time Image Processing, 11(2):335–348, 2013.
6. P. Moghadam, J. A. Starzyk, and W. S. Wijesoma. Fast vanishing-point detection in unstructured environments. IEEE Transactions on Image Processing, 21(1):425–430, Jan 2012.
7. C. K. Chang, C. Siagian, and L. Itti. Mobile robot monocular vision navigation based on road region and boundary estimation. In Intelligent Robots and Systems (IROS), 2012 IEEE/RSJ International Conference on, pages 1043–1050, Oct 2012.
8. Volker Grabe, Heinrich H. BĂĽlthoff, Davide Scaramuzza, and Paolo Robuffo Giordano. Nonlinear ego-motion estimation from optical flow for online control of a quadrotor uav. The International Journal of Robotics Research, 2015.
9. G. C. Barcelo, G. Panahandeh, and M. Jansson. Image-based floor segmentation in visual inertial navigation. In 2013 IEEE International Instrumentation and Measurement Technology Conference (I2MTC), pages 1402–1407, May 2013.
10. I. Parra, M. A. Sotelo, D. F. Llorca, and M. Ocaña. Robust visual odometry for vehicle localization in urban environments. Robotica, 28:441–452, 5 2010.
11. Kurt Konolige, Motilal Agrawal, and Joan Sola. Large-scale visual odometry for rough terrain. In Makoto Kaneko and Yoshihiko Nakamura, editors,Robotics Research, volume 66 of Springer Tracts in Advanced Robotics, pages 201–212. Springer Berlin Heidelberg, 2011.
12. Y. Yu, H. Zhao, F. Davoine, J. Cui, and H. Zha. Monocular visual localization using road structural features. In Intelligent Vehicles Symposium Proceedings, 2014 IEEE, pages 693–699, June 2014.
13. J. Courbon, Y. Mezouar, N. Guenard, and P. Martinet. Visual navigation of a quadrotor aerial vehicle. In Intelligent Robots and Systems, 2009. IROS 2009. IEEE/RSJ International Conference on, pages 5315–5320, Oct 2009.
14. Mingyang Li and A.I. Mourikis. Improving the accuracy of ekf-based visual-inertial odometry. In Robotics and Automation (ICRA), 2012 IEEE International Conference on, pages 828–835, May 2012.
15. Francisco Amoros, Luis Paya, Oscar Reinoso, Walterio Mayol-Cuevas, and Andrew Calway. Global appearance applied to visual map building and path estimation using multiscale analysis. Mathematical Problems in Engineering, 2014.
16. N. Roy, P. Newman, and S. Srinivasa. Visual Route Recognition with a Handful of Bits, pages 504–. MIT Press, 2013.
17. N. Kita, F. Kanehiro, M. Morisawa, and K. Kaneko. Obstacle detection for a bipedal walking robot by a fisheye stereo. In System Integration (SII), 2013 IEEE/SICE International Symposium on, pages 119–125, Dec 2013.

18. You Li and Yassine Ruichek. Occupancy grid mapping in urban environments from a moving on-board stereo-vision system. Sensors, 14(6):10454, 2014.

19. H. Durrant-Whyte and T. Bailey. Simultaneous localization and mapping: part i. Robotics Automation Magazine, IEEE, 13(2):99 –110, june 2006.

20. H. Zhou, D. Zou, L. Pei, R. Ying, P. Liu, and W. Yu. Structslam: Visual slam with building structure lines. IEEE Transactions on Vehicular Technology, 64(4):1364–1375, April 2015.

21. S. Pi, B. He, S. Zhang, R. Nian, Y. Shen, and T. Yan. Stereo visual slam system in underwater environment. In OCEANS 2014 - TAIPEI, pages 1–5, April 2014.

22. Adam Schmidt. Computer Vision and Graphics: International Conference, ICCVG 2014, Warsaw, Poland, September 15-17, 2014. Proceedings, chapter The EKF-Based Visual SLAM System with Relative Map Orientation Measurements, pages 570–577. Springer International Publishing, Cham, 2014.

23. Rodrigo MunguÃ¬a, Sarquis Urzua, Yolanda Bolea, and Antoni Grau. Vision-based slam system for unmanned aerial vehicles. Sensors, 16(3):372, 2016.

24. M. Meilland and A. I. Comport. On unifying key-frame and voxel-based dense visual slam at large scales. In Intelligent Robots and Systems (IROS), 2013 IEEE/RSJ International Conference on, pages 3677–3683, Nov 2013.

25. Y. Lu and D. Song. Visual navigation using heterogeneous landmarks and unsupervised geometric constraints. IEEE Transactions on Robotics, 31(3):736–749, June 2015.

26. H. Strasdat, J.M.M. Montiel, and A.J. Davison. Real-time monocular slam: Why filter? In Robotics and Automation (ICRA), 2010 IEEE International Conference on, pages 2657 –2664, may 2010.

27. Rodrigo Munguia, Bernardino Castillo-Toledo, and Antoni Grau. A robust approach for a filter-based monocular simultaneous localization and mapping (slam) system. Sensors, 13(7):8501–8522, 2013.

28. T. Oskiper, S. Samarasekera, and R. Kumar. Multi-sensor navigation algorithm using monocular camera, imu and gps for large scale augmented reality. In Mixed and Augmented Reality (ISMAR), 2012 IEEE International Symposium on, pages 71–80, Nov 2012.

29. Wei Zheng, Fan Zhou, and Zengfu Wang. Robust and accurate monocular visual navigation combining imu for a quadrotor. Automatica Sinica, IEEE/CAA Journal of, 2(1):33–44, January 2015.

30. S. Lynen, M. W. Achtelik, S. Weiss, M. Chli, and R. Siegwart. A robust and modular multi-sensor fusion approach applied to mav navigation. InIntelligent Robots and Systems (IROS), 2013 IEEE/RSJ International Conference on, pages 3923–3929, Nov 2013.

31. S. Shen, Y. Mulgaonkar, N. Michael, and V. Kumar. Multi-sensor fusion for robust autonomous flight in indoor and outdoor environments with a rotorcraft mav. In Robotics and Automation (ICRA), 2014 IEEE International Conference on, pages 4974–4981, May 2014.

32. Elena López, Rafael Barea, Alejandro Gómez, Álvaro Saltos, Luis M. Bergasa, Eduardo J. Molinos, and Abdelkrim Nemra. Robot 2015: Second Iberian Robotics Conference: Advances in Robotics, Volume 1, chapter Indoor SLAM for Micro Aerial Vehicles Using Visual and Laser Sensor Fusion, pages 531–542. Springer International Publishing, Cham, 2016.

33. Shi and C. Tomasi. Good features to track. In Computer Vision and Pattern Recognition, 1994. Proceedings CVPR ’94., 1994 IEEE Computer Society Conference on, 1994.

34. J.Y. Bouguet. Camera calibration toolbox for matlab. In online, 2008.

35. Rodrigo Munguia and Antoni Grau. A practical method for implementing an attitude and heading reference system. International Journal of Advanced Robotic Systems, 11, 2014.

36. I. Skog, P. Handel, J.O. Nilsson, and J. Rantakokko. Zero-velocity detection, an algorithm evaluation. Biomedical Engineering, IEEE Transactions on, 57(11):2657–2666, Nov 2010.

37. D.D. Diel, P. DeBitetto, and S. Teller. Epipolar constraints for vision-aided inertial navigation. In Application of Computer Vision, 2005. WACV/MOTIONS ’05 Volume 1. Seventh IEEE Workshops on, volume 2, pages 221–228, Jan 2005.

38. R. Munguia and A. Grau. Concurrent initialization for bearing-only slam. Sensors, 10(3):1511–1534, 2010.

39. Jose-Luis Blanco, Francisco-Angel Moreno, and Javier Gonzalez. A collection of outdoor robotic datasets with centimeter-accuracy ground truth.Autonomous Robots, 27(4):327–351, November 2009.

CHAPTER 2

FAULT DIAGNOSIS FOR ACTUATORS IN A CLASS OF NONLINEAR SYSTEMS BASED ON AN ADAPTIVE FAULT DETECTION OBSERVER

Runxia Guo[1], Kai Guo[1], Quan Gan[1], Junwei Zhang[1], Jiankang Dong[1], and Lanping Bai[2]

[1]College of Electronics Information and Automation, Civil Aviation University of China, Tianjin 300300, China

[2]Process Engineering Department, Freescale Semiconductor (China) Limited, Tianjin 300300, China

ABSTRACT

The problem of actuators' fault diagnosis is pursued for a class of nonlinear control systems that are affected by bounded measurement noise and external disturbances. A novel fault diagnosis algorithm has been proposed by combining the idea of adaptive control theory and the approach of fault detection observer. The asymptotical stability of the fault detection observer is guaranteed by setting the adaptive adjusting law of the unknown fault vector. A theoretically rigorous proof of asymptotical stability has been given. Under the condition that random measurement noise generated by the sensors of control systems and external disturbances exist simultaneously, the designed fault diagnosis algorithm is able to successfully give specific estimated values of state variables and failures rather than just giving a simple fault warning. Moreover, the proposed algorithm is very simple and concise and is easy to be applied to practical engineering. Numerical experiments are carried out to evaluate the performance of the fault diagnosis algorithm. Experimental results show that the proposed diagnostic strategy has a satisfactory estimation effect.

INTRODUCTION

With the rapid development of modern technology, the engineering and technical personnel have put forward higher requirements for the performance and reliability of equipment. Due to the complexity of equipment and harsh work environment, in order to avoid serious losses, adopting fault diagnosis technology for real-time monitoring of equipment's status is necessary. In the past few decades, there has been a rapid development in fault diagnosis technology.

Generally, the fault diagnosis methods can be divided into model-based approaches [1–3], knowledge-based approaches [4–6], and the methods based on signal processing [7–9]. Among these troubleshooting approaches, knowledge-based fault diagnosis methods are strongly dependent on the diagnostic systems themselves, and the diagnostic efficiency is subjective to the integrity of the information about the diagnostic systems. The fault diagnosis algorithms based on signal processing, especially filtering theory, are easy to implement in practical applications, but the first-order linear truncation in extended Kalman filters (EKF) can lead to low precision and filters' divergence. The model-based fault diagnosis approaches can

Citation: Runxia Guo, Kai Guo, Quan Gan, Junwei Zhang, Jiankang Dong, and Lanping Bai, "Fault Diagnosis for Actuators in a Class of Nonlinear Systems Based on an Adaptive Fault Detection Observer," Mathematical Problems in Engineering, vol. 2016, Article ID 2618534, 12 pages, 2016. doi:10.1155/2016/2618534

make full use of the system's information for fault location, determination of the failures' type, and fault estimation to obtain higher accuracy. Therefore, the model-based fault diagnosis technology has been widely studied by scholars [10, 11]. The state observer technique is very important in model-based fault diagnosis, and it is able to give specific estimated values of both failures and state variables simultaneously, which is beneficial to the subsequent design of fault-tolerant control law. Consequently, fault diagnosis based on observer has been a research hotspot [12, 13]. The representative results are analyzed as follows.

A robust fault detection observer was designed for a Takagi-Sugeno (T-S) fuzzy model with sensor faults [14]. The approach used the technique of descriptor systems by considering sensor fault as an auxiliary state variable. The design of robust fault detection observer was formulated as an H_{-}/H infinity problem. A solution of the pursued problem based on nonquadratic Lyapunov functions was then given via a linear matrix inequality formulation. An example was presented to demonstrate the design conditions. The example showed that the proposed fault detection observer was effective.

The small-amplitude oscillatory failures in the electrical flight control system of an aircraft were specifically studied [15]. A nonlinear observer-based solution to detect oscillatory failures with small amplitude at a very early stage was presented. By narrowing the detection threshold, the proposed approach could detect a fault with small amplitude and achieve early fault warning but was not able to give a specific form of the failure.

The online fault estimation question was further pursued [16]. A nonlinear aircraft model with multiple control surfaces was considered and a fault detection and isolation (FDI) algorithm was proposed for the stuck fault detection. The estimation effect was perfect; however, the proposed algorithm was based on an ideal analytical model, without considering external disturbances and measurement noise.

Robust and accurate detection of failures in the actuators of a civil aircraft is a crucial issue. Actuator failures, if not accurately detected in time, can often lead to improper maintenance and may potentially lead to structural damage or waste of money. Motivated by the actual demand, the combination of the observer approach and the adaptive control theory is applied to carry out fault diagnosis for actuators in a class of nonlinear systems. The adaptive method has been applied in the development of fault detection observer. Adaptive control is one of the effective theories of nonlinear systems' design. By selecting adaptable compensators, the adaptive controller can achieve stable control of a nonlinear system whose parameters are not accurately known or affected by disturbances [17, 18]. The adaptive parameters' adjusting law of the unknown fault vector has been set up, guaranteeing the asymptotic stability of the observer.

The main contributions of this paper lie in three aspects. Firstly, different from the existing papers, the fault diagnosis algorithm in this paper is not based on ideal analytical models; in other words, the random measurement noise and external disturbances are taken into account simultaneously during the whole development process [19]. The designed fault diagnosis algorithm can successfully separate the measurement noise generated by the sensors, the unknown external disturbances, and the failures when they exist at the same time. Secondly, the fault diagnosis algorithm is able to give specific estimated values of both failures and state variables rather than just giving the estimated values of the state variables or a fault warning. This is very beneficial for in-depth fault analysis and taking appropriate troubleshooting actions. Thirdly, the proposed algorithm is very simple and concise, which can overcome the problem of "explosion of complexity" and is easy to be applied to practical engineering. All the above three aspects of research work are not perfect in the existing literatures.

The structure of this paper is organized as follows. An analytical model of nonlinear system with measurement noise and disturbances is presented in Section 2. Section 3 identifies the diagnostic targets and constraints under which the diagnosis algorithm is developed. A design scheme of the nonlinear adaptive fault detection observer is proposed for actuators' fault diagnosis. In addition, a theoretically rigorous proof of asymptotical stability has also been given. Experimental results are presented in Section 4. Conclusions and future works are discussed in Section 5.

NONLINEAR SYSTEM STATEMENT

Consider a fault-free nonlinear system:

$$\dot{x}(t) = g(x,t) + Bu(t) + Md(t),$$
$$y(t) = Cx(t) + Nn(t), \tag{1}$$

where $x(t) \in R^n$ denotes the state vector and it can be measured directly; $y(t) \in R^m$ and $u(t) \in R^r$ denote an output vector and a control input vector, respectively; $d(t) \in R^n$ denotes an external disturbance vector that is applied to the system's actuator; $n(t) \in R^n$ denotes a random measurement noise (band-limited white noise) vector that is generated by the system's sensors; B and C are appropriately dimensional constant matrices; M and N are full rank const. matrices; $g(x, t)$ is a nonlinear function of the state variables $x(t)$.

Remark 1. Nonlinear control system (1) describes the influence of disturbances on the state variables and the effect of measurement noise on the output, respectively.

When there are actuators' failures, the fault systemcan be expressed as

$$\dot{x}(t) = g(x,t) + B[u(t) + f(t)] + Md(t),$$
$$y(t) = Cx(t) + Nn(t), \tag{2}$$

where $f(t) \in R^r$ denotes an unknown time-varying fault vector.

DESIGN OF THE OBSERVER AND ADAPTIVE FAULT ESTIMATION ALGORITHM

Constraint Conditions

Before designing the fault diagnosis algorithm, some constraints need to be satisfied for fault system (2).

(i) (x, t) in (1) and (2) satisfies the Lipschitz condition about state vector (t); namely, there exists constant $\delta_1 > 0$, such that

$$\|g(x_1) - g(x_2)\| \leq \delta_1 \|x_1 - x_2\|. \quad (3)$$

Remark 2. To simplify writing, (t) is abbreviated to x. This kind of writing method is suitable for the subsequent content.

(ii) The unknown external time-varying disturbance vector (t) is bounded; that is,

$$\|Md\| \leq \mu_0, \quad (4)$$

where $\mu\, 0 \geq 0$ represents the maximum acceptable disturbance. The requirement that the power of external disturbances is bounded is obviously reasonable. Because the disturbances must exist actually in the running process of nonlinear system (1) or (2), in other words, if $\|Md\| \rightarrow +\infty$, state variables $x(t)$ and time-varying failures $f(t)$ will be "drowned" by disturbances. In this case, state estimation or fault estimation is impossible.

(iii) The sensor's measurement noise vector (t) is bounded; that is,

$$\|Nn\| \leq \mu_1, \quad (5)$$

where $\mu_1 \geq$ 0represents themaximumamplitude of the noise.

In general, measurement noise is mingled in the feedback signal coming from the sensors, which means that the noise signal exists for real. So the measurement noise should be bounded.

(iv) The time-varying fault vector f is a bounded vector with finite rate of change; that is,

$$\|f\| \leq \mu_2,$$

$$\|\dot{f}\| \leq \mu_3, \quad (6)$$

where $\mu_2 \geq 0$ and $\mu_3 \geq 0$ represent the maximum amplitude of the failures and the rate of change, respectively.

Constraint condition (6) has specified the applicable scope of the proposed algorithmin this paper; in other words, only bounded failures with finite rate of change can be given specific estimated values. If $\|f\| \rightarrow +\infty$ or $\|f\| \rightarrow +\infty$, specific estimated values of failures will not be obtained.

(v) The state vector x is also bounded with finite rate of change; that is,

$$\|x\| \leq \mu_4,$$

$$\|\dot{x}\| \leq \mu_5, \quad (7)$$

where $\mu_4 \geq 0$ and $\mu_5 \geq 0$ represent the maximum amplitude of the states and the rate of change, respectively.

Just as constraint condition (6), because the proposed algorithm will also give the specific estimated values of state variables, their values should also be bounded.

Fault Detection Observer

For fault system (2), the following observer is designed:

$$\dot{\hat{x}} = g(\hat{x}) + B\left[u + \hat{f}\right] + H(y - \hat{y}),$$

$$\hat{y} = C\hat{x}, \quad (8)$$

where $\hat{x} \in \Re^n$, $\hat{f} \in \Re^r$, and $\hat{y} \in \Re^m$ denote state estimation, fault estimation, and output estimation, respectively; H is a finite gain matrix of the designed observer.

Here, the deviation signal $\tilde{x} \in \Re^n$ is defined in the following form:

$$\tilde{x} = x - \hat{x}. \quad (9)$$

Remark 3. x denotes the real values of state variables in fault system (2), which is measureable. However, $\hat{x}$ comes from fault detection observer (8).When failures occur in actuators, namely, $\|f\| \neq 0$, x is not equal to $\hat{x}$; namely, $\|\tilde{x}\| \neq 0$. In other words, $\tilde{x}$ always contains the deviation information caused by actuators' failure f. If there are no failures,^x asymptotically approaches x and in this case, fault detection observer (8) becomes a standard state observer. So, (8) is named as "fault detection observer."

By fault system (2) and fault detection observer (8), we can obtain

$$\dot{\tilde{x}} = \dot{x} - \dot{\hat{x}}$$

$$= -HC\tilde{x} + (g - \hat{g}) + B(f - \hat{f}) + Ed - HNn, \quad (10)$$

where $\hat{g}$ denotes $g(\hat{x})$.

Adaptive Fault Estimation

Consider the system to be supervised given by (2); the proposed adaptive fault estimation algorithm for this fault system is given by

$$\dot{\hat{f}} = \Pi B^T W \tilde{x}, \quad (11)$$

where $\Pi = \text{diag}\{\Pi_{11}, \Pi_{22}, \ldots, \Pi_{rr}\}$ $\forall \Pi_{ii} > 0$, $i \in [1,r]$, denotes a convergence factor matrix, namely, an adaptive rate, which can influence the estimation performance; W is a symmetric positive definite matrix and the conditions that need to be satisfied for W will be discussed in detail in the following.

Remark 4. Matrix Π in (11) is discussed here. In the subsequent proof process, only the conditions for elementΠ_{ii} in the diagonal of matrix Π need to be met, that is, $\forall \Pi_{ii} > 0$, $\forall \Pi_{ii} < +\infty$, and $\forall \Pi_{ij} = 0$, $j \in [1, r]$ and $i \neq j$. These constraint conditions can ensure that the eigenvalues of matrix Π exist. In addition, it is easy to satisfy these constraint conditions, but the optimal value for Π_{ii} cannot be given here.

The error signal of fault estimation $\tilde{f} \in R^r$ is defined in the following form:

$$\tilde{f} = f - \hat{f}. \quad (12)$$

By constraint condition (6), it can be obtained that

$$\dot{\tilde{f}} = \dot{f} - \dot{\hat{f}} = \dot{f} - \Pi B^T W \tilde{x}. \quad (13)$$

Theorem 5. *If W in (11) and the symmetric positive definite matrix U are selected to satisfy the condition*

$$(-HC)^T W + W(-HC) + \delta WW + \delta I = -U, \tag{14}$$

where δ is a positive constant that satisfies Lipschitz condition (3) and I is an appropriately dimensional unit matrix, then the fault diagnosis algorithm based on fault detection observer (8) and adaptive fault estimation algorithm (11) can ensure that state estimation error $\tilde{x}$ and fault estimation error $\tilde{f}$ are driven asymptotically to bounded constants simultaneously, in the sense that

$$\lim_{t\to\infty} \|\tilde{x}(t)\| \le \tau_1,$$

$$\lim_{t\to\infty} \left\|\tilde{f}(t)\right\| \le \tau_2, \tag{15}$$

Where $\tau_1 > 0$ and $\tau_2 > 0$ denote two bounded positive constants

Proof. We now define a Lyapunov function (t) as

$$V(t) = \tilde{x}^T W\tilde{x} + \tilde{f}^T \Pi^{-1}\tilde{f}. \tag{16}$$

Remark 6. (t) evaluates deviation signal $\tilde{x}$ and fault estimation error $\tilde{f}$ comprehensively.

After taking the time derivative of (16) and substituting (10) and (13) into it, it can be obtained that

$$\dot{V} = \dot{\tilde{x}}^T W\tilde{x} + \tilde{x}^T W\dot{\tilde{x}} + \dot{\tilde{f}}^T \Pi^{-1}\tilde{f} + \tilde{f}^T \Pi^{-1}\dot{\tilde{f}}. \tag{17}$$

There are four terms in $\dot{V}$. In order to facilitate reading, they are expanded, respectively, into the following forms:

$$\begin{aligned}
\dot{\tilde{x}}^T W\tilde{x} &= \left[\tilde{x}^T(-HC)^T + (g-\hat{g})^T + \tilde{f}^T B^T + d^T E^T \right.\\
&\quad \left. - n^T (HN)^T\right] W\tilde{x},\\
\tilde{x}^T W\dot{\tilde{x}} &= \tilde{x}^T W\left[(-HC)\tilde{x} + (g-\hat{g}) + B\tilde{f} + Ed\right.\\
&\quad \left. - HNn\right],\\
\dot{\tilde{f}}^T \Pi^{-1}\tilde{f} &= \left(\dot{f} - \Pi B^T W\tilde{x}\right)^T \Pi^{-1}\tilde{f} = \dot{f}^T \Pi^{-1}\tilde{f}\\
&\quad - \tilde{x}^T W B\tilde{f},\\
\tilde{f}^T \Pi^{-1}\dot{\tilde{f}} &= \tilde{f}^T \Pi^{-1}\left(\dot{f} - \Pi B^T W\tilde{x}\right) = \tilde{f}^T \Pi^{-1}\dot{f}\\
&\quad - \tilde{f}^T B^T W\tilde{x}.
\end{aligned} \tag{18}$$

After substituting (18) into (17), $\dot{V}$ can be expressed as

$$\begin{aligned}
\dot{V} &= \tilde{x}^T\left[(-HC)^T W + W(-HC)\right]\tilde{x}\\
&\quad + \left[(g-\hat{g})^T W\tilde{x} + \tilde{x}^T W(g-\hat{g})\right]\\
&\quad + \left[d^T M^T W\tilde{x} + \tilde{x}^T W M d\right]\\
&\quad - \left[n^T N^T H^T W\tilde{x} + \tilde{x}^T W H N n\right] + 2\tilde{f}^T \Pi^{-1}\dot{f}.
\end{aligned} \tag{19}$$

After utilizing Lipschitz condition (3), the following inequality can be obtained:

$$\begin{aligned}
\dot{V} &\le \tilde{x}^T\left[(-HC)^T W + W(-HC)\right]\tilde{x} + 2\left\|\tilde{x}^T W\right\| \delta \|\tilde{x}\|\\
&\quad + 2\|Md\|\lambda_{\max}(W)\|\tilde{x}\|\\
&\quad + 2\|HNn\|\lambda_{\max}(W)\|\tilde{x}\| + 2\tilde{f}^T \Pi^{-1}\dot{f},
\end{aligned} \tag{20}$$

where $\lambda_{\max}(W)$ denotes the maximum eigenvalue of matrix W.

Next, the last term in (20) can be handled in accordance with the following form:

$$\begin{aligned}
2\tilde{f}^T \Pi^{-1}\dot{f} &\le \tilde{f}^T\tilde{f} + \dot{f}^T \Pi^{-1}\Pi^{-1}\dot{f}\\
&\le \left\|\tilde{f}\right\|^2 + \left\|\dot{f}\right\|^2 \lambda_{\max}\left(\Pi^{-1}\Pi^{-1}\right),
\end{aligned} \tag{21}$$

where $\lambda\max(\Pi^{-1}\Pi^{-1})$ denotes the maximum eigenvalue of matrix $\Pi^{-1}\Pi^{-1}$.

In addition, because the measurement noise vector (t) is bounded, where $\|Nn\| \le \mu_1$ has been declared in constraint conditions (5) and H is the finite gain matrix of the designed observer, $\|HNn\|$ will also be bounded. It can be defined as follows:

$$\|HNn\| \le \mu_6, \tag{22}$$

where μ_6 is a bounded positive constant.

Obviously, by combining conditions (21) and (22), (20) can be strengthened to

$$\begin{aligned}
\dot{V} &\le \tilde{x}^T\left[(-HC)^T W + W(-HC)\right]\tilde{x}\\
&\quad + \delta\left[\left\|\tilde{x}^T W\right\|^2 + \|\tilde{x}\|^2\right] + 2\mu_0\lambda_{\max}(W)\|\tilde{x}\|\\
&\quad + 2\mu_6\lambda_{\max}(W)\|\tilde{x}\| + \left\|\tilde{f}\right\|^2\\
&\quad + \left\|\dot{f}\right\|^2 \lambda_{\max}\left(\Pi^{-1}\Pi^{-1}\right)\\
&= \tilde{x}^T\left[(-HC)^T W + W(-HC) + \delta WW + \delta I\right]\tilde{x}\\
&\quad + 2(\mu_0 + \mu_6)\lambda_{\max}(W)\|\tilde{x}\| + \left\|\tilde{f}\right\|^2\\
&\quad + \left\|\dot{f}\right\|^2 \lambda_{\max}\left(\Pi^{-1}\Pi^{-1}\right),
\end{aligned} \tag{23}$$

where μ_0 has been defined in (4).

After utilizing constraint condition (6), the following inequality can be obtained easily:

$$\left\|\tilde{f}\right\| = \left\|f - \hat{f}\right\| \le \|f\| + \left\|\hat{f}\right\| \le 2\mu_2, \tag{24}$$

where μ_2 has been defined in (6).

After substituting (6), (14), and (24) into (23), the condition can be adjusted into the following expression

$$\begin{aligned}
\dot{V} &\le -\tilde{x}^T U\tilde{x} + 2\Upsilon\lambda_{\max}(W)\|\tilde{x}\| + 4\mu_2^2\\
&\quad + \mu_3^2\lambda_{\max}\left(\Pi^{-1}\Pi^{-1}\right)\\
&\le -\lambda_{\min}(U)\|\tilde{x}\|^2 + 2\Upsilon\lambda_{\max}(W)\|\tilde{x}\| + \Psi,
\end{aligned} \tag{25}$$

Where $\lambda\min(U)$ denotes the minimum eigenvalue of matrix U

Which satisfies Theorem 5, $\Upsilon = (\mu_0 + \mu_6)$ and $\Psi = 4\mu_2^2 + \mu_3^2\lambda_{\max}(\Pi^{-1}\Pi^{-1})$, and μ_3 has been defined in (6).

Next, in order to complete the stability analysis, two cases will be examined.

Case 1 ($\|\tilde{x}\| > \eta$). Consider $\eta = (2\Upsilon\lambda_{\max}(W) + \sqrt{4\Upsilon^2\lambda_{\max}^2(W) + 4\Psi\lambda_{\min}(U)})/2\lambda_{\min}(U)$.

In this case, $\dot{V} < 0$ is established and (8) is a stable observer of nonlinear fault system (2).

Case 2 ($\|\tilde{x}\| \le \eta$). In this case, $\dot{V} > 0$ is established and observer (8) will be divergent. This causes the increase of $\|\tilde{x}\|$, and let $\|\tilde{x}\| > \eta$. Next, it will satisfy Case 1, which brings $\|\tilde{x}\|$

to reduce, and so on; namely, the states error $\|\tilde{x}\|$ will remain at a certain level range.

Consolidating Cases 1 and 2, it can be seen that the observer is stable, and the designed fault detection observer is able to keep accurate tracking of fault system (2).

Remark 7. H in (14) is a finite gain matrix of fault detection observer (8), which just needs to satisfy the condition that ensures that $(-HC)$ is a stable matrix. For example, the following choice is reasonable:

$$H = \Theta W C^T, \tag{26}$$

where $\Theta > 0$ is an adjustable parameter.

EXPERIMENTS

Experiment 1. It is a comparative study between the proposed approach and the T-S fuzzy observer.

In order to fully evaluate the performance of the proposed fault diagnosis algorithm, an actuator motion model is selected and its parameters are analyzed as the following expression:

$$g(x) = \begin{bmatrix} x_2 \\ -0.5\sin(x_1) \end{bmatrix}, \quad B = \begin{bmatrix} 1.5 & 0 \\ 0 & 1 \end{bmatrix}, \quad C = M = N = \begin{bmatrix} 1 & 0 \\ 0 & 1 \end{bmatrix}. \tag{27}$$

Obviously, (x) satisfies the Lipschitz condition. The parameters in fault detection observer (8) are selected according to (14) with the following form:

$$\delta = 0.5, \quad W = H = \begin{bmatrix} 5.2 & -4.9 \\ -4.9 & 6.3 \end{bmatrix}, \quad U = \begin{bmatrix} 76.08 & -84.53 \\ -84.53 & 95.05 \end{bmatrix}. \tag{28}$$

By combining (27) and (28), the fault detection observer based on our proposed algorithm in this paper has been constructed as follows:

$$\begin{bmatrix} \dot{\hat{x}}_1 \\ \dot{\hat{x}}_2 \end{bmatrix} = \begin{bmatrix} \hat{x}_2 \\ -0.5\sin(\hat{x}_1) \end{bmatrix} + \begin{bmatrix} 1.5 & 0 \\ 0 & 1 \end{bmatrix}\begin{bmatrix} u_1 + \hat{f}_1 \\ u_2 + \hat{f}_2 \end{bmatrix} + \begin{bmatrix} 5.2 & -4.9 \\ -4.9 & 6.3 \end{bmatrix}\begin{bmatrix} y_1 - \hat{y}_1 \\ y_2 - \hat{y}_2 \end{bmatrix},$$
$$\begin{bmatrix} \hat{y}_1 \\ \hat{y}_2 \end{bmatrix} = \begin{bmatrix} 1 & 0 \\ 0 & 1 \end{bmatrix}\begin{bmatrix} \hat{x}_1 \\ \hat{x}_2 \end{bmatrix}. \tag{29}$$

The convergence factor matrix Π in adaptive fault estimation algorithm (11) is selected carefully as

$$\Pi = \Pi_1 = \begin{bmatrix} 1.53 & 0 \\ 0 & 6.0 \end{bmatrix}. \tag{30}$$

The observer based on Takagi-Sugeno (T-S) fuzzy models is another important approach for fault detection. So it is necessary to make a comparative study between the proposed approach and T-S fuzzy observer. Here, a reduced-order T-S fuzzy observer scheme has been cited and its design process is presented briefly (the specific technical details can be found in [20]).

The T-S fuzzy system fault model that contains actuators' failures is given in the following form:

$$\dot{x} = \sum_{i=1}^{l} h_i(x)\left(A_i x + B_i(u + f)\right), \quad y = Cx, \tag{31}$$

where l is the number of fuzzy rules; $h_i(x)$ is a membership function and $h_i(x) \in [0,1]$; $\sum_{i=1}^{l} h_i(x) = 1$; A_i and B_i are appropriately dimensional constant matrices; x, y, u, f, and C have been defined in (1) and (2) separately.

The fault detection observer based on T-S fuzzy model is designed:

$$\dot{\hat{x}} = \sum_{i=1}^{l} h_i(\hat{x})\left(A_i \hat{x} + B_i(u + \hat{f})\right), \quad \hat{y} = C\hat{x}, \tag{32}$$

where $\hat{x}$, $\hat{y}$, $\hat{f}$, and C have been defined in (8).

A T-S fuzzy adaptive fault estimation algorithm for fault system (2) is given by

$$\dot{\hat{f}} = \sum_{i=1}^{l} h_i(\hat{x})\Gamma\tilde{x}. \tag{33}$$

Assumed in the diagnosis process, there are no disturbances (internal or external); namely, $d = n = [0]$, but there are failures described as

$$f = \begin{bmatrix} 0.1\cos(3.14\cdot t) \\ 0.4\sin(3.14\cdot t) \end{bmatrix} \tag{34}$$

The control input vector adopts constants with the following form:

$$u = \begin{bmatrix} 0.02 \\ 0.6 \end{bmatrix}. \tag{35}$$

Initial state vector x and estimation vector $\hat{x}$ are selected as follows:

$$x = \begin{bmatrix} 15.0 \\ 5.0 \end{bmatrix}, \quad \hat{x} = \begin{bmatrix} 0.0 \\ 0.0 \end{bmatrix}. \tag{36}$$

The simulation process lasts 40 seconds. The effects of state estimation and fault estimation based on the proposed algorithm in this paper are shown in Figures 1, 2, 3, and 4.The effects of state estimation and fault estimation based on T-S fuzzy model are shown in Figures 5, 6, 7, and 8.

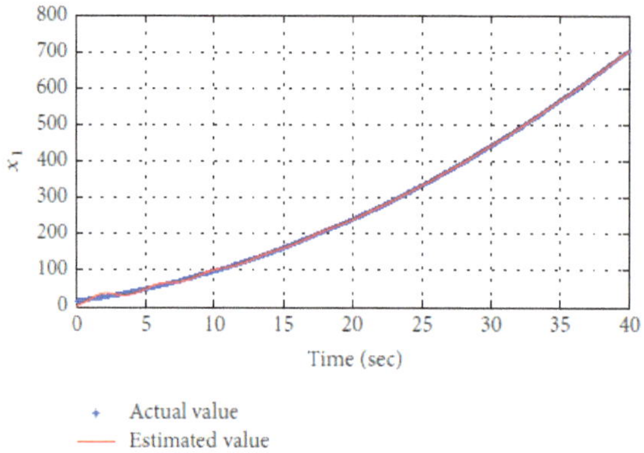

Figure 1: Estimation of x_1 based on the proposed approach.

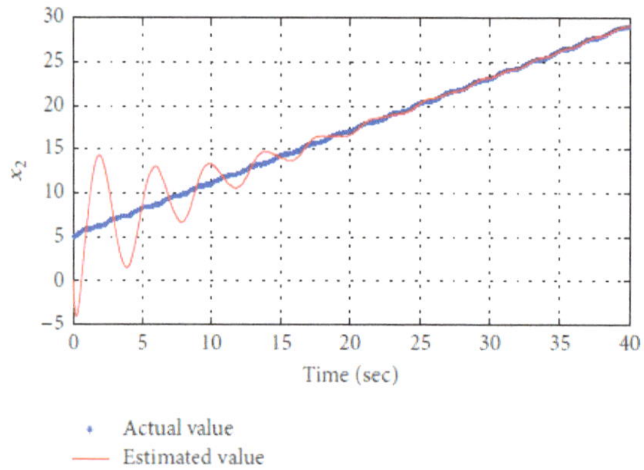

Figure 2: Estimation of x_2 based on the proposed approach.

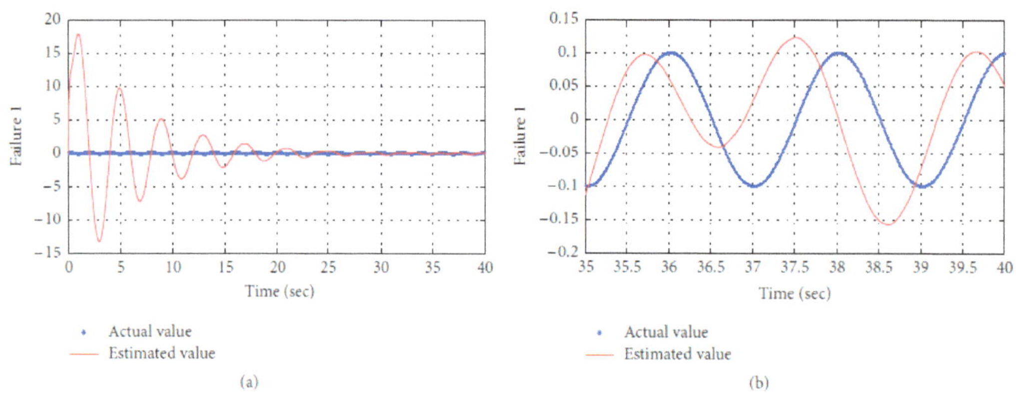

Figure 3: Estimation of failure 1 based on the proposed approach: (a) estimation of failure 1 and (b) enlarged view.

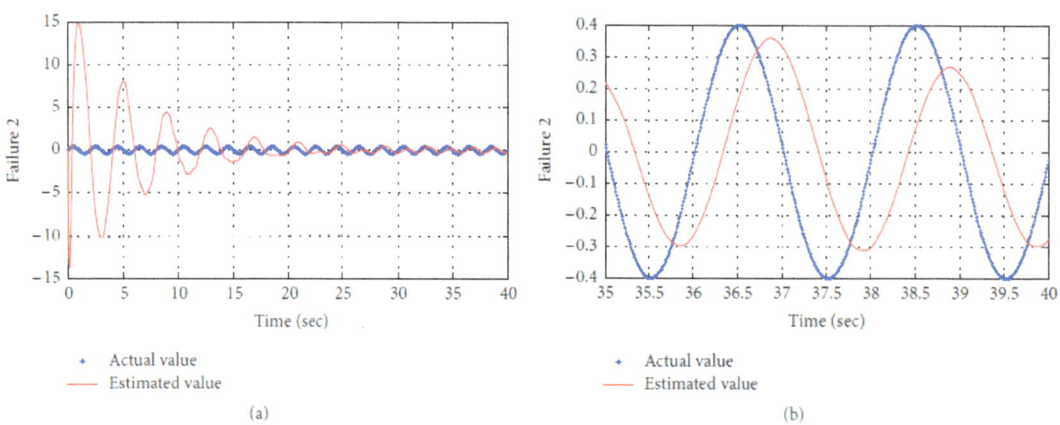

Figure 4: Estimation of failure 2 based on the proposed approach: (a) estimation of failure 2 and (b) enlarged view.

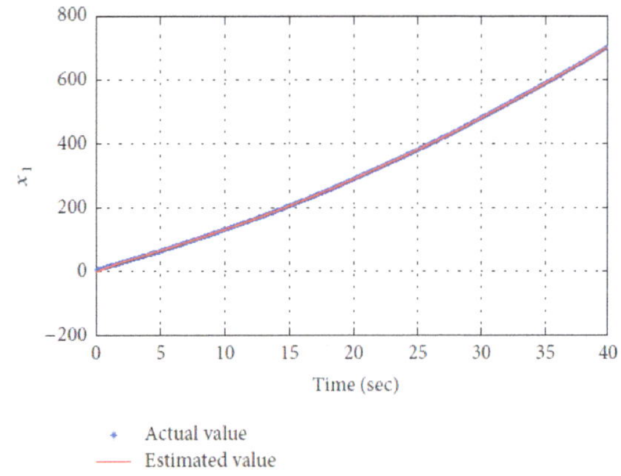

Figure 5: Estimation of x_1 based on the T-S fuzzy models.

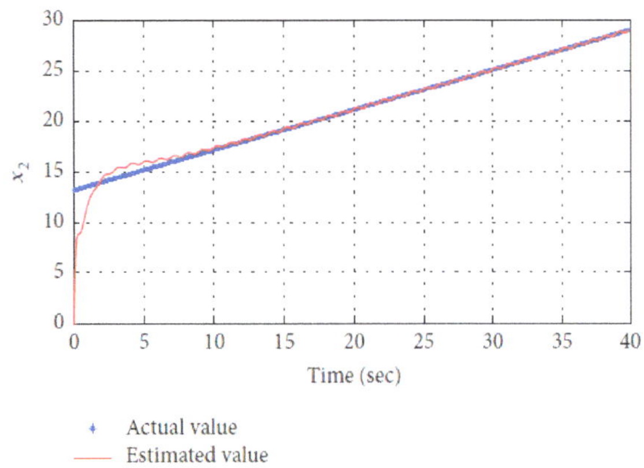

Figure 6: Estimation of x_2 based on the T-S fuzzy models.

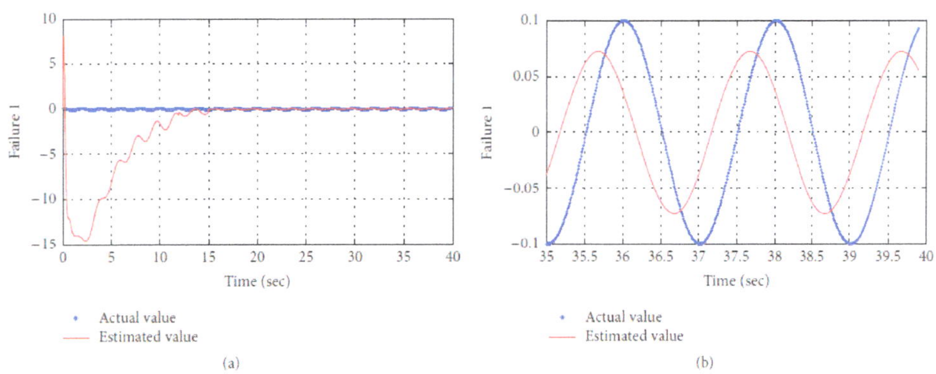

Figure 7: Estimation of failure 1 based on the T-S fuzzy models: (a) estimation of failure 1 and (b) enlarged view.

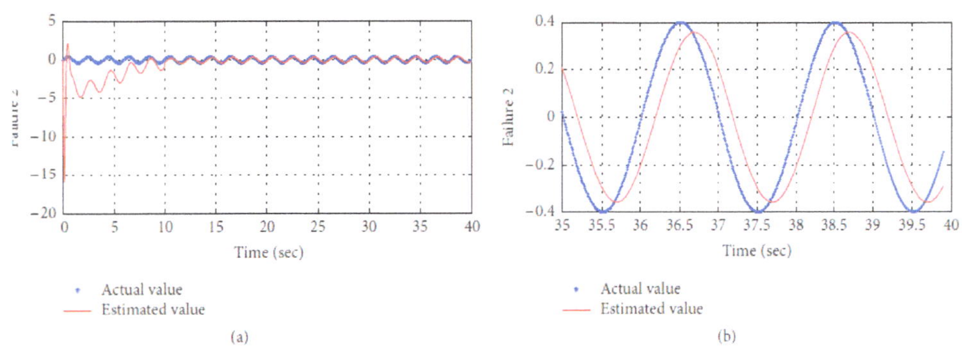

Figure 8: Estimation of failure 2 based on the T-S fuzzy models: (a) estimation of failure 2 and (b) enlarged view.

From Experiment 1, the following can be seen:

(1) There is no obvious difference between the performances of the proposed algorithm in this paper and the T-S fuzzy model. Both of the two algorithms can accurately estimate the state variables and the failures, and only small errors exist. The adjustment time of T-S fuzzy model is slightly shorter, but it is not obvious (see Figure 2 versus Figure 6, Figure 3 versus Figure 7, and Figure 4 versus Figure 8). In addition, from the point of view of the design process in two observers, the proposed algorithm is more concise.

(2) The T-S fuzzy observer proposed by [20] did not consider the influence of time-varying disturbances (random measurement noise, external disturbances, etc.) on estimation performance and the solution to the ant disturbance problem was also not given. However, the proposed algorithm in this paper has excellent ant disturbance performance, which is the most important advantage compared with the existing approaches. In order to fully illustrate this point, Experiment 2 is further carried out.

Experiment 2. It is the performance test with the presence of disturbances and failures.

Assuming that in the diagnosis process there are external time-varying disturbances as shown in (37) and time-varying failures as shown in (34) at the same time,

$$d = \begin{bmatrix} 0.08\cos(31.4\cdot t) \\ 0.3\sin(31.4\cdot t) \end{bmatrix}. \tag{37}$$

In addition, sensor variance is equal to 1; namely,

$$\begin{aligned} \mathrm{PSD}(n) &= 0.01, \\ \mathrm{mean}(n) &= 0, \\ \mathrm{var}(n) &= 1. \end{aligned} \tag{38}$$

The control input vector also uses (35); initial state vector x and estimation vector $\hat{x}$ use (36).

In order to thoroughly analyze and evaluate the convergence factors' impact on the performance of state estimation and fault estimation, the convergence factor matrix Π in adaptive fault estimation algorithm (11) is selected as (30) and (39), respectively:

$$\Pi = \Pi_2 = 0.5 * \Pi_1 = \begin{bmatrix} 0.765 & 0 \\ 0 & 3.0 \end{bmatrix}. \tag{39}$$

The effects of estimation based on Π_1 are shown in Figures 9–12. The effects of estimation based on Π_2 are shown in Figures 13–16.

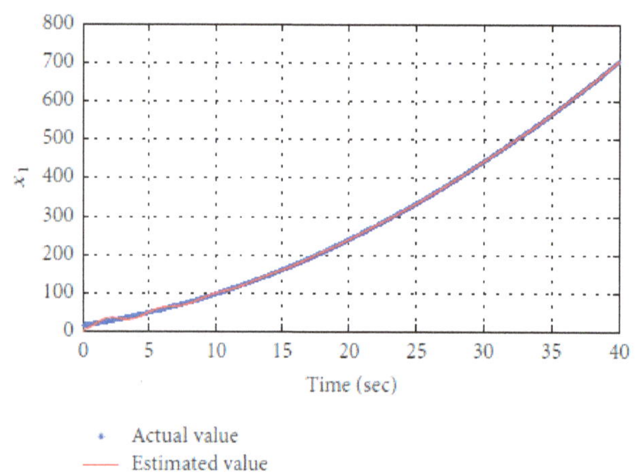

Figure 9: Estimation of x_1 with Π_1.

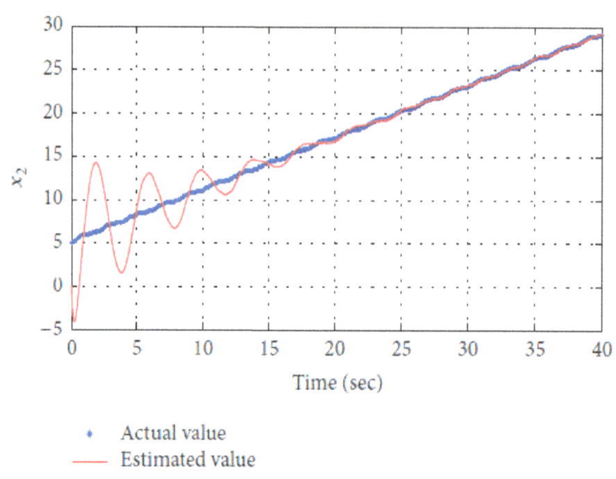

Figure 10: Estimation of x_2 with Π_1.

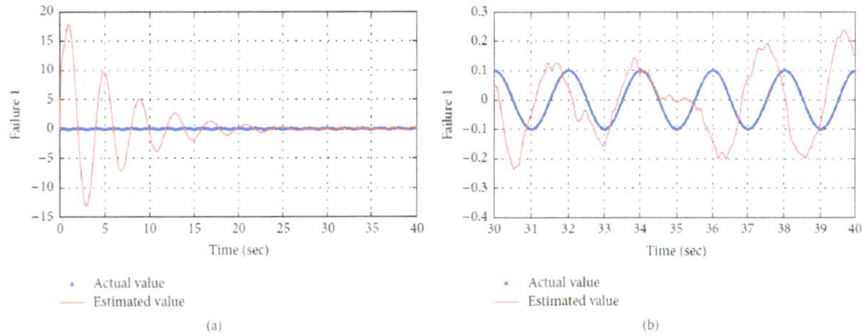

Figure 11: Estimation of failure 1 with Π_1: (a) estimation of failure 1 and (b) enlarged view.

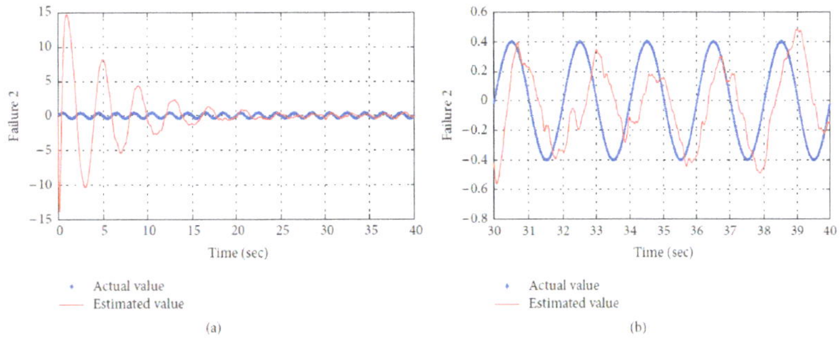

Figure 12: Estimation of failure 2 with Π_1: (a) estimation of failure 2 and (b) enlarged view.

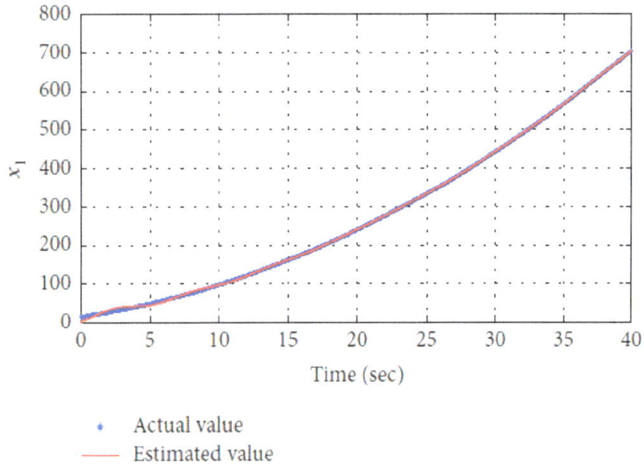

Figure 13: Estimation of x_1 with Π_2.

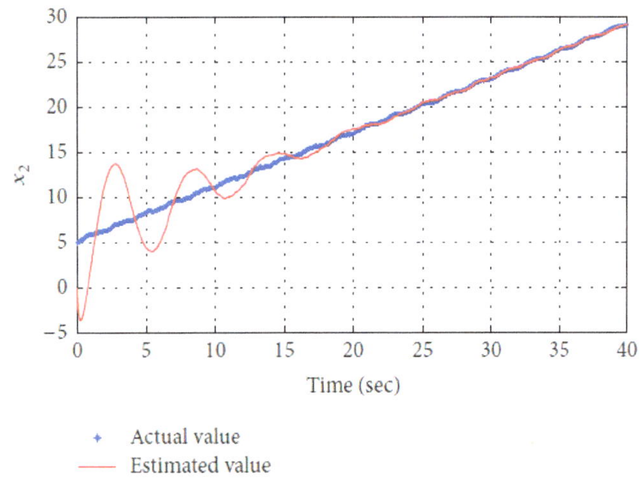

Figure 14: Estimation of x_2 with Π_2.

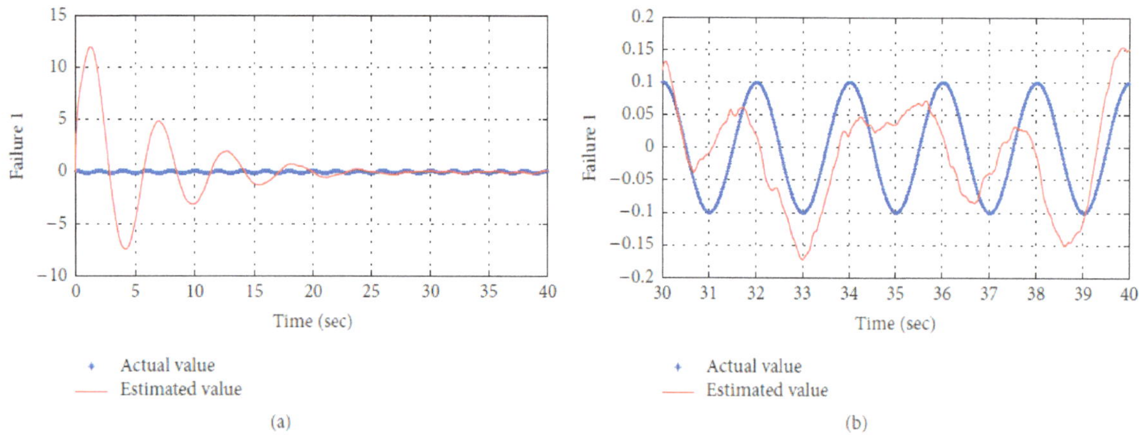

Figure 15: Estimation of failure 1 with Π_2: (a) estimation of failure 1 and (b) enlarged view.

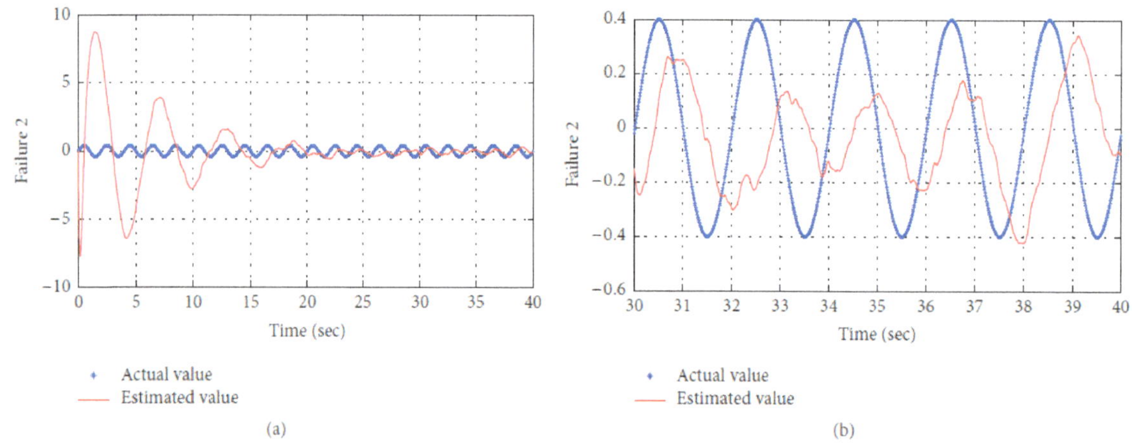

Figure 16: Estimation of failure 2 with Π_2: (a) estimation of failure 2 and (b) enlarged view.

For the proposed algorithm in this paper, the following can be concluded from Experiments 1 and 2:

(1) The performance of state estimation is perfect. Regardless of whether there are disturbances, the error signals of state estimation in stable state are very small; namely, $\lim_{t\to\infty}\|\tilde{x}\| \approx [\tilde{0}]$ (see Figures 1, 2, 9, 10,13, and 14). This shows that the designed fault detection observer (8) is not only stable but also very accurate. The precise state estimation provides the basis for the fault diagnosis of small-amplitude failures.(2) The external disturbances and random measurement noise have a very obvious influence on the results of fault estimation. From Experiment 2, it can be found that, with the presence of disturbances and noise, the values of fault estimation have obvious "burr" (see Figures 11(b), 12(b), 15(b), and 16(b)). At this time, the estimated failures will oscillate around the actual failures. However, when there are no disturbances or noise, the values of fault estimation are very "smooth" (see Figures 3 and 4). In actual operation, fault system

(2) has difficulty avoiding the influence of disturbances and noise, and therefore the results in Experiment 2 are more consistent with the actual situations.

(3) There are slight constant phase differences between actual values and estimated values in steady state (see Figures 3(b), 4(b), 11(b), 12(b), 15(b), and 16(b)). According to Theorem 5, the fault diagnosis algorithm based on fault detection observer (8) and fault estimation algorithm (11) can only ensure that state estimation error $\tilde{x}$ and fault estimation error $\tilde{f}$ are driven asymptotically to bounded positive constants, not to zero. In other words, the proposed algorithm is not an error-free estimation. So these simulation results are consistent with Lyapunov stability theory and this is why constant phase differences are generated. But the phase differences are not too large and are within an acceptable range.

(4) The convergence factor matrix Π has a significant impact on the performance of fault estimation. In Experiment 2, through the comparison of $\Pi_1 = \begin{bmatrix} 1.53 & 0 \\ 0 & 6.0 \end{bmatrix}$ and $\Pi_2 = \begin{bmatrix} 0.765 & 0 \\ 0 & 3.0 \end{bmatrix}$, it can be found that if is selected, the estimated failures in stable status have smaller amplitude errors (see Figure 11(b) versus Figure 15(b) and Figure 12(b) versus Figure 16(b)). Therefore, an appropriate increase of convergence factors can improve the accuracy of fault estimation and reduce the estimation errors. However, it is also found through experiments that too large values of convergence factors will cause the observer's divergence; therefore, careful selection of convergence factors is always required.

CONCLUSION

A nonlinear fault detection observer and an adaptive fault estimation algorithm have been proposed for the diagnosis of a class of nonlinear systems that have actuators' failures. Satisfactory experimental results have been achieved. The most important contribution of this paper is to present a novel diagnostic strategy that is able to give specific estimated values of state variables and failures when the external disturbances and measurement noise exist simultaneously. It has been proved that the fault detection observer, obtained via a Lyapunov function, is useful to achieve asymptotical stability in the case where unknown bounded disturbances and noise exist.

In this paper, we only discuss the problem of detection and estimation of actuators' failures. Future works will be focused on three aspects: the first aspect is the selection of convergence factors Π in (11). A guided approach of parameters' selection should be studied to guarantee that the proposed algorithm has faster estimation speed and smaller estimation errors; the second aspect is diagnosis algorithm's spreading. We hope that not only actuators' failures but also sensors' failures can be detected and estimated simultaneously; the third aspect is the practical use. This paper focuses on the theoretical research and how to apply the given results to engineering practice more effectively is also a major task to be completed in the future.

COMPETING INTERESTS

The authors declare that there are no competing interests regarding the publication of this paper.

ACKNOWLEDGMENTS

This work was jointly funded by the National Natural Science Foundation of China and the Civil Aviation Administration of China (Grant no. U1433103), the Natural Science Foundation of Tianjin (Grant no. 13JCYBJC39000), and the Basic R&D Operating Expenses of CAUC with no. 3122014B002 and no. 3122014D022.

REFERENCES

1. M. Chadli, S. Aouaouda, H. R. Karimi, and P. Shi, "Robust fault tolerant tracking controller design for a VTOL aircraft," Journal of the Franklin Institute, vol. 350, no. 9, pp. 2627–2645, 2013.
2. S. Aouaouda, M. Chadli, M. T. Khadir, and T. Bouarar, "Robust fault tolerant tracking controller design for unknown inputs T–S models with unmeasurable premise variables," Journal of Process Control, vol. 22, no. 5, pp. 861–872, 2012.
3. S. Aouaouda, M. Chadli, and H.-R. Karimi, "Robust static output-feedback controller design against sensor failure for vehicle dynamics," IET Control Theory and Applications, vol. 8, no. 9, pp. 728–737, 2014
4. H. R. Karimi, M. Zapateiro, and N. Luo, "A linear matrix inequality approach to robust fault detection filter design of linear systems with mixed time-varying delays and nonlinear perturbations," Journal of the Franklin Institute, vol. 347, no. 6, pp. 957–973, 2010.
5. Y. Y. Xu, S. C. Tong, and Y. M. Li, "Prescribed performance fuzzy adaptive fault-tolerant control of non-linear systems with actuator faults," IET Control Theory and Applications, vol. 8, no. 6, pp. 420–431, 2014.
6. Bregon, G. Biswas, B. Pulido, C. Alonso-Gonzalez, and H. Khorasgani, "A common framework for compilation techniques applied to diagnosis of linear dynamic systems," IEEE Transactions on Systems, Man, and Cybernetics: Systems, vol. 44, no. 7, pp. 863–876, 2014.
7. H. Z. Hosseinabadi, B. Nazari, R. Amirfattahi, H. R. Mirdamadi, and A. R. Sadri, "Wavelet network approach for structural damage identification using guided ultrasonic waves," IEEE Transactions on Instrumentation and Measurement, vol. 63, no. 7, pp. 1680–1692, 2014.
8. J. Li and G.-H. Yang, "Fault detection filter design for discrete-time switched linear systems with mode-dependent average dwell-time," International Journal of Adaptive Control and Signal Processing, vol. 28, no. 1, pp. 77–95, 2014.
9. S. Yin and X. Zhu, "Intelligent particle filter and its application to fault detection of nonlinear system,"IEEE Transactions on Industrial Electronics, vol. 62, no. 6, pp. 3852–3861, 2015.
10. L.-Y. Meng and B. Jiang, "Fault diagnosis based on the nonlinear adaptive observer," Systems Engineering and Electronics, vol. 30, no. 7, pp. 1317–1319, 2008.
11. S. Kelkar and R. Kamal, "Adaptive fault diagnosis algorithm for controller area network," IEEE Transactions on Industrial Electronics, vol. 61, no. 10, pp. 5527–5537, 2014.
12. S. Aouaouda, M. Chadli, P. Shi, and H. R. Karimi, "Discrete-time H-/H∞ sensor fault detection observer design for nonlinear systems with parameter uncertainty," International Journal of Robust and Nonlinear Control, vol. 25, no. 3, pp. 339–361, 2015.
13. S. Aouaouda, M. Chadli, V. Cocquempot, and K. M. Tarek, "Multi-objective H-/H∞ fault detection observer design for Takagi-Sugeno fuzzy systems with unmeasurable premise variables: descriptor approach," International Journal of Adaptive Control and Signal Processing, vol. 27, no. 12, pp. 1031–1047, 2013.
14. M. Chadli, A. Abdo, and S. X. Ding, "H-/H∞ fault detection filter design for discrete-time Takagi-Sugeno fuzzy system," Automatica, vol. 49, no. 7, pp. 1996–2005, 2013.
15. E. Alcorta-García, A. Zolghadri, and P. Goupil, "A nonlinear observer-based strategy for aircraft oscillatory failure detection: A380 case study," IEEE Transactions on Aerospace and Electronic Systems, vol. 47, no. 4, pp. 2792–2806, 2011.
16. Y. Han, S. Oh, B. Choi, D. Kwak, H. J. Kim, and Y. Kim, "Fault detection and identification of aircraft control surface using adaptive observer and input bias estimator," IET Control Theory and Applications, vol. 6, no. 10, pp. 1367–1387, 2012.
17. M. Rodrigues, H. Hamdi, D. Theilliol, C. Mechmeche, and N. Benhadj Braiek, "Actuator fault estimation based adaptive polytopic observer for a class of LPV descriptor systems," International Journal of Robust and Nonlinear Control, vol. 25, no. 5, pp. 673–688, 2015.

18. J. Zhang, A. K. Swain, and S. K. Nguang, "Robust H∞ adaptive descriptor observer design for fault estimation of uncertain nonlinear systems," Journal of the Franklin Institute, vol. 351, no. 11, pp. 5162–5181, 2014.

19. R. X. Guo, K. Guo, J. K. Dong, and Y. Zhu, "Time-varying and anti-disturbance fault diagnosis for a class of nonlinear systems," Proceedings of the Institution of Mechanical Engineers, Part I: Journal of Systems and Control Engineering, vol. 229, no. 7, pp. 573–586, 2015.

20. D. Krokavec and A. Filasová, "A reduced-order TS fuzzy observer scheme with application to actuator faults reconstruction," Mathematical Problems in Engineering, vol. 2012, Article ID 951953, 25 pages, 2012.

CHAPTER 3

FAULT DIAGNOSIS FOR ROTATING MACHINERY USING VIBRATION MEASUREMENT DEEP STATISTICAL FEATURE LEARNING

Chuan Li [1], René-Vinicio Sánchez [2], Grover Zurita [2], Mariela Cerrada [2] and Diego Cabrera [2]

[1]School of Mechanical Engineering, Dongguan University of Technology, Dongguan 523808, China

[2]Department of Mechanical Engineering, Universidad Politécnica Salesiana, Cuenca 010105, Ecuador

ABSTRACT

Fault diagnosis is important for the maintenance of rotating machinery. The detection of faults and fault patterns is a challenging part of machinery fault diagnosis. To tackle this problem, a model for deep statistical feature learning from vibration measurements of rotating machinery is presented in this paper. Vibration sensor signals collected from rotating mechanical systems are represented in the time, frequency, and time-frequency domains, each of which is then used to produce a statistical feature set. For learning statistical features, real-value Gaussian-Bernoulli restricted Boltzmann machines (GRBMs) are stacked to develop a Gaussian-Bernoulli deep Boltzmann machine (GDBM). The suggested approach is applied as a deep statistical feature learning tool for both gearbox and bearing systems. The fault classification performances in experiments using this approach are 95.17% for the gearbox, and 91.75% for the bearing system. The proposed approach is compared to such standard methods as a support vector machine, GRBM and a combination model. In experiments, the best fault classification rate was detected using the proposed model. The results show that deep learning with statistical feature extraction has an essential improvement potential for diagnosing rotating machinery faults.

INTRODUCTION

As one of the fundamental types of mechanical system, rotating machinery is widely applied in various fields. As a result of relative motion between mating surfaces, components of rotating machinery are prone to suffer from damage [1]. Effective fault diagnosis is thus important for maintaining the health of rotating machinery. One of the most challenging fault diagnosis tasks is the detection of faults and fault patterns, if any.

Different methods have been developed for fault diagnosis in rotating components such as gearboxes and bearings [2,3,4]. Gao et al. [5,6] systematically reviewed the fault diagnosis with model-based, signal-based, knowledge-based, and hybrid/active approaches. The most successful methods have three main steps: determining the fault symptoms, extracting the sensitive features, and classifying the condition patterns. Various fault symptoms, including vibration measurements [7], thermal features [8], acoustic signals [9], oil debris [10],

Citation: Chuan Li, René-Vinicio Sánchez , Grover Zurita , Mariela Cerrada and Diego Cabrera, Fault Diagnosis for Rotating Machinery Using Vibration Measurement Deep Statistical Feature Learning, doi:10.3390/s16060895

and other process parameters have been used as indices of the health of rotating systems. Vibration sensor signals have been proven effective for monitoring the health of rotating machinery.

Even in the vibration sensor category, different features sensitive to fault detection have been extracted in recent years. Most of these feature extractions are performed in the time domain, frequency domain, and time-frequency domain. To extract a fault feature in the time domain, Raad et al. [11] proposed using cyclostationarity as an indicator to diagnose gears. A diagnostic feature was introduced by Bartelmus and Zimroz [12] to monitor planetary gearboxes in time-varying operating conditions. The fault features are sometimes very sensitive in the frequency domain. Spectral kurtosis is one of the most popular fault features in the frequency domain [13]. Based on frequency domain kurtosis, an optimal mathematical morphology demodulation method was proposed for the diagnosis of bearing defects [14]. Compared to feature extraction in the time and frequency domains, time-frequency domain features have attracted much attention in both academia and industry. Continuous wavelet transform (CWT) [15], discrete wavelet transform (DWT) [16], wavelet packet transform (WPT) [17], second generation wavelet transform [18], comblet transform [19], and other time-frequency tools [20,21] have been successfully used to generate fault-sensitive features. In addition to feature extraction in a single domain, researchers have proposed detecting machinery faults in different domains. Lei et al. [22] proposed two diagnostic parameters from an examination of the vibration characteristics of planetary gearboxes in both the time and the frequency domains.

Based on the extracted fault features, different classifiers have been used to distinguish the healthy condition from different fault patterns. A multi-stage feature selected by genetic algorithms was proposed by Cerrada et al. [23] for the fault diagnosis of gearboxes. An intelligent diagnosis model jointly using a wavelet support vector machine (SVM) and immune genetic algorithm (IGA) was introduced for gearbox fault diagnosis [24]. Discriminative subspace learning has been used to diagnose faults in bearings [25]. Tayarani-Bathaie et al. [26] introduced a dynamic neural network to diagnose gas turbine faults. An artificial neural network and empirical mode decomposition have been applied to automatic bearing fault diagnosis using vibration signals [27]. It is clear that the SVM family has achieved good results in comparison with peer classifiers. Recently, deep learning has gained much attention in the classification community. Tamilselvan and Wang [28] introduced deep belief learning based health-state classification for failure diagnosis in datasets including iris, wine, Wisconsin breast cancer diagnosis, Escherichia coli and others. Tran et al. [29] used deep belief networks for the diagnosis of reciprocating compressor valves.

In this paper, we present a deep statistical feature learning approach for fault diagnosis in rotating machinery. The purpose of this paper is to use deep statistical feature learning as an integrated feature optimization and classification tool to improve fault diagnosis capability. For deep learning of statistical features with unknown value boundaries, a Gaussian-Bernoulli deep Boltzmann machine (GDBM) based on Gaussian-Bernoulli restricted Boltzmann machines (GRBMs) is proposed for the automatic learning of fault-sensitive features. The influences of different domains and typical rotating mechanical systems on fault classification are investigated. Deep learning is an effective learning framework for simultaneous statistical feature representation and classification, and the GRBM is a promising tool for dealing with unknown-boundary problems within the deep learning framework.

The remainder of this paper is structured as follows: the statistical features of the machinery vibration measurements are introduced in Section 2, and feature learning using the unsupervised GRBM and the supervised GDBM are also proposed in this section. In Section 3, fault diagnosis experiments for a gearbox and bearings are reported. The results of the experiments and discussions of the results are presented in Section 4. Conclusions are given in Section 5.

METHODOLOGIES

The GDBM is applied as a deep statistical feature learning tool for fault diagnosis in this paper. The methodologies used are introduced in this section. In Section 2.1, some classical statistical features are calculated from the time, frequency, and time-frequency domains of the vibration measurements. As the GDBM is constructed by stacking several GRBMs, and the GRBM is an improved version of the restricted Boltzmann machine (RBM), in Section 2.2 the basics of the GRBM are introduced. The statistical features calculated in the first subsection are used as the fault features represented by the unsupervised GRBM. As deep learning is an effective learning framework for simultaneous statistical feature representation and classification, the GDBM is constructed in Section 2.3. More details can be found in the following sections.

Statistical Features of the Vibration Sensor Signals

For a vibration measurement x(t) of the rotating machinery, its spectral representation X(f) can be calculated by:

$$X(f) = \hat{x}(f) = \int_{-\infty}^{+\infty} x(t) e^{-2\pi j f t} dt \tag{1}$$

where the hat "^" stands for the Fourier transform, t the time and f the frequency. For engineering applications, the collected vibration data are discrete values. Hence, the discrete version of Equation (1) (i.e., the discrete Fourier transform, DFT) should be used for the vibration data. There are several ways to calculate the DFT, among which the fast Fourier transform (FFT) is an efficient solution.

The time domain measurement x(t) and the frequency domain spectrum X(f) are capable of describing the machinery vibration in terms of time and frequency separately. For jointly representing the machinery vibration, the wavelet transform provides a powerful mathematical tool for signal processing and analysis. As mentioned in the Introduction, the CWT, DWT and WPT are in general the most popular categories in the wavelet transform family. Although different wavelet transforms have been successively applied in the fault diagnosis community, this paper uses the WPT to generate the time-frequency statistical features because it has comparatively low dimensions of the decomposition numbers and enhanced signal decomposition capability in the high frequency region.

The WPT is an extension of the typical DWT, in which detailed information is further decomposed by the WPT in the high frequency region. In other words, the WPT decomposes x(t) into a set of wavelet packet (WP) nodes through a series of low-pass and high-pass filters recursively.

With the integral scale parameter j and translation parameter k ($k = 0, \ldots, 2^j - 1$; $j = 0, \ldots, J$, which is the number of the decomposition levels), a WP function $T^n_{j,k}(t)$ is defined by:

$$T^n_{j,k}(t) = 2^{j/2} T^n(2^j t - k) \tag{2}$$

where n = 0, 1, … is the oscillation parameter [30]. The first two WP functions with j = k = 0 are the scaling function $\varphi(t)$ and the mother wavelet function $\psi(t)$, respectively. The remaining WP functions for n = 2, 3, … can be given by the WPT as:

$$T^{2n}(t) = \sqrt{2}\sum_k h(k) T^n_{1,k}(2t - k) \text{ and } T^{2n+1}(t) = \sqrt{2}\sum_k g(k) T^n_{1,k}(2t - k) \tag{3}$$

where the low-pass filter h(k) and the high-pass filter g(k) have the following forms:

$$h(k) = \frac{1}{\sqrt{2}} < \varphi(t), \varphi(2t - k) > \text{ and } g(k) = \frac{1}{\sqrt{2}} < \psi(t), \psi(2t - k) > \tag{4}$$

where <*,*> represents the inner product operator. The WP coefficients $P^n_{j,k}$ are therefore the inner product between the signal and the WP functions, i.e.:

$$P^n_{j,k} = < x(t), T^n_{j,k} > = \int_{-\infty}^{\infty} x(t) T^n_{j,k}(t) d(t) \tag{5}$$

In this way, the signal x(t) is decomposed by the WPT into J levels. At the j-th (j = 0, …, J) level, there are 2^j packets with the order $n = 1, 2, \ldots, 2^j$. For simplicity, we index the WP node as (j, n) whose coefficients are given by $P^n_{j,k}$.

According to the above analyses, the vibration measurement of the rotating machinery can be represented in the time domain, the frequency domain and the time-frequency domain. This can be formulated by:

$$M(p,q) = \begin{cases} x(t); & p \in R^1, q = t \in R^{n_0}, \text{time domain} \\ X(f); & p \in R^1, q = f \in R^{[n_0/2]}, \text{frequency domain} \\ [P^1_{1,k}, P^2_{1,k}, \ldots, P^{2^j}_{j,k}]; & p \in R^{2^{j+1}-1}, q \in R^{n_0/2^j}, \text{time-frequency domain} \end{cases} \tag{6}$$

where n_0 is the length of x(t).

As the three representations M(p,q) are usually very long, statistical features can be used as healthy condition indicators for rotating machinery. Statistical features have been approved as simple and effective features in fault diagnostics [17]. Based on the aforementioned studies, one can use the following statistical features for the vibration signals:

$$\begin{aligned} &F_{1,p}(M) = \frac{\int_{-\infty}^{\infty} [M-\mu]^4 P(M) dM}{\sigma^4}, F_{2,p}(M) = \frac{\int_{-\infty}^{\infty} [M-\mu]^3 P(M) dM}{\sigma^3}, \\ &F_{3,p}(M) = \frac{\max|M|}{\sqrt{\frac{1}{N}\sum_{q=1}^{N} M^2}}, F_{4,p}(M) = \frac{\max|M|}{\left(\frac{1}{N}\sum_{q=1}^{N}\sqrt{|M|}\right)^2}, F_{5,p}(M) = \frac{\sqrt{\frac{1}{N}\sum_{q=1}^{N} M^2}}{\frac{1}{N}\sum_{q=1}^{N}|M|}, \\ &F_{6,p}(M) = \frac{\max|M|}{\frac{1}{N}\sum_{q=1}^{N}|M|}, F_{7,p}(M) = \int_{-\infty}^{\infty} [M-\mu]^2 P(M), F_{8,p}(M) = \left(\frac{1}{N}\sum_{q=1}^{N}\sqrt{|M|}\right)^2, \text{ and} \\ &F_{9,p}(M) = \frac{1}{N}\sum_{q=1}^{N}|M| \end{aligned} \tag{7}$$

where N is the length of q for M(p,q), P(.) is the probability density [31], μ is the mean value, σ is standard deviation, and$F_{1,p}$, …, $F_{9,p}$ stand for kurtosis, skewness factor, crest factor, clearance factor, shape factor, impulse indicator, variance, denominator of clearance factor (the square of the averaged square roots of absolute amplitude), and mean of absolute amplitude values of the p-th vector of M(p,q), respectively [32]. Note that there are nine statistical features for the time domain representation M(p,q) = x(t), 9 for the frequency domain representation M(p,q) = X(f), and $9(2^{j+1} - 1)$ for the time-frequency domain representation $M(p,q) = [P^1_{1,k}, P^2_{1,k}, \ldots, P^{2^j}_{j,k}]$. The feature set F is therefore given by:

$$F = \begin{cases} [F_{1,1}(M), \ldots, F_{9,1}(M)]; & \text{time domain} \\ [F_{1,1}(M), \ldots, F_{9,1}(M)]; & \text{frequency domain} \\ [F_{1,1}(M), \ldots, F_{9,1}(M), F_{1,2}(M), \ldots, F_{9,2}(M), \ldots, F_{9,2^{j+1}-1}(M)]; & \text{time-frequency domain} \end{cases} \tag{8}$$

Statistical Feature Representation by Unsupervised Boltzmann Machines

After determining the statistical features in the time domain, the frequency domain and the time-frequency domain, in this subsection the unsupervised Boltzmann machine is proposed for feature representation.

The deep learning is a promising branch of the machine learning. It was developed to simulate the working mechanism of the brain to make sense of such data as images, sounds, and texts. The composed single layer GRBM model is the core to construct the deep learning (GDBM) frameworks in this work, and is originated from restricted Boltzmann machine (RBM).

The Boltzmann machine is a log-linear energy based model, where the energy function is linear in its free parameters. To restrict the Boltzmann machines to those without visible-visible and hidden-hidden connections, the RBM was proposed by Hinton, the father of deep learning, to form deep learning networks [33].

Conventional RBMs define the state of each visible and hidden neuron as binary codes (0 or 1). For real-valued data, the RBM has to normalize the input variables into [0, 1] with treating them as probabilities. For regular cases where the real values data have limited values, e.g., [0, 255] for pixels in the image processing, the RBM works well [34]. However, our statistical features scatter in different ranges. For example, the minimal value for F_1 is 0, but that for F_5 will be a negative number. This means that the conventional RBM is difficult to cope with our statistical features for the fault diagnosis.

To accommodate the real-valued data, the binary visible neurons can be replaced by the Gaussian ones to generate the Gaussian-Bernoulli RBM (GRBM). Although with real-valued neurons, the GRBM exhibits same structure compared to its RBM counterpart as shown in Figure 1.

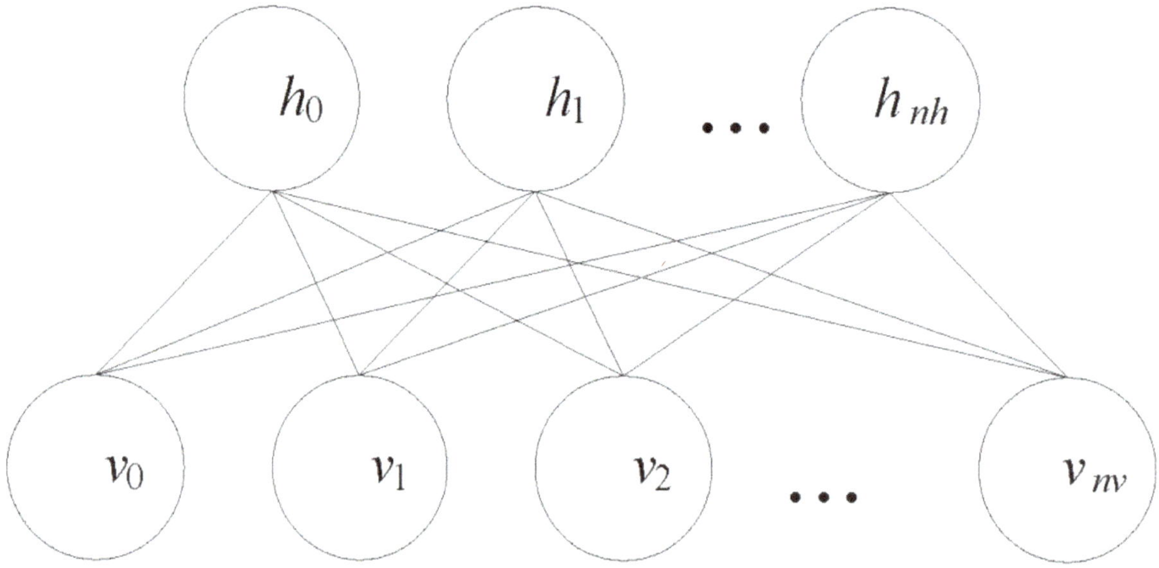

Figure 1: Illustration of the network connections with a GRBM. Note the GRBM exhibits same structure compared to its RBM counterpart.

For the GRBM shown in Figure 1, the energy function E(**v**, **h**) is given by:

$$E(\mathbf{v},\mathbf{h}|\theta) = \sum_{i=1}^{n_v} \frac{(v_i - b_i)^2}{2\sigma_i^2} - \sum_{i=1}^{n_v}\sum_{j=1}^{n_h} W_{ij} h_j \frac{v_i}{\sigma_i^2} - \sum_{j=1}^{n_h} c_j h_j \tag{9}$$

where **v** and **h** denote the visible and the hidden neurons, b_i and c_i stand for the offsets of the visible layers, w_{ij} represents the weights for the connection matrix, σ_i is the standard deviation associated with a Gaussian visible neuron v_i, and θ is the Gaussian parameter [35]. The traditional gradient-based training of the GRBM has difficulty learning σ_i, which is constrained to be positive. Hence, some algorithms fix σ_i as unity. With the improved energy function, Cho et al. [35] proposed conditional probabilities for the visible and the hidden neurons as follows:

$$p(v_i = v|\mathbf{h}) = (v|b_i + \sum_{j=1}^{n_h} h_j w_{ij}, \sigma_i^2) \tag{10}$$

and

$$p(v_i = v|\mathbf{h}) = (v|b_i + \sum_{j=1}^{n_h} h_j w_{ij}, \sigma_i^2) \text{ and } p(h_j = 1|\mathbf{v}) = S(c_j + \sum_{i=1}^{n_v} w_{ij} v_i / \sigma_i^2) \tag{11}$$

where $N(.|\mu,\sigma^2)$ is the Gaussian probability density function with mean μ and variance σ^2, and S(.) is a sigmoid function. The upgraded gradients with respect to the GRBM parameters are given by:

$$\nabla T_{ij} = \left\langle v_i h_j / \sigma_i^2 \right\rangle_d - \left\langle v_i h_j / \sigma_i^2 \right\rangle_m, \ \nabla b_i = \left\langle v_i / \sigma_i^2 \right\rangle_d - \left\langle v_i / \sigma_i^2 \right\rangle_m, \ \nabla c_j = \left\langle h_j \right\rangle_d - \left\langle h_j \right\rangle_m,$$
$$\nabla \log\sigma_i^2 = \exp(-\log\sigma_i^2)\left(\left\langle (v_i - b_i)^2/2 - \sum_{j=1}^{n_h} v_i h_j w_{ij} \right\rangle_d - \left\langle (v_i - b_i)^2/2 - \sum_{j=1}^{n_h} v_i h_j w_{ij} \right\rangle_m\right) \tag{12}$$

where $<.>_d$ and $<.>_m$ represent the expectation computed over the data and the model distributions, respectively.

When applying the GRBM for the unsupervised learning of the statistical features, the feature set F should be used as **v**and the GRBM results GR(F) = **h**. In this way, the n_v statistical features are represented by n_h neurons [36]. For condition monitoring and fault type classification, GRBM representations can be input to a classifier such as a support vector machine (SVM), decision tree, or random forest.

When applying the SVM as a classifier for fault diagnosis in rotating machinery, one should choose a multi-class SVM. The classical SVM is a binary classifier. Different methods have been proposed for using classical SVMs to compose multi-class SVMs. A pairwise coupling strategy was introduced by Hastie and Tibshirani [37] to perform multi-class classification by combining posterior probabilities provided by individual binary SVM classifiers.

Deep Statistical Feature Learning and Classification

After determining the statistical features in the time domain, the frequency domain and the time-frequency domain, in this subsection the unsupervised Boltzmann machine is proposed for feature representation.

In a common sense, an unsupervised mono-layer GRBM is inferior to a supervised multi-layer deep model. To stack several GRBMs on top of each other, a Gaussian-Bernoulli deep Boltzmann machine (GDBM) can be constructed for deep statistical feature learning of the machinery vibration signals. As an extension of the classical deep Boltzmann machine (DBM), the GDBM was introduced by Cho et al. [36]. Unlike other RBM-based deep models such as the deep belief network and the deep autoencoder, each neuron in the intermediate layers of the GDBM connects with both top-down and bottom-up information.

The GDBM structure used in this paper is shown in Figure 2a. The suggested GRBM is composed of three GRBMs (i.e., $GRBM_1$, $GRBM_2$, and $GRBM_3$). Each GRBM consists of one visible layer and one hidden layer, and the hidden layer of the previous GRBM is just the visible layer of the next GRBM. In this way, the first layer (data layer) and the second layer (hidden layer 1) forms the $GRBM_1$, the second layer and the third layer (hidden layer 2) forms the $GRBM_2$, the third layer and the last layer (output layer) forms the $GRBM_3$, and the three GRBMs are stacked together to form the GDBM.

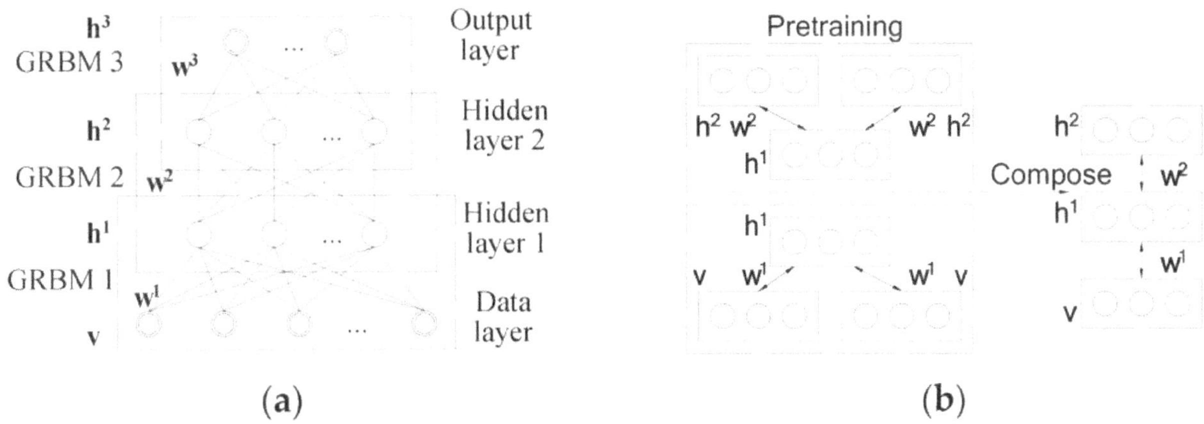

Figure 2: Schematic of the three-layer GDBM: (**a**) network structure; and (**b**) pretraining and composition of the GDBM.

The GDBM and its constituting GRBMs can be pretrained using a greedy, layer-by-layer unsupervised learning algorithm [37]. During the pretraining period as shown in Figure 2b, special attention should be paid to the GDBM as the neurons in the intermediate layers receive information both from the upper and the lower layers. To cope with this particularity, Salakhutdinov [38] halved the pretrained weights in the intermediate layers and duplicated the visible and topmost layers for the pretraining. With this idea, Equation (10) should be revisited to calculate the energy of the visible layer for the GRBM as:

$$E(\mathbf{v},\mathbf{h}^{(1)}|\theta) = \sum_{i=1}^{n_v}\frac{(v_i-b_i)^2}{2\sigma_i^2 N_v} - \sum_{i=1}^{n_v}\sum_{j=1}^{n_h} w_{ij}h_j^{(1)}\frac{v_i}{\sigma_i^2 N_v} - \sum_{j=1}^{n_h} c_j h_j^{(1)} \tag{13}$$

where $N_v = 2$ corresponds to the duplication of the visible layer. Similarly, the energy for the topmost $GRBM_L$ during the pretraining is given by:

$$E(\mathbf{h}^{(L-1)},\mathbf{h}^{(L)}|\theta) = -\sum_{j=1}^{n_h} N_v c_j h_j^{(L)} - \sum_{i=1}^{n_v}\sum_{j=1}^{n_h} N_v w_{ij}^{(L-1)} h_i^{(L-1)} h_j^{(L)} - \sum_{i=1}^{n_v} b_i h_i^{(L-1)} \tag{14}$$

The aforementioned pretraining is an unsupervised, bottom-up procedure for the GDBM. This means that it cannot be applied for the classification after the pretraining. Compared to conventional unsupervised learning, fortunately, the GDBM requires an extra supervised, top-down fine-tuning procedure [39,40]. At the fine-tuning procedure, the output layer is replaced by a multilayer perceptron (MLP) with sigmoid functions. To fit the fault classification task, all the weights **w** can be discriminatively fine-tuned using a back-propagation (BP) algorithm [41]. The supervised BP method uses labeled data as an extra MLP layer of variables to train the GDBM model for the classification. Unlike the unsupervised training process considering one GRBM at a time, the BP training considers all the layers in a GDBM simultaneously, which is in the same way as for the standard feed forward neural networks [42]. In this way, the GDBM can be regarded as an improvement of the MLP, or neural networks. It is capable of dealing with the classification for nonlinear, abnormal (non-Gaussian) data using a "deeper" fashion [43]. Of course, this "deeper" learning is much more time-consuming than the conventional ones.

Having introduced the GDBM and its constituting components, the GRBMs, the procedure of applying the GDBM based classification for the fault diagnosis of the rotating machines is shown in Figure 3 and is summarized as follows:

Step 1. Collect the vibration signals x(t), define the fault patterns and the diagnosis problems;

Step 2. Calculate the statistical feature set F according to Equation (8);

Step 3. Develop the GDBM model with the stack of the GRBMs according to Figure 1 and Figure 2;

Step 4. Pretrain the GDBM model and its constituting GRBMs using the layer-by-layer unsupervised learning algorithm from the training dataset;

Step 5. Fine-tune the GDBM weights using the BP algorithm from the training dataset; and

Step 6. Diagnose the rotating machinery condition using the trained GDBM model.

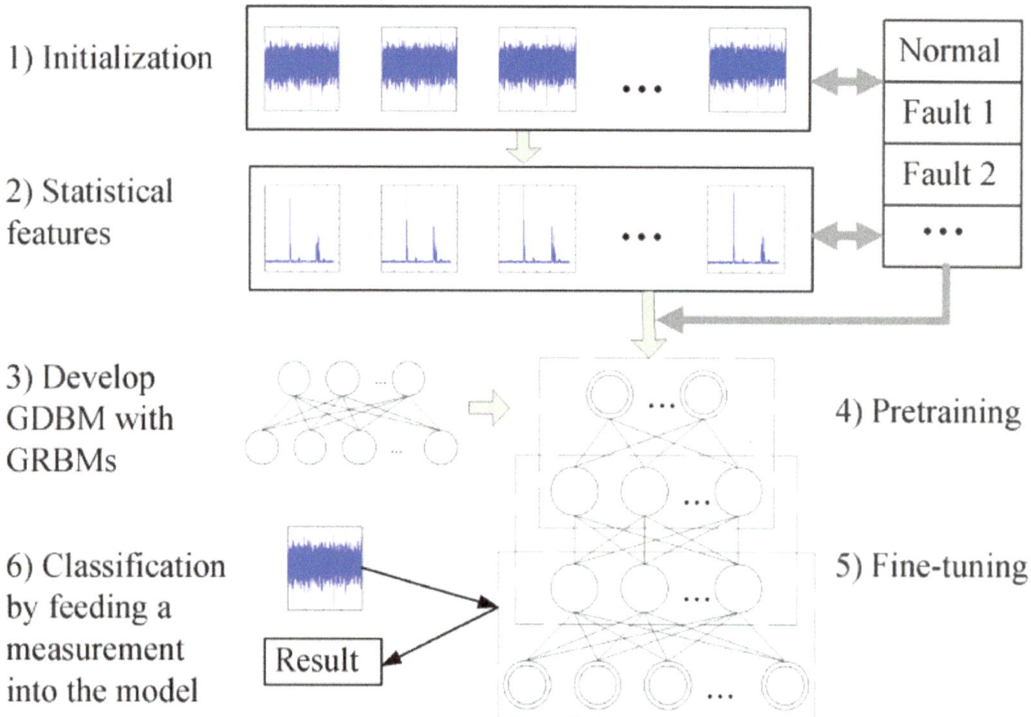

Figure 3: Flowchart of the deep statistical feature learning technique for the fault diagnosis of the rotating machinery.

DATA COLLECTION EXPERIMENTS FOR THE FAULT DIAGNOSIS

To validate the effectiveness of deep statistical feature learning for fault diagnosis, the proposed deep learning was applied to diagnose the health of two rotating mechanical systems. The experimental setups and procedures are detailed in the following two subsections.

Experimental Procedure for Gearbox Fault Diagnosis

The first experiments were carried out on a gearbox fault diagnosis system. As shown in Figure 4a, the output of a motor (3~, 2.0 HP, Siemens, Munich, Germany) was connected to the input shaft of a gearbox (fabricated by the lab of the Universidad Politécnica Salesiana, Cuenca, Ecuador) via a coupling. A 53-tooth pinion was installed on the input shaft of the gearbox, whose output shaft has an 80-tooth gear. An electromagnetic torque break (8.83 kW, Rosati, Monsano, Italy) was used as a load to connect with the output shaft of the gearbox via a belt transmission. The torque break was controlled by a controller (GEN 100-15-IS510, TDK-Lambda, Tokyo, Japan) which enabled the load to be adjusted manually. An accelerometer (ICP 353C03, PCB, Depew, NY, USA) was mounted on top of the gearbox to collect the vibration signals, which were sent to a laptop (Pavilion g4-2055la, HP, Palo Alto, CA, USA) through a data acquisition system (cDAQ-9234, NI, Austin, TX, USA). The laptop controlled an inverter (VLT 1.5 kW, Danfoss, Nordberg, Denmark) for adjusting the motor's rotation speed, which was monitored by a tachometer (VLS5/T/LSR optical sensor, Compact, Bolton, UK).

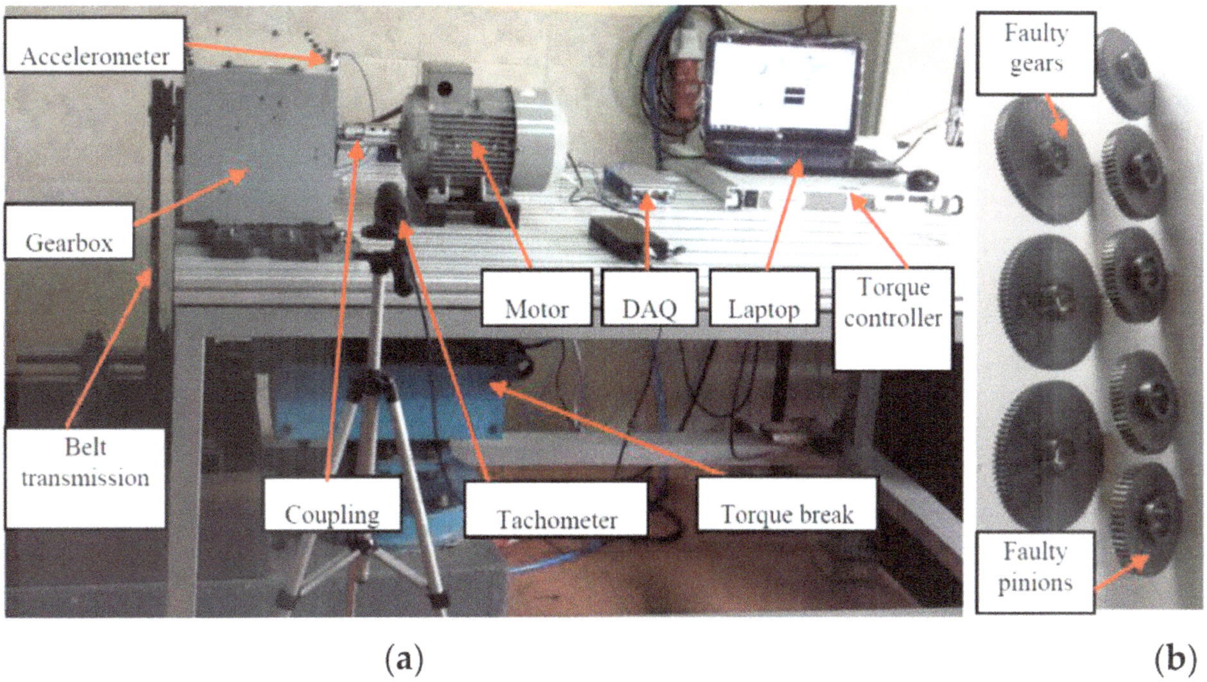

Figure 4: Gearbox fault diagnosis configurations: (**a**) experimental set-up; and (**b**) three different faulty gears and five different faulty pinions.

In the gearbox fault diagnosis experiments, in addition to one normal pinion and one normal gear, three different faulty gears and five different faulty pinions (shown in Figure 4b) were used to configure different condition patterns for the gearbox. The 10 different patterns shown in Table 1 were set for the collection of vibration signals.

Table 1: Experimental configurations of different condition patterns.

Experimental Setup	Pattern Label	Component 1	Component 2	Load
Gearbox (component 1-pinion; component 2-gear)	A	Normal	Normal	zero, small, great
	B	Chaffing tooth	Normal	zero, small, great
	C	Worn tooth	Normal	zero, small, great
	D	Chipped tooth 25%	Normal	zero, small, great
	E	Chipped tooth 50%	Normal	zero, small, great
	F	Missing tooth	Normal	zero, small, great
	G	Normal	Chipped tooth 25%	zero, small, great
	H	Normal	Chipped tooth 50%	zero, small, great
	I	Normal	Missing tooth	zero, small, great
	J	Chipped tooth 25%	Chipped tooth 25%	zero, small, great
Bearing (component 1-bearing 1; component 2-bearing 2)	1	Normal	Normal	Zero, 1, 2 flywheel(s)
	2	Normal	Inner race fault	Zero, 1, 2 flywheel(s)
	3	Normal	Outer race fault	Zero, 1, 2 flywheel(s)
	4	Normal	Ball fault	Zero, 1, 2 flywheel(s)
	5	Outer race fault	Inner race fault	Zero, 1, 2 flywheel(s)
	6	Ball fault	Inner race fault	Zero, 1, 2 flywheel(s)
	7	Ball fault	Outer race fault	Zero, 1, 2 flywheel(s)

To challenge the fault diagnosis performance, three different load conditions (no load, small load, and large load), were manually set for each pattern. For each pattern and load condition, we collected 24 signals, each of which covered 0.4096 s, with a sampling frequency of 10 kHz. The experiments were repeated five times, so 3600 vibration signals corresponding to 10 condition patterns (with three different loads) were recorded. Each vibration signal was used to generate the temporal, spectral, and WPT representations M(p,q) given by Equations (6)–(8) were then used to generate the feature set F for the vibration signals. The 3600 feature sets were divided into a training dataset with 2400 samples and the testing dataset with 1200 samples.

The unsupervised GRBM and the supervised GDBM were applied to learn the statistical features of the vibration signals. The statistical features represented by the unsupervised GRBM required an additional classifier for the pattern classification. Considering its excellent classification capability, the SVM was used as the classifier for the GRBM representations. For the GDBM, supervised deep learning as shown in Figure 3 was applied for the healthy condition pattern classification of the gearbox.

Experimental Procedure for Bearing Fault Diagnosis

To further challenge the deep statistical feature learning for fault diagnosis, we also carried out bearing fault diagnosis experiments. The gear fault patterns (displayed in Table 1) occupied areas of great damage, which introduced greater changes in the vibration measurements [44]. Compared to the vibration signal of the gear fault, an incipient bearing fault often has a smaller damage surface and thus generates weak vibration changes [45].

As shown in Figure 5a, a rolling element bearing test rig was constructed in the Universidad Politécnica Salesiana of

Ecuador to collect the vibration measurements for different healthy conditions. The test rig was driven by a motor (3~, 2.0 HP, Siemens) controlled by an inverter (VLT 1.5 kW, Danfoss). The rotating speed of the motor was monitored by a tachometer (VLS5/T/LSR optical sensor, Compact). A steel shaft (φ30 mm) was connected to the motor via a coupling. The two ends of the shaft were supported by two bearings (bearing 1 and bearing 2, 1207 EKTN9/C3, SKF, Goteborg, Sweden). An accelerometer (ICP 353C03, PCB) was mounted on the housing (SNL 507-606, SKF) of bearing 2 for measuring the vibration signals, which were collected by a data acquisition box (cDAQ-9234, NI) that communicated with a laptop (Pavilion g4-2055la, HP). Two flywheels were installed on the shaft as the load of the system.

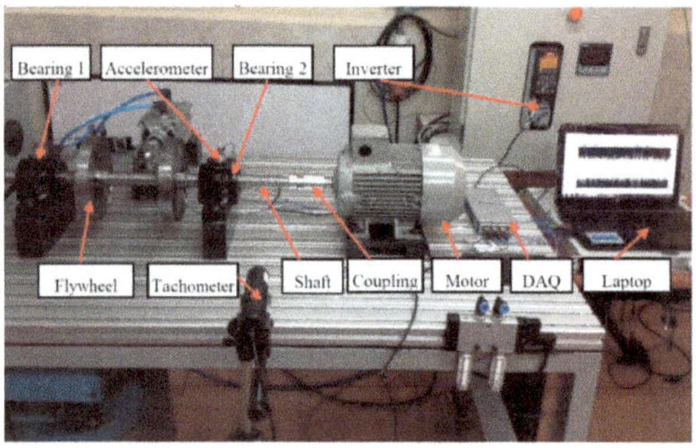

(a)

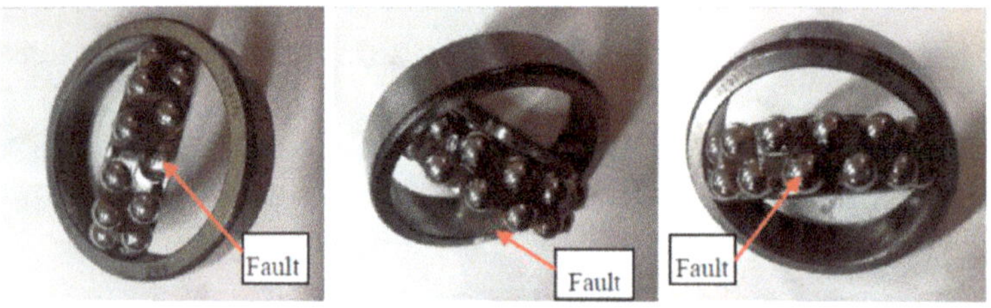

(b)

Figure 5: Fault diagnosis configurations for the rolling element bearings: (**a**) experimental set-up; and (**b**) 3 different faulty bearings with an inner race fault (**left**), an outer race fault (**middle**) and a ball fault (**right**), respectively.

In addition to the normal bearings, as shown in Figure 5b, three different faulty bearings with an inner race fault, an outer race fault and a ball fault, were used in the experiments. Using combinations of bearings in different conditions, seven healthy condition patterns were set, as shown in Table 1. For each experiment with each pattern, there were respectively 0, 1 and 2 flywheels used as the load. For each pattern and load configuration, 48 signals were collected for 0.4096 s. Each experiment was repeated five times. This means that 5040 signals were finally obtained. The sampling frequency for the bearing fault diagnosis was also set at 10 kHz.

Similar to the procedure described in the previous section, the statistical features were produced from the raw data of the bearing vibration signals. The unsupervised GRBM and the supervised GDBM were again applied for the fault diagnosis of the bearing system. The results of all the experiments are detailed in the next section.

RESULTS AND DISCUSSION

Gearbox Fault Diagnosis Results

Figure 6a,b plot the time domain waveform and statistical features for the first signal collected from the gearbox experimental setup. As the signal covered 0.4096 s with the sampling frequency of 10 kHz, the length of the discrete time signal is 4096. For all the collected 3600 signals, their time domain waveforms and statistical features are shown in Figure 6c,d, respectively.

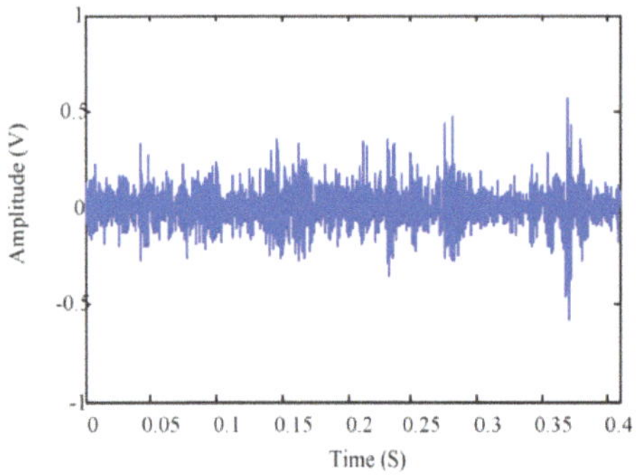

(a)

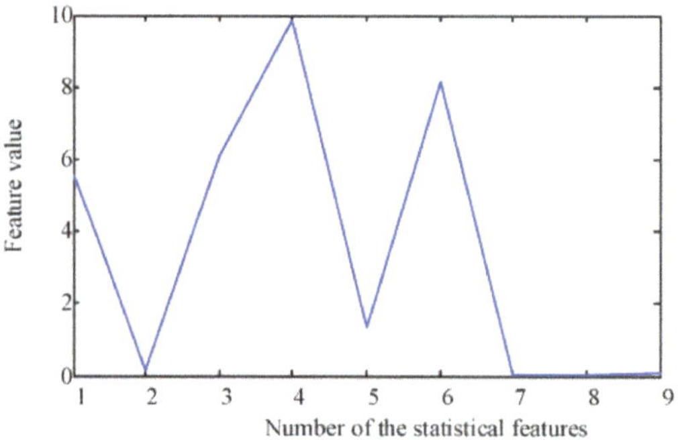

(b)

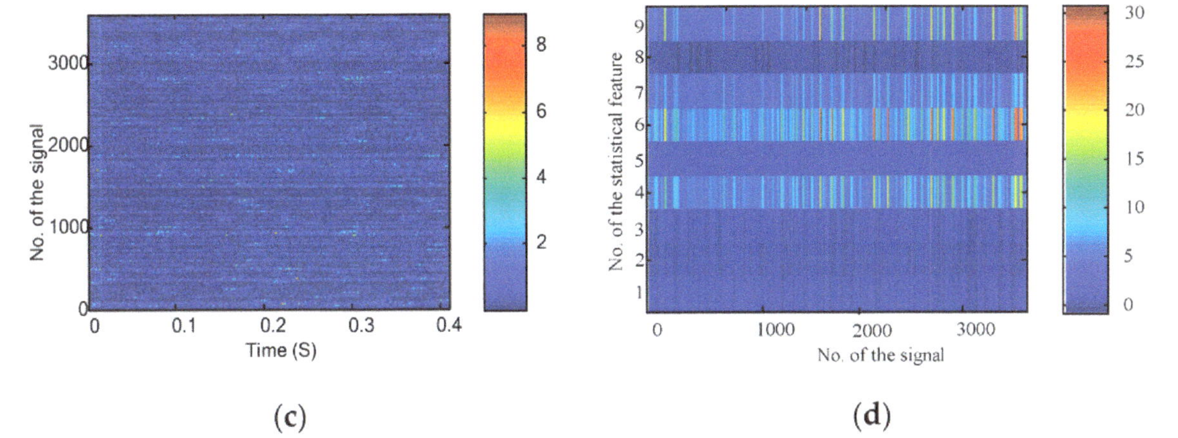

Figure 6: Time domain features for the gearbox fault diagnosis: (**a**) time domain waveform of the first signal; (**b**) time domain statistical features of the first signal; (**c**) time domain waveforms of the 3600 collected signals; and (**d**) time domain statistical features of the 3600 collected signals.

Vibration signals were then transformed into the frequency domain. The frequency domain representation and statistical features for the first signal are shown in Figure 7a,b, respectively. As the sampling frequency was 10 kHz, the effective frequency band in Figure 7a is [0, 5000] Hz. However, there are only 4096 points for the temporal waveform. This means that there are only 2048 frequency points ranging between [0, 5000] Hz. For all the collected 3600 signals, their frequency domain representations and statistical features are shown in Figure 7c,d, respectively.

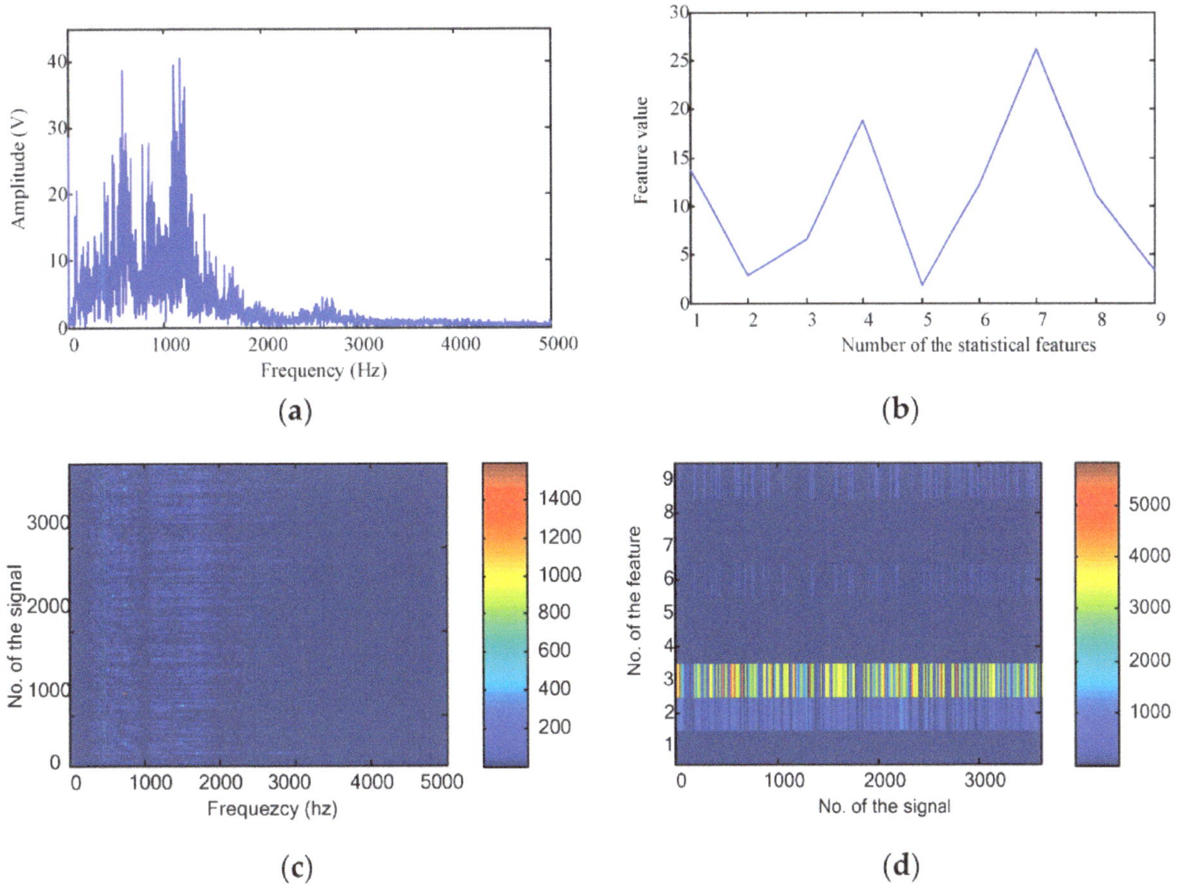

Figure 7: Frequency domain features for the gearbox fault diagnosis: (**a**) frequency domain representation of the first signal; (**b**) frequency domain statistical features of the first signal; (**c**) frequency domain representations of all the collected 3600 signals; and (**d**) frequency domain statistical features of all the collected 3600 signals.

For generating the time-frequency domain representations, the WPT was applied to decompose the raw data up to four levels. There are 2, 4, 8 and 16 nodes for each level. Put all the nodes together, the WPT presentation and statistical features are displayed in Figure 8a,b, respectively. As the length of the raw signal is 4096 points, numbers of data points for a node at the four levels are 2052, 1030, 519 and 264, respectively. In this way, the number of data points as shown in Figure 8a is 16,600. For the WPT, there are 30 nodes each of which has nine features. This generates 270 features for the first signal as shown in Figure 8b.

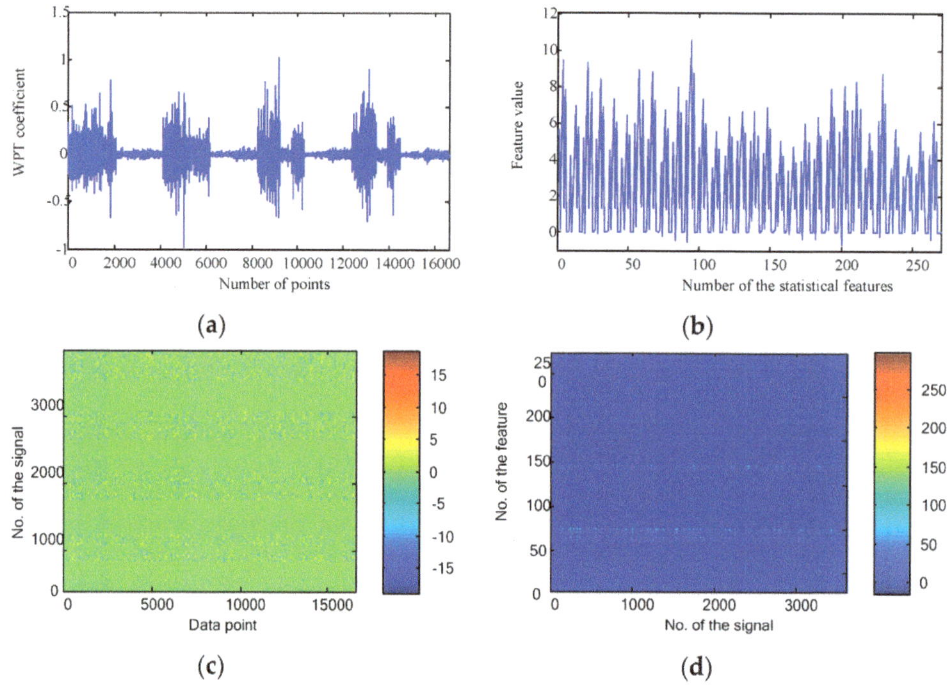

Figure 8: Time-frequency domain features for the gearbox fault diagnosis: (**a**) WPT representation of the first signal; (**b**) time-frequency domain statistical features of the first signal; (**c**) time-frequency domain representations of all the collected 3600 signals; and (**d**) time-frequency domain statistical features of all the collected 3600 signals.

All the 3600 data have been disordered for the experiments. Among all the 3600 samples for each data, 2400 samples were random chosen as the training dataset F. To represent the statistical feature set F, we first applied the mon-layer GRBM with parameters as: number of the neurons in the hidden layer = 200, number of the learning epochs = 150, the initial learning rate = 0.001, its upper-bound = 0.001, and the weight decay = 0.005. As unsupervised learning of the GRBM does not have the classification function, a multi-class SVM classifier was applied to obtain the first fault diagnosis model (# 1 peer model). The reason for us to implement the model is to show the performance of the present deep learning. For # 1 peer model, the GRBM acts as the second feature representation tool (statistical features given by Equation (7) is the first one) for the vibration measurements. The outputs of the GRBM were fed into the SVM classifier. The supervised GDBM was subsequently applied for the same dataset F with parameters as: number of the neurons in the hidden layer 1 = 200, number of the neurons in the hidden layer 2 = 200, number of the pretraining epochs (for each constituent and the model) = 150, number of the fine-tuning epochs = 150, the initial learning rate = 0.001, its upper-bound = 0.001, and the weight decay = 0.005. In this way, we obtained the second fault diagnosis model (the proposed GDBM model). For comparison, the SVM classifiers for the original statistical features M(p,q), and the combination of M(p,q) and the GRBM representation were respectively developed as the third fault diagnosis model (#2 peer model) and the fourth one (#3 peer model). All the algorithms were realized using Matlab®. One may note that in this work we have not employed more "shallow" learning models such as the decision tree, the random forest, and the neural network. The reason is that the SVM has been proven the prominent representative which outperforms most of the "shallow" learning members.

With the trained models, the remaining 1200 samples (in the time, frequency, and time-frequency domains, respectively) were used to test the classification performances, which are displayed in Table 2.

Table 2: Fault classification rates for the testing dataset (%), where N represents the device, d denotes the domain of the feature.

Device (*N*)	Domain (*d*)	Fault Diagnosis Model				
		#1 Peer	GDBM	#2 Peer	#3 Peer	Average [a]
Gearbox	Time domain	26.08	62.58	60.83	35.83	46.33
	Frequency domain	52.67	91.75	52.83	79.42	69.17
	Time-frequency domain	45.25	95.17	69.50	78.42	72.09
	Average [b]	41.33	83.17	61.05	64.56	62.53
Bearing	Time domain	18.52	60.63	59.58	41.96	45.17
	Frequency domain	39.95	87.57	80.74	82.91	72.79
	Time-frequency domain	58.84	91.75	81.53	82.70	78.71
	Average [b]	39.10	79.98	73.95	69.19	65.508

[a] the average value of the left four models; [b] the average value of the above three domains.

From the diagnosis results shown in Table 2, it is clear that the classification rates for the time-frequency domain statistical features are higher (72.09% on average) than those for the time and frequency domains. This is due to the joint time and frequency representation of the WPT. When comparing the statistical features of the time and frequency domains, the time domain features are always the worst. Among all the models, deep statistical feature learning via the GDBM exhibits the best classification rate for the same data (62.58%, 91.75%, 95.17%, and 83.17% for the time, frequency, time-frequency domain statistical features, and the average, respectively). The best classification rate of 95.17% is seen with the GDBM model and time-frequency statistical features. Compared to supervised learning methods (e.g., the GDBM), the unsupervised GRBM displays the lowest classification rates (26.67%, 52.67%, 42.25%, and 41.33% for the time, frequency, time-frequency domain statistical features, and the average, respectively). Nevertheless, it should be noted that the GRBM used in this paper is an unsupervised algorithm, which shows that there is still some potential for fault diagnosis, if a fine-tuning procedure can be introduced for its learning process. As one of the most important "shallow" learning approaches, the SVM exhibited good classification results for the gearbox system. This result is similar to that of existing studies (e.g., [46]). When the GRBM representations are combined with the original statistical features M(p,q), a small increase in the classification rates can be seen (from 52.83% to 79.42% for the frequency domain, 69.50% to 78.42% for the time-frequency domain statistical features, and 61.05% to 64.56% on average). However, due to the "shallow" learning limit, it is very difficult to further improve the classification rate for the SVM. Our results indicate that deep statistical feature learning has the best performance for gearbox fault diagnosis. It should be noted that deep learning is much more time-consuming than classical learning methods.

Bearing Fault Diagnosis Results

For the bearing fault diagnosis experiments, 5040 vibration signals and their statistical features in the time domain are plotted in Figure 9a,b. The Fourier transform were then used to generate the frequency data and their statistical features as shown in Figure 9c,d, respectively. The time-frequency representation produced by the 4-level WPT and their statistical features are shown in Figure 9e,f, respectively.

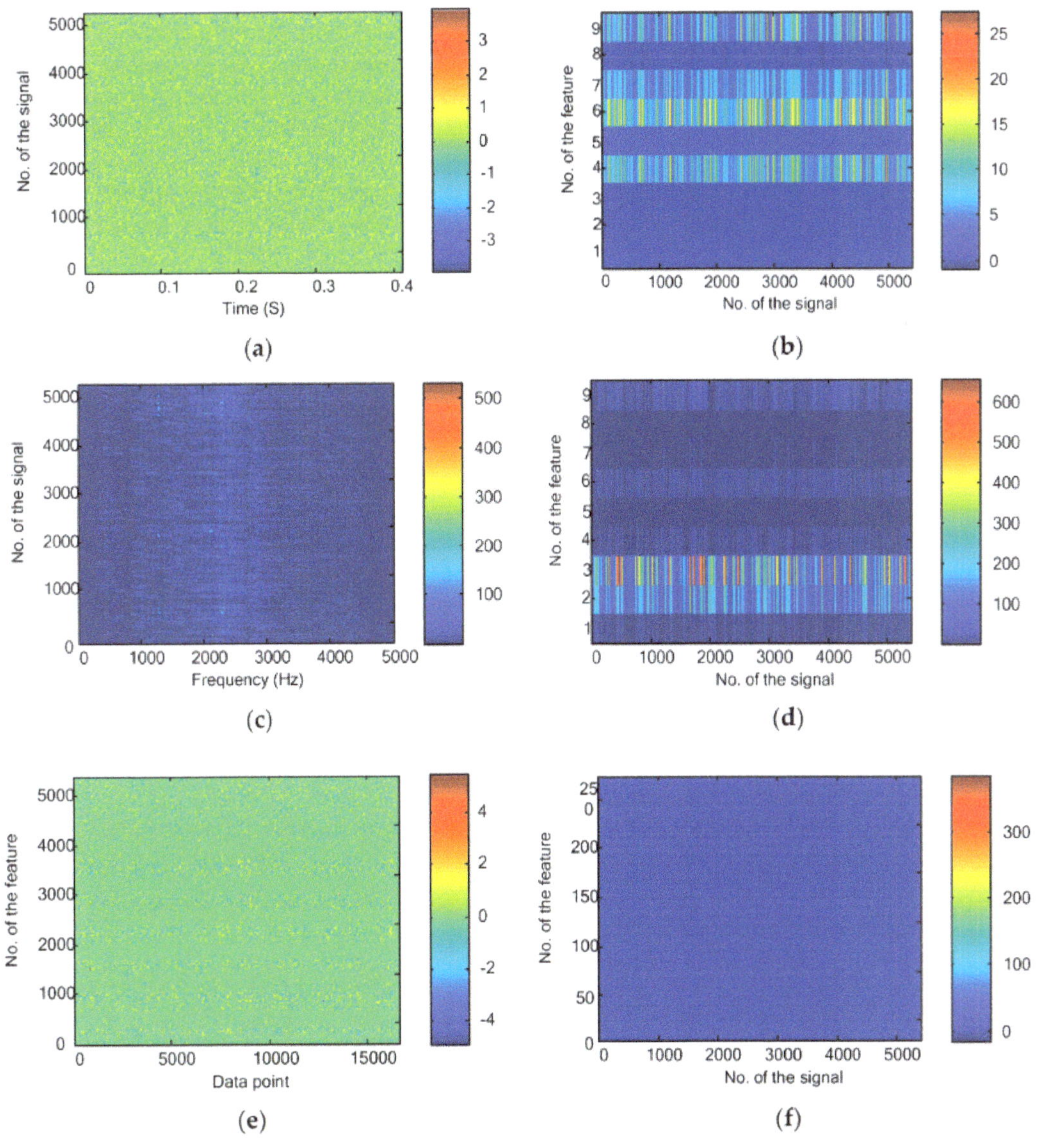

Figure 9: Bearing fault diagnosis experiments: (**a**) the time domain signals; (**b**) the time domain statistical features; (**c**) the frequency domain representations; (**d**) the frequency domain statistical features; (**e**) the WPT results; and (**f**) the time-frequency domain statistical features.

Of the 5040 samples, 3150 of the bearing system vibration signals were randomly chosen as the training dataset F. Similar modeling procedures to the gearbox fault diagnosis were repeated to develop the bearing fault diagnosis models. For comparisons, the same parameters are used in this subsection for the four models (i.e., No. 1: GRBM, No. 2: GDBM, No. 3: SVM, and No. 4: GRBM-SVM). After obtaining the trained models, the remaining 1890 samples (in the time, frequency, and time-frequency domains) were applied to test the classification performance for the bearing fault diagnosis. The results are displayed in Table 2.

A comparison of the feature performances in the different domains in Table 2 suggests that the time–frequency domain features exhibit the best performances (58.84%, 91.75%, 81.53% and 82.70% with the GRBM, GDBM, SVM and GRBM-SVM models, respectively), and the time domain features have the lowest classification rate (45.17% on average for all models). Compared with the gearbox fault diagnosis, the fault features for the rolling element bearings are more evident in the frequency domain, especially in the high frequency resonance band [47]. However, the model comparison results for the bearing fault diagnosis are almost same as those for the gearbox fault diagnosis. Among all the peer models, the deep statistical feature learning model (the GDBM) has the best classification rate (60.63% for the time domain, 87.57% for the frequency domain, 91.75% for the time-frequency domain, and 79.98% on average). This again validates the effectiveness of deep statistical feature learning for fault diagnosis in rotating machinery. Nevertheless, the improvement in fault diagnosis performance with deep learning is at the cost of complexity. The present GDBM is the most complex model with the largest number of parameters that must be estimated from the sample. It is an intrinsic drawback that "deeper" learning requires much more time than "shallow" learning does. As pointed out by LeCun et al. [48], the advent of fast graphics processing units (GPUs), which are convenient to program, allowed researchers to train deep networks 10 or 20 times faster. This indicates that parallel computation is helpful for reducing computation time. However, parallel computation is beyond the scope of this study. All the programs in this work were executed on a laptop. This resulted in much more computation time (hour-level) for the presented GDBM than its "shallow" counterparts (usually second- or minute- level on a laptop).

Remarks

Based on the fault diagnosis results as shown in the previous two subsections, one can see that the deep statistical feature learning holds the best classification performance comparing to the peer models. During the experiments, there were some very aberrant values (outliers) collected from the experimental setups, because the outliers are always unavoidable for real applications. It is obvious that the outliers may lead to deterioration of the fault diagnosis. However, we did not remove those outliers from the dataset, even if the removal of the outliers may increase the classification rates.

It should be noted that the given parameters also play important roles to the GDBM model. As indicated by Cho et al. [35], the training procedure of the GDBM can easily run into problems without careful selection of the learning parameters. Upon determining the network structure for different layers, therefore, the learning epochs for the pretraining and the fine-tuning will be directly related to the classification performance. In this subsection we will discuss the influence of the epochs for the pretraining and the fine-tuning procedure. We first adjusted the number of pretraining epochs (for the GRBMs and the presented model) with all the other parameters fixed. Figure 10a plots the change of the fault classification rates in response to the increase of the pretraining epochs for the WPT features. For the fault diagnosis of both the gearbox and the bearing systems, the number of pretraining epochs does not influence the classification very much. This means that even for a small number (e.g., 10) of the epochs, the pretraining can achieve good effect.

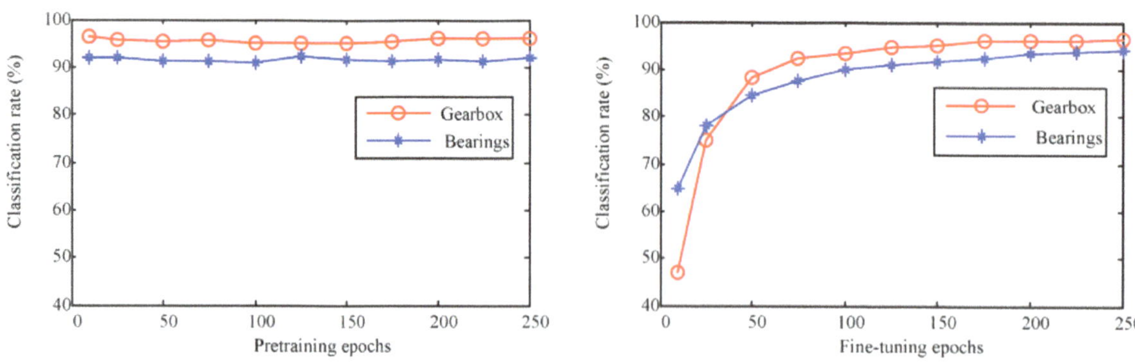

(a) (b)

Figure 10: Relationship between the classification rate and the number of the modeling epochs: (**a**) classification rates v.s. pretraining epochs; and (**b**) classification rates vs.fine-tuning epochs of the time-frequency domain GDBM models.

The pretraining epochs were subsequently set at 150 to adjust the number of the fine-tuning epochs between 10 and 250. The fault classification rates for the two mechanical systems were displayed in Figure 10b. With the increase of the fine-tuning epochs, the classification rates for both experiments improve accordingly. For the gearbox diagnosis experiments, the improvement goes slowly after 125 epochs. As for the bearing systems, the classification rate increases still evidently before 200 epochs. Figure 10a,b prove that the deep statistical feature learning using the GDBM is not very sensitive to the learning parameters. Even though, a careful selection of the model parameters will be helpful in improving the fault pattern classification for the rotating machinery. This means that the proposed method has an essential improvement potential for the fault diagnosis of the rotating machinery.

Figure 11a–c plots the comparisons between the real fault patterns and the classified patterns for the bearing fault diagnosis in the time, the frequency and the time-frequency domains, respectively. It is shown that some signals are correctly classified by the proposed method in some domains, but are misclassified in other domains. Let's take 10 vibration signals (#286~#295) of the bearing fault diagnosis as an example. The time domain diagnosis misclassified #295 signal, while both the frequency domain and the time-frequency domain obtained right classification. The frequency domain diagnosis misclassified #290 signal, but all the other two domains are correct. The time-frequency domain diagnosis misclassified #286 signal, while all the rest domains are right. If one considers the three domains simultaneously, therefore, all the 10 vibration signals (#286~#295) can be right diagnosed. This shows that the combination of the diagnosis results may contribute better classification rates. Though out of our scope in this paper, this discussion encourages us that further potentials can be explored for the proposed fault diagnosis approach.

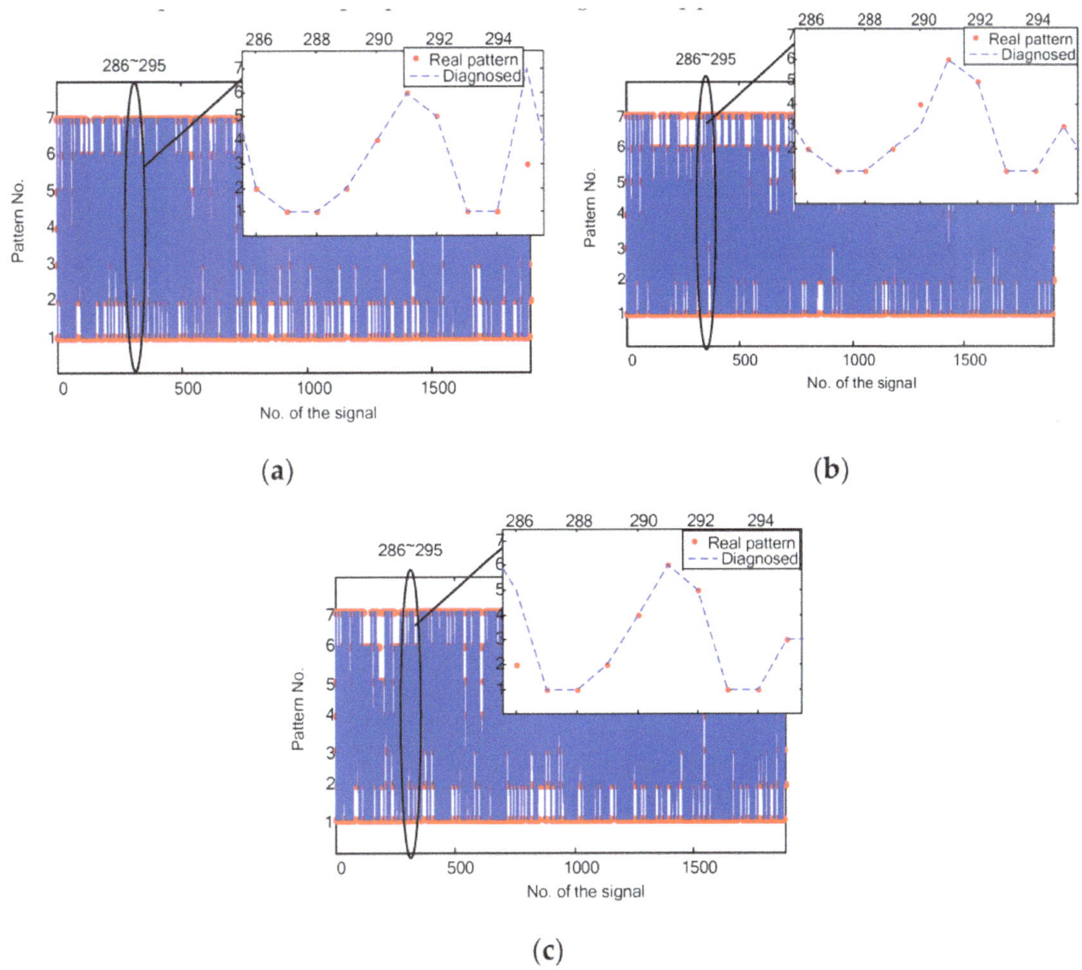

Figure 11: Bearing fault diagnosis results in different domains: (**a**) the time domain; (**b**) the frequency domain; and (**c**) the time-frequency domain.

CONCLUSIONS

In this paper, a deep statistical feature learning for vibration measurement has been proposed to diagnose fault patterns in rotating machinery. The statistical feature set was first extracted from the time, frequency, and time-frequency domains of the vibration signals. The real-valued RBMs were then stacked to develop a GDBM to accommodate statistical feature learning. Two typical rotating machinery systems (a gearbox and bearing test rigs), were constructed to validate the proposed approach, which was used for fault classification in the three-domain feature sets. The results show that deep statistical feature learning is capable of classifying fault patterns at higher rates than other models. Compared with the unsupervised GRBM, the SVM and the combined SVM and GRBM models, the deep statistical feature learning by the GDBM consistently had clearly better performances. This means that deep learning with statistical feature representation is a feasible update of conventional methods. The results also reveal that the statistical features in the time, frequency and time–frequency domains have different representation capabilities for fault patterns. Our further work will focus on optimizing the statistical features in different domains for different diagnostic applications.

ACKNOWLEDGMENTS

This work is supported in part by the Secretariat for Higher Education, Science, Technology and Innovation of the Republic of Ecuador (GIDTEC project No. 009-004-2015-07-16), the National Natural Science Foundation of China (Grant No. 51375517), and the Project of Chongqing Science & Technology Commission (Grant No. cstc2014gjhz70002). The valuable comments and suggestions from the two anonymous reviewers are very much appreciated.

AUTHOR CONTRIBUTIONS

Chuan Li designed the experiments and wrote the paper; Vinicio Sanchez contributed experimental set-up and materials; Grover Zurita and Mariela Cerrada analyzed the data; Diego Cabrera performed the experiments.

CONFLICTS OF INTEREST

The authors declare no conflict of interest.

REFERENCES

1. Lei, Y.; Lin, J.; Zuo, M.J.; He, Z. Condition monitoring and fault diagnosis of planetary gearboxes: A review. Measurement 2014, 48, 292–305.
2. Li, C.; Liang, M. Time-frequency signal analysis for gearbox fault diagnosis using a generalized synchrosqueezing transform. Mech. Syst. Sig. Process. 2012, 26, 205–217.
3. Batista, L.; Badri, B.; Sabourin, R.; Thomas, M. A classifier fusion system for bearing fault diagnosis. Expert Syst. Appl.2013, 40, 6788–6797.
4. Li, C.; Liang, M.; Wang, T. Criterion fusion for spectral segmentation and its application to optimal demodulation of bearing vibration signals. Mech. Syst. Sig. Process. 2015, 64–65, 132–148.
5. Gao, Z.; Cecati, C.; Ding, S.X. A survey of fault diagnosis and fault-tolerant techniques—Part I: Fault diagnosis with model-based and signal-based approaches. IEEE Trans. Ind. Electron. 2015, 62, 3757–3767.
6. Gao, Z.; Cecati, C.; Ding, S.X. A survey of fault diagnosis and fault-tolerant techniques—Part II: Fault diagnosis with knowledge-based and hybrid/active-based approaches. IEEE Trans. Ind. Electron. 2015, 62, 3768–3774.
7. Qin, Q.; Jiang, Z.N.; Feng, K.; He, W. A novel scheme for fault detection of reciprocating compressor valves based on basis pursuit wave matching and support vector machine. Measurement 2010, 45, 897–908.
8. Wong, W.K.; Loo, C.K.; Lim, W.S.; Tan, P.N. Thermal condition monitoring system using log-polar mapping, quaternion correlation and max-product fuzzy neural network classification. Neurocomputing 2010, 74, 164–177.
9. Arumugam, V.; Sidharth, A.A.P.; Santulli, C. Failure modes characterization of impacted carbon fibre reinforced plastics laminates under compression loading using acoustic emission. J. Compos. Mater. 2014, 48, 3457–3468.
10. Li, C.; Peng, J.; Liang, M. Enhancement of the wear particle monitoring capability of oil debris sensors using a maximal overlap discrete wavelet transform with optimal decomposition depth. Sensors 2014, 14, 6207–6228.
11. Raad, A.; Antoni, J.; Sidahmed, M. Indicators of cyclostationarity: Theory and application to gear fault monitoring.Mech. Syst. Sig. Process. 2008, 22, 574–587.
12. Bartelmus, W.; Zimroz, R. A new feature for monitoring the condition of gearboxes in non-stationary operating conditions. Mech. Syst. Sig. Process. 2009, 23, 1528–1534.
13. Randall, R.B.; Antoni, J. Rolling element bearing diagnostics—A tutorial. Mech. Syst. Sig. Process. 2011, 25, 485–520.
14. Li, C.; Liang, M. Continuous-scale mathematical morphology-based optimal scale band demodulation of impulsive feature for bearing defect diagnosis. J. Sound Vib. 2012, 331, 5864–5879.
15. Zuo, M.J.; Lin, J.; Fan, X. Feature separation using ICA for a one-dimensional time series and its application in fault detection. J. Sound Vib. 2005, 287, 614–624.
16. Kumar, R.; Singh, M. Outer race defect width measurement in taper roller bearing using discrete wavelet transform of vibration signal. Measurement 2013, 46, 537–545.
17. Shen, C.; Wang, D.; Kong, F.; Tse, P.W. Fault diagnosis of rotating machinery based on the statistical parameters of wavelet packet paving and a generic support vector regressive classifier. Measurement 2013, 46, 1551–1564.
18. Cao, H.; Chen, X.; Zi, Y.; Ding, F.; Chen, H.; Tan, J.; He, Z. End milling tool breakage detection using lifting scheme and Mahalanobis distance. Int. J. Mach. Tools Manuf. 2008, 48, 141–151.
19. Wang, D.; Tse, P.W.; Guo, W.; Miao, Q. Support vector data description for fusion of multiple health indicators for enhancing gearbox fault diagnosis and prognosis. Meas. Sci. Technol. 2011, 22, 025102.
20. Li, C.; Cabrera, D.; Valente de Oliveira, J.; Sanchez, R.V.; Cerrada, M.; Zurita, G. Extracting repetitive transients for rotating machinery diagnosis using multiscale clustered grey infogram. Mech. Syst. Sig. Process. 2016, 76–77, 157–173.
21. Li, C.; Sanchez, V.; Zurita, G.; Lozada, M.C.; Cabrera, D. Rolling element bearing defect detection using the generalized synchrosqueezing transform guided by time-frequency ridge enhancement. ISA Trans. 2016, 60, 274–284.
22. Lei, Y.; Kong, D.; Lin, J.; Zuo, M.J. Fault detection of planetary gearboxes using new diagnostic parameters. Meas. Sci. Technol. 2012, 23, 055605.
23. Cerrada, M.; Sánchez, R.V.; Cabrera, D.; Zurita, G.; Li, C. Multi-stage feature selection by using genetic algorithms for fault diagnosis in gearboxes based on vibration signal. Sensors 2015, 15, 23903–23926.
24. Chen, F.; Tang, B.; Chen, R. A novel fault diagnosis model for gearbox based on wavelet support vector machine with immune genetic algorithm. Measurement 2013, 46, 220–232.
25. Zhao, M.; Jin, X.; Zhang, Z.; Li, B. Fault diagnosis of rolling element bearings via discriminative subspace learning: visualization and classification. Expert Syst. Appl. 2014, 41, 3391–3401.
26. Tayarani-Bathaie, S.S.; Vanini, Z.N.S.; Khorasani, K. Dynamic neural network-based fault diagnosis of gas turbine engines. Neurocomputing 2014, 125, 153–165.

27. Ali, J.B.; Fnaiech, N.; Saidi, L.; Chebel-Morello, B.; Fnaiech, F. Application of empirical mode decomposition and artificial neural network for automatic bearing fault diagnosis based on vibration signals. Appl. Acoust. 2015, 89, 16–27.
28. Tamilselvan, P.; Wang, P. Failure diagnosis using deep belief learning based health state classification. Reliab. Eng. Syst. Saf. 2013, 115, 124–135.
29. Tran, V.T.; Thobiani, F.A.; Ball, A. An approach to fault diagnosis of reciprocating compressor valves using Teager-Kaiser energy operator and deep belief networks. Expert Syst. Appl. 2014, 41, 4113–4122.
30. Wang, Y.; Xu, G.; Liang, L.; Jiang, K. Detection of weak transient signals based on wavelet packet transform and manifold learning for rolling element bearing fault diagnosis. Mech. Syst. Sig. Process. 2015, 54–55, 259–276.
31. Randall, R.B. Vibration-Based Condition Monitoring: Industrial, Aerospace and Automotive Applications; John Wiley & Sons: Chichester, UK, 2011.
32. Li, C.; Sanchez, R.V.; Zurita, G.; Cerrada, M.; Cabrera, D.; Vásquez, R.E. Multimodal deep support vector classification with homologous features and its application to gearbox fault diagnosis. Neurocomputing 2015, 168, 119–127.
33. Hinton, G.E.; Salakhutdinov, R.R. Reducing the dimensionality of data with neural networks. Science 2006, 313, 504–507.
34. Hjelma, D.; Calhouna, V.; Salakhutdinov, R.; Allena, E.; Adali, T.; Plisa, S. Restricted Boltzmann machines for neuroimaging: An application in identifying intrinsic networks. Neuroimage 2014, 96, 245–260.
35. Cho, K.H.; Ilin, A.; Raiko, T. Improved Learning of Gaussian-Bernoulli Restricted Boltzmann Machines. In Artificial Neural Networks and Machine Learning—ICANN 2011; Springer Berlin Heidelberg: Berlin, Germany, 2011; Volume 6791, pp. 10–17.
36. Salakhutdinov, R. Learning Deep Generative Models. Ph.D. Thesis, University of Toronto, Toronto, ON, Canada, 2009.
37. Hastie, T.; Tibshirani, R. Classification by Pairwise Coupling. In Advances in Neural Information Processing Systems 10; MIT Press: Cambridge, MA, USA, 1998.
38. Cho, K.H.; Raiko, T.; Ilin, A. Gaussian-Bernoulli deep Boltzmann machine. In Proceedings of the 2013 International Joint Conference on Neural Networks, Dallas, TX, USA, 4–9 August 2013; pp. 1–7.
39. Schmidhuber, J. Deep learning in neural networks: An overview. Neural Netw. 2015, 61, 85–117.
40. Salakhutdinov, R. Learning in Markov Random Fields Using Tempered Transitions. In Advances in Neural Information Processing Systems 22; MIT Press: Cambridge, MA, USA, 2009; pp. 1598–1606.
41. Bishop, C.M. Neural Networks for Pattern Recognition; Clarendon Press: Oxford, UK, 1996.
42. Nabney, I.T. NETLAB: Algorithms for Pattern Recognition; Springer-Verlag: London, UK, 2001.
43. Hinton, G.E.; Osindero, S.; The, Y.W. A fast learning algorithm for deep belief nets. Neural Comput. 2006, 18, 1527–1554.
44. Khomfoi, S.; Tolbert, L.M. Fault diagnostic system for a multilevel inverter using a neural network. IEEE Trans. Power Electron. 2007, 22, 1062–1069.
45. Bartkowiak, A.; Zimroz, R. Outliers analysis and one class classification approach for planetary gearbox diagnosis. J. Phys. Conf. Ser. 2011, 305, 012031.
46. Souza, D.L.; Granzotto, M.H.; Almeida, G.M.; Oliveira-Lopes, L.C. Fault detection and diagnosis using support vector machines—A SVC and SVR comparison. J. Saf. Eng. 2014, 3, 18–29.
47. Hou, S.; Li, Y.; Wang, Z. A resonance demodulation method based on harmonic wavelet transform for rolling bearing fault diagnosis. Struct. Health Monit. 2010, 9, 297–308.
48. LeCun, Y.; Bengio, Y.; Hinton, G.E. Deep learning. Nature 2015, 521, 436–444.

CHAPTER 4 MEASUREMENT OF OUTFLOW FACILITY USING IPERFUSION

Joseph M. Sherwood[1], Ester Reina-Torres[1], Jacques A. Bertrand[1], Barnaby Rowe[2], Darryl R. Overby[1]

[1] Department of Bioengineering, Faculty of Engineering, Imperial College London, London, United Kingdom,

[2] Department of Physics and Astronomy, Faculty of Mathematical and Physical Sciences, University College London, London, United Kingdom

ABSTRACT

Elevated intraocular pressure (IOP) is the predominant risk factor for glaucoma, and reducing IOP is the only successful strategy to prevent further glaucomatous vision loss. IOP is determined by the balance between the rates of aqueous humour secretion and outflow, and a pathological reduction in the hydraulic conductance of outflow, known as outflow facility, is responsible for IOP elevation in glaucoma. Mouse models are often used to investigate the mechanisms controlling outflow facility, but the diminutive size of the mouse eye makes measurement of outflow technically challenging. In this study, we present a new approach to measure and analyse outflow facility using iPerfusion™, which incorporates an actuated pressure reservoir, thermal flow sensor, differential pressure measurement and an automated computerised interface. In enucleated eyes from C57BL/6J mice, the flow-pressure relationship is highly non-linear and is well represented by an empirical power law model that describes the pressure dependence of outflow facility. At zero pressure, the measured flow is indistinguishable from zero, confirming the absence of any significant pressure independent flow in enucleated eyes. Comparison with the commonly used 2-parameter linear outflow model reveals that inappropriate application of a linear fit to a non-linear flow-pressure relationship introduces considerable errors in the estimation of outflow facility and leads to the false impression of pressure-independent outflow. Data from a population of enucleated eyes from C57BL/6J mice show that outflow facility is best described by a lognormal distribution, with 6-fold variability between individuals, but with relatively tight correlation of facility between fellow eyes. iPerfusion represents a platform technology to accurately and robustly characterise the flow-pressure relationship in enucleated mouse eyes for the purpose of glaucoma research and with minor modifications, may be applied in vivo to mice, as well as to eyes from other species or different biofluidic systems

Citation: Sherwood JM, Reina-Torres E, Bertrand JA, Rowe B, Overby DR (2016) Measurement of Outflow Facility Using iPerfusion. PLoS ONE 11(3): e0150694. doi:10.1371/journal.pone.0150694

INTRODUCTION

Intraocular pressure (IOP) is regulated by the balance between the secretion of aqueous humour (AH) and its outflow across the hydrodynamic resistance of the conventional outflow pathway. An increase in outflow resistance can lead to ocular hypertension, which is a major risk factor in glaucoma; a disease for which a reduction of IOP is the sole therapeutic target for sight preservation [1–3]. The physiology of AH outflow is complex, involving flow through both the trabecular meshwork and Schlemm's canal (conventional outflow pathway) and the ciliary muscle, choroid and sclera (unconventional outflow pathway). Although the conventional pathway is the predominant outflow route in humans and increased resistance of this pathway is largely responsible for the elevated IOP observed in glaucoma [4], current medical treatments generally either promote unconventional outflow or reduce AH secretion [5, 6]. An increasing amount of research is therefore being aimed at developing treatments that target the root cause of ocular hypertension, namely increased resistance of the conventional outflow pathway [7, 8].

Mouse eyes are becoming an increasingly popular model for investigating the mechanisms of outflow and the hydrodynamic conductance of the conventional outflow pathway, known as outflow facility (the reciprocal of hydrodynamic resistance). Recent studies have shown that this pathway is anatomically similar in mice to that in human eyes [9, 10] and exhibits a comparable pharmacological response to drugs that both increase and decrease outflow facility [11–14]. In addition, the ability to genetically manipulate mice enables investigation of the genomic factors that regulate outflow physiology [15–20]. The considerable disadvantage of mouse eyes is their diminutive size, being approximately 3% of the volume of AH compared to human eyes [21,22], which corresponds to extremely low flow rates, on the order of 50 nl/min at physiological IOP. This makes the assessment of outflow facility extremely sensitive to uncertainties in the methods of measurement and data analysis.

The present study describes iPerfusion™, a new paradigm for the measurement of outflow facility by ocular perfusion, incorporating measurement techniques, data analysis and presentation methods. Although the present study focuses on enucleated mouse eyes, the system can be scaled up for facility measurements in any species, and with minor modifications may be applied in vivo. To provide context, a brief review of existing techniques is provided in the following section.

Existing Techniques to Measure Outflow Facility

The commonly used approach for measuring outflow facility is based on mass conservation of the flow entering and exiting the eye during an in vivo perfusion according to:

$$Q_{in} + Q = C(P - P_e) + Q_0 \qquad (1)$$

which is known as the modified Goldmann equation [23]. Q_{in} is the rate of AH secretion, Q is the flow rate into the eye from the perfusion apparatus and Q_0 is the pressure-independent outflow. P is the intraocular pressure and P_e is the pressure in the episcleral vessels (into which the AH drains). In this form, C is the total outflow facility, comprising both conventional outflow and any pressure-dependent components of unconventional outflow and AH secretion (pseudofacility) [23]. Herein we will use the term 'facility' to indicate C, for simplicity. In order to calculate facility, Q_0 and Q_{in}, P_e and C itself are often assumed to be pressure independent (thereby tacitly assuming a linear Q – P relationship). Under these assumptions, two measurements of P and Q are thus sufficient to estimate the facility according to the two-step perfusion protocol [24]:

$$C_{lin} = \frac{Q_{II} - Q_I}{P_{II} - P_I} \qquad (2)$$

where the subscripts I and II denote measurements at two different pressures, and C_{lin} is a pressure independent facility, based on the assumption of a linear Q – P response. Alternatively, for the case of enucleated eyes, Q_{in} and P_e are zero, hence Eq 1 reduces to

$$Q = C_{lin}P + Q_0 \qquad (3)$$

In order to provide a more robust method than using Eq 2, it has become common in mouse eye perfusions to measure multiple (3–5) points and fit a straight line to the Q – P relationship, assuming that the facility, given by the slope of the line, C_{lin}, is pressure independent. The unconventional outflow is sometimes estimated from the intercept using this approach [11–13,20, 21, 25], although others note that the validity of the method is questionable [26–28]. Fitting a straight line to the data to estimate facility is only appropriate if C is independent of pressure and Q_0 has a finite value. The validity of these assumptions in the context of enucleated mouse eyes will be investigated in the present study.

A number of methodologies for assessing the flow rate into mouse eyes from perfusion systems have been reported in the literature. These may be grouped into (1) syringe pump based systems, in which the flow into the eye is prescribed and the pressure in the eye measured, (2) pressure-decay based systems, in which the flow rate and pressure are inferred based on the change in height of a small diameter fluid reservoir, and (3) constant pressure systems, in which a large diameter fluid reservoir provides a relatively constant eye pressure and the flow rate is measured, for example, based on the pressure drop along a capillary tube of known resistance.

Syringe Pump Based Systems.

In these systems, a syringe pump is used to deliver a nominal flow rate, and the pressure in the eye is monitored. Syringe pumps function via a lead screw driven by stepper motor, and at the low flow rates required for mouse eye perfusions, the finite steps may introduce pulsations in the flow. These pulsations are proportional to the square of the inner diameter of the syringe and the step size, and thus use of appropriately small syringe sizes and high resolution pumps is critical. The simplest pump based approach is to apply a constant flow rate and allow the system to naturally reach a steady state pressure for each flow rate [12], although this may take a very long time (30–40 min) [16]. In order to decrease the time to reach a stable condition, active control of the flow rate can be used to regulate the measured pressure, often with a pre-pressurisation stage at each step to further improve the transient response of the system [11,16, 21, 25, 29, 30].

In pump based systems, a number of flow or pressure steps are analysed, and a straight line is fitted to the steady state values. The selection of steady state is often not clearly defined and is prone to subjectivity, particularly in active control systems. Furthermore, very few studies using pump based systems to measure outflow facility validate the pump output or elucidate the control algorithm used to obtain and regulate the desired pressure (in systems with active control), which complicates verification of the results by other research groups.

Pressure-decay Based Systems.

In pressure-decay based systems, a small diameter (D) reservoir open to atmospheric pressure is used to apply a hydrostatic pressure to the eye. As fluid flows from the reservoir into the eye, the reservoir height decreases over time, t. In the classic 'two-step' perfusion method developed for primates [24], the flow rate is measured based upon the change of mass, m, of the fluid reservoir, but this would not be suitable for mouse eyes due to the low flow rates (over a 10 minute period under physiological conditions, the change in mass for an average mouse eye would be less than 1 mg). Aihara et al. [21], estimated the flow rate by observing the change in height of the fluid reservoir, h_r, and related this value to the volume perfused into the eye over a certain time period. Camras et al. [31] instead related the pressure gradient (based on a straight line fit to P vs t) to the change in height, so as to estimate a flow rate. These three approaches are mathematically equivalent and can be written as:

$$Q = \frac{1}{\rho}\frac{dm}{dt} = \frac{\pi D^2}{4}\frac{dh}{dt} = \frac{\pi D^2}{4\rho g}\frac{dP}{dt} \qquad (4)$$

where ρ is the density of the fluid in the reservoir and g is the gravitational acceleration. In all three forms of Eq 4 (based on change in weight, height or pressure), the flow rate is inferred based on the derivative of a measured parameter, generally estimated from two time points. Consequently, instantaneous measurements of the flow rate are not possible, and the methods yield only an average value over the evaluated time period, typically on the order of minutes. Furthermore, as the reservoir height is used to infer both the flow rate and the pressure, a trade-off exists wherein a higher resolution in measurement of the flow rate, achieved by reducing D or increasing the time period, is accompanied by a greater change in the applied pressure. Additional sources of error in these methods include estimation of the height of the reservoir, non-linearity in the pressure sensor or weight measurement, inaccuracies in specification of D (as the flow rate scales with D^2) and evaporation. For these reasons, along with the uncertainty in differentiating inherently noisy experimental data [32], it is preferential to acquire data on a parameter that is proportional to the flow rate itself [33].

Constant Pressure Systems.

In constant pressure systems, a reservoir with a large diameter can be used to apply a relatively constant pressure, whilst measuring the flow rate directly. Typically, the flow rate is calculated based on the pressure drop along a capillary tube of known diameter, located upstream of the eye, using the Hagen-Poiseuille law. Ethier et al. [33] used a differential pressure sensor to measure the pressure drop across a 150 μm capillary tube, with a second pressure sensor to measure the IOP. The pressure drop across a capillary tube is highly dependent on temperature (water viscosity decreases by more than 30% as temperature increases from 20 to 37°C, ≈2% per degree), hence in their system Ethier et al. [33] immersed the capillary tubing in a temperature controlled water bath. The system was tested in the range of facilities relevant to human eyes (≈200 nl/min/mmHg). Kee et al. [34] proposed a modified version of this system in which a number of components added for accuracy were omitted in the interest of simplicity. They tested their system using rat eyes (C ≈ 50 nl/min/mmHg). This approach has not been adopted for studies in mouse eyes, which have considerably lower facilities (C ≈ 5 nl/min/mmHg). In addition to the requirement for temperature regulation and accurate pressure measurement, a small diameter capillary would be needed to generate an accurately measurable pressure drop (≈75 μm or less). From a practical perspective, such small diameter capillaries are prone to blockages and biofilm formation, which would drastically alter the hydrodynamic resistance of the tube and hence the estimation of the flow rate.

In the present study, we use a large diameter reservoir and commercially available thermal flow sensor, which enables direct, repeatable, and accurate measurement of the flow rate. The flow rate is calculated based on thermal gradients along a capillary tube, induced by a small heating element. Knowledge of the applied pressure, the resistance of the flow sensor and the flow rate through the sensor provides redundant information, which can be used to monitor the functioning of the flow sensor itself (for example, to detect blockages).

In summary, measurements of outflow facility in mice are technically challenging and extremely sensitive to small errors. Although a number of techniques have been described in the literature, none have been sufficiently validated and all have considerable limitations. To overcome these challenges, we have developed the iPerfusion system.

METHODS

iPerfusion comprises hardware, control software, protocol, data analysis and statistical methodologies, and graphical representations, that together maximise the scientific value of each eye perfused. This section is subdivided into data acquisition (including validation of the system) and statistical analysis.

Data acquisition

Experimental Setup.

The hardware consists of three main components; an actuated reservoir, a thermal flow sensor and a differential pressure transducer (Fig 1a). On either side of the pressure transducer, a manifold (not shown) is used to control the flow path, in order to switch between configurations required for sensor calibration and system validation. The pressure applied to the system is controlled using a 10 ml reservoir, filled with water. The reservoir is coupled to a vertically mounted linear actuator (L35; Nanotec, Germany). The fluid passes from the actuated reservoir through the flow sensor (SLG64-0075; Sensirion AG, Switzerland), which comprises a 75 μmdiameter glass capillary, onto which two temperature sensing elements are bonded on either side of a heating element. Based on knowledge of the flow characteristics (Poiseuille) and the heat transfer properties of water, the flow rate can be calculated in the range −5000 to 5000nl/min. The pressure difference across the outflow pathway is measured using a silicon micromachined wet-wet differential pressure transducer (0–50 mmHg; PX409; Omegadyne, USA). Supporting Information 1 provides analysis of the accuracy of the flow and pressure sensors. The results show that the uncertainties, in the range appropriate for mouse eye perfusions, are given by ±4 nl/min and ±0.024 mmHg (two standard deviations (SD)) for the flow and pressure sensors respectively.

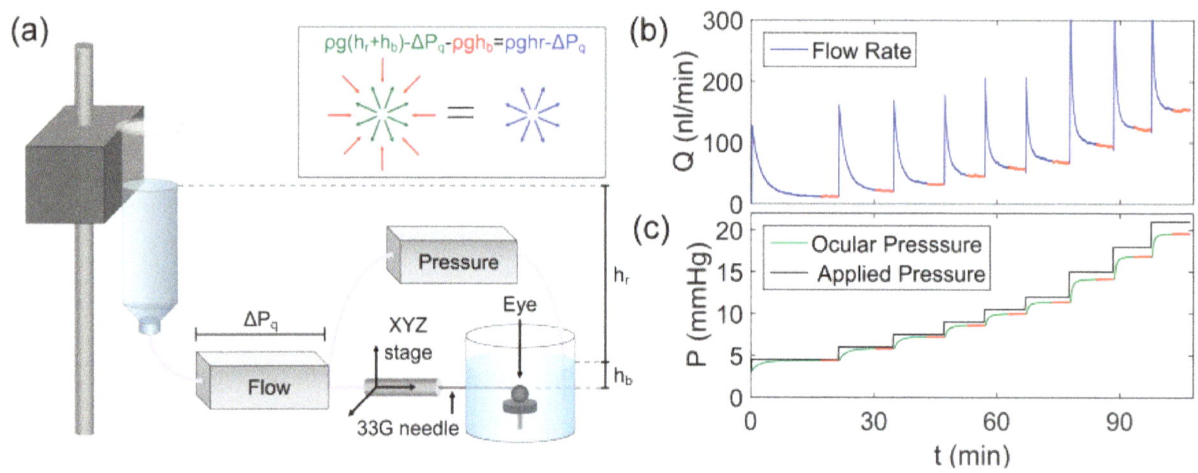

Figure1: The iPerfusion system. (a) Schematic of the experimental setup. Inset shows internal (green), external (red) and resultant (blue) pressures acting on the eye. Flow (b) and pressure (c) traces from a sample mouse eye perfusion. Red highlighted regions show steady-state periods, over which data were averaged. http://dx.doi.org/10.1371/journal.pone.0150694.g001

The eye is completely submerged in a relatively large volume of phosphate buffered saline (PBS; 20–40 ml), which is regulated at 35 ± 0.5°C. Submersion ensures that the temperature and hydration of the eye remain constant throughout the experiment. The submersion depth has no effect on the pressure drop across the outflow pathway for the following reason. As shown in Fig 1a, hydrostatic pressure acting on the outside of the eye is given by $\rho g h_b$, whereh_b is the submersion depth of the eye. For the case of zero flow into the eye, the pressure inside the eye is given by $\rho g(h_b + h_r)$, where h_r is the height of the reservoir relative to the surface of the fluid in the eye bath. The difference between the internal and external pressures, is given by $\rho g h_r$, which we refer to as the applied pressure, P_a. As the flow sensor consists of a small diameter capillary, it has a hydrodynamic resistance that results in a pressure drop, $\Delta P_q = Q(t)/C_q$ across the sensor, in the general case of non-zero flow. The measured pressure in the eye, P(t) is therefore related to the applied pressure, P_a, according to

$$P(t) = \rho g(h_r + h_b) - \Delta P_q(t) - \rho g h_b = P_a - Q(t)/C_q \quad \textbf{(5)}$$

which is independent of submersion depth. Due to its location downstream of the flow sensor, the wet-wet differential pressure transducer directly measures P(t), and therefore would not be affected by changes in h_b, for example due to evaporation from the heated eye bath or reservoir.

For the present study, two duplicates of the experimental setup were used, so that both eyes from a given mouse could be measured simultaneously, eliminating the influence of post-mortem time between paired eyes. The time between enucleation and cannulation was generally less than 30 minutes.

Experimental Protocol.

The basic protocol for each perfusion comprises three sections: calibration/testing of the system, cannulation of the eye and acclimatisation, and a multi-step perfusion regimen.

For each day of experiments, the pressure sensor was calibrated using an automated 8-point calibration with the automated linear actuator, which has a resolution of 1.25 μm per step. Subsequently, the resistance of the flow sensor was measured to ensure that it was blockage free. Prior to each perfusion, the resistance of a glass capillary of known resistance (comparable to a mouse eye) was measured to confirm that the system was functioning properly. The perfusion tubing (downstream of the flow sensor) and needle were then filled with perfusate and the resistance of the needle was measured before cannulating the eye. Note that, as the fluid in the system was water, there was a water-perfusate interface in the perfusion tubing. However, given that the volume of the tubing (≈2ml) was large compared to the total volume perfused in an experiment (≈25μl), neither advection nor diffusion could have altered the perfusate entering the eye. If any of the readings from the tests were unexpected, for example if the system time-response was too long or measured resistances were not correct, leaks/bubbles/blockages were identified and removed before the ocular perfusion commenced.

All animal experiments were done ex vivo in accordance with the Animals (Scientific Procedure) Act with the authority of a UK Home Office project licence (PPL 70/7306). All perfusions were carried out on 10–14 week old male C57BL/6J (B6) mice (Charles River UK, Ltd.), that were euthanised via cervical dislocation. All mice were fed ad libitum and housed in clear cages at 21°C with a 12 hour light-dark cycle (lights on at 7AM).

After enucleation, the eyes were stored in PBS at room temperature to await perfusion. Each eye was then affixed to a platform inside the heated bath with a small amount of cyanoacrylate glue (Loctite, UK). An XYZ micromanipulator (World Precision Instruments, USA) was used to cannulate the eye via the anterior chamber with a 33 gauge needle (Nanofil, World Precision Instruments, USA) under a stereomicroscope. The control perfusate was DBG: Dulbecco's PBS containing divalent cations, supplemented with 5.5 mM glucose and passed through a 0.2μm filter. After cannulation, the bath was filled with PBS to fully immerse the eye and the temperature was raised to 35°C. The applied pressure was held at 9 mmHg for a period of 30–45 minutes to allow the eye to acclimatise to the pressure and temperature environment.

After acclimatisation, a nine-step perfusion protocol was carried out, consisting of applied pressures of 4.5, 6, 7.5, 9, 10.5, 12, 15, 18 and 21 mmHg. A sample perfusion tracing is shown in Fig 1b and 1c.

In order to avoid a subjective element in the definition of steady state at each pressure step, a steady state condition was defined and automatically monitored by the perfusion software. A parameter $\Gamma(t) = Q(t)/P(t)$ was continuously evaluated and $d\Gamma/dt$ was estimated by linear regression over a moving window of 5 minutes. When $|d\Gamma/dt|$ was continuously less than 0.1nl/min/mmHg/min for one minute, the system was considered to be at steady state. The measured flow and pressure were then averaged over the four previous minutes to yield Q_j and P_j respectively, for pressure step j. The red lines in Fig 1b and 1c indicate these averaging periods. This approach ensures that the measured facility changes by less than 0.4nl/min/mmHg over the averaging period. For the eyes in the present study, the median change in facility over the averaging window was 0.16 nl/min/mmHg.

System Validation.

In order to ascertain the accuracy of the pressure and flow sensors, independent tests were carried out and are described in. A series of in vitro tests were then carried out to investigate the accuracy of the system in measuring facility (hydrodynamic conductance). A number of lengths of thick walled borosilicate glass capillaries (CM Scientific, UK) with inner diameter of 51 ± 5 μm were used to create resistances comparable to the outflow resistances measured in mouse eyes. The capillaries were cut to lengths, L_c, of approximately 50, 100 and 150 mm. Additional lengths of 250 mm and 350 mm were achieved by adding capillaries in series. For each length of capillary, a 7 step perfusion (steps of 4 mmHg starting at 4 mmHg) was carried out three times. For each pressure step j, the uncertainty in the pressure was negligibly small, and the uncertainty in the flow rate (given as a variance), s^2_{Qj}, was calculated by adding the variance of the measured flow rate over a 30 second window, s^2_{Qave}, and the uncertainty of the flow sensor, s^2_{Qsens}. The value of the hydrodynamic conductance of each length of capillary, C, was calculated for each test by fitting the line Q =CP to the Q_j and P_j values, using weighted regression with weights defined according to $1/s^2_{Qj}$.

Data based on regression analysis are presented as the mean ± the margin of error at a 95% confidence level, (ME_{95}), which is defined as the half-width of the 95% confidence interval (CI). When analysing the variability within a sample, two standard deviations will be reported, which indicate the range of values in which 95% of the population are expected to lie: we will refer to this as 'two-sigma' herein.

Fig 2 shows the measured C with the 95% CI for each of the repeated experiments at each capillary length. The red line and thin grey lines show the best fit and 95% confidence bounds, respectively, to the equation $C = 1/(R_L L_c+R_0)$, where R_L is the resistance per unit length and R_0 represents the resistance of the system downstream of the flow sensor. This model is used as hydrodynamic resistance (the reciprocal of conductance) is proportional to length. The model was fitted using weighted regression, with weights defined according to $1/s^2_C$, where s_C is assumed to be equivalent to the 68% ME on C. The resistance per unit length was found to be R_L = 0.627 ± 0.029 mmHg/μl/min/mm (mean ± ME_{95}), while R_0 = 1.50 ± 2.25 mmHg/μl/minwas not significantly different from zero. To confirm that the system was measuring correct values in absolute terms, the diameter of the capillary based on Poiseuille flow (at 22°C) was calculated from R_L and its confidence intervals. The predicted diameter was 52.8 ± 0.6μm, which is within the manufacturer's specification (see above). The root mean square deviation from the best fit was 0.8 nl/min/mmHg, and thus a conservative estimate of two-sigma for the measurement of facility could be given as ±1.6 nl/min/mmHg over the range of facilities relevant for mouse eyes.

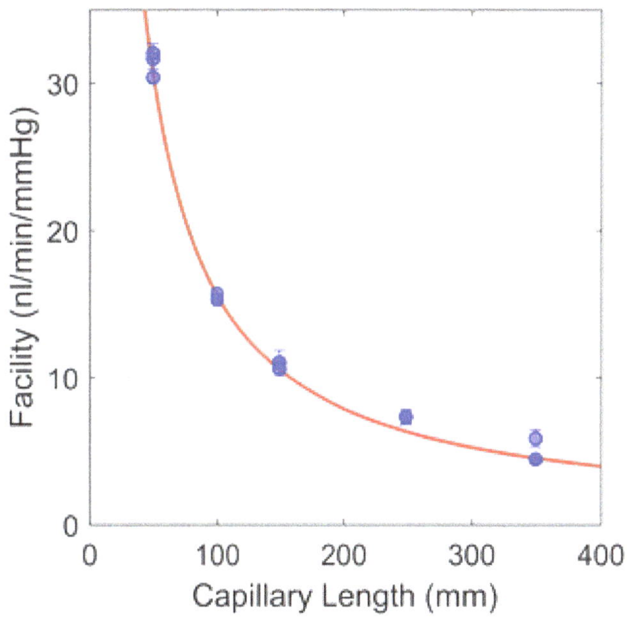

Figure 2: In vitro validation of the system. Data points show the 'facility' (hydraulic conductance) of various lengths of glass capillary with 95% confidence intervals. The fit in red is a linear relationship between hydraulic resistance and length, with 95% confidence bounds shown in grey. Deviations from the fit can be used to estimate the accuracy of facility measurement usingiPerfusion. http://dx.doi.org/10.1371/journal.pone.0150694.g002

These in vitro capillary tests provide a frame of reference to establish that facility measurements can be made accurately and reproducibly within the range appropriate for mouse eyes using iPerfusion. It must be stressed, however, that the accuracy when measuring outflow facility in real eyes will not necessarily be the same as that measured with in vitrocapillaries. While some sources of uncertainty overlap, such as the flow rate measurement, the measurement of biological tissues will inevitably involve additional, and likely greater, uncertainties, such as those associated with tissue variability and with interfacing mechanical and biological systems.

Statistical Analysis

In ocular perfusion studies, statistical analysis is generally carried out using hypothesis testing, where the statistical significance of an experimental treatment is assessed using, for example, Student's two sample t-test (paired or unpaired). However, due to biological variability and the limitations of any experimental measurement, each data point analysed will have some degree of uncertainty associated with it. The standard t-test, implicitly assumes that the measurement uncertainty is equal for all samples. The validity of the statistical analysis can be improved by accounting for uncertainty in the acquired data. To this end, we have developed a statistical approach, including the 'weighted t-test', a modification of Student's t-test that includes variable uncertainties between data points.

Here we present an overview of the statistical analysis; a complete description of the mathematics is given in. Although this approach has been developed for processing eye perfusion data, the weighted t-test is applicable to any statistical analysis in which a t-test might normally be used, but in which individual measurement uncertainties differ between cases. The analysis method additionally provides useful estimates of the sample variability and treatment effect.

A nomenclature listing all mathematical terms used in the analysis is provided

Statistical Distributions.

Any parametric statistical analysis requires an assumption of the underlying probability distribution. In order to use the t-test and many of the tools utilised in the following analysis, the data should be sampled from a normal distribution. However, as reviewed by Limpert et al. [35,36], for many real world applications, a normal distribution is often not appropriate. We posit that this is also the case for facility, for the following reasons.

For the outflow facility of mouse eyes reported in the literature, the standard deviation is often of similar magnitude to the mean [17, 20, 21, 31], such that range of facilities for a sample of mice (the lower limit of which can be estimated as the mean minus two standard deviations), would predict some very low, unphysiological, facilities according to the normal distribution. Furthermore, facility is inherently multiplicative, rather than additive (for example, doubling the resistance for a given flow rate would double the pressure drop across the outflow pathway). From these points, we conclude that outflow facility should be treated as a lognormally distributed variable, as will be confirmed in the Results section. Therefore, by analysing the logarithm of C, $Y = \ln(C)$, we may use statistical tools based on the normal distribution.

Unlike facility, the measured flow rate and pressure data are normally distributed variables, as uncertainties in these measurements occur due to electrical and external noise. The calculation of facility will be carried out by fitting a model to the measured flow rate and pressure data using weighted regression analysis, which requires normally distributed data. However, where the output of the regression analysis yields a lognormally distributed variable, such as facility, further consideration is required.

Consider fitting Eq 3 to the measured pressure and flow data, for a given eye, for which the best estimate of $C_{lin} \pm s_{Clin}$ would be obtained. If the variable C_{lin} was normally distributed, it could be described according to $C_{lin} \pm s_{Clin}$. However, as discussed, it is not C_{lin}, but $Y_{lin} = \ln(C_{lin})$ that is normally distributed, and we therefore require $Y_{lin} \pm s_{Ylin}$. As $\ln(C_{lin} - s_{Clin})$ and $\ln(C_{lin}+s_{Clin})$ are not equidistant from $\ln(C_{lin})$, it is not possible to state $\pm s_{Ylin}$ as a function of s_{Clin}. Rather than fitting Eq 3, it is possible to directly calculate s_{Ylin} from the regression, by fitting the model

$$Q = e^{Y_{lin}} P + Q_0 \tag{6}$$

which yields $Y_{lin} \pm s_{Ylin}$, where the value of C_{lin} is equal to $e^{Y_{lin}}$.

With regards to the Q_0 term, it is worth noting that as Q_0 can take on negative values, it cannot be lognormally distributed. It is therefore reasonable to assume that Q_0 is normally distributed, and the values of $Q_0 \pm s_{Q0}$ will be appropriate for further analysis.

Sources of Uncertainty.

There are five sources of uncertainty that contribute to variability in ocular perfusion measurements. We use the term s^2 to indicate an uncertainty in the form of a sample variance. Variances are used as they may be added to combine sources of uncertainty.

Measurement of pressure and flow rate: due to non-linearity, irrepeatability and hysteresis present in all transducers, each measurement has an associated degree of uncertainty that should be accounted for in the analysis, s^2_{sens} (sensor uncertainty). Additionally, there is a component of uncertainty due to averaging the signal over a period of time, s^2_{ave} (averaging uncertainty). For the flow measurement (subscript Q), the total uncertainty for each pressure step j, can be calculated according to

$$s^2_{Q,j} = s^2_{Qsens} + s^2_{Qave,j}$$

For the present system, we found that for the pressure measurements (subscript P) the total uncertainty $s^2_{P,j}$, is negligibly small. Thus from the raw data we obtain the steady state values P_j, Q_j and the uncertainty in the flow rate, $s^2_{Q,j}$.

Fitting a model to the data: the facility is calculated by fitting a model to the flow rate and pressure data using weighted least-squares regression between Q_j and P_j, with the weights defined as $1/s^2_{Q,j}$. For each eye, i, each of the regression parameters in the model has an associated variance based on its confidence interval, $s^2_{reg,i}$ (regression uncertainty).

Intra-individual variability: when using a paired experimental design, it is generally implicitly assumed that in the absence of a treatment, no difference would be observed. However, in complex measurements, a certain amount of deviation from the ideal case is to be expected. To quantify this, we introduce the intra-individual variability between contralateral eyes, s^2_{con}. This parameter inherently includes both the intrinsic biological variability in outflow facility between contralateral eyes and the added uncertainty associated with interfacing the eye with the perfusion system (cannulation), as numerically these are inseparable. For a given experimental system and sample population, the value of s^2_{con}, can be estimated by carrying out paired perfusions on untreated contralateral eyes, as described in detail in the Results section and in. In order to do this, it is assumed that the total variability in the difference in facility between the paired eyes, s^2_{dif}, is made up of contributions from the average uncertainty in the regression, $\overline{s^2_{reg}}$, and s^2_{con}. Therefore, the intra-individual variability is given by

$$s^2_{con} = s^2_{dif} - 2\overline{s^2_{reg}}$$

where the factor 2 arises due to regression uncertainty from both eyes of each pair. Although s^2_{con} does not numerically affect the outcome of the weighted t-test, knowledge of its value is necessary for estimating the variability of a given treatment in paired analyses, assists in the visualisation of uncertainties for pairs of eyes, and is beneficial for understanding the limitations of a perfusion system.

Inter-individual variability: when using an unpaired experimental design there will be a variability in outflow facility between individuals for a given population, s^2_{pop}, termed the inter-individual variability. The value of s^2_{pop} can be estimated by calculating the total variance of measured facilities over a given sample, s^2_{tot}, and assuming that it comprises both s^2_{pop} and $\overline{s^2_{reg}}$. The inter-individual variability is then given by

$$s^2_{pop} = s^2_{tot} - \overline{s^2_{reg}}$$

As with the s^2_{con} term, this inherently includes an uncertainty associated with interfacing the eye with the perfusion system.

Treatment variability: in experiments where the effect of a treatment is evaluated using either paired or unpaired analysis, there will be a variability in the efficacy of the treatment, s^2_{tre}. This parameter can be estimated as the difference in the variability between the treated and control samples. Large values of s^2_{tre} imply that the treatment effect varies considerably between individuals, for example if there are subpopulations of 'responders' and 'non-responders'. Calculation of very small or negative values of s^2_{tre} would indicate that the additional uncertainty due to the treatment is negligible compared to the inherent variability present in untreated eyes. Equations for calculating s^2_{tre} are provided in the next section, and the derivations are presented in

Central tendency and measures of spread using weighting.

Typically, the arithmetic mean, standard deviation and standard error on the mean (SEM) are used as measures of the central tendency and spread of the data. However, these values do not incorporate the variable uncertainty for each measurement. In order to account for measurement uncertainty and thus provide improved estimates of the sample statistics, the weighted arithmetic mean (WAM) is used. To calculate the WAM, each value is weighted relative to the reciprocal variance of the total uncertainty for each pair or individual eye (see below for details). Thus data points with large uncertainty contribute less to the weighted average, making the calculation more robust. The weighted standard error on the mean (WSEM) can be calculated as the square root of the variance on the WAM, and the square root of the unbiased weighted sample variance gives the weighted standard deviation (WSD). A brief description is given in the following section, refer to for details and derivation.

Paired data: for paired data, we first calculate the difference in the parameter of interest for each pair, and then calculate the average of the differences. For facility, we must use the log transform Y = ln(C). The difference for a pair of eyes p is then given by $Z_p = Y_{2,p} - Y_{1,p}$, where the subscripts 1 and 2 refer to the control and treated eyes respectively. The weight for each Z_p is given by the reciprocal of the total uncertainty for each pair

$$s^2_{Z_p} = s^2_{tot,Z} + s^2_{reg,p,1} + s^2_{reg,p,2} - 2\overline{s^2_{reg}}$$

where $s^2_{tot,Z}$ is the unweighted variance of Z_p. The WAM, $\overline{Z}$, can be calculated as the weighted arithmetic mean of Z_p. The variance of $\overline{Z}$ is $s^2_{\overline{Z}}$ and the unbiased weighted sample variance is s^2_Z. The treatment variability s^2_{tre} is estimated using

$$s^2_{tre} = s^2_Z - s^2_{con} - 2\overline{s^2_{reg}}$$

Unpaired data: for unpaired data, we calculate the averages for each population A and B, and then calculate the difference between the averages. For each eye, i, the weight is given by the reciprocal of the total uncertainty for each eye

$$s^2_{Y_i} = s^2_{tot} + s^2_{reg,i} - \overline{s^2_{reg}}$$

where s^2_{tot} is the unweighted variance of Y_i. $\overline{Y}$, $s^2_{\overline{Y}}$ and s^2_Y can then be calculated for each population sample. $\overline{Z}$ is then calculated as the difference between the WAM of the two population samples, $\overline{Z} = \overline{Y}_B - \overline{Y}_A$. The variance of $\overline{Z}$ can be calculated by adding the variances from the two population sample, $s^2_{\overline{Z}} = s^2_{\overline{Y}_A} + s^2_{\overline{Y}_B}$. The treatment variability s^2_{tre} is estimated by assuming s^2_{pop} is equal for both samples, and thus where the treated population sample has a larger variance, it can be attributed to the treatment, hence:

$$s^2_{tre} = s^2_{Y_B} - s^2_{Y_A} - \left(\overline{s^2_{reg,B}} - \overline{s^2_{reg,A}}\right)$$

using the values of $\overline{s^2_{reg}}$ for each population.

Lognormal variables: for a given population, the variable $\overline{Y}$ corresponds to the average logarithm of the outflow facility (expressed in dimensionless units). As it is preferable to report results in units of facility, we must take the exponential of $\overline{Y}$, yielding the geometric mean, $\overline{C}^* = e^{\overline{Y}}$, with $s^*_{\overline{C}} = e^{s_{\overline{Y}}}$. Likewise, $s^*_C = e^{s_Y}$. Note that the * denotes the use of geometric variables. When moving from the additive (log) domain, in which we describe $\overline{Y} \pm s_{\overline{Y}}$, into the multiplicative (linear) domain, the 'plus or minus' (±) term no longer applies, and is replaced by 'times or divide by' ($^\times/$) [36]. Therefore, $\overline{Y} \pm s_{\overline{Y}}$ is expressed as $\overline{C}^* {}^\times/ s^*_{\overline{C}}$. Note that when using $\pm 2s_{\overline{Y}}$, the lognormal equivalent becomes ${}^\times/ s^{*2}_{\overline{C}}$, as $e^{2s_{\overline{Y}}} = (e^{s_{\overline{Y}}})^2 = s^{*2}_{\overline{C}}$.

For both paired and unpaired data, the variable $\overline{Z}$ represents the average difference in the logarithm of the facility, Y. A more intuitive variable is the average fold change given by $\overline{D}^* = e^{\overline{Z}}$. This can be described easily for unpaired data wherein, $\overline{Z} = \overline{Y}_B - \overline{Y}_A$. Converting into the multiplicative domain yields $\overline{D}^* = e^{\overline{Z}} = \overline{C}^*_B / \overline{C}^*_A$, which is the proportional change in the average facility. Therefore $s^*_{\overline{D}} = e^{s_{\overline{Z}}}$ describes the spread in the estimate of $\overline{D}^*$, and hence how well the average fold change is known.

Reporting Statistical Results.

For each reported statistic, we provide a weighted mean (geometric or arithmetic for lognormal or normal variables respectively), along with the 95% confidence interval on the weighted mean. In addition we provide an indication of the spread or variability in the population. For discussion of why we report the statistics in this way,

Statistical results for a lognormally distributed variable, such as facility, will be reported as

$$\overline{C^*} \ {}^{\times}/ \ ME_{\overline{C^*},95}\left(s_{\text{pop}}^{*2}\right)$$

where $ME_{\overline{C^*},95}$ is the margin of error given by the half-width of the 95% confidence interval on the estimate of $\overline{C^*}$. The confidence interval on $\overline{C^*}$ can be calculated as $\left[\overline{C^*}/ME_{\overline{C^*},95}, \overline{C^*} \times ME_{\overline{C^*},95}\right]$. The term s_{pop}^{*2} corresponds to two-sigma, an estimate of the range that would encompass 95% of the facility values within the population ($s_{\text{pop}}^{*2} = e^{2s_{\text{pop}}}$). Note that the margin of error and two-sigma values are multiplicative.

Statistical results for a normally distributed variable, such as Z, will be reported as

$$\overline{Z} \pm ME_{\overline{Z},95}\left(2s_{\text{tre}}\right)$$

where $2s_{\text{tre}}$ is two-sigma, gives an estimate of the range that encompasses 95% of the treatment variability. The proportional change will be reported as

$$\overline{D^*} \ {}^{\times}/ \ ME_{\overline{D^*},95}\left(s_{\text{tre}}^{*2}\right)$$

In addition to $\overline{D^*}$, it may be preferable to report the average percentage change, given by $(\overline{D^*} - 1) \times 100\%$. The confidence bounds for the percentage change are given by $(\overline{D^*} \ {}^{\times}/ \ ME_{\overline{D^*},95} - 1) \times 100\%$ and thus it is not possible to state them in terms of a simple ${}^{\times}/$. This is also the case for the treatment effect. Thus, when reporting a percentage change, we will provide the confidence interval, and/or the range bounded by two-sigma directly. For example, a proportional change of $\overline{D^*} = 1.5$ with $ME_{\overline{D^*},95} = 1.1$ and $s_{\text{tre}}^{*2} = 1.25$ could be reported as a 50% increase, with confidence interval [36,65%] and a treatment effect between 20% and 88%.

Hypothesis Testing.

Student's t-test yields a p-value, which is an estimate of the probability that the null hypothesis ($\overline{Z} = 0$) is true, based on a t-statistic and an estimate of the number of degrees of freedom, ν. The 'weighted t-test' described in is similar but incorporates uncertainties by calculating the t-statistic as the ratio of the weighted mean to the square root of its weighted variance, $t = \overline{Z}/s_{\overline{Z}}$. Due to the weighting of each data point, ν will be less than that normally used in a standard t-test (for example $\psi - 1$ for a paired t-test with ψ pairs). For the weighted statistics used in this study, ν can be approximated using the Welch-Satterthwaite equation

Selection Criteria.

Three stages of selection criteria must be passed in order for a given perfusion to be included in further analysis. These approaches were designed to be as robust and as objective as possible.

A benefit of the iPerfusion system is that it enables continuous monitoring of the pressure and flow rate throughout the perfusion. Therefore, complications due to a poor cannulation, such as the cornea blocking the needle or poor sealing around the needle tip, become readily apparent. Cases in which the tracings deviate significantly from the prototypical tracing shown in Fig 1b and 1c, were thus excluded from further analysis. This process was carried out prior to calculating the facility, in order to minimise bias.

In response to a step increase in applied pressure, the measured pressure typically displays an exponential decay-type response, settling to a constant pressure over several minutes. In some eyes, typically at higher pressures, the applied pressure did not reach an apparent steady state, but slowly decreased over a longer time. To detect and eliminate these cases, the time constant of the pressure decay was estimated by fitting a single exponential, and any pressure trace that varied by more than 0.2 mmHgafter 6 time constants (at which time the system should be within 0.3% of its steady state value), was considered not to have a definable steady state condition. In such cases, the non-steady state pressure step and all subsequent pressure steps were eliminated from the regression analysis. Cases with fewer than four stable pressure steps were omitted on the basis of being insufficient for regression analysis using a model with two free parameters. The median number of pressure steps used per eye in the present study was 8.

For paired data, putative outliers in the observed Z values were identified. A non-parametric approach was used, whereby the median med(Z) was calculated as a robust estimate of the central value. The median absolute deviation (MAD)

$$\text{MAD} = 1.4826 \text{ med}\left|Z_p - \text{med}(Z)\right| \qquad \textbf{(7)}$$

was used as a robust estimator of the sample standard deviation (the 1.4826 is dependent on the assumption that the underlying distribution is normal). Outliers were classified as data points for which the measured value including the uncertainty exceeded 2.5 MAD from the median [37]. Thus, data points greater than the median were omitted from the analysis if Z_p minus the regression uncertainty exceeded the median plus 2.5 MAD, and vice versa for data points below the median. This approach is similar to using 2.5 standard deviations from the mean, but is more robust by accounting for uncertainties on the measurements and because the estimates of the median and MAD are relatively insensitive to outliers. None of the paired measurement in this study that passed the first two criteria were classified as outliers.

RESULTS AND DISCUSSION

A Power Law Model for the Flow-Pressure Relationship

Fig 3a shows a sample flow vs. pressure plot (for the same perfusion as the traces shown inFig 1b and 1c). The linear fit to the data (Eq 6) with 95% confidence bounds are shown in blue. The raw data, shown in black, appear to be fairly well represented by the linear fit, aside from a small systematic deviation. The slope of the line, C_{lin} predicts a facility for this eye of 9.1 ${}^{\times}/$ 1.06nl/min/mmHg. The other free-parameter in the model, Q_0, yields a value of −27.8 ± 5.7 nl/min. Considering Eq 3, the value of Q_0 implies that if the pressure in the eye were zero, fluid would leave the eye at a rate of 27.8 nl/min. For an in vivo eye, some reverse flow would be expected due to AH secretion, but AH secretion is expected to be negligible in an enucleated eye.

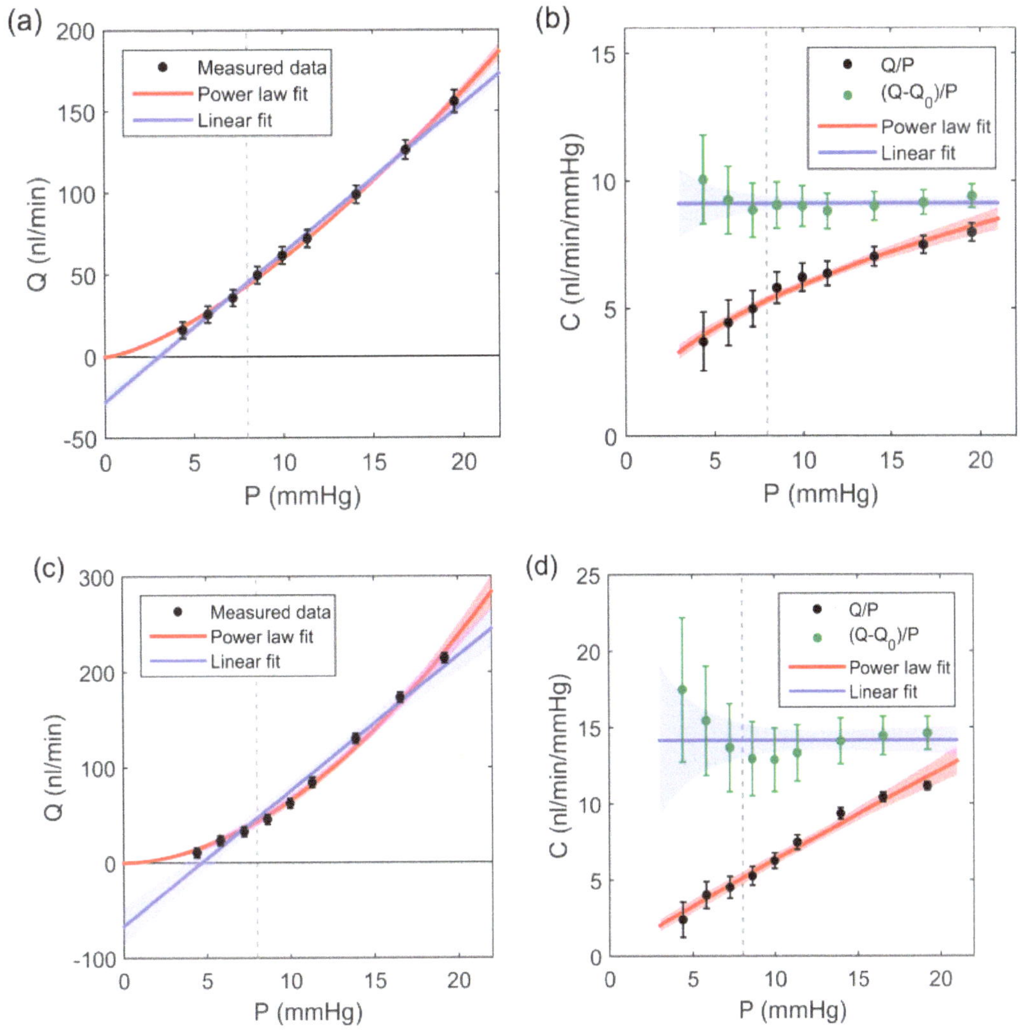

Figure 3: Selecting an appropriate model for the flow-pressure relationship. (a) A sample flow-pressure curve for the enucleated mouse eye perfusion shown in Fig 1b and 1c. Points show measured data with 95% confidence intervals. Blue: linear fit (Eq 3), C = 9.1 nl/min/mmHg, $Q_0 = -27.8$ nl/min. Red: power law (Eq 9), $C_r = 5.4$nl/min/mmHg, $\beta = 0.44$. Shaded regions show 95% confidence bounds. (b) The facility as calculated by the linear and power law models. Black markers show Q/P, which is independent of the fit, and green markers show facility as calculated according to Eq 8, $(Q - Q_0)/P$, showing the large influence of the model on the calculated facility. (c) and (d) show equivalent plots for a more non-linear case. Linear fit, C = 14.2 nl/min/mmHg,$Q_0 = -66.2$ nl/min. Power law, $C_r = 5.4$ nl/min/mmHg, $\beta = 0.85$. http://dx.doi.org/10.1371/journal.pone.0150694.g003

In some outflow studies [12, 25], a non-zero value of Q_0 was interpreted as 'unconventional' or pressure-independent outflow. In an enucleated eye, however, there is no clear physical mechanism that could generate a pressure-independent outflow. The osmotic driving force for uveovortex outflow [38], typically associated with pressure-independent outflow, would be eliminated by severing the ocular blood vessels during enucleation. We thereby posit that the term Q_0 should be identically zero in enucleated eyes. In order to test this hypothesis, we measured the flow rate at an applied pressure of 0 mmHg in six enucleated mouse eyes, following a standard perfusion. For intraocular pressures of 0.036 ± 0.068 mmHg, the average flow rate was 1.1 ± 6.5 nl/min (WAM ± 2WSD). These values are negligibly different from zero. We thus propose that any flow-pressure model for enucleated mouse eyes should not include aQ_0 term, such that there is no flow at zero pressure. As a corollary, including a non-zero Q_0term in Eq 3 may introduce errors in the estimation of facility.

To investigate the influence of a non-zero Q_0 on the estimation of facility, consider Eq 3rearranged in terms of facility at the jth pressure step.

$$C_{\mathrm{lin},j} = \frac{(Q_j - Q_0)}{P_j} \tag{8}$$

The values of $C_{\mathrm{lin},j}$ are represented by the green data points in Fig 3b, with the blue line representing the constant value of C_{lin} as predicted by fitting a linear model to the data. Shaded regions represent the 95% confidence interval on the fit. However, as $Q_0 = 0$ for enucleated mouse eyes, the facility for the jth pressure step is not given by Eq 8 but rather by$C_j = Q_j/P_j$, as indicated by the black data points in Fig 3b. At a physiological pressure drop across the outflow pathway of 8 mmHg (represented by the vertical dashed lines), the facility predicted when including a non-zero Q_0 value is almost twice that predicted when $Q_0 = 0$, demonstrating that the estimation of facility is highly sensitive to assumptions regarding Q_0. As illustrated in Fig 3c and 1d, the relative influence of Q_0 on the estimate of facility is increased for more non-linear cases. Note that the absence of a Q_0 term in enucleated eyes does not negate the presence of unconventional (non-trabecular) outflow, but does require that any unconventional outflow that exists be pressure dependent.

We thus require an alternative to the linear model, which allows both a pressure-dependent facility and a zero-intercept. We propose a simple power law model:

$$Q(P) = C_r \left(\frac{P}{P_r}\right)^{\beta} P \tag{9}$$

where P_r is a reference pressure (defined to be 8 mmHg), at which C_r is the facility. FollowingEq 6, the model actually used to fit the measured data is

$$Q(P) = e^{Y_r} \left(\frac{P}{P_r}\right)^{\beta} P \tag{10}$$

where $Y_r = \log(C_r)$. This fit is shown in red in Fig 3. The facility is given by

$$C(P) = C_r \left(\frac{P}{P_r}\right)^{\beta} \tag{11}$$

The power exponent β characterises the non-linearity of the flow-pressure relationship and can be interpreted as an index of the combined sources of non-linearity affecting the flow-pressure relationship through the outflow pathway. When $\beta = 0$, the facility is independent of pressure ($C = C_r$ for all P) and the flow-pressure relationship is entirely linear. For $\beta < 0$, the facility decreases with increasing pressure, as may occur due to collapse of Schlemm's canal, which increases outflow resistance with increasing IOP [39, 40]. For $\beta > 0$, the facility increases with increasing pressure. Evidence suggests that 'anterior chamber deepening' may artificially increase outflow facility as IOP increases by applying traction to the trabecular meshwork [41]. Anterior chamber deepening typically occurs during ocular perfusion via the anterior chamber, whenever the pressure in the anterior chamber exceeds that in the posterior chamber [42].

For the case shown in Fig 3a and 3b, $C_r = 5.4 \,{}^{\times}\!/\, 1.04$ nl/min/mmHg and $\beta = 0.44 \pm 0.06$, which represents a modest non-linearity. Recall that for this case the linear model predicted $C_{lin} = 9.1 \,{}^{\times}\!/\, 1.06$ nl/min/mmHg. For the more non-linear case shown in Fig 3c and 3d, $\beta = 0.85 \pm 0.16$ and $C_r = 5.4 \,{}^{\times}\!/\, 1.12$ nl/min/mmHg. For this case, the linear model predicts $C_{lin} = 14.2 \,{}^{\times}\!/\, 1.13$nl/min/mmHg, which is almost three times the facility estimated by the power law model at physiological pressure drop.

In summary, our data reveal that the flow-pressure relationship is non-linear in enucleated mouse eyes and can be represented by a power law model that captures the pressure-dependence of outflow facility. Ignoring this non-linearity and fitting a linear flow-pressure relationship to the data introduces an error in the estimation of outflow facility, even when the non-linearity appears rather subtle as in the case of Fig 3a. This underscores the importance of specifying an appropriate model to interpret flow-pressure data and to estimate outflow facility.

Comparison of Regression Parameters for the Power Law and Linear Models

In order to assess the power-law model and compare it to the commonly used linear model over a population of eyes, we evaluated the regression parameters from 66 independent, untreated eyes perfused with DBG. For each eye, the data were fit to both the linear and power law models.

As shown in Fig 4a, no correlation was observed between β and $\ln(C_r)$, indicating that the regression parameters were independent of one another for the power law model. For the linear model, however, there was a strong correlation between $\ln(C_{lin})$ and Q_0 ($p < 10^{-6}$; Fig 4b), where larger negative values of Q_0 coincide with larger positive values of $\ln(C_{lin})$. Based on current understanding of AH dynamics, there is no physiological precedent for a relationship between pressure-independent outflow and outflow facility, suggesting that the observed correlation is a result of insufficiencies in the linear model, rather than a physiological link between the two parameters. Similarly, there is no reason to suspect a relationship between βand $\ln(C_r)$, and no such relationship was observed using the power law model (Fig 4a).

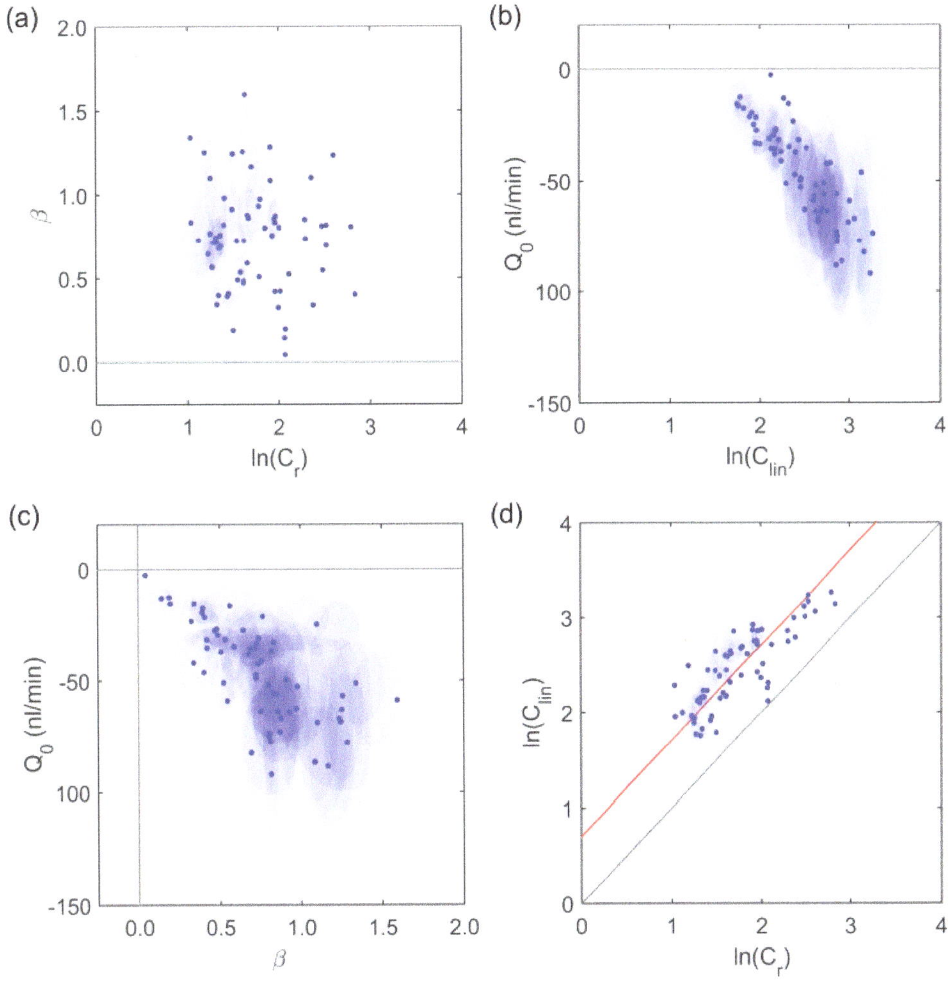

Figure 4: Comparing linear and power-law models. Outer ellipses show 95% CI on each parameter. (a) Power law exponent, β, against the logarithm of reference facility for the power law model: no correlation is observed ($p = 0.49$). (b) Pressure independent flow, Q_0, against the logarithm of facility for the linear fit: a strong correlation is observed ($p < 10^{-6}$), suggesting that the linear model is inappropriate. (c) Pressure-independent flow against power law exponent: a strong correlation is observed: $p < 10^{-6}$, indicating that non-zero Q_0 values are a result of the non-linearity in the Q − P relationship. (d) Comparison between log facility predicted by linear and power law models. Red line shows average over-prediction of ≈103% by the linear model. http://dx.doi.org/10.1371/journal.pone.0150694.g004

Fig 4c compares Q_0 to β. When the non-linearity of the flow-pressure relationship was negligible (small β), so was the Q_0 value, whereas for highly non-linear cases (large β), larger values of Q_0 were found ($p < 10^{-6}$). This suggests that the non-linearity of the flow-pressure relationship introduces an artificial non-zero value of Q_0, as demonstrated by the data in Fig 3a and 3c.

As Q_0 was negative in all cases, outflow facility as predicted by the linear model, C_{lin} was consistently overestimated. Fig 4d compares the facility calculated by the two models, showing that the linear model overestimated the facility with an average increase of ≈103%, as indicated by the red line, with a maximum of 268%. The error appears to be relatively systematic, implying that in studies using the linear model, changes in facility are still measurable; however, the scatter in the data implies that a larger number of samples would be required to achieve sufficient statistical power, compared to the power law model.

Population Distributions

In this section, the population characteristics of the 66 independent B6 mice are investigated.Fig 5a shows a histogram of the C_r values. The histogram is skewed towards lower values ofC_r, although no mice had a facility below 2.5 nl/min/mmHg, and a number of mice exhibited relatively high facilities. Under the assumption of a lognormal distribution, the reference facility is best described in terms of the geometric mean and geometric standard deviation, yielding $C_r = 5.89 \times\!/\ 1.12(2.50)$ nl/min/mmHg $(\text{mean} \times\!/\ ME_{95}(s^{*2}_{pop}))$. Thus approximately 95% of the eyes measured lie in the range 2.4–14.8 nl/min/mmHg. The red curve in Fig 5a represents the predicted population distribution based on these values. As shown in the histogram, 3 eyes did have facilities greater than the upper confidence bound, which for 66 eyes could be expected. At the lower end of the scale, the distribution rapidly tails off and predicts that very few eyes would yield facilities below 2 nl/min/mmHg (1 in 109). Using the common assumption of normally distributed facilities, the population distribution would be described by $C_r = 6.61 \pm 0.85(7.01)$ nl/min/mmHg, as shown in blue in Fig 5a. The normal distribution predicts facilities of less than 2 nl/min/mmHg in 1 out of 11 mice, and negative facilities in 1 out of 34 B6 mice, as indicated by the blue shaded regions in Fig 5a. As negative facilities cannot exist, and the very low predicted facilities are not in agreement with the current measurements, C_r cannot be well described by a normal distribution.

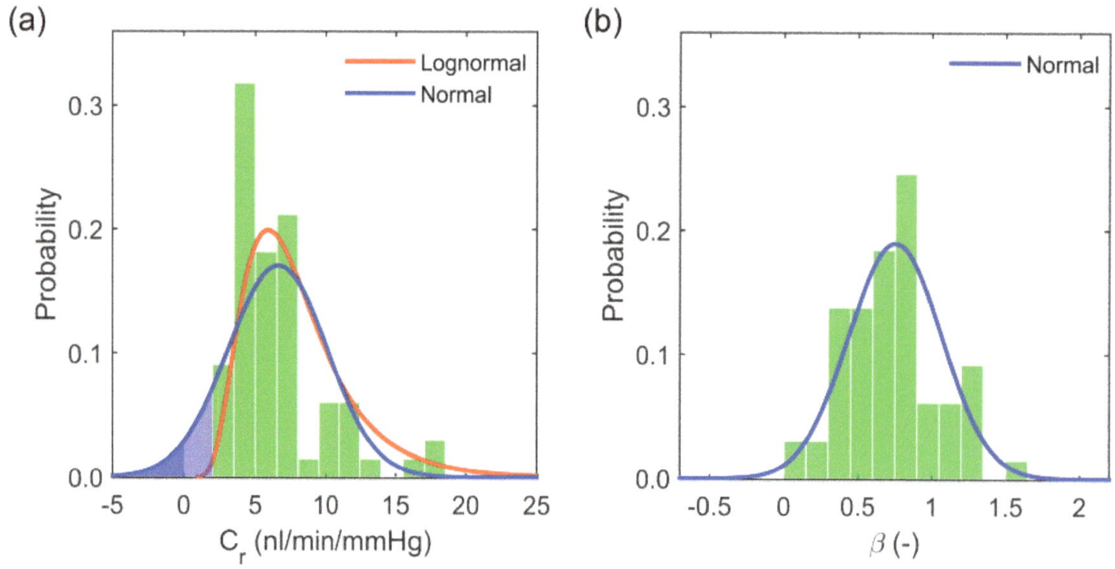

Figure 5: Population distributions of the regression parameters for the power law model. (a) Reference facility C_r: histogram of 66 eyes with overlaid normal (blue) and lognormal (red) distributions. Shaded blue regions show predicted facilities below 2 nl/min/mmHgand below 0 nl/min/mmHg. (b) β: histogram from 66 eyes with overlaid normal distribution. http://dx.doi.org/10.1371/journal.pone.0150694.g005

In order to quantify how well the alternative distributions represent the data, we used the Shapiro-Wilk test [43], which evaluates the probability that null hypothesis that the data are sampled from a normal distribution is true for the observed data. Small p-values thus indicate that the data are unlikely to be sampled from a normal distribution. For the facility, C_r, the Shapiro-Wilk test yields $p = 2.5 \times 10^{-6}$. For a lognormally distributed parameter such as C_r, its logarithm $Y = \ln(C_r)$ should be normally distributed. The Shapiro-Wilk test for $\ln(C_r)$ yielded $p = 0.016$. The Anderson-Darling test (reported to be the second most powerful normality test after the Shapiro-Wilk test [43]) yielded similar results. Although the tests both reject the normality of the $\ln(C_r)$ distribution at the standard threshold of $p = 0.05$, the p-value for $\ln(C_r)$ is significantly larger than for the distribution of C_r, indicating that the lognormal distribution is a considerable improvement.

The predicted population range in facility is rather large, considering that the mice should be genetically identical and their environment relatively constant. In order to investigate whether external factors contributed to the spread of C_r, we examined the relationship between C_r and a number of possible confounding variables using Pearson's correlation coefficient, r. The time of enucleation did not correlate with C_r ($r = 0.014$), suggesting that diurnal oscillations do not appear to be a cause of the large range in baseline facilities for the population. The time between enucleation and cannulation ($r = -0.036$) and the age of the mouse ($r = 0.067$) also showed no correlation with the measured facility. No effect of the day of the week on which the perfusion was carried out was detected using a one-way ANOVA ($p = 0.10$).

For the non-linearity parameter, β, the population distribution is shown in Fig 5b. In this case, the data are reasonably well represented by the normal distribution. The Shapiro-Wilk and Anderson-Darling tests yielded $p = 0.044$ and $p = 0.086$ respectively, indicating that it is reasonable to assume a normal distribution for this parameter. The distribution of β is described by $\beta = 0.752 \pm 0.079(0.631)$.

Analysis of Paired Measurements

Reference Facility.

The large range of facilities observed in the population highlights the benefits of using paired analysis, which considerably reduces the parameter space by providing a contralateral control eye for each treated eye. In paired analyses, it is often tacitly assumed that if no modification was applied, then the control and experimental cases would yield identical values; that is the pairs are 'ideal'. However, with such difficult measurements, and particularly with biological tissue, there will always be an inherent degree of uncertainty, and thus the extent of the 'pairedness' should be considered. In order to analyse the intra-individual variability between untreated contralateral eyes, $\Psi_{con} = 10$ pairs of eyes were perfused simultaneously with DBG on duplicate systems.

For each pair, p, the value of Z_p (the difference in the log-transformed facility between the treated and control eye) was calculated. Fig 6a shows a 'paired facility plot' for the data. The xand y-axes are the log-transformed facility of the two contralateral eyes, therefore each pair yields a single point on the graph, with the blue ellipses indicating the 95% confidence intervals based on the output of the weighted regression fit of Eq 9. It can be seen that, even when accounting for the confidence intervals from the regression, not all data points overlap the unity line (shown in blue), which would indicate a measurement that is equal for both untreated eyes. Therefore, an additional source of uncertainty must exist, which can be attributed to the intra-individual variability, s^2_{con}.

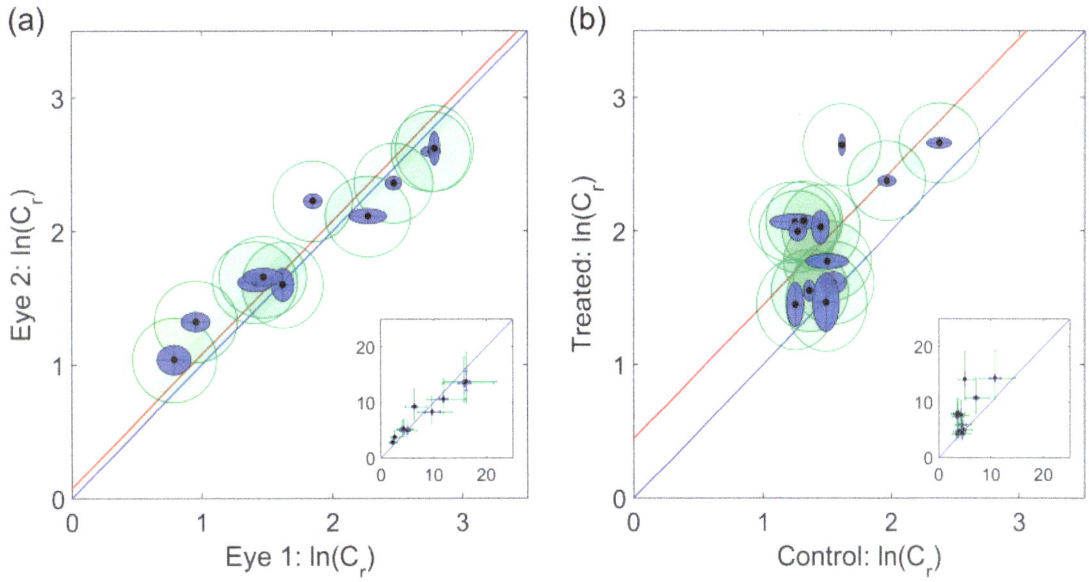

Figure 6: Comparison of facility for paired eyes using 'Paired Facility plots'. Each data point represents one individual, with the log-transformed facility of each eye defining the co-ordinates. Filled ellipses indicate 95% confidence intervals from the regression fitting of Eq 9 ($1.96s_{reg}$) and outer ellipses indicate additional uncertainty due to intra-individual variability (s_{con}). Inset: facility in nl/min/mmHg. Unity line is shown in blue. Red line shows average difference $\overline{Z}$, with its confidence intervals in grey. (a) Untreated contralateral pairs. No pairs exhibited a significant difference between contralateral eyes when accounting for uncertainty. (b) Effect of PDA205 treatment compared to control eyes. A significant increase in the facility following treatment is observed, although there is considerable variability (s^2_{tre}) introduced by the drug. http://dx.doi.org/10.1371/journal.pone.0150694.g006

In order to estimate s^2_{con}, we assume that the total deviations from the unity line, s^2_{dif}, result from both the intra-individual variability and the regression uncertainty, such that $s^2_{dif} = s^2_{con} + 2\overline{s^2_{reg}}$. The value of s^2_{dif} is estimated using the mean squared difference, which would be equivalent to the unbiased variance of Z_p, if $\overline{Z}$ were zero. Using the average $\overline{s^2_{reg}}$, we can estimate s^2_{con} according to

$$s^2_{con} = \frac{1}{\Psi_{con}} \sum_{p=1}^{\Psi_{con}} Z_p^2 - 2\overline{s^2_{reg}} \qquad \textbf{(12)}$$

Please for further details. For the B6 mice analysed here, $s^2_{con} = 0.046$, giving two-sigma of $2s_{con} = 0.430$, as a difference in the log domain. This corresponds to a 53% increase or a 35% decrease in facility, for a given pair. The green outer ellipses in Fig 6a show the uncertainty including s_{con}. For each of the pairs, the outer ellipses cross the unity line, and thus the facilities for the two eyes from each mouse are not significantly different from one another when accounting for intra-individual variability. Analysing these data in terms of the difference in log-transformed facility between contralateral eyes yields $\overline{Z} = 0.075 \pm 0.160$, which is not significantly different from zero. The red line in the figure shows $\overline{Z}$, and the grey shaded area shows the confidence interval on $\overline{Z}$ ($\overline{Z} \pm ME_{\overline{Z}.95}$), which overlaps the blue unity line, and thus $\overline{Z}$ is not statistically different from zero. See Table 1 for a summary of the key statistical parameters.

Table 1: Summary of the results of the untreated and PDA205 analyses. All terms are dimensionless. $p(\overline{Z} = 0)$ is the probability that the observed $\overline{Z}$ is not significantly different from zero, according to the weighted t-test described in. http://dx.doi.org/10.1371/journal.pone.0150694.t001

Case	$\overline{Z}$	$ME_{\overline{Z}.95}$	$2s_{tre}$	%	$\overline{D}^*$	$ME_{\overline{D}^*.95}$	s^2_{tre}	$p(\overline{Z} = 0)$
Untreated Pairs	0.075	0.160	-	8	1.08	1.17	-	0.3159
PDA205 Paired	0.446	0.212	0.487	56	1.56	1.24	1.63	0.0007
PDA205 Unpaired	0.440	0.322	0.543	55	1.55	1.38	1.72	0.0100

doi:10.1371/journal.pone.0150694.t001

Although variability between contralateral eyes appears to be fairly large, it is still possible to use the system to conclusively measure differences in facility. To verify this, 12 pairs of eyes were perfused intracamerally with 10 nM 3,7-dithia prostaglandin E_1 (PDA205) or vehicle control in DBG. PDA205 (Allergan Inc., Irvine CA) was stored as a lyophilised powder at −80°C, dissolved in ethanol as a 10 mM stock solution, stored at −20°C and protected from light at all times. Working stock was made up daily by diluting the stock solution in DBG. The control eye received DBG with an equivalent concentration of vehicle (17 µM ethanol). PDA205 is an EP_4agonist and has been shown previously to increase outflow facility in enucleated mouse eyes [11], enucleated human eyes [44] and living monkey eyes [45].

The results of the PDA205 experiments are presented as a paired facility plot in Fig 6b. The x-axis is the log-transformed facility of the control eyes, and the y-axis is the log-transformed facility of the treated eyes. For data points lying above the blue unity line, an increased facility was measured in response to the treatment. Although PDA205 is known to increase outflow facility, of the 12 pairs examined here, only 5 pairs independently exhibited a significant difference in facility in response to treatment (green ellipses did not cross the unity line). However, all pairs did show some increase in facility, and thus a statistically significant increase was observed overall. For these data, $\overline{Z} = 0.446 \pm 0.212(0.487)$. This gives a proportional change of $\overline{D}^* = 1.56 \times / 1.24(1.63)$. These data imply an overall percentage increase in facility of 56% in response to PDA205 (p = 0.0008). The two-sigma on the percentage facility increase for individual pairs, however, spanned a wide range from -4% to 154%. One could thus conclude from these data that PDA205 increases outflow facility on average, but that the effect of the drug varies considerably between individuals.

Non-linearity Parameter.

For the 10 untreated pairs, the power law exponent, β was not correlated between paired eyes (r = 0.246). However, when one case in which the maximum pressure was less than 9 mmHg(and thus β may not be well characterised) was removed, the correlation coefficient increased to r = 0.799 (p = 0.0097), implying that the pressure-dependence of facility may be paired between contralateral eyes. Applying Eq 12 to the nine β values yielded a two-sigma of $2s_{con,\beta} = 0.322$, approximately half the average value for the population.

Analysis of Unpaired Measurements

In some cases, it is not possible to perform paired measurements between contralateral eyes. In order to demonstrate the processing for unpaired data, the PDA205 data set was also analysed as if the control and treated eyes were obtained from different individuals. Fig 7shows a modified form of a violin plot of the data, which we refer to as a 'cello' plot. The individual C_r values for each eye are shown by the markers, with the 95% confidence intervals on the regression uncertainty shown by the error bars. As described above, there will be an additional uncertainty on C_r related to the cannulation, but we cannot quantify this uncertainty independent of the population variability, and it is therefore not indicated in the cello plot. The shaded region in the background indicates the predicted lognormal distribution of C_r, as described by the calculated $\overline{C_r^*}$ and $s^*_{C_r}$, and the width is scaled such that area of each shaded region is equal for the control and treated eyes. The geometric mean $\overline{C_r^*}$ is indicated by the thick central white line, and the thin white lines indicate two-sigma, within which 95% of eyes from this population are expected to lie. The grey area in the centre of the data set indicates the 95% confidence interval on $\overline{C_r^*}$. As compared to a box plot, the cello plot shows all individual data points and their uncertainty, as well as the predicted distribution and the confidence interval on the mean, providing a complete overview of the data. Consistent with the paired analysis, the cello plot in Fig 7 shows an increase in C_r in response to treatment with PDA205 using unpaired analysis.

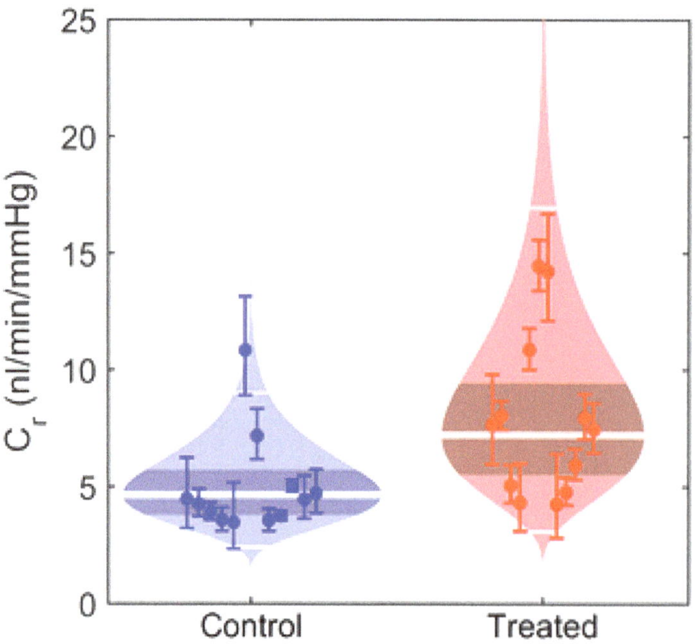

Figure 7: Comparison of facility for unpaired eyes using the 'Cello plot'. Unpaired analysis of facility for PDA205 treated and control eyes. Each data point shows the reference facility, C_r, with the error bars showing 95% confidence intervals from the regression fitting of Eq 10 ($1.96s_{reg}$). Shaded regions show best estimates of the sample distributions, with the geometric mean and two-sigma shown by the thick and thin horizontal lines respectively. Dark central bands show 95% CI on the mean values. http://dx.doi.org/10.1371/journal.pone.0150694.g007

The statistics for the unpaired analysis yield a similar proportional increase in response to PDA205 to that calculated with the paired analysis, $\overline{D}^* = 1.55$ (see Table 1). However, the CI on the mean proportional change increases significantly for the unpaired analysis, as evidenced by $ME_{\overline{D}^*,95}$ increasing from 1.24 to 1.38. Correspondingly, the calculated p-value is more than an order of magnitude greater when the data were analysed as unpaired relative to paired, although both p-values would still be considered significant.

PDA205 has a relatively large facility effect, however, and thus treatments having smaller effect sizes may not achieve sufficient statistical power to be detected by unpaired analysis without using prohibitively large number of animals. Note also that the value of s^{*2}_{tre} increases, but by relatively little, from 1.63 for paired analysis to 1.72 for unpaired analysis, indicating that both methods predict a similar variability in the treatment effect.

Recommendations

The measurement of outflow facility in mice, or indeed any animal, is technically challenging. A number of different approaches have been reported in the literature for acquiring pressure and flow data during ocular perfusions, to which a model is fitted in order to estimate the facility. Subsequently, statistical analysis is carried out on the facilities of treated and control eyes, in order to establish whether observed trends are significant. Based on findings from this study, we make the following recommendations regarding measurement of outflow facility in enucleated mouse eyes.

System Validation.

In order to analyse the uncertainty, one must first establish that the pressure and flow can be measured with sufficient accuracy and resolution in the physiological range of interest. As the facility is the main parameter of interest, acceptable facility measurements should be demonstrated in the appropriate range, which can be done using glass capillaries. Similar validation would be necessary for techniques to measure other parameters in live animals, such as AH secretion, episcleral venous pressure or unconventional outflow.

Steady State Condition.

Due to the compliant response of the eye to a change in pressure or flow, an accurate measurement of facility can only be acquired when the system has reached steady state. An objective steady state condition should be defined and its implications for the measurement of facility should be discussed.

Pressure Steps.

The pressure steps used in the present protocol have been selected so as to provide high resolution around the physiological range of pressure drops across the outflow pathway (≈8mmHg), as well as data at elevated intraocular pressures. Both the range and number of pressure steps are arbitrary and can be tailored to a particular experiment, but should be held constant as much as possible for comparing cases within a given experiment. Increasing the number of pressure steps improves confidence intervals on the facility estimate and thereby effectively increases measurement accuracy, but also increases the duration of the experiment.

Model Fitting.

The present results demonstrate that the flow-pressure relationship is non-linear in enucleated mouse eyes, and the standard linear fit (and assumption of a facility that is independent of pressure) is therefore not appropriate. Applying a linear fit leads to the appearance of pressure-independent outflow, but no such outflow is present in enucleated mouse eyes. Furthermore, using the linear model to estimate facility, even when the flow-pressure relationship appears relatively linear, may result in errors up to 250%. For these reasons, we recommend using the power law model proposed in Eq 9. We also advocate showing sample flow-pressure relationships, along with the fit and 95% confidence bounds, to demonstrate the quality of the perfusion data.

Statistical Distribution.

The present data have shown that facility is better described by a lognormal distribution than the commonly assumed normal distribution. In order to make use of the standard statistical tools that require normality, the log-transformed facility should be obtained by fitting Eq 10 to the flow-pressure data, and all statistical analysis should be conducted in the log domain.

Data Analysis.

Ocular perfusion measurements, particularly those in mouse eyes, include several sources of uncertainty that may vary between individual samples. Additionally, the average treatment size and variability in the treatment effect are of importance. For these reasons, we developed statistical tools, including the weighted t-test, which can be applied to any perfusion data, and introduced the paired XY plot and cello plot. Use of the data analysis methods reported here would improve the transparency of data in future studies and enable improved comparability between studies.

Negative Results.

It is not possible to conclude from an experiment that a given experimental condition has no effect. However, using the statistics developed in this paper, an appropriate CI on $\overline{Z}$ can be determined, from which one can obtain an estimate of the minimum effect size that could have been detected. For example, in the paired analysis of the PDA205 data set, $ME_{\overline{Z},95} = 0.214$. This implies that, for the twelve pairs we measured, the minimum net change we would have been able to statistically resolve would be ±0.214, a fold change of approximately ±20%.

Limitations

A number of outstanding issues remain in the development of methods to accurately measure outflow facility in mice. One issue is that, when using a single needle to perfuse with compounds via the anterior chamber, it is unclear whether the compound has acted on the entire outflow pathway before facility measurements are initiated. Considering an average B6 mouse with a facility of 6 nl/min/mmmHg at a physiological pressure drop of 8 mmHg, it would take more than two hours to turnover the anterior chamber volume, which is approximately 6μl[21]. This problem is exacerbated by segmental outflow that may deliver the compound non-uniformly to the trabecular meshwork [46, 47]. Variability in the uniformity of drug delivery may elevate the apparent value of s^2_{tre} and increase the number of mice required to achieve sufficient statistical power. A possible solution is to fully exchange the contents of the anterior chamber with the treatment solution prior to the start of the stepping protocol, but this would require a second perfusion needle, which is technically challenging and would likely introduce additional variability into the facility measurements. Alternatively, it may be possible to pre-treat the eyes prior to the perfusion either topically or intracamerally, assuming that this delivery did not interfere with the perfusion and was compatible with the time scale of the facility effect.

A second question is the mechanism of the pressure-dependent increase in outflow facility. In other species such as primates, outflow facility typically decreases with increasing pressure [40], which has been attributed to pressure-induced collapse of Schlemm's canal [39]. If perfusions are carried out ex vivo via the anterior chamber, however, the resulting pressure difference generated across the iris exerts traction on the outflow pathway and increases outflow facility; this phenomenon is known as 'anterior chamber deepening' [48] and is an experimental artefact that can be overcome by perfusing via the posterior chamber or by creating a shunt across the iris (iridectomy) [41]. The eyes in this study were perfused via the anterior chamber without an iridectomy, and thus it is possible that anterior chamber deepening may have contributed to the pressure-dependent increase in outflow facility and to the non-linearity of the flow-pressure relationship. It is conceivable that pressure dependence in both the conventional and unconventional outflow pathways could contribute to the observed non-linearity. However, it is not possible to distinguish reliably between alternative anatomical routes based on perfusion data alone. Unconventional outflow is commonly regarded to be relatively pressure insensitive [22, 49] and therefore increasing unconventional outflow facility seems an unlikely candidate for the observed pressure-dependent increase in total outflow facility.

Finally, we accept that as the perfusions were conducted on post-mortem enucleated eyes, the current experiments do not fully capture the mechanisms that may influence outflow facility regulation in vivo. This includes the potential effect of innervation, vascular regulation or other local or systemic factors that may influence outflow facility. As discussed below, in vivomeasurements have their own limitations regarding other aspects of AH dynamics that may change during the perfusion and thereby complicate measurements of facility in living animals, but some of these in vivo limitations may be overcome using the iPerfusion system. Studies to establish in vivo techniques using iPerfusion are currently underway.

Implications for *in vivo* Perfusions

The main advantage of perfusing enucleated eyes, relative to eyes maintained in vivo or in situ, is control over the parameter space. In enucleated eyes, the variables Q_{in}, P_e and Q_0 in Eq 1are zero, and the temperature, hydration state of the eye and external pressure can be accurately controlled by fully immersing the eye in isotonic saline. Conversely, for in vivoexperiments, Q_{in} and Q_0 are unknown, the downstream pressure in the conventional outflow pathway, P_e, is determined by episcleral venous pressure, and the hydration state of the eye is harder to control. Furthermore, anaesthesia may alter these parameters during a perfusion. For in situ measurements in euthanised animals, the values of Q_{in}, Q_0 and P_e would decay to zero eventually, but the time scales of this decay are unknown and would likely differ between parameters. Although a range of complementary methods may be used to estimate these variables, the compound uncertainties in each variable would weaken confidence on the estimates on the final parameters. These added complications can be justified by the fact that ultimately in vivo perfusions are a direct measurement of the entire system of ocular fluid dynamics, rather than a disconnected component as measured ex vivo. However, the present findings suggest that some of the common assumptions made in the analysis of in vivo data may not be appropriate.

In living animals, AH secretion, Q_{in}, pressure-independent outflow, Q_0, and episcleral venous pressure, P_e, are all expected to be non-zero, and these terms would need to be added to an analogous form of Eq 9 that is appropriate for in vivo pefusions. If one assumed that Q_0, $Q_{in}$and P_e were independent of both time and pressure, then a potential model for the flow-pressure relationship in living mouse eyes would be given by $Q(P) = C_r\left(\frac{P-P_e}{P_r-P_e}\right)^{\beta}(P-P_e)+Q_0-Q_{in}$, where the total pressure-independent flow in the system is $Q_0 - Q_{in}$, and would be negative indicating a net flow from the eye into the perfusion system at $P = P_e$. Due to the non-linearity of facility, $Q_{in} - Q_0$ could not be accurately estimated by extrapolation, but would need to be measured directly when $P = P_e$. Additionally, knowledge of P_e is necessary to estimate C, which would not be the case if C was independent of pressure. Methods have been reported to estimate P_e and Q_{in} in mouse eyes, but the accuracy of such measurements requires validation [12, 13, 21].

An additional and considerable complication in the context of in vivo perfusions is the change in IOP over time under anaesthesia [31, 50–52]. Episcleral venous pressure is directly correlated with mean arterial pressure [53] and will thus decrease over time, as well as being influenced by topical anaesthesia [54]. AH secretion, Q_{in}, is also reduced under anaesthesia [55], despite its mechanism being predominantly independent of pressure. The change of these variables in time, introduces additional uncertainty in the estimation of outflow facility. The two-step constant pressure perfusion commonly used in monkeys to account for time-varying parameters inherently assumes a facility that is independent of pressure [24], and so may not be completely appropriate if in vivo mouse eyes exhibit a non-linear flow-pressure relationship, as observedex vivo.

CONCLUSIONS

In this study, we described iPerfusion, a constant pressure perfusion system that provides faster and more accurate measurements of outflow facility in enucleated mouse eyes, relative to perfusion systems previously described in the literature. We fully characterised the iPerfusionsystem in the physiological range of facility values relevant for mice, including in vitro tests andex vivo studies. We developed a statistical analysis and the weighted t-test to account for uncertainties in the perfusion measurements and in the estimation of facility, and devised graphical formats to present the data in a manner that appropriately displays the uncertainty and spread in the measurements.

Using iPerfusion, we demonstrated that there is negligible pressure-independent outflow in fully hydrated enucleated eyes from young B6 mice. Using a 9-step perfusion protocol, we showed that the flow-pressure relationship was consistently non-linear and as a consequence, inappropriate application of a linear model to the data led to considerable overestimation of outflow facility. In contrast to the linear model, the flow-pressure relationship was well described by a simple power law model that captured the pressure-dependence of outflow facility. Comparing the linear and power law models revealed that the appearance of pressure-independent outflow when analysed using the linear model is most likely attributable to the non-linearity in the flow-pressure relationship and the inappropriate assumption of pressure-independent outflow facility.

Analysis of 66 mice revealed that baseline facility is better described by a lognormal distribution than a normal distribution. Hence, statistical analysis of facility should be carried out in the log domain, by using log-transformed values. The distribution of baseline facility values spanned nearly a 6-fold range, but facility was relatively tightly correlated between contralateral eyes. This large variability between individuals was not attributable to any apparent experimental condition, and given the correlation between contralateral eyes, it appears that this spread may have a physiological origin. As IOP variability within the B6 population has been reported to be 14.6 ± 5.2 (mean ± 2SD) [56], such a large spread in facility is unexpected and may suggest that facility is somehow regulated to maintain a fairly constant IOP, despite other systemic variables that may affect AH dynamics. Further research is needed to investigate the basis of the wide distribution of facility between individuals, but the fact that the spread is so large emphasises the advantage of using a paired experimental design.

With minor modifications, the iPerfusion system may be translated to other species and to in vivo ocular perfusion, as well as perfusion of other biofluidic systems.

ACKNOWLEDGMENTS

We would like to thank Professors C. Ross Ethier (Georgia Institute of Technology, US) and Kim Parker (Imperial College London, UK) for their helpful discussions on the data analysis.

AUTHOR CONTRIBUTIONS

Conceived and designed the experiments: JMS ERT JB DRO. Performed the experiments: JMS ERT JB. Analyzed the data: JMS BR DRO. Contributed reagents/materials/analysis tools: JMS ERT JB BR DRO. Wrote the paper: JMS ERT JB BR DRO.

REFERENCES

1. The AGIS investigators. The Advanced Glaucoma Intervention Study (AGIS): 7. The Relationship Between Control of Intraocular Pressure and Visual Field Deterioration. American Journal of Opthalmology. 2000; 130(4):429–440. doi: 10.1016/S0002-9394(00)00538-9.
2. Heijl A, Leske MC, Bengtsson B, Hyman L, Bengtsson B, Hussein M. Reduction of intraocular pressure and glaucoma progression: results from the Early Manifest Glaucoma Trial. Archives of Ophthalmology. 2002;120(10):1268–1279. doi: 10.1001/archopht.120.10.1268. pmid:12365904
3. Leske MC, Heijl A, Hussein M, Bengtsson B, Hyman L, Komaroff E, et al. Factors for glaucoma progression and the effect of treatment—The Early Manifest Glaucoma Trial. Archives of Ophthalmology. 2003;121(1):48–56. doi: 10.1001/archopht.121.1.48. pmid:12523884
4. Grant WM. Clinical measurements of aqueous outflow. AMA Archives of Ophthalmology. 1951;46(2):113–131. doi: 10.1001/archopht.1951.01700020119001. pmid:14856471
5. Quigley HA. Glaucoma. Lancet. 2011;377(9774):1367–1377. doi: 10.1016/S0140-6736(10)61423-7. pmid:21453963
6. Weinreb RN, Khaw PT. Primary open-angle glaucoma. The Lancet. 2004;363(9422):1711–1720. doi: 10.1016/S0140-6736(04)16257-0.
7. Kaufman PL, Rasmussen CA. Advances in glaucoma treatment and management: outflow drugs. Investigative Ophthalmology & Visual Science. 2012;53(5):2495–2500. doi: 10.1167/iovs.12-9483m.
8. Stamer WD, Acott TS. Current understanding of conventional outflow dysfunction in glaucoma. Current Opinion in Ophthalmology. 2012;23(2):135–143. pmid:22262082 doi: 10.1097/icu.0b013e32834ff23e
9. Overby DR, Bertrand J, Schicht M, Paulsen F, Stamer WD, Lütjen-Drecoll E. The structure of the trabecular meshwork, its connections to the ciliary muscle, and the effect of pilocarpine on outflow facility in mice. Investigative Ophthalmology & Visual Science. 2014;55(6):3727–3736. doi: 10.1167/iovs.13-13699.
10. Smith RS, Zabaleta A, Savinova O V, John SW. The mouse anterior chamber angle and trabecular meshwork develop without cell death. Biomed Central Developmental Biology. 2001;1:3. doi: 10.1186/1471-213X-1-3. pmid:11228591
11. Boussommier-Calleja A, Bertrand J, Woodward DF, Ethier CR, Stamer WD, Overby DR. Pharmacologic manipulation of conventional outflow facility in ex vivo mouse eyes. Investigative Ophthalmology & Visual Science. 2012;53(9):5838–5845. doi: 10.1167/iovs.12-9923.
12. Crowston JG, Aihara M, Lindsey JD, Weinreb RN. Effect of latanoprost on outflow facility in the mouse. Investigative Ophthalmology & Visual Science. 2004;45(7):2240–2245. doi: 10.1167/iovs.03-0990.
13. Millar JC, Clark AF, Pang IH. Assessment of aqueous humor dynamics in the mouse by a novel method of constant-flow infusion. Investigative Ophthalmology & Visual Science. 2011;52(2):685–694. doi: 10.1167/iovs.10-6069.
14. Whitlock NA, McKnight B, Corcoran KN, Rodriguez LA, Rice DS. Increased intraocular pressure in mice treated with dexamethasone. Investigative Ophthalmology & Visual Science. 2010;51(12):6496–6503. doi: 10.1167/iovs.10-5430.
15. Dai Y, Lindsey JD, Duong-Polk X, Nguyen D, Hofer A, Weinreb RN. Outflow facility in mice with a targeted type I collagen mutation. Investigative Opthalmology & Visual Science. 2009;50(12):5749–5753. doi: 10.1167/iovs.08-3367.
16. Kumar S, Shah S, Tang HM, Smith M, Borrás T, Danias J. Tissue plasminogen activator in trabecular meshwork attenuates steroid induced outflow resistance in mice. PLoS ONE. 2013;8(8):e72447. doi: 10.1371/journal.pone.0072447. pmid:23977299
17. Shepard AR, Millar JC, Pang IH, Jacobson N, Wang WH, Clark AF. Adenoviral gene transfer of active human transforming growth factor-β2 elevates intraocular pressure and reduces outflow facility in rodent eyes.

Investigative Opthalmology & Visual Science. 2010;51(4):2067–10. doi: 10.1167/iovs.09-4567.

18. Stamer WD, Lei Y, Boussommier-Calleja A, Overby DR, Ethier CR. eNOS, a pressure-dependent regulator of intraocular pressure. Investigative Ophthalmology & Visual Science. 2011;52(13):9438–9444. doi: 10.1167/iovs.11-7839.

19. Zhang D, Vetrivel L, Verkman AS. Aquaporin deletion in mice reduces intraocular pressure and aqueous fluid production. The Journal of General Physiology. 2002;119(6):561–569. doi: 10.1085/jgp.20028597. pmid:12034763

20. Zhang Y, Davidson BR, Stamer WD, Barton JK, Marmorstein LY, Marmorstein AD. Enhanced inflow and outflow rates despite lower IOP in bestrophin-2-deficient mice. Investigative Ophthalmology & Visual Science. 2009;50(2):765–770. doi: 10.1167/iovs.08-2501.

21. Aihara M, Lindsey JD, Weinreb RN. Aqueous humor dynamics in mice. Investigative Ophthalmology & Visual Science. 2003;44(12):5168–5173. doi: 10.1167/iovs.03-0504.

22. Gabelt BT, Kaufman PL. Aqueous humor hydrodynamics. In: Kaufman PL, Alm A, editors. Adler's Physiology of the Eye. Elsevier Health Sciences; 2011. p. 237–289.

23. Brubaker RF. Goldmann's equation and clinical measures of aqueous dynamics. Experimental Eye Research. 2004;78(3):633–637. pmid:15106943 doi: 10.1016/j.exer.2003.07.002

24. Barany EH. Simultaneous measurement of changing intraocular pressure and outflow facility in the vervet monkey by constant pressure infusion. Investigative Ophthalmology & Visual Science. 1964;3:135–143.

25. Lei Y, Overby DR, Boussommier-Calleja A, Stamer WD, Ethier CR. Outflow physiology of the mouse eye: pressure dependence and washout. Investigative Ophthalmology & Visual Science. 2011;52(3):1865–1871. doi: 10.1167/iovs.10-6019.

26. Boussommier-Calleja A, Overby DR. The influence of genetic background on conventional outflow facility in mice. Investigative Ophthalmology & Visual Science. 2013;54(13):8251–8258. doi: 10.1167/iovs.13-13025.

27. Johnson M, Erickson K. Mechanisms and routes of aqueous humor drainage. In: Albert DM, Jakobiec FA, editors. Principles and Practice of Opthalmology. Saunders; 2000. p. 2577–2595.

28. Yablonski ME. Letter to the editor: some thoughts on the pressure dependence of uveoscleral flow. Current Eye Research. 2003;12(1):90–92.

29. Ko MK, Yelenskiy A, Gonzalez JMJ, Tan JCH. Feedback-controlled constant-pressure anterior chamber perfusion in live mice. Molecular Vision. 2014;20:163–170. pmid:24520185

30. Mermoud A, Baerveldt G, Minckler DS, Prata JA, Rao NA. Aqueous humor dynamics in rats. Graefe's Archive for Clinical and Experimental Ophthalmology. 1996;234 Suppl 1:S198–203. doi: 10.1007/BF02343072. pmid:8871174

31. Camras LJ, Sufficool KE, Camras CB, Fan S, Liu H, Toris CB. Duration of anesthesia affects intraocular pressure, but not outflow facility in mice. Current Eye Research. 2010;35(9):819–827. doi: 10.3109/02713683.2010.494241. pmid:20795864

32. Wood GA. Data smoothing and differentiation procedures in biomechanics. Exercise and Sport Sciences Reviews. 1982;10(1):308–362. pmid:6749517 doi: 10.1249/00003677-198201000-00010

33. Ethier CR, Ajersch P, Pirog R. An improved ocular perfusion system. Current Eye Research. 1993;12(8):765–770. doi: 10.3109/02713689308995773. pmid:8222738

34. Kee C, Hong TM, Choi K. A sensitive ocular perfusion apparatus measuring outflow facility. Current Eye Research. 1997;16(12):1198–1201. doi: 10.1076/ceyr.16.12.1198.5023. pmid:9426951

35. Limpert E, Stahel W, Abbt M. Log-normal distributions across the sciences: keys and clues. BioScience. 2001;51(5):341–352. doi: 10.1641/0006-3568(2001)051%5B0341:LNDATS%5D2.0.CO;2.

36. Limpert E, Stahel WA. Problems with using the normal distribution–and ways to improve quality and efficiency of data analysis. PLoS ONE. 2011;6(7):e21403. doi: 10.1371/journal.pone.0021403. pmid:21779325

37. Leys C, Ley C, Klein O, Bernard P, Licata L. Detecting outliers: do not use standard deviation around the mean, use absolute deviation around the median. Journal of Experimental Social Psychology. 2013;49(4):764–766. doi: 10.1016/j.jesp.2013.03.013.

38. Pederson JE, Gaasterland DE, MacLellan HM. Uveoscleral aqueous outflow in the rhesus monkey: importance of uveal reabsorption. Investigative Ophthalmology & Visual Science. 1977;16(11):1008–1007.

39. Brubaker RF. The effect of intraocular pressure on conventional outflow resistance in the enucleated human eye. Investigative Ophthalmology & Visual Science. 1975;14(4):286–292.

40. Johnson MC, Kamm RD. The role of Schlemm's canal in aqueous outflow from the human eye. Investigative Ophthalmology & Visual Science. 1983;24(3):320–325.

41. Boussommier-Calleja A, Li G, Wilson A, Ziskind T, Elena O, Ashpole NE et al. Physical factors affecting outflow facility measurements in mice. Investigative Ophthalmology & Visual Science. 2015;56(13):8331–8339. doi: 10.1167/iovs.15-17106.

42. Moses RA. The effect of intraocular pressure on resistance to outflow. Survey of Opthalmology. 1977;22(2):88–100. doi: 10.1016/0039-6257(77)90088-1.

43. Razali NM, Wah YB. Power comparisons of Shapiro-Wilk, Kolmogorov-Smirnov, Lilliefors and Anderson-Darling tests. Journal of Statistical Modeling and Analytics. 2011;2(1):21–33.

44. Millard LH, Woodward DF, Stamer WD. The role of the prostaglandin EP_4 receptor in the regulation of human outflow facility. Investigative Opthalmology & Visual Science. 2011;52(6):3506–3508. doi: 10.1167/iovs.10-6510.

45. Woodward DF, Nilsson SFE, Toris CB, Kharlamb AB, Nieves AL, Krauss AHP. Prostanoid EP_4 receptor stimulation produces ocular hypotension by a mechanism that does not appear to involve uveoscleral outflow. Investigative Opthalmology & Visual Science. 2009;50(7):3320–3328. doi: 10.1167/iovs.08-3031.

46. Chang JYH, Folz SJ, Laryea SN, Overby DR. Multiscale analysis of segmental outflow patterns in human trabecular meshwork with changing intraocular pressure. Journal of Ocular Pharmacology and Therapeutics. 2014;30(2–3):213–223. doi: 10.1089/jop.2013.0182. pmid:24456518

47. Swaminathan SS, Oh DJ, Kang MH, Rhee DJ. Aqueous outflow: segmental and distal flow. Journal of Cataract and Refractive Surgery. 2014;40(8):1263–1272. doi: 10.1016/j.jcrs.2014.06.020. pmid:25088623

48. Ellingsen BA, Grant WM. The relationship of pressure and aqueous outflow in enucleated human eyes. Investigative Ophthalmology & Visual Science. 1971;10(6):430–437.

49. Alm A, Nilsson SFE. Uveoscleral outflow—A review. Experimental Eye Research. 2009;88:760–768. pmid:19150349 doi: 10.1016/j.exer.2008.12.012

50. Avila MY, Carre DA, Stone RA, Civan MM. Reliable measurement of mouse intraocular pressure by a servo-null micropipette system. Investigative Ophthalmology & Visual Science. 2001;42(8):1841–1846.

51. Burke JA, Potter DE. The ocular effects of xylazine in rabbits, cats, and monkeys. Journal of ocular pharmacology. 1986;2(1):9–21. doi: 10.1089/jop.1986.2.9. pmid:3503099

52. Cone FE, Steinhart MR, Oglesby EN, Kalesnykas G, Pease ME, Quigley HA. The effects of anesthesia, mouse strain and age on intraocular pressure and an improved murine model of experimental glaucoma. Experimental Eye Research. 2012;99:27–35. doi: 10.1016/j.exer.2012.04.006. pmid:22554836

53. Reitsamer HA, Kiel AW. A rabbit model to study orbital venous pressure, intraocular pressure, and ocular hemodynamics simultaneously. Investigative Ophthalmology & Visual Science. 2002;43(12):3728–3734.

54. Zamora DO, Kiel JW. Topical proparacaine and episcleral venous pressure in the rabbit. Investigative Ophthalmology & Visual Science. 2009;50(6):2949–2952. doi: 10.1167/iovs.08-3048.

55. Sperber GO, Bill A. A method for near-continuous determination of aqueous humor flow; effects of anaesthetics, temperature and indomethacin. Exp Eye Res. 1984;39(4):435–453. doi: 10.1016/0014-4835(84)90044-7. pmid:6389166

56. Yun H, Lathrop KL, Yang E, Sun M, Kagemann L, Fu V, et al. A laser-induced mouse model with long-term intraocular pressure elevation. PLoS ONE. 2014;9(9):e107446. doi: 10.1371/journal.pone.0107446. pmid:25216052

CHAPTER 5 GLOBAL MONITORING OF FLUIDIZED-BED PROCESSES BY MEANS OF MICROWAVE CAVITY RESONANCES

Johan Nohlert [a], Livia Cerullo [a], Johan Winges [a], Thomas Rylander [a], Tomas McKelvey [a], Anders Holmgren [b], Lubomir Gradinarsky [b], Staffan Folestad [b], Mats Viberg [a], Anders Rasmuson [c]

[a] Department of Signals and Systems, Chalmers University of Technology, SE-41296 Göteborg, Sweden

[b] AstraZeneca R&D, Pharmaceutical Development, SE-431 83 Mölndal, Sweden

[c] Department of Chemical and Biological Engineering, Chalmers University of Technology, SE-41296 Göteborg, Sweden

ABSTRACT

We present an electromagnetic measurement system for monitoring of the effective permittivity in closed metal vessels, which are commonly used in the process industry. The measurement system exploits the process vessel as a microwave cavity resonator and the relative change in its complex resonance frequencies is related to the complex effective permittivity inside the vessel. Also, thermal expansion of the process vessel is taken into account and we compensate for its influence on the resonance frequencies by means of a priori information derived from a set of temperature measurements. The sensitivities, that relate the process state to the measured resonance frequencies, are computed by means of a detailed finite element model. The usefulness of the proposed measurement system is successfully demonstrated for a pharmaceutical fluidized-bed process, where the water and solid contents inside the process vessel is of interest.

INTRODUCTION

In the pharmaceutical industry, monitoring and control of manufacturing processes are important in order to achieve high-quality products to a low production-cost. A common step in the manufacturing process for many pharmaceutical products is coating and drying of particles in fluidized-bed processes. Measurement techniques that are capable of monitoring the material distribution and moisture content in this type of process are important tools for control, optimization and improved understanding of the process.

Currently available methods for monitoring of moisture content in pharmaceutical fluidized-beds include spectroscopic methods such as near infra-red (NIR) and Raman spectroscopy [1] and [2]. Although these methods yield important information on moisture content as well as physical and chemical properties of the material, they mainly provide local information for the region in the very proximity of the measurement probe, which may be insufficient or misleading for the estimation of the global process state. Recently, Buschmüller et al. [3] developed a stray field microwave resonator technique for in-line moisture measurements during fluidized-bed drying. However, their approach is based on local fields in the vicinity of the resonator probe and, therefore, it does not yield global information on the process state.

Citation: Johan Nohlert, Livia Cerullo, Johan Winges, Thomas Rylander, Tomas McKelvey, Anders Holmgren, Lubomir Gradinarsky, Staffan Folestad, Mats Viberg, Anders Rasmuson, Global monitoring of fluidized-bed processes by means of microwave cavity resonances, Measurement, Volume 55, September 2014, Pages 520-535, ISSN 0263-2241, http://dx.doi.org/10.1016/j.measurement.2014.05.034

At low frequencies, electric capacitance tomography (ECT) can be applied to regions with high particle density [4]. Although ECT yields important information on the particle distribution on a global scale, it suffers from limited spatial resolution and it is not suited for the estimation of dielectric losses, which leads to difficulties in distinguishing between material density and moisture content.

In this article, we present a microwave-based measurement technique that exploits electromagnetic eigenmodes that are supported by the process vessel bounded by its metal walls. The process vessel is considered as a cavity resonator and the permittivity distribution inside the vessel is estimated based on measurements of the complex resonance frequencies of the cavity. This technique resembles the well-established cavity perturbation measurement technique, which is regularly used for accurate measurements of dielectric and magnetic properties of materials at microwave frequencies [5] and [6]. In contrast to traditional cavity-perturbation measurements, our technique exploits several different eigenmodes, which yields three-dimensional spatial resolution and measurement information at multiple frequencies.

The process vessel is equipped with a set of probes that are connected to a network analyzer. Given the scattering parameters measured with the network analyzer, we estimate the complex resonance frequencies of the process vessel, where the real part corresponds to the frequency of oscillation and the imaginary part corresponds to the damping. We exploit a linearization of the electromagnetic response of the process vessel and it yields a relation between (i) a perturbation of the resonance frequencies and (ii) a perturbation of the complex permittivity inside the process vessel. We parameterize the permittivity with space dependent basis functions and determine the coefficients by means of solving an estimation problem based on an over-determined system of linear equations, where the right-hand side contains the measured deviations in the resonance frequencies as compared to the empty vessel.

MEASUREMENT SYSTEM

Process equipment

Fig. 1 shows a schematic picture of a fluidized-bed process vessel of Wurster type, which is the type of process we consider in this article. This process is used for coating and drying of pharmaceutical particles, such as pellets or granules, and the particular vessel we consider is a custom built lab-scale device for batches consisting of up to 1 kg of particles. The particles are dielectric objects of size that normally range from 250 μm to 1 mm and in this article, we consider particles that mainly consist of microcrystalline cellulose (MCC).

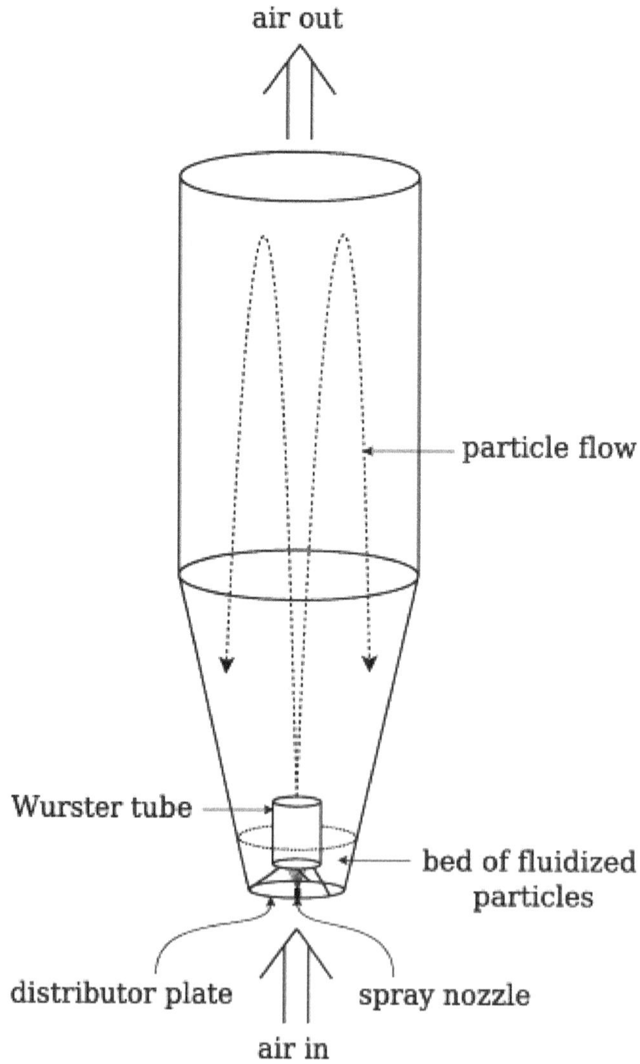

Figure 1: Schematic picture of the process vessel.

The bottom plate of the process vessel, also referred to as the distributor plate, is perforated with a large number of small holes. An airflow is applied through the distributor plate, which fluidizes the particles and creates a fluidized bed. Above the distributor plate resides a metal tube, which is referred to as the Wurster tube. At the center of the distributor plate below the Wurster tube, a pneumatic spray nozzle injects a rapidly moving mixture of air and liquid that accelerates the particles in its immediate vicinity in an upward motion. The particles move through the Wurster tube and create a fountain-like flow pattern above the Wurster tube. As the particles exit the fountain region, they fall back into the fluidized bed outside the Wurster tube.

The liquid injected through the spray nozzle is a solution that is sprayed onto the surface of the particles. The solvent evaporates mainly inside and immediately above the Wurster tube, which results in a deposited layer of the solute on the surface of the particles [7]. A common solvent is water and there is a large variety of solutes such as active pharmaceutical ingredients (API), polymers used for film coating and various filler materials. Thus, this type of process allows for the manufacturing of multi-layered particles. Here, the different layers provide designed functions such as protection from humidity during storage, delay of API release in the digestive system and the actual delivery of the API.

The water and its distribution inside the vessel for the duration of the process influence the process state and, consequently, the quality of the final product. For example, API that are sensitive to long-term exposure to moisture may be coated with a film that protects the API from ambient humidity. In such cases, it is important to establish that the particles are sufficiently dry before this film is applied. In addition, a too fast drying rate may have negative impact on the coating quality. A more extreme situation is the undesirable event of excessive amounts of water inside the vessel, which can cause extensive particle agglomeration and eventually process break-down. If such a condition can be detected, it is feasible to change the process parameters in order to counter-act agglomeration. The water inside the vessel can occur in a number of different forms: (i) water vapor; (ii) sorbed water inside the particles; (iii) free water on the surface of the particles; (iv) water droplets in the air; (v) condensed water droplets on the vessel walls; and (vi) water in agglomerates or dense gatherings of particles.

Measurement Principle

From an electromagnetic point of view, the process vessel that we consider is a rather good approximation of a microwave cavity resonator. The geometry of the vessel, which is used in the electromagnetic model described later in the article, is shown in Fig. 2. We exploit two magnetic-field probes to excite eigenmodes supported by the microwave resonator and the corresponding eigenfrequencies are measured with the same set of probes. A network analyzer is connected to the two probes and the 2×2 scattering matrix is measured as a function of frequency. The measured scattering parameters are used to extract the complex resonance frequencies of a selection of the lowest eigenmodes supported by the process vessel.

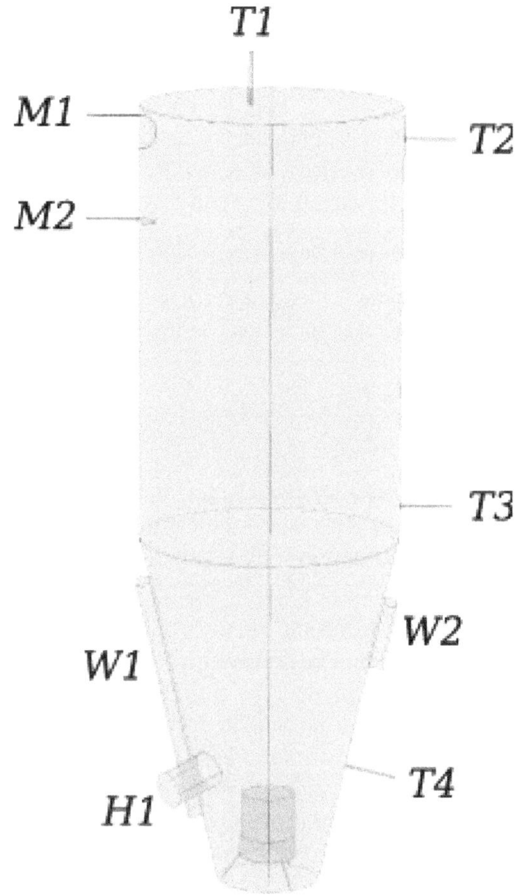

Figure 2: The process vessel including its two rectangular windows (W1 and W2) and the cylindrical tube used for loading of particles (H1). This figure also illustrates four temperature probes (T1–T4) and two microwave probes (M1 and M2) although these are not part of the electromagnetic model.

It should be emphasized that although these probes are rather small in comparison to the microwave resonator, i.e., the process vessel, the electromagnetic field for basically all eigenmodes extends throughout most of the process vessel's volume. The eigenfrequencies are influenced by the geometry of the process vessel and, more importantly for this application, the material and its distribution inside the process vessel. Consequently, important information on the distribution of particles and their moisture content can be derived from the measured eigenfrequencies.

Homogenization of a Mixture of Air and Particles

The mixture of particles and air influences the electromagnetic fields inside the process vessel. The typical length-scales associated with the lowest eigenmodes are comparable to the size of the cavity and, as a consequence, the particles are extremely small in relation to the spatial variation of the electromagnetic field. Thus, it is beneficial to represent the mixture of particles and air in terms of an effective permittivity, where the possible presence of other small dielectric objects, such as liquid droplets, contributes to the effective permittivity in the same manner as the particles. Given the material parameters and geometry of the particles together with their spatial distribution, the effective permittivity can be calculated by means of electromagnetic mixing formulas [8]. Here, we focus on the well-known Maxwell–Garnett mixing formula

$$\frac{\epsilon_{\text{eff}} - \epsilon_{\text{b}}}{\epsilon_{\text{eff}} + 2\epsilon_{\text{b}}} = \frac{1}{3\epsilon_{\text{b}}} \sum_{n=1}^{N} \vartheta_n \frac{\alpha_n}{V_n}, \tag{1}$$

where ϵ_{eff} is the effective permittivity and ϵ_{b} is the permittivity of the background medium for a mixture with N different types of inclusions indexed by n . In this article, we assume that ϵ_{b} is real, i.e., we have a lossless background medium. For the inclusions of type n , we have the polarizability α_n, the inclusion volume V_n and the volume fraction ϑ_n. For homogeneous spherical inclusions with permittivity ϵ_i, we have the polarizability

$$\alpha = 3\epsilon_{\text{b}} V \frac{\epsilon_{\text{i}} - \epsilon_{\text{b}}}{\epsilon_{\text{i}} + 2\epsilon_{\text{b}}}, \tag{2}$$

where V is the volume of the particle. Layered spherical particles with a core and an outer shell with different permittivities can also be incorporated into Eq. (1) by using the associated polarizability for this type of inclusions [8].

A dispersive inclusion material yields a dispersive behavior of the effective permittivity, although the dispersion relation of the mixture may be different from that of the inclusions. For example, air with inclusions that are described as a Debye medium yields an effective permittivity that also behaves like a Debye medium, but with different values of the Debye parameters [9]. Furthermore, air and inclusion particles with non-zero conductivity yields Debye-characteristics of the effective permittivity, due to the Maxwell–Wagner effect [10].

Density Independent Moisture Measurements

A feasible way to extract information about moisture content of the particles from the estimated complex effective permittivity $\epsilon_{\text{eff}} = \epsilon'_{\text{eff}} - j\epsilon''_{\text{eff}}$ in the process vessel, is the approach of density-independent measurement techniques [10]. The main idea of this approach is that if the material density in a given volume changes, both ϵ'_{eff} and ϵ''_{eff} will be affected in a similar way, since they both depend on the total amount of polarizable material in the volume.

In a situation where a certain material is dispersed in a background medium with permittivity ϵ_{b}, the ratio

$$\eta = \frac{\epsilon''_{\text{eff}}}{\epsilon'_{\text{eff}} - \epsilon_{\text{b}}}, \tag{3}$$

is an example of a density-independent function that has shown to be more or less independent on the density for many material mixtures [11] and [12]. This is not surprising, and it can be shown using the Maxwell–Garnett mixing formula applied to lossy spherical inclusions dispersed in a lossless background medium, that η is independent on the volume fraction ϑ in the low-density limit. Therefore, it is expected that measurements of ϵ''_{eff} and ϵ'_{eff} taken at different material densities would fall on a straight line that intersects the origin if ϵ''_{eff} is plotted versus $\epsilon'_{\text{eff}} - \epsilon_{\text{b}}$, given that the particles have constant permittivity. The slope η of this line usually depends on the moisture content of the inclusion material in a characteristic manner. The slope η may therefore be used as a density-independent metric of the moisture content, if it is known how η depends on the moisture content for the material considered. Such relations are usually obtained from calibration procedures which involve measurements on material samples with known moisture content. It should be emphasized that η may depend on many other factors as well, such as the physical state of the water (e.g. whether it is free or bound in the material) and the temperature. In the case of free water, its distribution in the material may also be of importance.

In practical measurement situations, the estimated effective permittivity ϵ_{eff} may be biased by, e.g., temperature variations of the metal parts of the cavity, or by changes in the permittivity of the background medium (which is usually air) caused by, e.g., a varying humidity. In such cases, there may be non-zero residual components of ϵ''_{eff} and $\epsilon'_{\text{eff}} - \epsilon_{\text{b}}$ at zero material density due to biased estimates of ϵ_{eff}, which makes a direct use of the ratio η unreliable. We therefore make an attempt to extend the classical density-independent measurement approach to situations where an uncontrolled bias of the estimated ϵ_{eff} is present. Our approach is based on the covariance between ϵ''_{eff} and $\epsilon'_{\text{eff}} - \epsilon_{\text{b}}$, which is computed for a fairly large number of measurement samples. The material (in our case the particles in the process) is re-distributed between different sample points which yields a variation in the material density and hence a variation in ϵ''_{eff} and ϵ'_{eff}, which is used to calculate the covariance. As a motivation to this approach, we consider a situation where lossy spherical inclusions with permittivity $\epsilon_{\text{i}} = \epsilon'_{\text{i}} - j\epsilon''_{\text{i}}$ are dispersed in a background medium with permittivity ϵ_{b} with volume fraction ϑ . If the dispersion is dilute ($\vartheta \ll 1$) and if the loss factor of the inclusion material is low ($\epsilon''_{\text{i}}/\epsilon'_{\text{i}} \leqslant 0.1$), the Maxwell–Garnett mixing formula yields the following expressions for the real and imaginary parts of the effective permittivity

$$\epsilon'_{\text{eff}}/\epsilon_{\text{b}} - 1 \simeq 3\frac{\epsilon'_{\text{i}}/\epsilon_{\text{b}} - 1}{\epsilon'_{\text{i}}/\epsilon_{\text{b}} + 2}\vartheta \tag{4}$$

$$\epsilon''_{\text{eff}}/\epsilon_{\text{b}} \simeq 9\frac{\epsilon''_{\text{i}}/\epsilon_{\text{b}}}{(\epsilon'_{\text{i}}/\epsilon_{\text{b}} + 2)^2}\vartheta. \tag{5}$$

If we now assume that ϑ, ϵ'_{i} and ϵ''_{i} vary randomly around their respective mean values, this will obviously yield variations in ϵ_{eff}. Hence, by studying the covariance between the real and imaginary part of ϵ_{eff}, one may extract important information about the parameters ϑ, ϵ'_{i} and ϵ''_{i} and their variabilities. In particular, we notice from Eqs. (4) and (5) that both the real and imaginary part of ϵ_{eff} depend linearly on the volume fraction ϑ , which explains the density-independent nature of the ratio η and the reason why ϵ''_{eff} and $\epsilon'_{\text{eff}} - \epsilon_{\text{b}}$ obtained from different volume fractions tend to fall on a straight line. Deviations from this line can be explained by means of variations in ϵ'_{i} and ϵ''_{i}, since they influence the real and imaginary part of ϵ_{eff} in different ways, as seen from Eqs. (4) and (5).

Experimental Setup

The experimental setup is shown in Fig. 3a and it includes the process vessel, the process machine and the two-port network analyzer used to perform the microwave measurements.

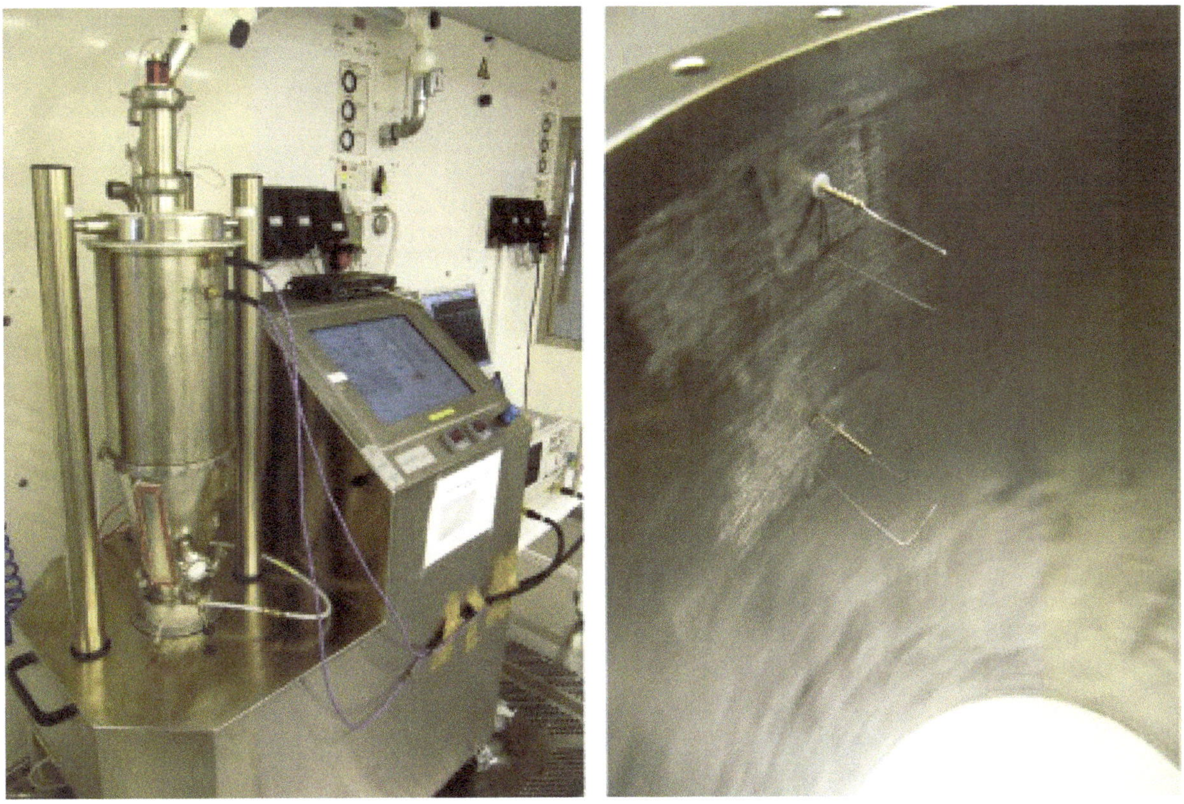

(a) Experimental setup (b) H-probes

Figure 3: Photographs of (a) the process vessel in the experimental setup and (b) the two H-probes.

Reduction of Undesired Losses

The original process vessel has a couple of apertures, such as windows that are used for visual inspection of the process, and such apertures damp the microwave cavity by means of radiation losses. We are interested in measuring the damping associated with the moisture content of the particles and, thus, we wish to reduce other contributions to the total damping of the cavity. Therefore, we shield the process vessel by covering the apertures W1, W2 and H1 in Fig. 2 by aluminum foil, and use a perforated plate to shield the microwave cavity from the air outtake in the upper part of the process vessel. In addition, the shielding mitigates possible problems with external electromagnetic sources that may disturb the measurements. With these arrangements we obtain unloaded Q -values of the cavity which are in the order of 10^3–10^4 for the lowest eigenmodes.

Choice of Probes and Their Design

Pharmaceutical mixtures processed in steel containers often become electrically charged due to the triboelectric effect [13] and [14]. Consequently, the statically charged particles yield a quasi-static electric field as they move inside the vessel. In order to protect the network analyzer from induced voltages of large magnitude and electric discharges that may occur in the process vessel, we exploit magnetic field probes (also referred to as H-probes) that are mainly sensitive to the time variations of the magnetic flux that passes through the loop of the H-probe.

Fig. 3b shows the process vessel equipped with two H-probes located in the upper part of the cavity for measurement of the azimuthal and vertical components of the magnetic field. At these locations, we can efficiently couple to the magnetic field of the lowest eigenmodes that we use for sensing. In addition, the size of each loop is chosen such that we get sufficiently good coupling to all of these modes. For industrial usage, it may not be feasible to let the probe extend into the process vessel. A possible remedy is to create a shallow dent in the cavity wall and to cover this dent by a dielectric "radome".

Measurement Settings

The network analyzer (Agilent E8361A) is controlled from a PC via GPIB interface, in order to perform repeated scans of a frequency band that includes the resonance frequencies of the sensing modes (0.7–1.7 GHz). This frequency band takes approximately 2 s to scan (for excitation on one of the two ports), using totally 6400 frequency points and the IF bandwidth set to 5 kHz, which means that it takes 1–10 ms to scan the frequency band occupied by a single resonance. The time between two consecutive frequency sweeps is set to 10 s, where a substantial part of this time is required for data transfer over the GPIB interface. In the post-processing of the measurement data, we use a statistical approach which benefits from a large number of measurement samples. Therefore, a measurement setup that allows for faster sampling of the process would be beneficial for this measurement technique.

ELECTROMAGNETIC MODEL

For the analysis of the resonance frequencies and the corresponding eigenmodes, we use an idealized model of the process vessel: (i) all the metal parts of the process vessel are assumed to be perfect electric conductors (PEC); (ii) all apertures are completely sealed such that the interior of the process vessel is completely enclosed by a perfect electric conductor; and (iii) the probes are excluded from the model. Further, we assume that the material inside the process vessel can be described by relatively simple dispersion relations for the frequency band of interest and, as a starting point, we consider a frequency-independent permittivity that may vary with respect to space. Also, we assume that the magnetization for the media involved is negligible and use the permeability $\mu=\mu_0$.

Thus, we consider the eigenvalue problem

$$\nabla \times \boldsymbol{E} = -j\omega \boldsymbol{B} \text{ in } \Omega, \tag{6}$$

$$\nabla \times \frac{1}{\mu_0}\boldsymbol{B} = (j\omega\epsilon + \sigma)\boldsymbol{E} \text{ in } \Omega, \tag{7}$$

$$\hat{\boldsymbol{n}} \times \boldsymbol{E} = 0 \text{ on } \Gamma. \tag{8}$$

We seek the eigenmodes $(\boldsymbol{E}_m, \boldsymbol{B}_m)$ and the corresponding eigenfrequencies ω_m that satisfy the eigenvalue problem (6), (7) and (8), where the index m labels the resonances supported by the process vessel.

Weak Formulation

The eigenvalue problem (6), (7) and (8) is expressed in the weak form: Find$(\boldsymbol{E},\boldsymbol{B},\omega)\in H_0(\text{curl};\Omega)\times H(\text{div};\Omega)\times C$ such that

$$c_{FL}(\boldsymbol{E},\boldsymbol{v})=-j\omega m_1(\boldsymbol{B},\boldsymbol{v}), \tag{9}$$

$$c_{AL}(\boldsymbol{B},\boldsymbol{w})=j\omega\mu_0\epsilon_0 m_{\epsilon r}(\boldsymbol{E},\boldsymbol{w})+\mu_0 m_\sigma(\boldsymbol{E},\boldsymbol{w}), \tag{10}$$

for all $(\boldsymbol{v},\boldsymbol{w})\in H(\text{div};\Omega)\times H_0(\text{curl};\Omega)$. Here, we express the real (absolute) permittivity $\epsilon=\epsilon_0\epsilon_r$ in terms of the corresponding relative permittivity. Further, we have the bilinear forms

$$c_{FL}(\boldsymbol{u},\boldsymbol{v}) = \int_\Omega (\nabla \times \boldsymbol{u}) \cdot \boldsymbol{v} d\Omega, \tag{11}$$

$$c_{AL}(\boldsymbol{u},\boldsymbol{v}) = \int_\Omega \boldsymbol{u} \cdot (\nabla \times \boldsymbol{v}) d\Omega, \tag{12}$$

$$m_1(\boldsymbol{u},\boldsymbol{v}) = \int_\Omega \boldsymbol{u} \cdot \boldsymbol{v} d\Omega, \tag{13}$$

$$m_{\epsilon_r}(\boldsymbol{u},\boldsymbol{v}) = \int_\Omega \epsilon_r \boldsymbol{u} \cdot \boldsymbol{v} d\Omega, \tag{14}$$

$$m_\sigma(\boldsymbol{u},\boldsymbol{v}) = \int_\Omega \sigma \boldsymbol{u} \cdot \boldsymbol{v} d\Omega. \tag{15}$$

Finally, we have the spaces

$$H(\text{curl};\Omega) = \{\boldsymbol{u} : \boldsymbol{u} \in L_2(\Omega) \text{ and } \nabla \times \boldsymbol{u} \in L_2(\Omega)\}, \tag{16}$$

$$H(\text{div};\Omega) = \{\boldsymbol{u} : \boldsymbol{u} \in L_2(\Omega) \text{ and } \nabla \cdot \boldsymbol{u} \in L_2(\Omega)\}, \tag{17}$$

and $H_0(\text{curl};\Omega)$ is the subspace of $H(\text{curl};\Omega)$ that satisfies the boundary condition(8).

Finite Element Approximation

We expand the electric field in terms of curl-conforming elements and the magnetic flux density in divergence-conforming elements, see Ref. [15] for further details. This yields the discrete eigenvalue problem

$$\begin{bmatrix} \mathbf{0} & -\mathbf{C} \\ \mathbf{C}^T & -Z_0\mathbf{M}_\sigma \end{bmatrix} \begin{bmatrix} c_0\mathbf{b} \\ \mathbf{e} \end{bmatrix} = \frac{j\omega}{c_0} \begin{bmatrix} \mathbf{M}_1 & \mathbf{0} \\ \mathbf{0} & \mathbf{M}_{\epsilon_r} \end{bmatrix} \begin{bmatrix} c_0\mathbf{b} \\ \mathbf{e} \end{bmatrix}, \tag{18}$$

where we have scaled the magnetic flux density with the speed of light and Ampère's law with the impedance of vacuum. This form is useful since it yields an eigenvalue problem that is linear in terms of the eigenvalue $j\omega/c_0$. Reference [16] contains further details on the treatment of eigenvalue problems by means of the finite element method.

Elimination of the magnetic flux density from Eq. (18) yields the corresponding eigenvalue problem

$$\left(\mathbf{C}^T\mathbf{M}_1^{-1}\mathbf{C} + j\omega\mu_0\mathbf{M}_\sigma - \omega^2\mu_0\epsilon_0\mathbf{M}_{\epsilon_r}\right)\mathbf{e} = \mathbf{0}, \tag{19}$$

formulated in terms of only the electric field. This eigenvalue problem is nonlinear since it is quadratic in terms of the eigenfrequency ω.

For problems with conducting media, it is computationally advantageous to solve Eq.(18). Such situations may occur in pharmaceutical processes that feature solutions with ions. For dry processes, the non-conducting background medium with possibly conducting inclusions, displays a more complicated dispersion for the effective permittivity. Given the somewhat limited frequency-band used for the measurements in this article, we choose to approximate the effective permittivity by a frequency-independent function $\epsilon(\boldsymbol{r})=\epsilon'(\boldsymbol{r})-j\epsilon''(\boldsymbol{r})$. This yields the eigenvalue problem

$$\left(\mathbf{C}^T\mathbf{M}_1^{-1}\mathbf{C} - \omega^2\mu_0(\mathbf{M}_{\epsilon'} - j\mathbf{M}_{\epsilon''})\right)\mathbf{e} = \mathbf{0}, \tag{20}$$

which is not particularly attractive for direct computation but useful for the sensitivity analysis that follows below.

Losses due to Finite Conductivity in the Cavity Walls

It is possible to approximately determine the losses due to cavity walls of finite conductivity based on the electromagnetic field solution computed for a PEC boundary. This type of approach yields the quality factor associated with a particular eigenmode and, consequently, its damping [17]. The losses are given by

$$P_{\text{wall}} = \text{Re}\left\{\frac{1}{2}\int_\Gamma Z_\Gamma |\hat{\boldsymbol{n}} \times \frac{\boldsymbol{B}_m}{\mu_0}|^2 d\Gamma\right\}, \tag{21}$$

where $Z_\Gamma=(1+j)/(\sigma\delta_m)$ is the surface impedance for the wall of finite conductivity. Here, $\delta_m=(2/(\omega_m\mu_0\sigma))^{1/2}$ is the penetration depth for the eigenfrequency ω_m and the magnetic flux density $\boldsymbol{B}_m$ of the corresponding eigenmode is computed based on the approximation that the wall conductivity is infinite. Finally, the quality factor is given by $Q_{\text{wall},m}=\omega_m W_m/P_{\text{wall},m}$ where W_m is the time-average energy stored in the vessel by the eigenmode $(\boldsymbol{E}_m,\boldsymbol{B}_m)$, i.e.,

$$W_m = \frac{1}{4}\int_\Omega \epsilon|\boldsymbol{E}_m|^2 d\Omega + \frac{1}{4}\int_\Omega \frac{1}{\mu_0}|\boldsymbol{B}_m|^2 d\Omega. \tag{22}$$

ESTIMATION PROCEDURE

Given the measured scattering parameters, we estimate the complex resonance frequencies by means of a subspace-based multivariable system identification algorithm[18]. This measurement procedure is repeated in a regular fashion throughout the entire duration of the process time, and in the following, we refer to the estimated resonance frequencies as the measured resonance frequencies. First, the vessel is empty and the measured resonance frequencies are denoted $\bar{\omega}_m^{\text{ref}}$, where the overline denotes a measured quantity. Next, the process vessel is loaded with particles and the subsequent measured resonance frequencies $\bar{\omega}_m$ are used to form the deviations in the resonance frequencies $\Delta\bar{\omega}_m = \bar{\omega}_m - \bar{\omega}_m^{\text{ref}}$. As the process evolves, we exploit the relative deviation $\Delta\bar{\omega}_m/\bar{\omega}_m^{\text{ref}}$ to estimate the state of the process and, later in this section, we present formulas that relate a perturbation in resonance frequency to the corresponding perturbations in (i) the material parameters and (ii) the shape of the vessel due to thermal expansion.

The estimation of the material and shape parameters is based on minimization of the misfit between the measured and computed relative deviations in resonance frequencies. Thus, we wish to minimize the following expression

$$\sum_m \left| \frac{\Delta\omega_m}{\omega_m^{\text{ref}}} - \frac{\Delta\bar{\omega}_m}{\bar{\omega}_m^{\text{ref}}} \right|^2, \tag{23}$$

where $\Delta\omega_m = \omega_m - \omega_m^{\text{ref}}$ is the deviation in the computed resonance frequencies ω_m as compared to the corresponding (computed) result ω_m^{ref} for the empty reference case. The computed resonance frequencies ω_m are functions of the material and shape parameters, and this dependency may be expressed according to

$$\omega_m = \omega_m|_{\text{LP}} + \sum_n \frac{\partial\omega_m}{\partial p_n}|_{\text{LP}}\delta p_n, \tag{24}$$

for small deviations around a certain linearization point, which is denoted by the subscript LP. The empty cavity may be used as linearization point, but if the permittivity inside the vessel deviates substantially from vacuum, it is beneficial to use a linearization point with higher permittivity in certain regions of the vessel. The perturbations δp_n in the parameters with respect to the linearization point, are stored in a vector **x** and we compute an estimate $\hat{\mathbf{x}}$ of **x** by means of the regularized approach

$$\hat{\mathbf{x}} = \arg\min_{\mathbf{x}} ||\mathbf{A}\mathbf{x} - \mathbf{b}||^2 + \gamma||\mathbf{L}(\mathbf{x} - \tilde{\mathbf{x}})||^2. \tag{25}$$

Here, the elements of the matrix **A** and the right-hand-side vector **b** are given by

$$A_{mn} = \frac{1}{\omega_m^{\text{ref}}}\frac{\partial\omega_m}{\partial p_n}|_{\text{LP}} \tag{26}$$

$$b_m = \frac{\Delta\bar{\omega}_m}{\bar{\omega}_m^{\text{ref}}} - \frac{\omega_m|_{\text{LP}} - \omega_m^{\text{ref}}}{\omega_m^{\text{ref}}}. \tag{27}$$

The first term in Eq. (25) resembles a system of linear equations which is over-determined, i.e., the number of rows in **A** exceeds the number of columns. Hence, the residual $||\mathbf{A}\mathbf{x} - \mathbf{b}||$ can be used as a metric of how well the column space of **A** is able to represent the data vector **b**. The second term in Eq. (25) is a regularization term, which penalizes solutions $\hat{\mathbf{x}}$ away from a given vector $\tilde{\mathbf{x}}$ using a weighted Euclidean norm. The matrix **L** in the regularization term is diagonal with all diagonal elements equal to unity, except for the elements that multiply the shape parameters, which are substantially larger. The balance between the two terms in Eq. (25), i.e., a solution which makes $\mathbf{A}\hat{\mathbf{x}} \approx \mathbf{b}$ and a solution $\hat{\mathbf{x}} \approx \tilde{\mathbf{x}}$, is controlled by the regularization parameter γ. From a Bayesian point of view we can interpret Eq. (25) as the posterior mean estimate of **x** under the assumption that the measurement errors in **b** are Gaussian distributed, have zero mean and covariance matrix $\varsigma^2\mathbf{I}$ and the parameter vector **x** has a Gaussian prior distribution with mean value $\tilde{\mathbf{x}}$ and covariance matrix$(\varsigma^2/\gamma)(\mathbf{L}^T\mathbf{L})^{-1}$[19].

For normal process conditions, the losses in the process vessel are small and, therefore, we compute the sensitivity matrix **A** for linearization points that corresponds to the vessel without losses. Consequently, the permittivity is represented by $\epsilon = \epsilon'_{\text{LP}} + \delta\epsilon' - j\delta\epsilon''$, where ϵ'_{LP} is the permittivity at the linearization point. Here, the perturbations with respect to the linearization point are denoted $\delta\epsilon'$ and $\delta\epsilon''$, which are estimated by means of Eq.(25).

Thus, we use the eigenvalue problem (20) with $\epsilon'(\mathbf{r}) \geqslant \epsilon_0$ and $\epsilon''(\mathbf{r}) = 0$ to compute the eigenfrequencies and eigenmodes at the linearization point, which reduces to a special case of Eq. (18). The computed eigenfrequencies and eigenmodes are also used to compute the sensitivities as described later in this section. The down-bed region with the fluidized particles yields an effective permittivity that deviates substantially from vacuum. Therefore, we use a linearization point

$$\epsilon'_{\text{LP}}(\mathbf{r}) = \begin{cases} \epsilon'_{\text{bed}} & \text{in the bed region,} \\ \epsilon_0 & \text{outside the bed region,} \end{cases} \tag{28}$$

where $\epsilon'_{\text{bed}} \geqslant \epsilon_0$. We solve the eigenvalue problem for a range of bed heights h_{bed} and bed permittivities ϵ'_{bed}, where the solutions from the eigenvalue problem together with the corresponding sensitivities are stored in a database. It is noticed that the resonance frequencies and their sensitivities do not change significantly as h_{bed} decreases, if ϵ'_{bed} is increased to an appropriate extent. This is reasonable since it merely corresponds to a redistribution of the bed material to a smaller region. During the evolution of the process, we continuously seek the linearization point ϵ'_{LP} that yields the best fit with the corresponding deviations in the measured resonance frequencies, where the bed height is fixed at a representative value $h_{\text{bed}} = 5$ cm.

Hence, this estimation procedure can be used for a process where the particles grow significantly in size, by means of changing the linearization point ϵ'_{LP}. To summarize, Eq.(25) yields reliable estimates of the perturbations $\delta\epsilon'_{\text{eff}}$ and $\delta\epsilon''_{\text{eff}}$ with respect to the linearization point. This approach is sufficient for most normal process states.

Sensitivity Analysis

The sensitivity analysis is based on the eigensolution $(\boldsymbol{E}_m, \boldsymbol{B}_m, \omega_m)$ computed for a linearization point with $\epsilon'(\boldsymbol{r}) > \epsilon_0$ and no losses, i.e., $\epsilon''(\boldsymbol{r}) = 0$ and $\sigma(\boldsymbol{r}) = 0$. Given this linearization point, we formulate the sensitivities that relate a small perturbation $\delta\omega_m$ in the resonance frequency ω_m to perturbations in (i) the material parameters $\delta\epsilon'$ and $\delta\epsilon''$ and (ii) the normal displacement δn of the cavity's boundary. Fig. 4a shows a subdivision of the cavity's volume that is used for the parameterization of the material andFig. 4b shows the cavity's wall with a set of basis functions used for the representation of the normal displacement, where both parameterizations are axisymmetric.

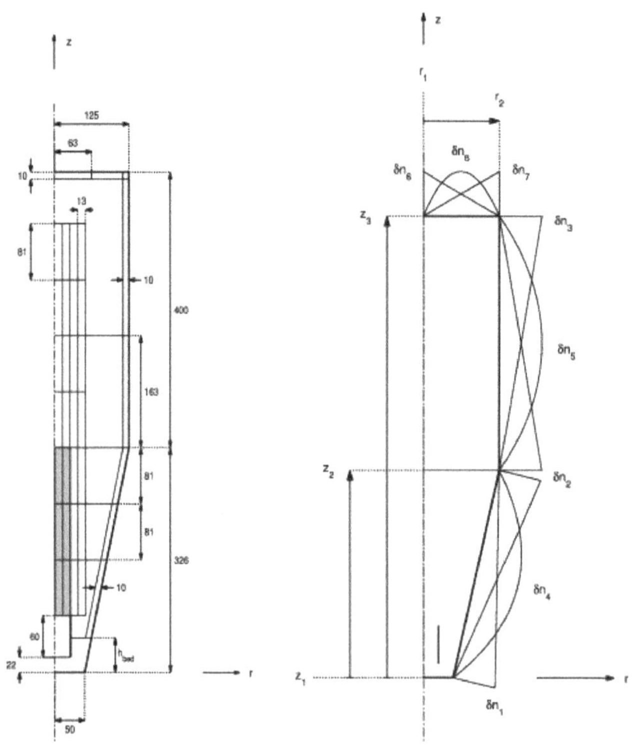

(a) Material representation (b) Shape representation

Figure 4: Two-dimensional projection of the process vessel showing (a) the basic subregions that may be combined in different ways to obtain the material basis functions φ_n and (b) basis functions used to represent shape deformations. All dimensions are in millimeters.

Material Sensitivities

Based on the eigenvalue problem (20), the sensitivity of the resonance frequency ω_m with respect to changes in the material parameters $p = \epsilon'$ and ϵ'' is expressed as

$$\frac{\delta\omega_m}{\omega_m} = -\frac{m_{\delta\epsilon'}(\boldsymbol{E}_m, \boldsymbol{E}_m) - jm_{\delta\epsilon''}(\boldsymbol{E}_m, \boldsymbol{E}_m)}{2m_{\epsilon'}(\boldsymbol{E}_m, \boldsymbol{E}_m)}, \tag{29}$$

where we have

$$m_{\delta p}(\boldsymbol{u}, \boldsymbol{v}) = \int_\Omega \delta p \boldsymbol{u} \cdot \boldsymbol{v} \, d\Omega, \tag{30}$$

for a perturbation δp with respect to the material parameter p [17] and [20]. From Eq.(29), we notice that $\delta\epsilon'$ affects only the real part of the eigenfrequencies, and that $\delta\epsilon''$ only affects the imaginary part. The discrete version of this linearization is

$$\frac{\delta\omega_m}{\omega_m} = -\frac{\mathbf{e}_m^T[\mathbf{M}_{\delta\epsilon'} - j\mathbf{M}_{\delta\epsilon''}]\mathbf{e}_m}{2\mathbf{e}_m^T\mathbf{M}_{\epsilon'}\mathbf{e}_m}. \tag{31}$$

Representation of Material Perturbations

The perturbation δp of the material parameter p is represented in terms of a set of N space dependent basis functions $\varphi_n(\boldsymbol{r})$, which are multiplied with the coefficients δp_n, according to

$$\delta p(\mathbf{r}) = \sum_{n=1}^{N} \delta p_n \varphi_n(\mathbf{r}). \tag{32}$$

Here, we subdivide the vessel's volume Ω into regions Ω_n such that Ω_n do not overlap and $\bigcup_{n=1}^{N} \Omega_n = \Omega$. We exploit piecewise constant basis functions $\varphi_n(\boldsymbol{r})$ that are equal to one if $\boldsymbol{r} \in \Omega_n$ and zero elsewhere. In particular, we associate the basis functions with three regions: (i) the bed region of height h_{bed} with a relatively high density of particles; (ii) a cylinder shaped volume that coincides with the core of the fountain; and (iii) the rest of the process vessel. These regions are formed by combining the basic subregions shown in Fig. 4a.

The residual $\|\mathbf{A}\hat{\mathbf{x}} - \mathbf{b}\|$ in the solution to the estimation problem (25) is used as a metric of how well these basis functions are able to represent the actual permittivity distribution in the experiment. For instance, the height and radius of the fountain region is chosen such that the residual is minimized. Based on measurement results presented later in this article, we find that the fountain region indicated by gray color in Fig. 4a yields the best agreement with the measured resonance frequencies.

Shape Sensitivities

For a normal displacement δn of the vessel's boundary Γ, the resulting perturbation in the resonance frequency is given by [17] and [20]

$$\frac{\delta\omega_m}{\omega_m} = -\frac{d_{\delta n}(\boldsymbol{B}_m, \boldsymbol{B}_m)/\mu_0 - d_{\delta n}(\epsilon'\boldsymbol{E}_m, \boldsymbol{E}_m)}{2W_m}, \tag{33}$$

where W_m is the time-average energy in the cavity given by Eq. (22), and

$$d_{\delta n}(\boldsymbol{u}, \boldsymbol{v}) = \int_\Gamma \boldsymbol{u} \cdot \boldsymbol{v}^* \delta n \, dS. \tag{34}$$

A positive deformation $\delta n > 0$ here corresponds to that the volume of the cavity increases. We notice that shape deformations only affect the real part of the eigenfrequencies.

The corresponding discrete formulation of Eq. (33) is

$$\frac{\delta\omega_m}{\omega_m} = -\frac{\frac{1}{\mu_0}\mathbf{b}^H\mathbf{D}_{\delta n}\mathbf{b} - \mathbf{e}^H\mathbf{D}_{\epsilon'\delta n}\mathbf{e}}{\frac{1}{\mu_0}\mathbf{b}^H\mathbf{M}_1\mathbf{b} + \mathbf{e}^H\mathbf{M}_{\epsilon'}\mathbf{e}}, \tag{35}$$

where the superscript ^{H} indicates conjugate transpose.

Representation of Shape Perturbations

We use an axisymmetric representation of the normal displacement δn by means of a quadratic hierarchical basis for three different parts of the vessel's boundary: (i) the upper lid; (ii) the cylindrical surface; and (iii) the conical surface. Fig. 4b shows the basis functions in relation to the geometry of the vessel. On each of the three parts of the boundary, we exploit the (local) basis functions

$\psi_1(u)=1-u$ (36)

$\psi_2(u)=u$ (37)

$\psi_3(u)=4u(1-u)$ (38)

where the part of the boundary is parameterized by the local coordinate u such that the interval $0 \leqslant u \leqslant 1$ corresponds to the entire part of the boundary.

In order to constrain the number of degrees of freedom associated with the cavity's shape to a number that is practical, and to suppress non-physical shape deformations, we form two new basis functions by linearly combining $\delta n_1 - \delta n_8$ according to

$$\delta n_{lid}=0.97\delta n_6+0.24\delta n_8 \quad (39)$$

$$\delta n_{cone/cyl}=0.69\delta n_2+0.16\delta n_3+0.71\delta n_4. \quad (40)$$

Thus, δn_{lid} describes a smooth deformation of the top lid and $\delta n_{cone/cyl}$ describes a joint deformation of the cylindrical and conical parts of the vessel. The coefficient values in Eqs. (39) and (40) are chosen such that the residual $\|\mathbf{A}\hat{\mathbf{x}} - \mathbf{b}\|$ is minimized, given only these two independent degrees of freedom for the shape.

Temperature Measurements

Before the vessel is loaded with particles, it is normally pre-heated. We exploit this part of the process to gain information that is used to compute the a priori information $\tilde{\mathbf{x}}$ for later parts of the process. During the pre-heating of the vessel, the vessel is empty and we perform simultaneous and synchronized measurements on (i) the resonance frequencies and (ii) the temperature registered by the set of temperature probes shown in Fig. 2. Given the deviations $\mathbf{y}_{\delta T}$ in the temperatures (relative to the initial temperatures) and the shape perturbations $\hat{\mathbf{x}}_{\delta n}$ estimated from the resonance frequencies for the empty cavity, we use the relatively simple model $\hat{\mathbf{x}}_{\delta n} = \mathbf{B}\mathbf{y}_{\delta T}$, where the matrix $\mathbf{B}$ is determined by means of linear regression from the data acquired during the pre-heating interval. This model is used to obtain a priori estimates $\tilde{\mathbf{x}}_{\delta n} = \mathbf{B}\mathbf{y}_{\delta T}$ from the measured temperatures during later parts of the process, after the particles have been loaded. In the following, we exploit the a priori information

$$\tilde{\mathbf{x}} = \begin{bmatrix} \tilde{\mathbf{x}}_{\delta\epsilon_{eff}} \\ \tilde{\mathbf{x}}_{\delta n} \end{bmatrix} = \begin{bmatrix} 0 \\ \mathbf{B}\mathbf{y}_{\delta T} \end{bmatrix}, \quad (41)$$

in the estimation problem (25), where $\tilde{\mathbf{x}}_{\delta\epsilon_{eff}}$ corresponds to all the material parameters that are incorporated in the estimation procedure.

Tests demonstrate that about 95% of the perturbations in the resonance frequencies due to thermal expansion of the vessel during heating and cooling can be accounted for by this model, when the vessel is empty. This approach is less successful once the vessel is loaded with particles, since the presence of particles yields a different temperature distribution in the vessel. However, it is presumed that it is feasible to improve the a priori information, should the metal temperature be measured at more places on the vessel's wall and top lid.

RESULTS

In the measurement results presented in the following, we perform repeated microwave measurements on the process vessel as the process is running. Each process experiment typically takes a few hours in total. The external process parameters such as the flow-rate, temperature and humidity of the fluidizing air, the pressure and flow in the pneumatic spray-nozzle and the flow-rate of the spraying liquid, are controlled from the user interface of the process machine. These external process parameters together with measured temperatures at the locations indicated in Fig. 2, are logged by the process machine.

Cavity without Particles

Among the eigenmodes hosted by the process vessel we have identified eight modes that are feasible to use for sensing. These modes are listed in Table 1 together with their resonance frequencies and unloaded Q -values. All these modes have similarities with the TE and TM modes hosted by a circular cylindrical cavity. In particular, the modes$m=1,2,5$ and 7 resemble the modes $TE_{111}, TE_{112}, TE_{113}$, and TE_{114} of a circular cylindrical cavity, and the modes $m=3,4$ and 6 resemble the modes TM_{010}, TM_{011} and TM_{012} respectively. The mode $m=7$ (TE_{114}) has a strong electric field in the lower part of the vessel, which makes it well-suited for probing the permittivity in the down-bed region. The mode $m=8$ resembles TE_{011} which has an electric field that is zero on all cavity walls, which implies that its resonance frequency is insensitive to dielectric material adhering to the cavity walls. This mode also has a notably high Q-value.

Table 1: Computed and measured values of the resonance frequencies and unloaded Q-values of the lowest eigenmodes in the process vessel. The variabilities (here represented by one standard deviation) in the measured quantities arise from disassembling and reassembling the complete vessel.

m	Computed		Measured	
	f_0 (MHz)	Q	f_0 (MHz)	Q
1	759	5100	761 ± 0.07	5400 ± 100
2	894	5800	894 ± 0.05	5100 ± 400
3	928	5900	930 ± 0.02	4400 ± 400
4	1008	6000	1009 ± 0.03	3200 ± 800
5	1051	6100	1051 ± 0.06	4700 ± 200
6	1149	6300	1150 ± 0.03	3400 ± 700
7	1207	5600	1206 ± 0.08	3500 ± 300
8	1495	9800	1498 ± 0.05	10,300 ± 400

The loaded Q -values in the experiment can be calculated from the real and imaginary part of the measured complex resonance frequency ω according to

$$Q_l = \frac{\mathrm{Re}(\omega)}{2 \cdot \mathrm{Im}(\omega)}, \quad (42)$$

for the weakly damped cavity modes [17]. Since the cavity is loaded by two probes with different coupling coefficients, the following expression is used to calculate the unloaded Q-value

$$Q_u = Q_l\left(\frac{2}{|S_{11}| + |S_{22}|}\right), \quad (43)$$

where the reflection coefficients S_{11} and S_{22} are evaluated at each resonance frequency [21].

We calculate the *Q*-value from the electromagnetic model using the method described in Section 3.3, and here we assume that the cavity walls have a conductivity $\sigma_c = 10^6\ \mathrm{S/m}$, which should be representative for the stainless steel that the vessel is made of.

As can be seen in Table 1, the measured and computed resonance frequencies agree well. There is also a generally good agreement between measured and computed *Q*-values although, for some modes, the measured *Q*-value is significantly lower than the computed. The latter may be a result of insufficient electric contact in the joint between the conical and cylindrical part of the cavity, or at the aluminum foil that covers the windows. As the process vessel is disassembled and reassembled, a certain variability in the measured resonance frequencies and *Q*-values is introduced, which is presented inTable 1. This variability is significant in relation to the perturbations in resonance frequencies caused by changes in the state of the process. Therefore, the particles are inserted into the process vessel through a small hole (indicated by H1 in Fig. 2) after the necessary reference measurements on the empty cavity have been collected, in order to eliminate this variability. Investigations show that the variability introduced from inserting the particles in this manner is negligible.

Effects from air Humidity and Spray

In a Wurster-process, several different components that influence the resonance frequencies are present simultaneously. The most important of these components are (i) particles, (ii) the temperature of the vessel (iii) liquid droplets from the spray nozzle and (iv) water vapor in the fluidizing air. The temperature of the vessel affects the resonance frequencies mainly via perturbations of the cavity's shape, whereas the other components are assumed to influence the resonance frequencies via the permittivity. In order to assess how each of these components contribute to the microwave response, they have been studied separately. Some of these components, such as the spraying liquid and the temperature, are inherently interconnected, i.e., the spraying rate cannot be changed without also affecting the temperature.

The humidity of the fluidizing air can be controlled independently using the process machine and humidities ranging from 0 to $20\ (\mathrm{g\ H_2O})/(\mathrm{kg\ air})$, corresponding to dew-points in the range from −30 °C to 25 °C, can be achieved. Our measurements show that the humidity influences only the real part of the permittivity of the air and hence the resonance frequencies, to a degree that is clearly detectable.

Water sprayed from the pneumatic spray nozzle (also referred to as the atomizer) yields similar changes in the permittivity inside the cavity as the humidity does. Only ϵ'_{eff} is influenced by the spray, and if the same amount of water is added through the atomizer as via the fluidizing air, the change in ϵ'_{eff} is very similar. The contributions to the permittivity of the air from humidity in the fluidizing air and from water spray are additive, as long as the spraying rate is sufficiently low to allow for complete evaporation of the liquid droplets.

Measurements with Particles

In the following, we present measurement results for processes involving particles. The particles used are made of MCC (Cellets® brand) and have diameters in a distribution from 500 to 700 µm. A batch consisting of 200 g of particles is used in all of the following experiments. All process settings are identical in the following experiments, except for the type of spraying solution and the spraying rate. The fluidization flow is set to $30\ \mathrm{Nm^3/h}$ (where N denotes normal temperature and pressure) with an inlet temperature of 70 °C. The atomizer pressure is 2.0 bar and the associated atomizer flow is $2.0\ \mathrm{Nm^3/h}$.

The spraying liquid used in the experiments is water containing either (i) Mannitol or (ii) NaCl. The permittivity of these spraying liquids is measured using the dielectric probe kit 85070E from Agilent Technologies, and the results are shown in Fig. 5.

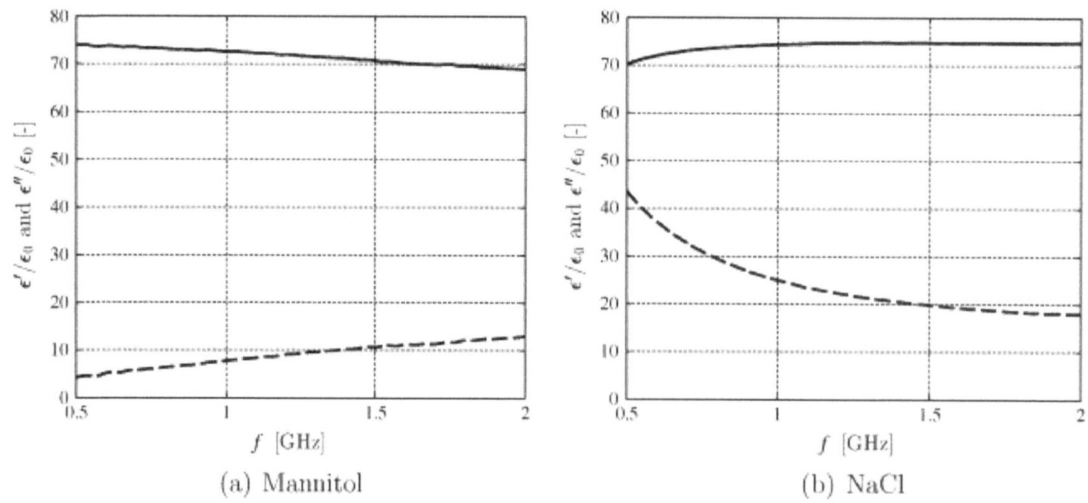

Figure 5: Relative permittivity of distilled water at room temperature that contains (a) Mannitol (13.5%) and Kollicoat IR® (1.5%), and (b) NaCl (0.5%). Solid curves represent ϵ'/ϵ_0 and dashed curves ϵ''/ϵ_0.

In the beginning of each experiment, the vessel is pre-heated and the particles are loaded, before the experiment-specific part of the process (e.g. coating) is initiated. In the results presented in the following, we omit the pre-heating interval and focus on the experiment-specific parts.

The average error between the model and measurement is computed as the root-mean-square of the difference between measurements (**b**) and model fit ($\mathbf{A}\hat{\mathbf{x}}$), averaged with respect to the number of sensing modes M. With particles present in the vessel, this average error is in the order of 10^{-5} during normal process conditions (no extensive agglomeration, etc.) which should be compared with the average magnitude of the measurement response ($||\mathbf{b}||/M$), which is in the order of 10^{-3}.

According to the sensitivity analysis in Section 4.1, perturbations in the real part of the resonance frequencies are related to both $\delta\epsilon'_{\text{eff}}$ and δn. Similarly, a perturbation in the imaginary part of the resonance frequency is related to $\delta\epsilon''_{\text{eff}}$. As described in Section 4.2, δn is to a large extent determined based on a priori information that exploits temperature measurements. An error in the a priori information may yield a corresponding error in $\delta\epsilon'_{\text{eff}}$, which varies on a timescale that is significantly slower than the fast timescales associated with the fluid-dynamics of the process. Based on these two time-scales, we make the assumption $\epsilon_{\text{eff}}(t)=\epsilon_{\text{eff,slow}}(t)+\epsilon_{\text{eff,fast}}(t)$, where $\epsilon_{\text{eff,fast}}(t)$ represents fast random fluctuations and $\epsilon_{\text{eff,slow}}(t)$ represents slow changes in the permittivity caused either by errors in the a priori information or by changes in the actual process state (such as particles growing in size or accumulation of moisture in the particles). We model $\epsilon_{\text{eff,slow}}(t)$ by a low-order polynomial with respect to time t, and this polynomial is determined by means of linear regression based on the estimated $\epsilon_{\text{eff}}(t)$. Furthermore, we compute the covariance matrix from the real and imaginary part of $\epsilon_{\text{eff,fast}}(t)$ according to $\mathbf{C}=\text{cov}(\mathbf{z}_1,\mathbf{z}_2)$ where $\mathbf{z}_1 = [\epsilon'_{\text{eff,fast}}(t_n)/\epsilon_0 - 1]^{\text{T}}$ and $\mathbf{z}_2 = [\epsilon''_{\text{eff,fast}}(t_n)/\epsilon_0]^{\text{T}}$ are vectors containing samples of $\epsilon_{\text{eff,fast}}$ taken at discrete time instants t_n. If the eigenvectors of $\mathbf{C}$ are scaled by the square root of its eigenvalues (i.e. the standard deviations), they define the major and minor semi-axes of an ellipse that represent the co-variation of the data. Such ellipses are included in the scatter-plots in the presentation of the following measurement results. Variations in $\epsilon_{\text{eff,fast}}(t)$ along the eigenvector associated with the largest eigenvalue of $\mathbf{C}$, are mainly attributed to fluctuations in the particle volume fraction ϑ. Therefore, we take the slope of this eigenvector (i.e. the major semi-axis of the ellipse) as a density-independent ratio which is equivalent to the ratio η in Section 2.2.2.

The relationship between this slope and underlying physical parameters such as the moisture content of the particles and the concentration of salt (NaCl) in the spraying liquid, have been studied by means of Monte-Carlo simulations. These simulations are based on measurements of the complex permittivity of MCC [11] and empirical models for the permittivity of water with and without salt [10], together with the mixing formulas described in Section 2.2.1. Thus, we generate random mixtures of MCC particles and air by using statistical normal-distributions to represent the volume fraction and water content of the particles. The water is assumed to be either absorbed uniformly in the MCC-particles or distributed in a uniform water-layer on the surface of the particles. The case with a pronounced water-layer may be representative also for situations where the surface of the particles have a substantially higher moisture content than the core. Based on these simulations, we observe the following trends:

1. An increase in the moisture content of the MCC particles (absorbed water) *increases* the slope, if the initial moisture content is less than around 7%. For higher moisture contents, the slope instead *decreases* with increasing moisture content.
2. Pure water distributed on the surface of the particles *decreases* the slope.
3. Water containing salt distributed on the surface of the particles *increases* the slope if the salt concentration exceeds 1%.
4. The slope is *not* affected by changes in the volume fraction, if the volume fraction is sufficiently low (around 5% or lower), whereas at higher volume fractions, the slope *increases* with increasing volume fraction.

Fast Coating Process

Here, we present results from a process where MCC particles are coated with a mixture of Mannitol (a sugar) and a polymer labeled Kollicoat IR® (ethylene glycol vinyl acetate grafted polymer), where the Mannitol and the Kollicoat IR are dissolved in de-ionized water. The dry content of the solution (the fraction that ideally stays on the surface of the particles) is 15 mass% (13.5% Mannitol and 1.5% Kollicoat IR). The complex relative permittivity for this solution is shown in Fig. 5a, and here we notice that the response is similar to that of pure water. In this experiment, a relatively high spraying rate is used to obtain fast build-up of the coating layer. First, the spraying rate is set to 11.3 g/min and kept constant during approximately 15 min. Next, the spraying rate is increased even further to 14.4 g/min, which eventually leads to extensive agglomeration and particles adhering to the walls of the vessel.

Fig. 6 shows the estimated permittivity in three different regions: (i) the down-bed region; (ii) the fountain region above the Wurster tube; and (iii) the rest of the volume above the down-bed region, as function of process time in hours. At $t = 0.9$ h, the particles are fluidized in the process with the fountain off, and the temperature is in equilibrium. The following changes in process settings are marked with dashed vertical lines: at $t = 1.04$ h the fountain is switched on; in the interval $1.2 \text{ h} < t < 1.26$ h the spraying rate is successively increased to 11.3 g/min; at $t = 1.51$ h the spraying rate is increased to 14.4 g/min. Some interesting features can be observed in these figures:

1. The estimated permittivities fluctuate substantially, which is reasonably caused by the continuous re-distribution of the particles in the process which occurs on a short time scale (<0.1 s).
2. As the fountain is switched on, the estimated permittivity increases clearly in the fountain and remaining volume regions, whereas it decreases in the bed region. This is reasonable since as the fountain is switched on, particles start to circulate in the upper part of the vessel which reduces the amount of particles in the down-bed region.
3. The estimated permittivity in the fountain decreases slightly as the spraying starts. This could be an artifact (incorrect estimation) resulting from insufficient compensation for temperature variations, but it could

also be due to a lower particle density in the fountain as the spraying starts. The estimated permittivity in the remaining volume increases as the spraying starts, reasonably due to increased air humidity inside the vessel.

4. During the interval 1.5 h < t < 1.75 h, extensive agglomeration occurs in the process and the particles start to build up dense layers on parts of the cavity walls, as a result of the high spraying rate (this could be observed visually after the experiment was finished). This process is clearly reflected in the estimated permittivities, which are changing dramatically during the agglomeration interval.

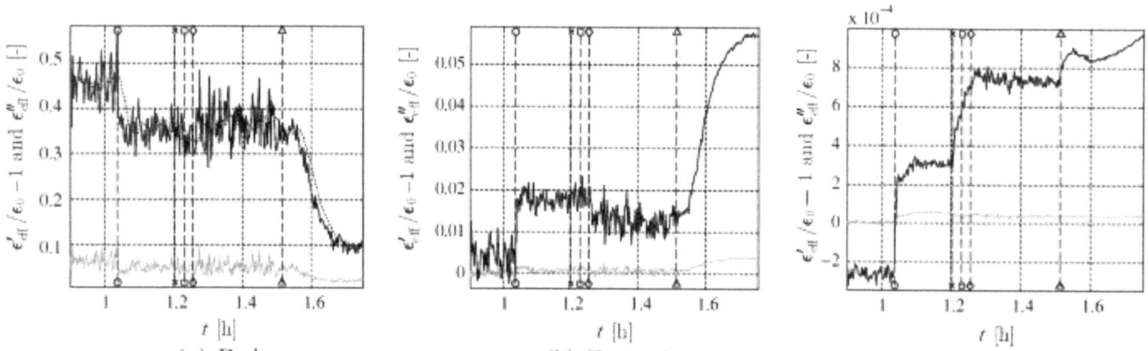

(a) Bed (b) Fountain (c) Volume

Figure 6: Estimated permittivity as a function of process time in the fast coating process. Black lines represent $\epsilon'_{eff}/\epsilon_0 - 1$ and gray lines $\epsilon''_{eff}/\epsilon_0$. The dashed black curve in (a) shows the permittivity in the bed region at the current linearization point.

In situations similar to the agglomeration interval described in item 4 above, i.e. at very moist process conditions that yields material build-up on the vessel walls, we repeatedly observe a larger discrepancy between the model and measurements, i.e. an increasing residual $\|\mathbf{Ax} - \mathbf{b}\|$. Thus, a model that allows only for a constant permittivity in the bed, fountain and remaining volume regions, is unable to accurately represent the material distributions formed during extensive agglomeration. This indicates the possibility of detecting anomalous process conditions based on the residual. Studies show however that it is possible to maintain a low residual also during extensive agglomeration if the cavity is divided into sub-regions that better represent the actual material distribution. The sub-regions in the direct vicinity of the cavity walls shown in Fig. 4a may be well-suited for this purpose.

Fig. 7 shows scatter plots of $\epsilon''_{eff}/\epsilon_0$ versus $\epsilon'_{eff}/\epsilon_0 - 1$ in the three regions before and after spraying starts. The ellipses in these scatter plots represent the co-variation of the data, where the length of each semi-axis is three standard deviations. All ellipses are highly elongated and the variations along the major semi-axis is assumed to be caused by variations in material density, according to the discussion on density-independent functions in Section 2.2.2. The slopes in the scatter plots are shown in Table 2. Here we notice that (i) the slope is higher in the bed region, as compared to the fountain and volume regions and (ii) the slope decreases in the fountain and in the volume after the spraying starts, but remain unchanged in the bed. According to the results from the Monte-Carlo simulations described previously, the decreasing slope in the fountain and volume regions could indicate that the moisture content at the surface of these particles increases during the spraying interval. The higher slope in the bed is in agreement with the Monte-Carlo simulations, since the volume fraction in the bed is substantially higher than in the other regions.

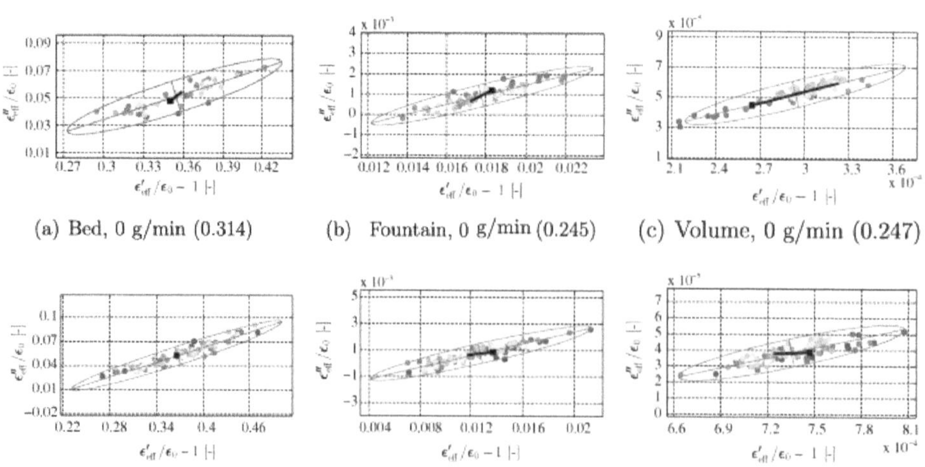

(a) Bed, 0 g/min (0.314) (b) Fountain, 0 g/min (0.245) (c) Volume, 0 g/min (0.247)

(d) Bed, 11.3 g/min (0.313) (e) Fountain, 11.3 g/min (0.228) (f) Volume, 11.3 g/min (0.211)

Figure 7: Scatter plots of $\epsilon''_{eff}/\epsilon_0$ versus $\epsilon'_{eff}/\epsilon_0 - 1$ in the fast coating process. Values inside parentheses are the slopes of the major semi-axis of the ellipses. The black lines starting with squares represent the slowly varying component $\epsilon_{eff,slow}$ of the estimated permittivity.

Table 2: Slopes of the major semi-axis in the scatter plots of ϵ''_{eff} versus $\epsilon'_{\text{eff}} - 1$.

	Bed	Fountain	Volume
Spraying off	0.314	0.245	0.247
Spraying 11.3 g/min	0.313	0.228	0.211

Slow Coating Process

In this experiment, MCC particles are coated with Mannitol and Kollicoat IR® with the mixing ratio described in Section 5.2.1, under relatively dry process conditions, i.e., a relatively low spraying rate of 6.6 g/ min is used.

Fig. 8 shows the estimated permittivity in the bed, fountain and remaining volume regions, as function of process time in hours. The time interval 0 h < t < 0.9 h, which has been omitted, includes pre-heating of the vessel and loading of particles, so at t = 0.9 h, the particle fountain is switched on and the temperature is close to equilibrium. The process conditions are changed at the time instants indicated by the dashed vertical lines: at t = 0.89 h spraying (coating) starts with a spraying rate of 4.6 g/min; at t = 0.96 h the spraying rate is increased to the final value 6.6 g/min; at t = 1.13 h the temperature is considered stable and at t = 1.78 h the spraying ends. In these figures, some interesting features can be observed:

1. The estimated permittivity in the bed increases slowly during the coating interval 0.90 h < t < 1.78 h, whereas the permittivities in the fountain and the remaining volume regions remain fairly constant. This may be explained by particles growing in size as a result of the coating, which results in more dielectric material that accumulates mainly in the bed region. The growth in particle size was verified by means of sieving and weighing the particles after the coating process was finished, and the mass-increase was found to be approximately 25%. A bed that grows in size in the experiment is interpreted by our model as an increase in the effective permittivity in the bed-region, which has a fixed size. This effect can however not be clearly observed in the fast coating experiment described in Section 5.2.1, which is somewhat surprising given that we use the same coating solution in both processes.
2. The permittivity in the remaining volume region (Fig. 8) increases as the spraying starts, it increases further as the spraying rate increases and drops as the spraying is stopped. This may be attributed to changes in the permittivity of the air, due to increased humidity caused by the evaporating water in the spray.

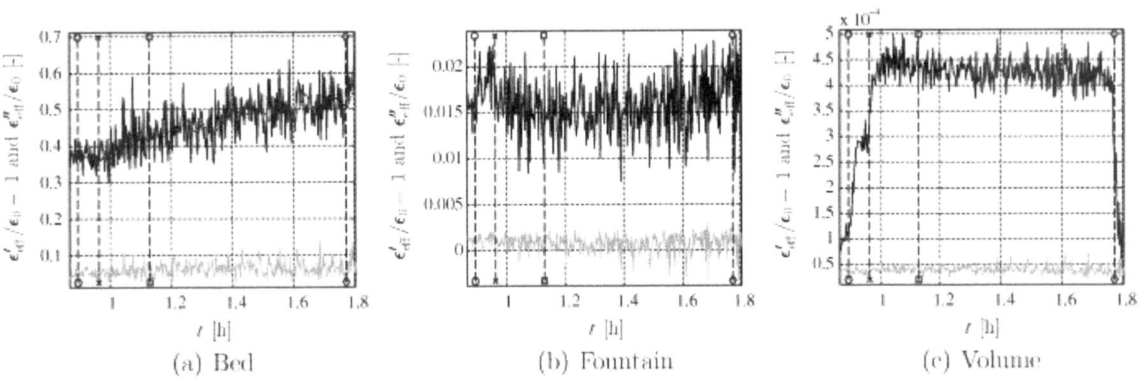

Figure 8: Estimated permittivity as a function of process time in the slow coating process. Black curves represent $\epsilon'_{\text{eff}}/\epsilon_0 - 1$ and gray curves $\epsilon''_{\text{eff}}/\epsilon_0$. The dashed black curve in (a) shows the permittivity in the bed region at the current linearization point.

Fig. 9 shows scatter plots of $\epsilon''_{\text{eff}}/\epsilon_0$ versus $\epsilon'_{\text{eff}}/\epsilon_0 - 1$ for the tree regions, during the coating interval. Here, we notice that the slope is higher in the bed as compared to the other two regions, which is reasonable given the higher volume fraction in the bed. We also notice how the slowly varying component $\epsilon_{\text{eff,slow}}$ shown by black lines in Fig. 9, varies substantially in the bed region, reasonably due to the growth in particle size.

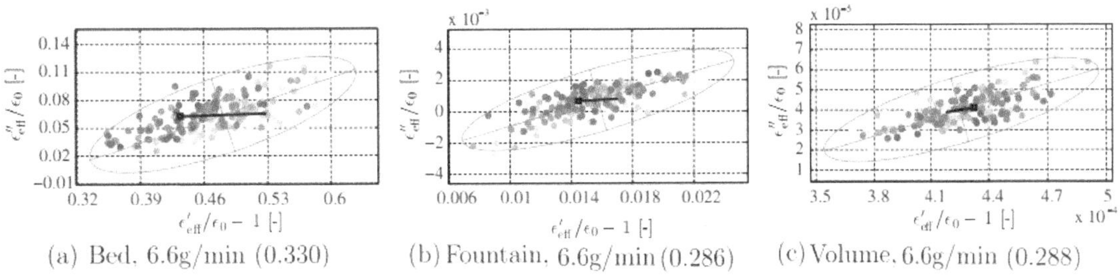

Figure 9: Scatter plots of $\epsilon''_{\text{eff}}/\epsilon_0$ versus $\epsilon'_{\text{eff}}/\epsilon_0 - 1$ in the slow coating process. Values inside parentheses are the slopes of the major semi-axis of the ellipses. The black lines starting with squares represent $\epsilon_{\text{eff,slow}}$, i.e. the slowly varying component of the estimated permittivity.

Particles Sprayed With Salt Water

In this experiment, MCC particles are sprayed with salt water (0.5 mass% NaCl dissolved in distilled water at room temperature). The complex relative permittivity for this solution is shown in Fig. 5b and here we notice a significant increase in ϵ'' at low frequencies due to the conductivity introduced by the salt. Two different spraying rates are studied, namely $11.6\,\mathrm{g/min}$ and $13.3\,\mathrm{g/min}$.

Fig. 10 shows the estimated permittivity in the bed, fountain and remaining volume regions, as function of process time in hours. At the first time instant shown in Fig. 10, the particles are fluidized, the fountain is on and the temperature is stable. The following changes in process settings are marked with dashed vertical lines: at t = 1.84 h the spraying starts (11.6 g/min); at t = 2 h the temperature is considered stable; at t = 2.24 h the spraying rate is increased to 13.3 g/min which yields very moist process conditions and presumed initiation of extensive agglomeration. The increase in spraying rate to 13.3 g/min is followed by clear changes in the estimated permittivities, possibly due to free water (with salt) on the surface of the particles in the fountain region and/or agglomeration with particles adhering to the walls of the vessel.

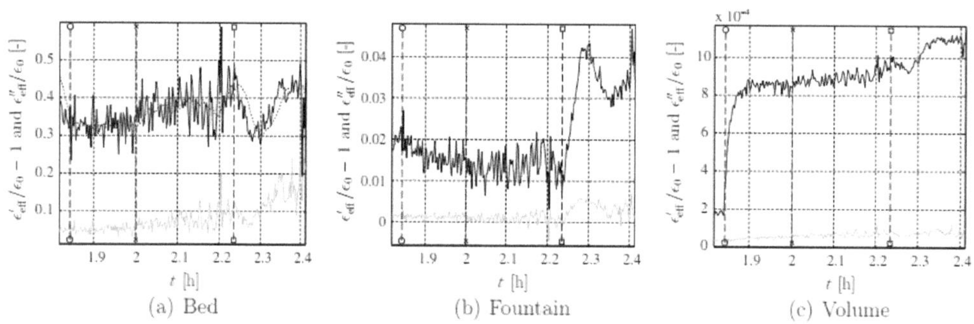

Figure 10: Estimated permittivity as a function of process time in the salt water experiment. Black curves represent $\epsilon'_{\mathrm{eff}}/\epsilon_0 - 1$ and gray curves $\epsilon''_{\mathrm{eff}}/\epsilon_0$. The dashed black curve in (a) shows the permittivity in the bed region at the current linearization point.

Fig. 11 shows scatter plots of $\epsilon''_{\mathrm{eff}}/\epsilon_0$ versus $\epsilon'_{\mathrm{eff}}/\epsilon_0 - 1$ in two intervals: (i) no spray and (ii) spraying at 11.6 g/min (which does *not* yield agglomeration). The slopes of the major axis for the three regions is also presented in Table 3. Here, it is interesting to notice how the slopes in all regions clearly increase after the spraying is switched on. This is expected, given the high dielectric losses of the salt water in the frequency range of the cavity resonances, and it is also in agreement with the Monte-Carlo simulations presented previously.

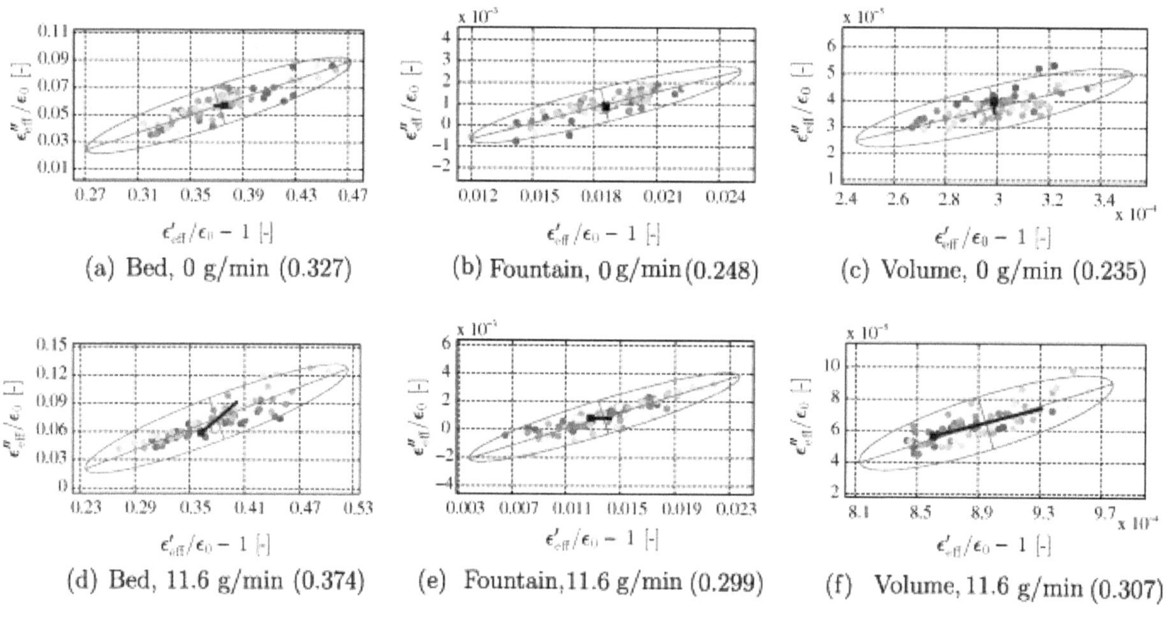

Figure 11: Scatter plots of $\epsilon''_{\mathrm{eff}}/\epsilon_0$ versus $\epsilon'_{\mathrm{eff}}/\epsilon_0 - 1$ in the salt water experiment. Values inside parentheses are the slopes of the major semi-axis of the ellipses. The black lines starting with squares represent the slowly varying component $\epsilon_{\mathrm{eff,slow}}$ of the estimated permittivity.

Table 3: Slopes of the major semi-axis in the scatter plots of $\epsilon''_{\mathrm{eff}}$ versus $\epsilon'_{\mathrm{eff}} - 1$ for the salt water experiment.

	Bed	Fountain	Volume
Spraying off	0.327	0.248	0.235
Spraying: 11.6 g/min	0.374	0.299	0.307

CONCLUSIONS

We have presented a microwave measurement technique for industrial processes contained in closed metal vessels, which exploits a number of electromagnetic resonances of the metal vessel to estimate the process state. In particular, we estimate the effective permittivity inside a vessel that hosts a pharmaceutical fluidized-bed process, which is used for particle coating. This process involves spraying a solution onto the particles, where water acts as the solvent for different types of coating substances. For this particular application, it is important to monitor the moisture content of the particles and the particle distribution. Electromagnetic waves interact strongly with water in a non-invasive and non-destructive manner, which makes this type of measurement technology very attractive. In addition, our measurement system is sensitive to changes in permittivity throughout the process vessel and, thus, it can be considered to perform a global measurement of the process state, which is challenging for more conventional techniques.

We exploit two magnetic field probes and measure the scattering parameters as function of frequency using a network analyzer. The complex resonance frequencies are estimated from the scattering parameters and these are related to a reference measurement of the empty vessel in order to achieve relative changes in the resonance frequencies. The perturbations in the resonance frequencies are related to (i) the perturbations in the complex effective permittivity inside the vessel and (ii) the thermal expansion of the vessel caused by variations in the temperature. We exploit an estimation algorithm with regularization to estimate the effective permittivity and the shape of the vessel, where the unknown effective permittivity and shape is parameterized with the aid of spatial basis functions. The real and imaginary part of the effective permittivity are assumed to be frequency-independent in the frequency band of the used resonance frequencies. We model the vessel by means of the finite element method and the computed eigenmodes and eigenfrequencies are used to calculate sensitivities with respect to changes in the material parameters and the shape of the vessel. The estimation incorporates a priori information that is collected from probes that measure the temperature of the vessel's metal walls and the air that flows through the vessel. This approach accounts for substantial parts of the temperature variations but it is believed that it can be significantly improved.

The measurement technique performs well for the estimation of the effective permittivity that is parameterized for a small number of sub-regions. The re-distribution of material occurring when the particle fountain is switched on, is clearly reflected in the estimated permittivity in all these sub-regions. The effect of particles that grow in size during coating processes can be observed in the estimated permittivities, and we have also shown that material build-up on the cavity walls can be clearly detected. We have related the real part of the effective permittivity to its imaginary part by means of scatter plots and covariance matrices, where measurements taken at different time instants fall approximately on a straight line. This line is defined by the eigenvector associated with the largest eigenvalue of the covariance matrix, and the slope of the line is affected by the liquid sprayed onto the particles in the process. In particular, the slope of this line clearly increases if the particles are sprayed with a liquid that features substantial dielectric losses.

ACKNOWLEDGEMENTS

Livia Cerullo was financially supported by AstraZeneca. In addition, Johan Winges was financially supported by the Swedish Research Council (dnr 2010-4627) in the project "Model-based Reconstruction and Classification Based on Near-Field Microwave Measurements". The computations were performed on resources at Chalmers Centre for Computational Science and Engineering (C3SE) provided by the Swedish National Infrastructure for Computing (SNIC).

REFERENCES

1. J. Rantanen, S. Lehtola, P. Rämet, J.-P. Mannermaa, J. Yliruusi, On-line monitoring of moisture content in an instrumented fluidized bed granulator with a multi-channel NIR moisture sensor, Powder Technol. 99 (1998) 163–170.
2. T.D. Beer, A. Burggraeve, M. Fonteyne, L. Saerens, J. Remon, C. Vervaet, Near infrared and Raman spectroscopy for the in-process monitoring of pharmaceutical production processes, Int. J. Pharm. 417 (2011) 32–47.
3. C. Buschmüller, W. Wiedey, C. Döscher, J. Dressler, J. Breitkreutz, Inline monitoring of granule moisture in fluidized-bed dryers using microwave resonance technology, Euro. J. Pharm. Biopharm. 69 (2008) 380–387.
4. V. Rimpiläinen, L.M. Heikkinen, M. Vauhkonen, Moisture distribution and hydrodynamics of wet granules during fluidized-bed drying characterized with volumetric electrical capacitance tomography, Chem. Eng. Sci. 75 (2012) 220–234.
5. M. Dressel, O. Klein, S. Donovan, G. Grüner, Microwave cavity perturbation technique: Part III: applications, Int. J. Infrared Millimeter Waves 14 (1993) 2489–2517.
6. B. Meng, J. Booske, R. Cooper, Extended cavity perturbation technique to determine the complex permittivity of dielectric materials, IEEE Trans. Microwave Theory Tech. 43 (1995) 2633–2636.
7. S. Karlsson, A. Rasmuson, B. van Wachem, I.N. Björn, CFD modeling of the Wurster bed coater, AIChE J. 55 (2009) 2578–2590.
8. Sihvola, Electromagnetic Mixing Formulas and Applications, The Institution of Engineering and Technology (2008).
9. G. Kristensson, S. Rikte, A. Sihvola, Mixing formulas in the time domain, J. Opt. Soc. Am. 15 (1998) 1411–1422.
10. E. Nyfors, P. Vainikainen, Industrial Microwave Sensors, Artech House, 1989.
11. L. Gradinarsky, H. Brage, B. Lagerholm, I.N. Björn, S. Folestad, In situ monitoring and control of moisture content in pharmaceutical powder processes using an open-ended coaxial probe, Measure. Sci. Technol. 17 (2006) 1847–1853.
12. S. Trabelsi, S.O. Nelson, Density-independent functions for on-line microwave moisture meters: a general discussion, Measure. Sci. Technol. 9 (1998) 570–578.
13. D.A. Engers, M.N. Fricke, R.P. Storey, A.W. Newman, K.R. Morris, Triboelectrification of pharmaceutically

relevant powders during low-shear tumble blending, J. Electrostat. 64 (2006) 826–835.

14. D.A. Engers, M.N. Fricke, A.W. Newman, K.R. Morris, Triboelectric charging and dielectric properties of pharmaceutically relevant mixtures, J. Electrostat. 65 (2007) 571–581.

15. J.C. Nédélec, Mixed finite elements in R3, Numer. Math. 35 (3) (1980) 315–341.

16. T. Rylander, A. Bondeson, P. Ingelström, Computational Electromagnetics, Springer, 2012.

17. D. Pozar, Microwave Engineering, second ed., Wiley, 1998.

18. T. McKelvey, H. Akçay, L. Ljung, Subspace-based multivariable system identification from frequency response data, IEEE Trans. Automatic Control 41 (1996) 960–979.

19. S.M. Kay, Fundamentals of Statistical Signal Processing: Estimation Theory, Prentice Hall, 1993.

20. I.W. Lee, D.O. Kim, G.H. Jung, Natural frequency, Natural frequency and mode shape sensitivities of damped systems: Part I, distinct natural frequencies, J. Sound Vib. 223 (1999) 399–412.

21. K. Leong, J. Mazierska, Measurements of unloaded Q-factor of transmission mode dielectric resonators, in: IEEE MTT-S International Microwave Symposium Digest 3, 1997.

CHAPTER 6

A PRACTICAL METHOD TO ASSESS THE PERFORMANCE OF GRAVIMETRIC FLOW TEST RIGS BY USING THE TIMING ERROR

L. Cordova and, T. Lederer

Heat and Vacuum Laboratory of the Physikalisch-Technische Bundesanstalt, Germany

ABSTRACT

This paper proposes an improved method to assess the performance of a gravimetric flow test rig. It addresses the study of two of the most important measurement systems: the weighing scale and all mass-related corrections, and the time measurement system. By performing a regular diverter test according to ISO 4185:1980 several times and under different conditions, it was possible to determine not only the statistical variances of the diverter system, but also the variances of the meter under test and of the flow test facility separately. The proposed method allows also under special conditions the determination of systematic errors in the mass determination. The method has been successfully applied at the large water flow calibration facilities of PTB in Braunschweig and Berlin and is recommended as a routine to test the performance of gravimetric flow test facilities.

INTRODUCTION AND MOTIVATION

For any flow laboratory the uncertainty estimation of its flow test rig used for research or traceability purposes and its validation is of central importance. The uncertainty estimation methods are universally accepted and the main components like the weighing system, the water density determination, the flow conditions and the repeatability of the meter under test (MUT) are well known and have been the subject of several studies and publications. Nevertheless, there are no standardized methods for the validation of the obtained uncertainty.

The most recognized way of validating declared uncertainties is through inter-laboratory comparisons. It is expected that the degree of equivalence, i.e. the difference between a laboratory's result and the comparison reference value, is to be smaller than the declared best measurement capability. In order for the reported data of each laboratory to be comparable and consistent, all participants have to measure exactly the same measurand, i.e. all relevant measurement conditions have to be identical at every laboratory. It is hard to achieve those preconditions. Since inter-laboratory comparisons often are the only way to detect biases, it is most likely that participants will have undetected sources of uncertainty, which makes it difficult, or even impossible to explain the encountered differences.

Citation: L. Cordova, T. Lederer, A practical method to assess the performance of gravimetric flow test rigs by using the timing error, Flow Measurement and Instrumentation, Volume 44, August 2015, Pages 61-70, ISSN 0955-5986, http://dx.doi.org/10.1016/j.flowmeasinst.2014.11.010.

One way of avoiding this problem is by taking advantage of an external reference. This could be achieved by performing more than one round in a comparison; the first round should then allow an initial view into the participant's performance giving them the possibility to roughly recognize their flaws and to correct them if possible. Subsequently, a second comparison round can be performed. The efforts needed to perform a double comparison round are huge, therefore a different option would be preferred. An ideal option should allow participants to check their systems qualitatively in order to detect major problems before the comparison is started.

In the following a method is presented to qualitatively detect if important components of the flow test rig are working properly. It is an extended version of the timing error detection test proposed by ISO 4185 [1]. It performs, from a metrological point of view, the quality control of the flow test rig performance. Additionally it suggests a way to quantify the repeatability of the test flow measurement device, and consequently of the test bench alone.

MODEL OF THE DIVERTER ERROR

The time (Δt_d) the diverter takes to change from one end position to the other may vary from a few milliseconds up to 1 s. During this time, the flow directed into the weighing tank increases from zero to the actual flow rate. The form of the incrementing flow rate withinΔt_d is not known. The magnitude of the ratio $\Delta t_d/t$ defines whether analysis is necessary.

There are two ways of estimating the timing error: through modeling and characterizing the blade dynamics and the water jet, or by modeling the effects that are caused indirectly by the diverter. The first method delivers more information on the system like the repeatability of the diverting phenomena itself. However, large modeling and measuring efforts are necessary to assess the flow profile measurements at the outlet of the nozzle, and to characterize the motion dynamics of the blades. This analysis was performed by Engel and Baade. [3]. The second method is empirical and was introduced by the international standard ISO 4185. The advantage of this method is its universal applicability, since no additional measurement systems are required.

Considering Fig. 1, to obtain an error-free measurement, the area defined by the testing time limits and the flow rate must be equal to the dashed area and the following must apply:

$$a+c=b+d \quad (1)$$

If the flow rate during both diverting motions is different, in the case that the diverting time Δt_d is large, it is advisable to additionally apply the following symmetry conditions:

$$a=b, \quad c=d \quad (2)$$

If only the first condition was used to adjust the trigger point, the measurement would be error-free only if the flow rate is exactly the same for both diverting actions. Even symmetric diverters like the uni-diverter from NIST [2] or the double wing diverter from AIST [5] have to ensure very stable flow rates at the beginning and at the end of the measurement. But if also condition (2) were fulfilled, the diverter would cancel its error automatically even if the flow rate differences are large. Adjusting the trigger point to fulfill condition (2) for every flow rate might, however, not always be possible.

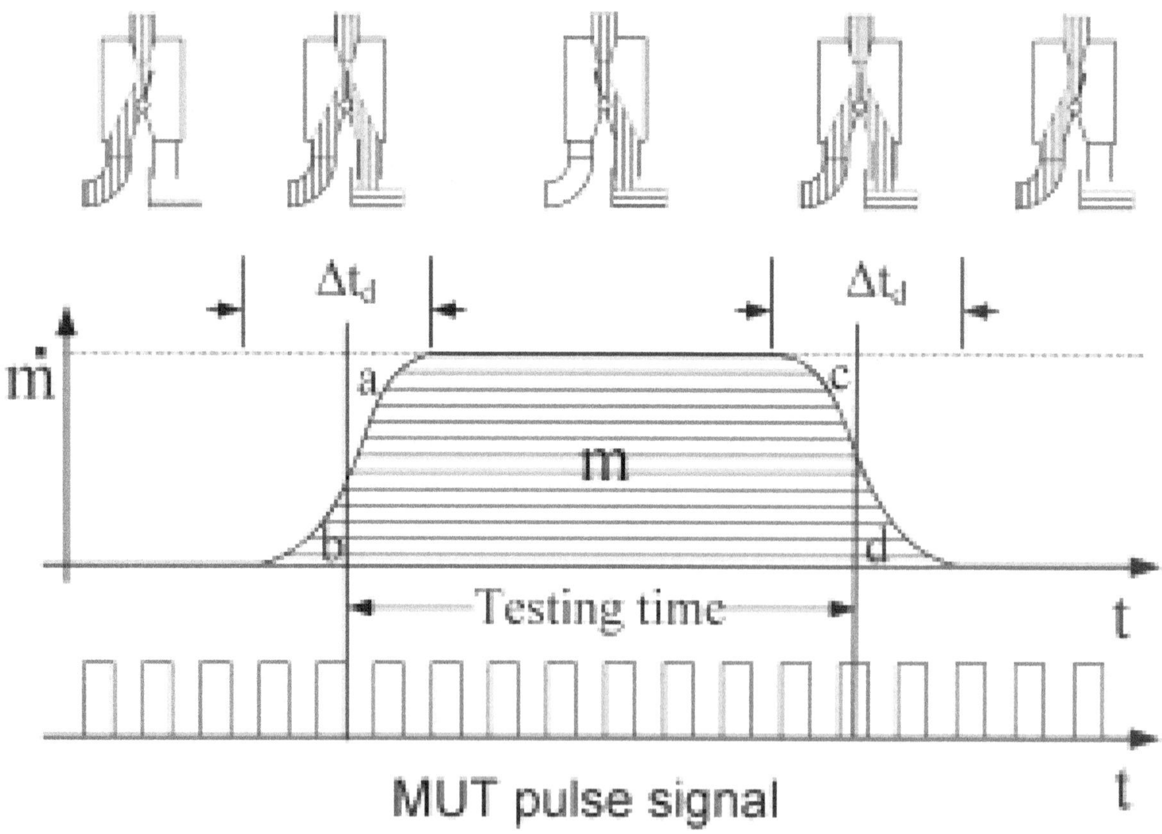

Figure.1: Diagram of the timing error concept. The curve shows the mass flow rate entering into the weighing tank. The dotted line represents the mass flow rate. The time the diverter takes to change its position is marked with Δt_d. The ideal testing time is defined as the time in which the dashed area becomes equal to the area defined by the testing time limits and the dotted line. The pulse trail shown below schematically represents the uninterrupted flow rate perceived by the meter under test.

The ISO 4185 Diverter Equation

Every time a diverting action is performed during a flow meter calibration, a constant time error ($-\Delta t$) is added to the time measurement. If a regular flow meter calibration is repeated including several diverting actions or "interruptions", it is possible to calculate the constant time error introduced by the diverting system.

For this, two measurements are necessary. The first measurement consists of only one filling interval, i.e. a regular calibration, and the second measurement is performed with $n$interruptions. The time and pulses are only counted while the diverter is directing the water into the weighing tank. For further details on the procedure refer to ISO 4185. The equation proposed by ISO 4185 to estimate the diverter error based on these two measurements is

$$\Delta t_{iso} = \frac{t'_c}{(n-1)}\left(\frac{\frac{m_i}{t'_i}}{q_i/q\frac{m_c}{t'_c}} - 1\right) \quad (3)$$

The symbols t , Δt, q and m, represent the time, the timing error of the diverter, the volumetric flow rate and the collected mass respectively; subindexes i and c refer to interrupted or continuous interval measurements respectively, and superindex $'$ refers to apparent magnitudes that are directly measured.

Eq. (3) contains simplifications that in some cases might introduce systematic errors. The unbiased formulation requires the introduction of two additional factors: kt to correct for different filling times, and kf to correct for different mass flow rates instead of volume flow rates. If no simplifications are introduced in the deduction, the resulting equation, rearranged to have the same form of the equation proposed by the ISO is given by

$$\Delta t = \frac{\Delta t + t'_c}{\left(\frac{n}{k_t} - 1\right)}\left(\frac{\frac{m_i}{t'_i}}{k_f\frac{m_c}{t'_c}} - 1\right) \quad (4)$$

$$k_t = t'_i/t'_c \quad (5)$$

$$k_f = \dot{m}_i/\dot{m}_c \quad (6)$$

Considering

$$k_m = \frac{m_i}{m_c} \quad (7)$$

it can be easily shown that the systematic error introduced by Eq. (3) is defined by

$$\frac{\Delta t}{\Delta t_{iso}} = \frac{n-1}{n - \frac{k_m}{k_f}} k_t \quad (8)$$

Note. For a detailed derivation of the timing error equations proposed by the ISO 4185:1980, refer to Appendix A and Appendix B.

If Eq. (3) is used iteratively to either calculate the real flow rates or to adjust the trigger point to eliminate the time measurement error to zero, then the bias has no considerable influences.

If the flow test rig calculates the flow meter correcting factor automatically, it might be more practical to use the following equivalent form (see Appendix C):

$$\frac{\Delta t}{t_c} = \frac{k'_i - k'_c}{\frac{n}{k_t}k'_i - k'_c} \quad (9)$$

where k'_c and k'_i are the apparent flow meter correction factors defined as

$$k' = \frac{\textit{Indication of the MUT}}{\textit{Indication of the reference}} \quad (10)$$

In order to determine the sensitivity to systematic errors, we would use Eq. (3) due to its simplicity. If we introduce the dimensionless variables including the time

$$\epsilon = \frac{\Delta t}{t_c} \quad (11)$$

we obtain

$$\epsilon = \frac{1}{n - k_t}\left(\frac{k_m}{k_f} - k_t\right) \quad (12)$$

The time measurement is typically measured with an uncertainty below 10^{-6}. For this reason, the term kt will be neglected for the following analysis:

$$\epsilon = \frac{1}{n-1}\left(\frac{k_m}{k_f} - 1\right) \quad (13)$$

Eq. (13) shows the robustness of this method. The measured variables are involved through measurement ratios. The use of ratios has the advantage of reducing statistical variability, and minimizing the effects of biases.

Assumptions

The aim of comparing flow rate facilities is to ensure that the calibration results of any given flow meter are independent of the laboratory used within the reproducibility of the flow meter. How to determine the reproducibility of the flow meter is a complicated and demanding metrological problem that has to be customized each time to the flow meter used. However, the most important parameters to check for in a gravimetric facility are always the measured mass including all corrections, and the measured time.

The diverter test mentioned in the section above has the potential to serve as a method not only to determine the timing error, but also to validate the flow test rig results. This is possible, since the requirements for the test to deliver consistent results overlap with the requirements of a gravimetric flow test rig. The assumptions to be fulfilled for a diverter test are the following:

Regarding the weighing systems:

- The weighing drift is negligible for the duration of the measurements.
- There is no time-dependent water loss or gain due to leaks, condensation or evaporation.
- The linearity error of the weighing scale in the working range is negligible.

Regarding the flow rate:

- The distribution of the error of the flow measurement device at the flow rate used is normally distributed.
- The flow measurement device error can be linearized in the flow rate region used.

- The flow rates during the initial and the final diverter motion are the same.
- The diverter motion does not influence the flow rate.

Regarding the actuator of the diverting system:

- The duration of the diverter motion has a normal distribution.
- The velocity of the diverter motion as a function of time, is nearly constant and normally distributed.

To make this concept clear, let us consider two sets of 10 timing-error determination tests. Each set is performed at the same temperature and flow rate, but using a different filling mass and/or a different number of filling interruptions. If any of the assumptions above happened not to be fulfilled, this would be detected through a bias in the timing error determination between the two sets as explained in the section below.

Systematic Influences

For a diverting system to work properly, it is required that the introduced timing error is normally distributed. Any correlation of the timing error to any quantity different than flow or temperature would be an indication of undetected systematic errors. By undertaking a sensitivity analysis on Eq. (12) we can estimate the magnitude of the introduced biases.

Mass Sensitivity

Let us assume that there is a bias in the mass determination. The reason for this could be an error in the buoyancy correction, a non-linearity effect between two mass calibration points or mass variations due to evaporation, condensation or leaks. If this bias is for example 1 kg for a filling mass of 15 000 kg and if we attempted to detected the bias through a flow meter calibration, in order to be able to detect its influence, it would be necessary for the repeatability of the flow meter to be better than 7×10^{-6} which is not possible. But by using this method, random variability is reduced through the use of measurement ratios making it possible to detect small biases.

One precondition has to be given: biases are only detected if they are dependent on time or on the total accumulated mass. If this is not the case, they will remain undetected. A bias on both mass measurements on the same direction would be canceled out and remain undetected. This is the advantage of the method: if all conditions are kept constant, biases are canceled out, and the timing error can be calculated accurately. However if slightly different conditions are used, small biases can be detected.

Based on Eq. (13), we can write:

$$\frac{\Delta t}{t}\approx\frac{1}{n-1}\left(\frac{1}{k_f}\frac{m_i}{m_c}-1\right) \tag{14}$$

Since we assume that $k_t\approx1$, we can write

$$\Delta t\approx\frac{1}{n-1}\left(\frac{1}{k_f}\frac{m_i}{m_c}-t\right) \tag{15}$$

and the sensitivity would finally be given by

$$\frac{\partial\Delta t}{\partial m}\approx\frac{1}{\dot{m}(n-1)} \tag{16}$$

We can see that the sensitivity of Δt to a mass variation is independent of the filling times and filling masses. Based on this equation, Fig. 2 can be drawn. The influence that 1 kg total bias on mass would have on the determination of the timing error is shown. For example, if a diverter test is performed at a flow rate of $200\ m^3/h$ with 10 filling intervals, if a mass bias of 1 kg exists, a systematic error of 2 ms would affect the value of Δt.

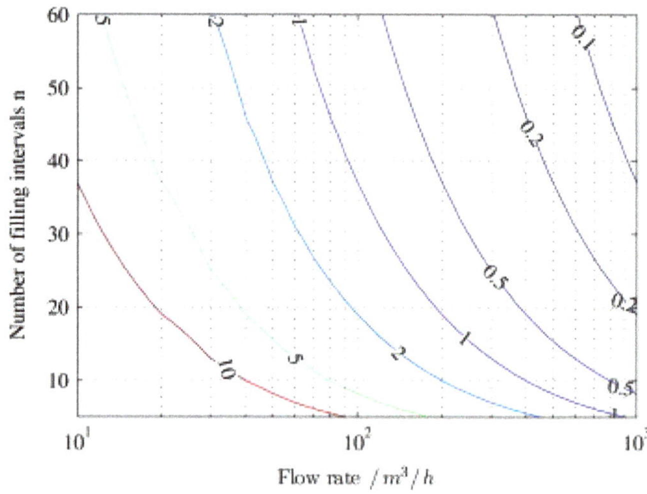

Figure 2: Sensitivity of the timing error to a systematic mass error of 1 kg, given in milliseconds.

Sensitivity to the Flow Measurement

For this case, the same condition applies: the bias has to be different on both performed measurements. Otherwise it would remain undetected. One thinkable phenomenon causing an asymmetrical bias on the flow rate measurement would be when the diverter action causes dynamic changes to the flow rate. This is the case where closed coupled valves are used. Depending on the nature of the flow rate variations and on the flow meter working principle the volume integrated by the flow meter will be more or less biased. The introduced bias has a reduced importance for a regular calibration, but it is important for the timing error determination. Note also that the flow measurement result and the fluid density measurement always appear together. Due to this, their effects cannot be separated.

Based again on Eq. (13) and assuming *km* to be approximately 1, we can write

$$\frac{\Delta t}{t}\approx\frac{1}{n-1}\left(\frac{\dot{m}_c}{\dot{m}_i}-1\right) \tag{17}$$

$$\frac{\partial\Delta t}{\partial\dot{m}_i}\approx\frac{t}{(n-1)}\frac{-\dot{m}_c}{\dot{m}_i^2} \tag{18}$$

assuming the flow rate bias is small we can finally write

$$\frac{\partial\Delta t}{\partial\dot{m}}\approx\frac{-t}{(n-1)\dot{m}} \tag{19}$$

Based on Eq. (19) we can draw Fig. 3.

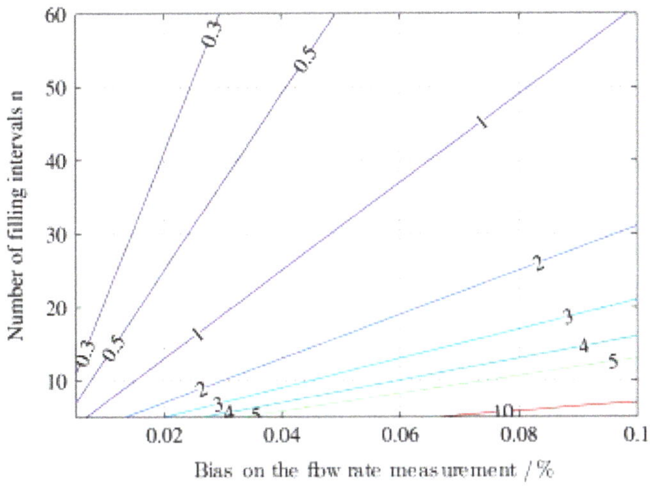

Figure 3: Sensitivity of the timing error to a systematic error in one flow rate measurement, calculated for 1 min filling time, given in milliseconds. A bias on one density measurement would produce the same influences.

From the deductions above, it can be concluded that if the diverter test is performed with slightly different filling times or masses, it is a sensitive test to determine the influence of biases that would remain undetected otherwise.

Variance of the Method

It is necessary to estimate the variance of the method to include it into the uncertainty budget. For this the experimental sample variance would suffice. From Eq. (13) we can see that the variance of the flow measurement, the variance of the mass measurement and the variance of the diverter itself have to be related.

Since time measurements are performed normally at an accuracy better than 10^{-6} s, we can assume $\sigma kt \approx 0$, justifying the use of *Eq.* (13). The associated variance of the method would be given by

$$\sigma_\epsilon^2 = \left(\frac{1}{n-1}\right)^2 \frac{1}{k_f^2}\sigma_{km}^2 + \left(\frac{1}{n-1}\right)^2 \frac{k_m^2}{k_f^4}\sigma_{kf}^2 \qquad (20)$$

As a first conclusion we can say intuitively, if the number of interruptions increases, the variance should decrease rapidly.

Analyzing Eq. (20), we can see that the variance of the diverter system itself σ_δ^2 does not apparently play a role. This assumption would be physically incorrect. The contribution ofσ_δ includes the effects of σkm, since every variation of the blade dynamics, or of the nozzle flow profile would be detected by the weighing system.

Note. The variance $\sigma_{\Delta t}$ describes the spread of the timing error determination method. The diverter variance σ_δ depends on different parameters; it represents an additional contribution to $\sigma_{\Delta t}$.

Given that $k_f = \dot{m}_i/\dot{m}_c$, and that the variance of the flow meter remains constant

$$\sigma_{\dot{m}i} = \sigma_{\dot{m}c} = \sigma_{\dot{m}} \qquad (21)$$

the value of σkf depends on the level of correlation r of $\sigma_{\dot{m}}$:

$$\sigma_{kf} = \sigma_{\dot{m}}\sqrt{2-2r} \qquad (22)$$

The results of an ideal measurement instrument are not autocorrelated $(r=0)$. This means that all relevant systematic effects are known and have been compensated, and that all remaining influences are of random nature. But in reality there is a small correlation remaining since not all influence quantities can be compensated. The level of correlation between $\dot{m}_i$ and $\dot{m}_c$ can be determined empirically. But in order to be able to estimate autocorrelation, it is necessary for the flow rate indications to be normalized to the same reference. Consequently, the correlation analysis cannot be performed based directly on the flow rate indications, but has to be based on the correction factors of the flow meters. Therefore, since the same reference is used, an additional correlation source is introduced. Nevertheless, it is considered that for low correlation factors, the estimation of the correlation factor through the correction factors is sufficiently accurate.

For *km* special treatment is necessary. As stated before, if the mass ratio measurement is unbiased, and the measurements have been performed under repeatability conditions,[1]the variations of *km* can be attributed only to the randomness of the diverter motion σ_δ. By rewriting *km* as a function of flow rate and time, we can include the effects of the randomness of the diverter motion:

$$k_m = \frac{\dot{m}_i\left(t_i' + \sum_{j=1}^{n}\Delta t_j\right)}{\dot{m}_c(t' + \Delta t)}$$

The mass flow rates are per definition of the true flow rates and can be assumed to be constant, if additionally $\dot{m}_i \approx \dot{m}_c$, we obtain

$$k_m = \frac{\left(t_i' + \sum_{j=1}^{n}\Delta t_j\right)}{(t' + \Delta t)} \qquad (23)$$

Note that $\lim_{n\to\infty}\sum_{j=1}^{n}\Delta t_j = n\Delta t$ but $\Delta t \neq \Delta t_j$. Considering that due to the averaging effect we have

$$\sigma_{\sum \Delta t_j} = \frac{\sigma_\delta}{\sqrt{n}} \qquad (24)$$

and given that it can be easily proven that $\sum_{j=1}^{n}\Delta t_j$ and Δt can be treated in this application as independent random numbers, the variance of σkm can be approximated by

$$\sigma_{km}^2 = \frac{(t_i' + n\Delta t)^2}{(t' + \Delta t)^4}\sigma_\delta^2 + \frac{n^2}{(t' + \Delta t)^2}\sigma_\delta^2\frac{1}{n} \qquad (25)$$

Assuming also that the filling times are the same, and that the timing error is small

$$\frac{(t_i' + n\Delta t)}{(t' + \Delta t)} \approx k_t \approx 1 \quad \text{and} \quad k_f \approx 1 \qquad (26)$$

the variance of consecutive mass measurements caused by the diverter variations would be given by

$$\sigma_{km}^2 = \frac{1+n}{t_c^2}\sigma_\delta^2 \qquad (27)$$

By introducing Eq. (27) and applying the conditions of (26) into Eq. (20), we obtain

$$\sigma_\epsilon^2 = \left(\frac{1}{n-1}\right)^2 \frac{1+n}{t_c^2}\sigma_\delta^2 + \left(\frac{1}{n-1}\right)^2 \sigma_{kf}^2 \qquad (28)$$

Reintroducing $\sigma_{\Delta t}$ and rearranging the terms gives

$$\sigma_{\Delta t}^2 = \frac{n+1}{(n-1)^2}\sigma_\delta^2 + \frac{t_c^2}{(n-1)^2}\sigma_{kf}^2 \quad (29)$$

As we can see the resulting value of $\sigma_{\Delta t}$ depends on the duration of the tests and on σkf. By observing Eq. (29) it becomes clear that it is possible to extract the values of σkf and σ_δ.

EXPERIMENTAL RESULTS

Exhaustive diverter tests have been performed at the heat meter testing laboratory of the PTB-Berlin (PTB-1) and at the flow test facility of the PTB-Braunschweig (PTB-2). Some of these results are published by Mathies and Lederer in [4] or by Engel and Baade [3]. The configuration of the diverter installed at the PTB-Berlin has some different characteristics compared to typical diverters for cold water. The reason is that water temperatures up to 90 °C are used to test heat flow meters. Due to the high evaporation rates expected at temperatures above 40 °C open water surfaces have to be avoided.

The most simple representation of a diverter consists of a blade with one angular degree of freedom within a containment with an entrance at the top, and two outputs at the bottom. The water beam coming from the test lines is directed through a nozzle directly to the rotation center of the blade. This type of diverter is used by PTB-2.

In the case of PTB-1 basically the same system is used, but instead of using only one diverter system, eight small diverting units are used simultaneously. To reduce the required blade size, instead of having the diverting motion actuated by rotation, the motion is performed by the nozzles through translation maintaining the angle between the blade and the flow direction constant. This eliminates the fluid forces acting on the blades. The eight diverter systems are coupled and positioned in a radial symmetry to reduce the required space as seen in Fig. 4.

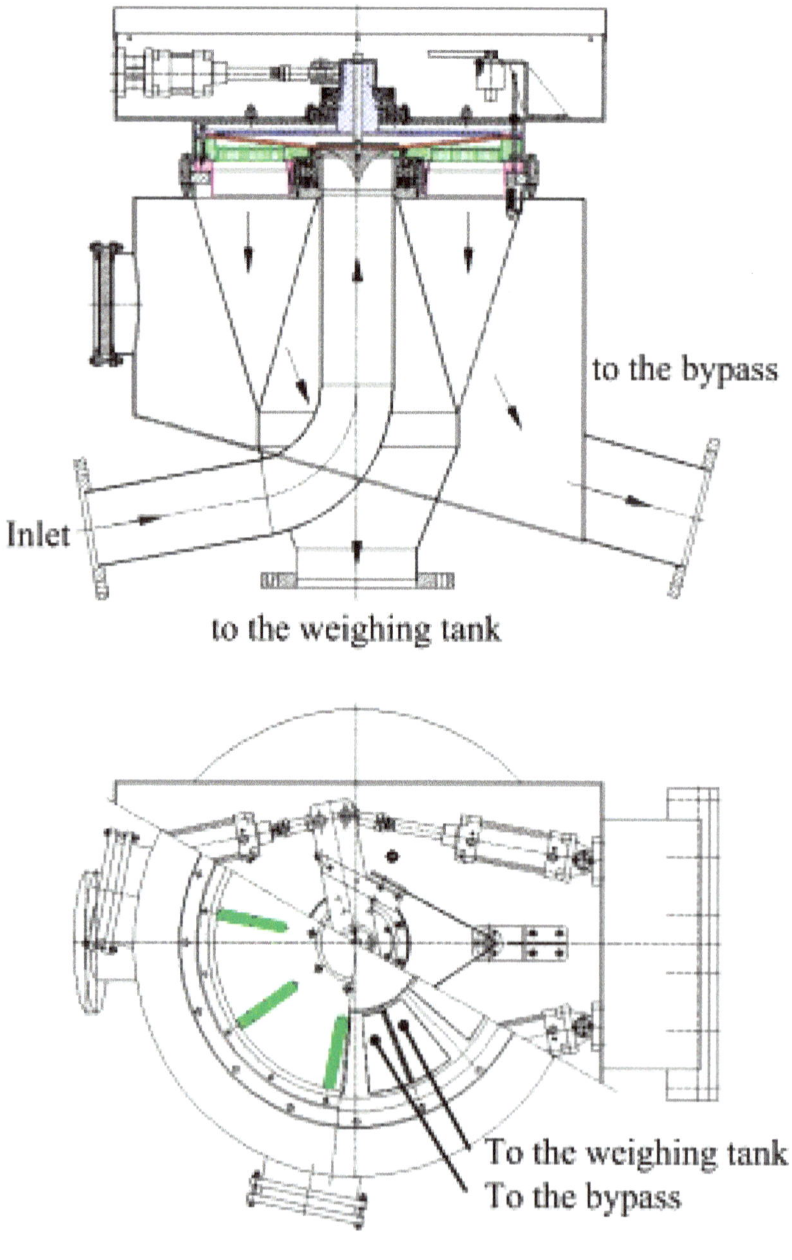

Figure 4: Diagram of the diverter installed at PTB-Berlin (PTB-1). The working principle is the same as for single blade diverters; the main difference is that instead of using only one blade, several radial-positioned blades are used. The fluid enters the diverting system at the bottom. Water is distributed into the 8 nozzles and directed to the blades. The actuating system consists of four pneumatic cylinders. Source: Alfons Witt.

Fig. 4 shows a diagram of the diverter installed at PTB in Berlin (PTB-1). Two diverter systems are used, one for the range up to 200 m³/h and the other for the flow range up to 1000 m³/h. The fluid is directed from the bottom into the diverter. There the flow is directed down to the eight nozzles. At each of the nozzles a flow conditioning system with 280 holes is installed. This is necessary in order to obtain a uniform block profile. The trigger point that starts the time measurement and the pulse counting on the MUT is defined geometrically, i.e. for example at 51% of the total way. It is the same point for both diverting directions. This means that an early trigger point in the first diverting action corresponds to a late triggering point in the final diverting action. In spite of being almost completely encapsulated, the diverter still works without influencing the flow. The actuators are pneumatic ensuring fast dynamics; the position is monitored using a radial incremental encoder.

The test conditions are summarized in Table 1. The flow meters used are of the electromagnetic type in both cases. For the PTB-1 100 pulses per liter were configured, and for the PTB-2 about 13 pulses per liter. In order to guarantee normality, rough errors have been filtered out using standardized methods prior to their inclusion in the analysis.

Table 1: Diverter test conditions.

Parameter	PTB-2	PTB-1
Temperature (°C)	20	50
Flowrate (m³/h)	90	199
Interruptions	7, 14, 26, 51	8, 17, 33, 54
Duration (s)	100	260
Mass (kg)	2500	14 000[a]
Repetitions	15	15–20
Timing error (ms)	4.6	8.6
Corre. *r* for *km*	0.076	0.170
Flow repeat.[b] (%)	0.008	0.016

[a] Some measurements have been performed at 19 900 kg.

[b] Refers to the repeatability of the *k*-factor of the reference flow meter.

As we can see in Fig. 5, the determination of the diverter error can be as accurate as 1 ms for n>50.

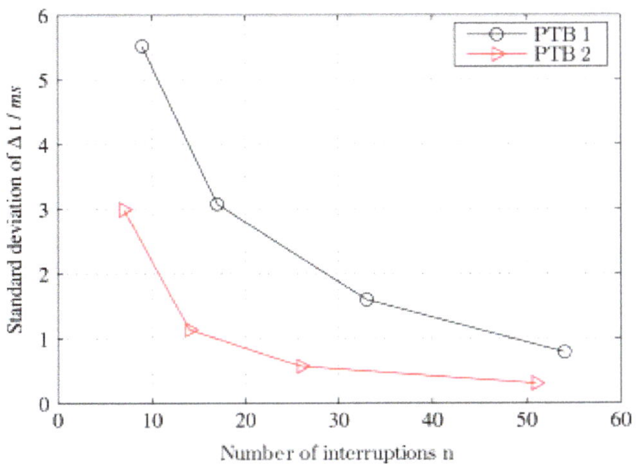

Figure 5: Standard deviation of the timing error for different amounts of filling interruptions during the tests in the PTB-Braunschweig (2) and PTB-Berlin (1). The standard deviation values are based on 15 repetitions, only the last point of PTB-1 is based on 20 repetitions.

Considering Fig. 6, PTB-2 does not present any anomaly. For PTB-1 a dependence of the timing error on the amount of interruptions seems to exist. This is physically not possible. The reason for this deviation can be explained through a systematic influence of the weighing system. Considering Section 2.2 as seen in Fig. 2 a bias of 1 kg would be necessary to produce a bias of about 2 ms at 200 m³/h and n=9 in the timing error, the actual bias is of about 1 ms, which would correspond to a bias on the mass measurement of about 0.5 kg. The weighing scale of PTB-1 is calibrated on a daily basis, and if required, before every measurement. The loads used for calibration are increased at 560 kg steps. The loads in between are validated, but not calibrated periodically using external weights. The tolerable mass bias (linearity error), due to this lack of calibration between points, is as large as 0.2 kg between two calibration steps. The diverter tests for$n$=9 and n =14 have been performed with masses $m_i \approx 13\ 700$ kg and $m_c \approx 14\ 300$ kg. The results strongly suggest that there has been a bias of different signs on both mass measurements. The loads used *mi* and *mc* are just between two calibrated points. A bias cannot be rejected.

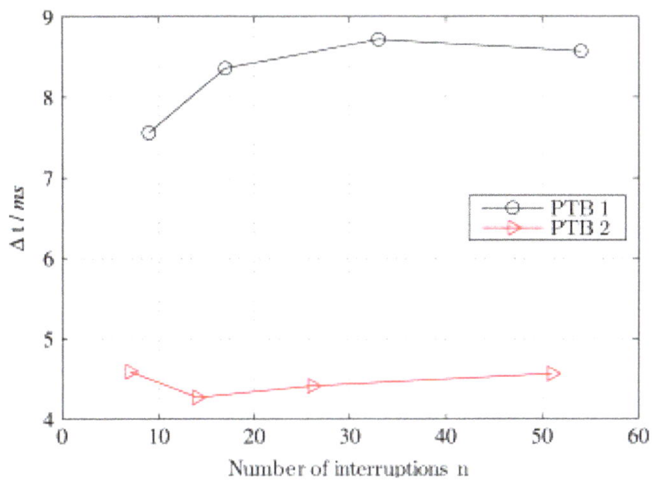

Figure 6: Timing error at both laboratories.

Given the nature of the sensitivity of the timing error to a mass bias, the measurements with large n provide a good estimation despite the small mass measurement errors. A mass bias of this magnitude is being considered already in the uncertainty budget of the calibration facility. [2]

Estimation of the Parameters

The parameters of Eq. (29) can be estimated through a weighted least square linear regression. For this purpose, the equation will be rearranged introducing the independent parameter λ and the dependent parameter φ :

$$\phi = \sigma_{\Delta t}^2 \frac{(n-1)^2}{(n+1)} \qquad (30)$$

$$\lambda = \frac{1}{(n+1)} \qquad (31)$$

By doing so we obtain

$$\phi = t^2 \sigma_{kf}^2 \lambda + \sigma_\delta^2 \qquad (32)$$

Now it is easier to fit the values of σ_δ and σkf in Eq. (32) based on the experimental data. Considering that for a normal distribution the variance of the sample variance is given by

$$\sigma^2_{\sigma\Delta t} = \sigma^2_{\Delta t}\sqrt{2/N} \quad (33)$$

as weights for the weighted least squares we will use the inverse variances of the sample variances.

The results of the regression are the slope S and the 0-crossing φ_0. Given these two parameters and considering Eq. (22) we can write

$$S = t^2\sigma^2_{kf} = 2\,t^2\sigma^2_{\dot{m}}(1-r) \quad (34)$$

$$\phi_0 = \sigma^2_\delta \quad (35)$$

In Fig. 7 and Fig. 8 we can see the obtained sample variances, the 0.95 confidence intervals of the sample variances, the weighted least square regression line, and the 2-sigma region for the zero-crossing (φ_0). Fig. 9 shows the fitting results back in the domain of n and $\sigma_{\Delta t}$.

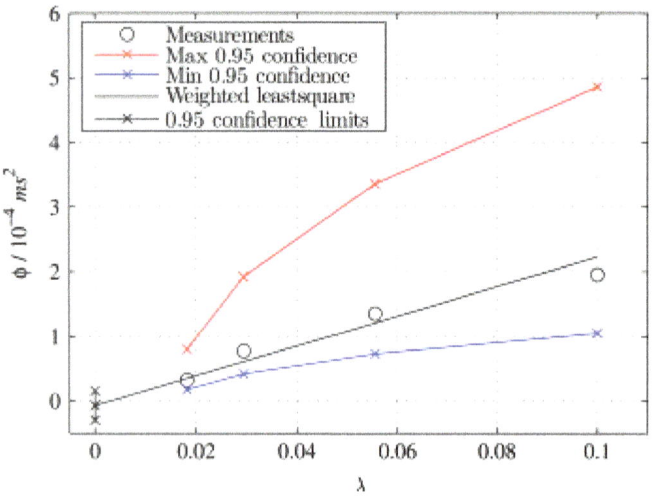

Figure 7: Linear fit function of the parameters φ and λ for the tests carried out at the PTB-Berlin.

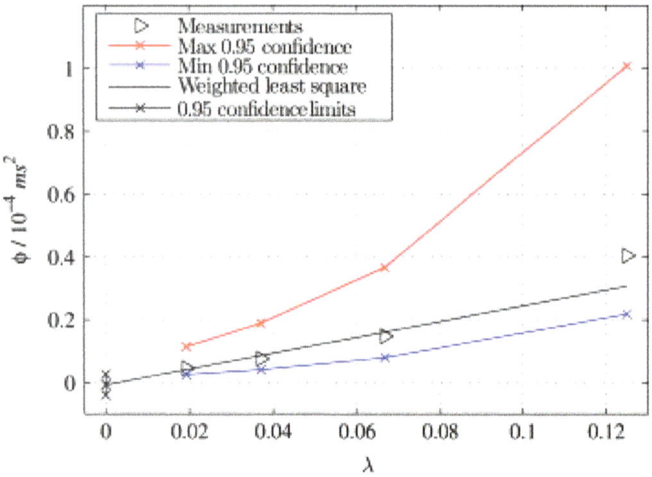

Figure 8: Linear fit function of the parameters φ and λ for the tests carried out at the PTB-Braunschweig.

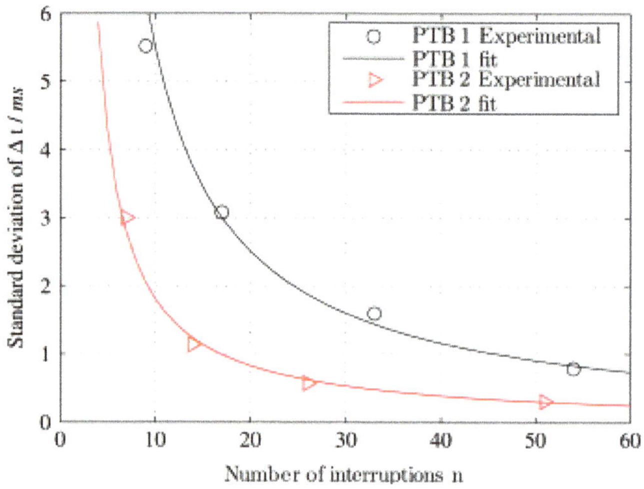

Figure 9: Measured and fitted standard deviation of the timing error for different amounts of filling interruptions during the tests in the PTB-Braunschweig (2) and PTB-Berlin (1).

Table 2 resumes the regression results shown in Fig. 7 and Fig. 8.

Table 2: Weighted least squares results with its corresponding standard errors.

Parameter	PTB-1	PTB-2
S	23.03E−04	2.511E−04
σ_S	4.17E−04	0.52E−04
φ_0	−7.07E−06	−0.67E−06
$\sigma_{\varphi 0}$	11.31E−06	1.64E−06
$\sigma_{\dot{m}}$	1.43E−04	1.16E−04
$\sigma_{\sigma\dot{m}}$	0.61E−04	0.53E−04

Considering Eqs. (34) and (35), it is clear that the slope S of the regression line, and the 0-crossing point φ_0 have to be positive to produce real results. However, the weighted least squares result in very small negative values for φ_0. This can be attributed to the uncertainty of the experiment itself. But considering for example the 0.95 confidence intervals of the sample variances for both experiments at PTB-1 and PTB-2, we can see on Figures 7 and 8 that for these experiments, a family of possible solutions that fit between both 0.95 limits, and that fulfill the condition of $\varphi_0 > 0$ does exist.

By applying a weighted least squares regression instead of regular least squares, the obtained results are not just single points, but rather result ranges defined by an upper and a lower limit. The real value is between those limits. These limits are defined by the uncertainty associated to the regression results. A simple strategy for obtaining a robust worst case estimation of the wanted parameter is to use those limits that result in the largest uncertainty. Please consider the following interesting numerical example extracted from [6] for selecting the appropriate limits. Considering the characteristics of the variance of the sample variance $\sigma_{\sigma}sv$, if normality is given, the 0.95 confidence region would be defined by

$$P\left\{\left|\sigma^2_{\sigma sv} - \sigma^2\right| \le 1.96\sigma^2\sqrt{2/N}\right\} = 0.95 \quad (36)$$

if we state a requirement of a maximum error of 10%, then we would obtain

$$\frac{|\sigma^2_{\sigma SV} - \sigma^2|}{\sigma^2} = 0.10 \quad (37)$$

$$\frac{1.96\sigma^2\sqrt{2/N}}{\sigma^2} = 0.10 \quad (38)$$

consequently, the required sample size would be

$N_{0.95} \approx 750$

By reducing the confidence down to 0.68, would require

$N_{0.68} \approx 200$

This lets us conclude that even the requirement of a confidence level of 0.68 for any regular measurements where the sample variance is directly estimated is hard to fulfill. Therefore, in the following analysis a confidence level of 0.68 will be considered to be sufficiently rigorous. Consequently, the expansion factor for the variance of the sample variance would be 1.

For the timing error, the worst case scenario would be given by

$$\max(\sigma_\delta) = \varphi_0 0.5 + \sigma_{\varphi 0} 0.5 \quad (39)$$

In this case, the value of φ_0 is negative, therefore we assume $\varphi_0^{0.5}$ approx 0. The extreme values for the repeatability of the flow meter are defined by

$$\max(\sigma_{\dot{m}}) = \sigma_{\dot{m}} + \sigma_{\sigma\dot{m}} \quad (40)$$

$$\min(\sigma_{\dot{m}}) = \sigma_{\dot{m}} - \sigma_{\sigma\dot{m}} \quad (41)$$

As seen in Table 3 the standard deviation of the diverting systems of the PTB-Braunschweig is 1.2 ms and for the PTB-Berlin 3.3 ms. Engel and Baade [3] have proven, using the modeling approach, that the uncertainty contribution of the timing error at the PTB-Braunschweig is not greater than 4 ms. This result can be considered compatible with the results found in this paper. In the case of the PTB-Berlin it is not possible to establish a full model of the diverting phenomena. The reason for this is that the flow at the entrance of the diverter will always be unknown since there is no possibility of measuring the velocity profiles. For this reason, the obtained value of 3.3 ms will be included in the uncertainty budget.

Table 3: Repeatability of the flow meter and of the diverter.

Parameter	PTB-1	PTB-2
$\max(\sigma_{\dot{m}})$	2.04E−04	1.69E−04
$\sigma_{\dot{m}}$	1.43E−04	1.17E−04
$\min(\sigma_{\dot{m}})$	0.82E−04	0.64E−04
$\max(\sigma_\delta)$	0.0034 s	0.0013 s

Repeatability of the Test Bench

The repeatability of the test bench alone σref can be expressed based on the repeatability of the calibration factor σcal of the flow meter, and on the repeatability of the flow meter alone $\sigma_{\dot{m}}$. By considering this we have

$$\sigma_{ref} = \sqrt{\sigma^2_{cal} - \sigma^2_{\dot{m}}} \quad (42)$$

If we considered the upper limits for the repeatability of the flow meter $\sigma_{\dot{m}}$, we would underestimate the repeatability of the test bench; therefore, the lower limit should be used in order to have a worst case estimation. The following results were obtained (Table 4).

Table 4: Fitting results for σref.

Laboratory	$\sigma_{\dot{m}}$	σcal	σref
PTB-1	0.82E−04	1.20E−04	0.87E−04
PTB-2	0.64E−04	0.90E−04	0.64E−04

The repeatability of the flow rate facilities, including all systems taking part in the measurement have been estimated to be at the most 0.64×10^{-4} and 0.87×10^{-4} for both facilities. This is the only method that provides a well-founded result to estimate the repeatability of the test bench alone.

DISCUSSION

In order to obtain the presented results about 120 flow meter calibrations were required for each facility. About 3000 diverting actions per diverter were performed. Considering the large effort required and the material fatigue induced on the diverter components, if the goal was only to determine the diverter induced timing error and its uncertainty, it is questionable whether the invested resources would always justify the results. Even more so if we consider that only one flow rate at one fluid temperature has been analyzed. Performing tests for several flow rates and temperatures should be analyzed well, since aging might cause a drift of the diverter characteristics, and the conclusions drawn might not be valid.

However, if the aim of the test is to obtain an estimation of the repeatability of the test bench, or the repeatability of one or of several flow meters installed during the tests, then the efforts might be seen as reasonable. One might consider avoiding measurements with a lower number of filling intervals since they do not weigh much during the least squares regression. But if we consider that systematic errors on the mass determination are detectable only using low *n*, these measurements are important as those with large*n*.Given that the method aims to obtain a worst case estimation, the fact that the regression analysis delivers a very small negative value does not invalidate the results, but underlines instead the importance of estimating also the variance of the regression results by using the weighted least squares analysis.

The estimation of systematic errors in the mass measurement through the diverter tests can give well-founded evidence of the presence of errors. The evidence of a small mass bias on the results of the PTB-1 confirm the sensitivity of the method. Nevertheless, it cannot be used as a general calibration replacement, since the reasons for a bias may be diverse, and require a more detailed study to be isolated and eliminated.

Only by using the weighted least squares method is it possible to estimate adequately the standard error of the regression, and consequently to establish boundaries on the obtained variances. The present approach is considered a good option to determine the repeatability of the flow test rig. Other methods based on correlation and variance subtraction as proposed by Hayward [7] and Poeschel [8] are considered difficult in practice given the fact that the variance of the sample variance requires hundreds of measurements to deliver accurate results.

CONCLUSIONS

It can be concluded that the proposed method is applicable to test the overall performance of a flow test rig, providing a tool to detect whether the mass determination is working properly. Additionally, it allows the estimation of the variances of the diverter system, and of the flow meter and the flow test rig independently.

It has been shown that small statistically relevant differences in the determined timing errors indicate if a bias has occurred. Furthermore, it can also be inferred that the weighted least squares regression would deliver plausible results only if all measurements have been performed under repeatability conditions.

The regression results give further insight into the diverter characteristics, into the flow meter used, and into the flow test rig itself. If this method is used periodically as part of a validating procedure, relevant biases that could affect the results, of interlaboratory comparisons for example, could be detected early and eliminated.

Considering that the diverting systems, like every mechanical system, are prone to aging, it is recommended to keep the amount of tests to a minimum.

ACKNOWLEDGMENTS

The measurement results used to undertake the analysis of the flow rate facility at PTB-Braunschweig were kindly shared by Dr. Rainer Engel from the Flow Laboratory of the PTB-Braunschweig. We acknowledge his cooperation gratefully.

APPENDIX A. ISO 4185 METHOD 2

Consider Fig. 10 and Fig. 11. The timing error shown in the horizontal axis can also be interpreted as an apparent flow rate error. The dashed area is the mass error that would originate through timing error Δt.

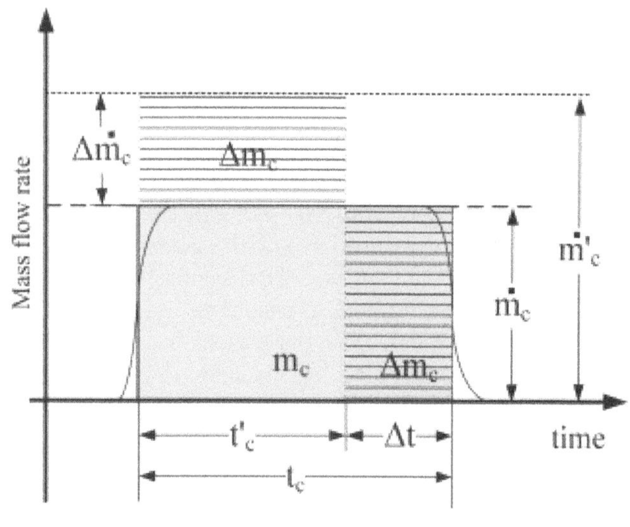

Figure 10: The figure shows that the diverter error in the time measurement shown in the horizontal axis can also be seen as an error on the vertical axis as a flow rate error.

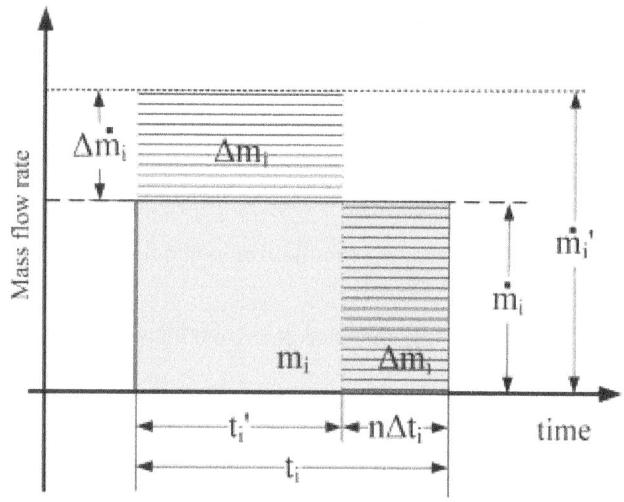

Figure 11: The figure shows that also for an interval filling measurement the diverter error can also be seen as an error on the vertical axis as a flow rate error. The diagram is simplified. Off-times, where water flows to the bypass, are not shown.

For a continuous test run:

$$\Delta m_c = \Delta \dot{m}_c \cdot t'_c \tag{43}$$

$$\Delta m_c = \Delta t \cdot \dot{m}_c \tag{44}$$

For a discontinuous test run:

$$\Delta m_i = n \cdot \Delta t \cdot \dot{m}_i \tag{45}$$

$$\Delta m_i = t'_i \cdot \Delta \dot{m}_i \tag{46}$$

eliminating Δm_c and Δm_i from Eqs. (43), (44), (45) and (46), we obtain the apparent flow rate variation:

$$\Delta \dot{m}_c = \frac{\Delta t}{t'_c}\dot{m}$$

$$\Delta \dot{m}_i = \frac{n \cdot \Delta t}{t'_i}\dot{m}_i$$

Considering a continuous test run we obtain the following relations:

$$\dot{m}'_c = \frac{m}{t'}$$

$$\dot{m}_c = \dot{m}'_c - \Delta \dot{m}_c$$

$$\dot{m}_c = \frac{m_c}{t'_c} - \frac{\Delta t}{t'_c}\dot{m}_c \tag{47}$$

solving for $\dot{m}$

$$\dot{m}_c = \frac{m_c}{t_c}\frac{1}{1+\dfrac{\Delta t}{t_c}} \tag{48}$$

Analogously, considering an interrupted test run we obtain

$$\dot{m}_i = \frac{m_i}{t'_i} - \frac{n \cdot \Delta t}{t'_i}\dot{m}_i \tag{49}$$

$$\dot{m}_i = \frac{m_i}{t_i}\frac{1}{1+\dfrac{n\Delta t}{t_i}} \tag{50}$$

Ideally, the continuous test run, and the one with interval filling periods should be performed under the same conditions, i.e. exactly the same flow rates should be installed and the same filling times should be aimed at. When this is however not possible or not wanted, a correcting factor *kq* can take into account the small flow rate variations. Similarly we apply *kt* to compensate for different filling times:

$$k_t = t_i'/t_c'$$
$$k_f = \dot{m}_i/\dot{m}_c \quad (4)$$

By inserting Eqs. (48) and (50) into *kf* and applying *km* and *kt*

$$k_f = \frac{k_m}{k_t}\frac{1+\frac{\Delta t}{t_c'}}{1+\frac{n\Delta t}{t_i'}} \quad (51)$$

rearranging the equation gives

$$\frac{k_m}{k_t k_f} = \frac{t_c' + \frac{n\Delta t}{t_i'}t_c'}{t_c' + \Delta t} \quad (52)$$

subtracting 1 yields

$$\frac{k_m}{k_t k_f} - 1 = \frac{t_c' + \frac{n\Delta t}{t_i'}t_c' - t_c' - \Delta t}{t_c' + \Delta t} \quad (53)$$

Turn MathJaxon

Factorizing Δt on the numerator

$$\frac{k_m}{k_t k_f} - 1 = \frac{\Delta t\left(\frac{n}{k_t} - 1\right)}{t_c' + \Delta t} \quad (54)$$

and solving for Δt on the numerator we finally obtain

$$\Delta t = \frac{\Delta t + t_c'}{\left(\frac{n}{k_t} - 1\right)}\left(\frac{\frac{m_i}{t_i'}}{k\frac{m}{t_c'}} - 1\right) \quad (55)$$

The simplifications made in the equation proposed by ISO 4185 might be adequate for most of the cases, but under certain circumstances they may introduce a considerable bias on the diverter error as seen. Eq. (4) is completely equivalent to the equations used for example in [2] and [4].

APPENDIX B. ISO 4185 METHOD 1

The second approach introduced by ISO 4185 has not been treated in this paper. However it was considered relevant to provide the way it works in order to allow a better judgment of the methods used to test the diverter system. This approach is also based on the comparison of two calibrations, but instead of varying the number of diverting actions, only the duration of the test is changed. Since the diverter error does not change, the effect on the resulting apparent flow rate allows the determination of the diverting error.

By rewriting Eq. (48) in terms of the apparent flow rate for two measurements in different conditions $\dot{m}_1'$ and $\dot{m}_2'$:

$$\dot{m}_1' = \left(\frac{\Delta t}{t_1}\dot{m}_1 + \dot{m}_1\right)$$

$$\dot{m}_2' = \left(\frac{\Delta t}{t_2}\dot{m}_2 + \dot{m}_2\right)$$

Subtracting term by term and reordering we obtain

$$\Delta t\left(\frac{\dot{m}_1}{t_1'} - \frac{\dot{m}_2}{t_2'}\right) = (\dot{m}_1' - \dot{m}_2') - (\dot{m}_1 - \dot{m}_2) \quad (55)$$

and dividing by $\dot{m}_2'$ we obtain a similar equation presented by ISO 4185.

$$\frac{\Delta t}{\dot{m}_2'}\left(\frac{\dot{m}_1}{t_1'} - \frac{\dot{m}_2}{t_2'}\right) = \frac{(\dot{m}_1' - \dot{m}_2') - (\dot{m}_1 - \dot{m}_2)}{\dot{m}_2'}$$

As seen in Eq. (55), this method is strongly dependent on the accuracy of the flow meters used. In order to reduce this random error, several measurements have to be made. Considering that Δt is the slope of the equation of a line drawn by Eq. (55), the international recommendations suggest estimating its value by means of a linear regression. This method has the advantage that no additional special measurements have to be performed. It is basically possible to include every measurement made by the test rig to confirm or to monitor the diverter error, provided that the reference flow meters used are stable and repeatable.

APPENDIX C. K-FACTOR FORMULATION

The apparent relative error of a flow meter for the continuous and the interrupted measurements are defined as

$$k_c' = \frac{\dot{m}_c'}{\frac{m_c}{t_c'}} \quad (56)$$

$$k_i' = \frac{\dot{m}_i'}{\frac{m_i}{t_i'}} \quad (57)$$

The real relative error would be given by

$$k_{cr} = \frac{\dot{m}_c'}{\frac{m_c}{t_c' + n\Delta t}} \quad (58)$$

$$k_{ir} = \frac{\dot{m}_i'}{\frac{m_i}{t_i' + \Delta t}} \quad (59)$$

By expressing Eqs. (58) and (59) as a function of k_c' and k_i' we can write

$$k_{cr} = k_c' + \frac{\Delta t}{\frac{m_c}{\dot{m}_c'}} \quad (60)$$

$$k_{ir} = k_i' + n\frac{\Delta t}{\frac{m_i}{\dot{m}_i'}} \quad (61)$$

Assuming that the relative error of the flow meter for the region near the working point is constant, then we can write

$$kcr = kir \quad (62)$$

Finally, solving for Δt gives

$$\frac{\Delta t}{t_c} = \frac{k_i' - k_c'}{\frac{n}{k_t}k_i' - k_c'} \quad (63)$$

REFERENCES

1. International Standard ISO 4185. Measurement of liquid flow in closed conduits—weighing method, first ed. ISO; Beuth Verlag, Berlin, 1980.
2. Marfenko Iryna V, Yeh Tsyh T, Wright John D. Diverter uncertainty less than 0.01 percent for water flow calibrations. In: International symposium for fluid flow

measurement conference proceedings. Queretaro; 2006.

3. Rainer Engel, Hans Baade Model-based flow diverter analysis for an improved uncertainty determination in liquid flow calibration facilities Meas Sci Technol, 21 (2) (2010)
4. Mathies N, Lederer T. Neue Methode zur Bestimmung des Zeitfehlers für Umschalteinrichtungen von gravimetrischen Messanlagen [New way of estimating the diverter timing error for gravimetric flow calibration systems] tm - Technisches Messen, vol. 74(6); June 2007.
5. T. Shimada, S. Oda, Y. Terao, M. Takamoto Development of a new diverter system for liquid flow calibration facilities Flow Meas Instrum, 14 (June (3)) (2003)
6. Yaakov-Bar-Shalom, X. Rong Li, Kirubarajan Thiagalingam Estimation with applications to tracking and navigation: theory algorithms and software John Wiley & Sons. Inc., Weinheim, Germany (2001), p. 107
7. Hayward ATJ. Measuring the repeatability of flow meters. NEL Report no. 636. National Engineering Laboratory; May 1977.
8. Poeschel W. Internal and external comparison measurements for ensuring accuracy and traceability in flowmeter calibration. In: International symposium on fluid flow measurement. Denver, USA; 1999

CHAPTER 7

MONITORING AND ANALYSIS OF AIR EMISSIONS BASED ON CONDITION MODELS DERIVED FROM PROCESS HISTORY

M. Liukkonen[1] and T. Hiltunen[2]

[1] Department of Environmental Science, University of Eastern Finland, P.O. Box 1627, Kuopio FIN-70211, Finland.

[2] Foster Wheeler Power Group, P.O. Box 201, Varkaus FIN-78201, Finland

ABSTRACT

Evaluation of online information on operating conditions is necessary when reducing air emissions in energy plants. In this respect, automated monitoring and control are of primary concern, particularly in biomass combustion. As monitoring of emissions in power plants is ever more challenging because of low-grade fuels and fuel mixtures, new monitoring applications are needed to extract essential information from the large amount of measurement data. The management of emissions in energy boilers lacks economically efficient, fast, and competent computational systems that could support decision-making regarding the improvement of emission efficiency. In this paper, a novel emission monitoring platform based on the self-organizing map method is presented. The system is capable, not only of visualizing the prevailing status of the process and detecting problem situations (i.e. increased emission release rates), but also of analyzing these situations automatically and presenting factors potentially affecting them. The system is demonstrated using measurement data from an industrial circulating fluidized bed boiler fired by forest residue as the primary fuel and coal as the supporting fuel.

INTRODUCTION

Air emissions are currently highly regulated, and it is obvious that the environmental legislation and regulations concerning energy plants will be tightened in the near future. Despite these regulations, energy producers have to be able to respond quickly to altering load demands (Wang, Huang, Liu, Yan, & Cen, 2008) and tolerate variations in fuel quality (Beér, 2000), which emphasizes the importance of prompt and precise operational control of combustion. Furthermore, since fuel costs cover the majority of the operating costs of energy boilers (Basu, 1999), the increasing price of electricity creates additional challenges, and producers may be obliged to favor inexpensive, poor-quality fuels like waste, recycled fuels, and low-grade coal. The state of affairs is complicated, because varying fuel quality may have unpredictable effects on combustion and, on the other hand, process modifications may be impracticable or too expensive.

The flip side of the problematic solid fuels is their influence on the nature and on the amount of air emissions released. It is possible that due to a change in the fuel quality, the combustion

Citation: M. Liukkonena & T. Hiltunenb, Monitoring and analysis of air emissions based on condition models derived from process history. Cogent Engineering Volume 3, Issue 1, 2016. DOI: 10.1080/23311916.2016.1174182.

conditions may temporarily become more favorable for the formation reactions of emissions, for which their formation is suddenly accelerated. Alternatively, low-grade fuels may include an enriched amount of fuel-bound emission precursors that are released and activated during combustion.

The technology of fluidized bed combustion (FBC) is especially suitable for converting solid fuels (coal, peat, biomass, etc.) into energy. In a circulating fluidized bed (CFB), the combustion occurs in a suspension (i.e. fluidized bed), which is typically a mixture of sand, fuel ash, and a possible sorbent material for sulfur capture. The bed material is fluidized by feeding in some air through the bottom of the combustion chamber (primary air). High fluidizing velocities are used in CFBs, so the bed particles are in a constant circulating movement within the system. Due to the large heat capacity of the bed, the combustion is generally stable, and supporting fuels such as oil or gas are usually needed only during the start-up. The intense turbulence of the CFB ensures good mixing and combustion. Combustion typically takes place at bed temperatures of ca. 850–900°C.

CFBs are extremely flexible with fuels, and they are able to process a wide variety of fuels without a noteworthy reduction in performance. This robustness enables co-combustion of a wide variety of solid fuels, which is a major advantage in the present energy market (Coda Zabetta, Hiltunen, Moulton, & Hotta, 2009). On the other hand, this can be considered a drawback as well, because changing fuel types and varying fuel quality typically inflict new challenges when it comes to controlled and stable combustion. For example, the use of alternative fuels, which are typically of lower quality, tends to worsen and complicate the NO_X emission performance of combustion systems (Hill & Douglas Smoot, 2000). Sulfur dioxide (SO_2), nitrogen oxides (NO_X), and nitrous oxide (N_2O) are the principal air pollutants released from CFB boilers fired by fossil fuels (Basu, 1999).

Assessment of online information on operating conditions is an essential task when reducing emissions in energy plants, and automated monitoring and control are of primary concern when preventing emissions, particularly in biomass combustion (Nussbaumer, 2003). The goal in traditional process monitoring is to detect abnormalities in process operation (MacGregor & Cinar, 2012), and the simplest ways of monitoring are to either observe the process performance by naked eye or to perceive the changes in it by using a preselected set of key measurements. Nevertheless, monitoring of industrial processes, including combustion and its emissions, is generally more demanding than this, which can be concluded from many facts. To begin with, emission monitoring with diagnosing abilities would be more useful than the general approach, because it is able to reveal the prevailing status of the process and present the potential reasons for possible problems. Moreover, automated computerized monitoring systems are usually needed by the industry for supporting decisions, because analyzing the prevailing process conditions based on an extensive data-set of measurements is arduous and difficult, even to a practiced specialist who has abundant former knowledge of the process. On the other hand, the utilization of automated monitoring systems makes it necessary to get reliable information on the process by systematizing and organizing measurements, which often leads to other dilemmas.

Principal component analysis and partial least squares are the customary methods of choice for monitoring systems with diagnosing abilities (MacGregor & Cinar, 2012; Qin, 2012). Nonetheless, when used in process monitoring, these two methods seem to be restricted by the lack of appropriate visualization properties for presenting the essential parts of extracted information in an understandable manner, such as the prevailing process status, and for analyzing the problem situations. As a result, the outputs of any systems based on these platforms are difficult and laborious to interpret. Monitoring based on a limited number of certain key measurements is sometimes feasible (Liukkonen, Juntunen, Laakso, & Hiltunen, 2013), but this kind of approach requires special knowledge of the process and is thus most applicable to cases in which the group of key variables is easily recognized and relatively small. Taken as a whole, there is room for new kinds of advanced emission monitoring and analysis systems.

The following topical issues related to modeling, management, and monitoring of NO_X emission in CFB boilers can be identified:

- Monitoring of NOX in CFBs lacks economically efficient, fast, and competent computational systems that could support decision-making for improving the emission efficiency.
- A large part of the monitoring applications designed for CFB, especially those based on physical models, are case-specific and entail arduous model development and special knowledge on the process, and new cost-efficient alternatives to this would be useful.
- As monitoring of CFB emissions is ever more challenging because of low-grade fuels and fuel mixtures, new monitoring applications are needed to extract essential information from the large amount of measurement data.
- In these complicated industrial environments, it is essential to find out reasons for problems and to know the prevailing status of the process, and the methods currently available do not directly provide these kinds of functionalities.

Complex and dynamic industrial processes are difficult in terms of online monitoring. For example, condition-dependent behavior is characteristic for industrial processes (Juntunen, Liukkonen, Lehtola, & Hiltunen, 2013a, 2013b; Liukkonen, Havia, Leinonen, & Hiltunen, 2011; Liukkonen, Heikkinen, et al., 2011; Liukkonen, Hiltunen, Hälikkä, & Hiltunen, 2011), which usually requires flexibility and adaptivity from practical software applications. In practice, this means that the process adheres to different regularities under different conditions, which means that these conditions have to be recognized by the application, so that it can adapt to the prevailing situation. A typical example on this is load-dependent behavior in energy and water processes (Liukkonen, Heikkinen, et al., 2011; Liukkonen & Hiltunen, 2014; Liukkonen, Laakso, & Hiltunen, 2013). Moreover, it is shown that problems originating from fluctuation can be significant in a dynamically behaving process (Liukkonen, Hiltunen, et al., 2011). These may occur in transition circumstances such as load change situations, and inflict undesired side effects such as increased emission release rates. These kinds of challenges are frequently faced by energy producers, who have to deal with uncontrollable factors such as varying fuel quality, which may pose rapidly or slowly arising problems including fouling, corrosion, and air emissions. For these reasons, it would be useful to be able to observe the combustion performance and process status more efficiently.

Modeling and model-based predictive systems are potential approaches for improving the control of complex industrial processes (Keskitalo, la Cour Jansen, & Leiviskä, 2010; Liukkonen, Hälikkä, Hiltunen, & Hiltunen, 2012, 2013; Sanjay & Prithvi,2014), and models based on process history (i.e. measurement data) are a potential approach to be used as the basis of advanced monitoring systems. For example, model-predictive monitoring of emissions has been of wider interest recently, and methods such as neural networks are widely used as the basis of predictive emission monitoring systems (Iliyas, Elshafei, Habib, & Adeniran, 2013; Lv, Liu, Yang, & Zeng, 2013; Shakil, Elshafei, Habib, & Maleki, 2009; Smrekar, Potočnik, & Senegačnik,2013). A descriptive data-based system for monitoring and evaluating dynamic industrial processes has to be able to process a large amount of multivariate data, extract the fundamental information from the data, and provide an understandable presentation of this information to be evaluated by experts. Additionally, these so-called intelligent monitoring platforms should be able to cope with multivariate and dynamic characteristics, which are typically inherent in industrial processes.

A promising platform for developing descriptive monitoring and diagnostics systems is the self-organizing map method (SOM) (Kohonen, 2001; Oja, Kaski, & Kohonen, 2002), its variants such as time adaptive SOM (Shah-Hosseini & Safabakhsh, 2000) and growing hierarchical SOM (Dittenbach, Merkl, & Rauber, 2000), the combination of SOM and k-means clustering (Chang & Chen, 2015; Liukkonen, Heikkinen, et al., 2011; Liukkonen, Laakso, et al., 2013), and the alternatives of SOM such as generative topographic maps (Bishop, Svensén, & Williams, 1998) and elastic maps (Gorban & Zinovyev, 2005).

The basic SOM is an artificial neural network based on unsupervised learning. The training of SOM results in a topological arrangement (lattice), of output neurons, each of which has a special property vector describing the input vectors associated with it. During training, these so-called reference vectors become weighted averages of the original data samples assimilated to it. In the lattice, the input vectors sharing common features end up into the same or neighboring neurons. As a result, the grid of neurons (i.e. the map) reflects variations in the statistics of the data-sets and selects common features, which approximate to the distribution of the data points. Recently, the SOM has been widely used in a wide variety of environmental monitoring systems (Chea, Grenouillet, & Lek, 2016; Strebel et al., 2013; Yu et al., 2014; Zhang, Liang, Wang, & Xu, 2015).

Nonetheless, the methods mentioned above do not provide a direct way of monitoring and diagnosing industrial processes as such. In the earlier study (Liukkonen & Hiltunen, 2014) we presented an approach for advanced monitoring of emissions in CFB boilers. In this approach, a model based on SOM and fuzzy c-means clusters is updated regularly to respond to the prevailing condition, the current condition of the process is monitored and evaluated, and an alert is presented whenever the preset limit for each emission component is exceeded. Although found suitable for monitoring emissions in a way that is not dependent on the load level, the method lacks the basic functionality required in online monitoring applications, and the structure based on clustering the map after every training stage may be unnecessarily complicated and heavy when it comes to flexible monitoring and simple visualization platforms. Thirdly, the approach is not fully capable of presenting an efficient diagnosis of the problem at hand, because the analysis is performed using the cluster centers, and not the prototype vectors of the model itself.

Keeping these limitations in mind, the objectives of this study were

- to determine if SOM-based models derived from process history could be applicable to automated, online emission monitoring systems,
- to design a flexible, adaptive, and visual decision-support system based on data-driven models for the improvement of emission efficiency in energy boilers,
- and to provide the system with automatable diagnostic abilities in problem situations.

The novelty of the SOM-based platform presented in the paper becomes from the fact that it can be utilized for decision-support in automated emission monitoring applications. It is also more practical in visualizing the dynamics and evolution of the process, and the structure of the platform is more simple, because the map units are used as conditional prototypes when determining the present condition of the process. Furthermore, the system is capable of presenting a diagnosis whenever a problem, i.e. an increase in the emission release rate, is detected. The environment in which the system is simulated involves measurement data from an industrial CFB boiler fired by forest residue as the primary fuel and coal as the supporting fuel.

METHODS AND DATA

The Basis of the Adaptive Emission Monitoring System

Self-organizing maps (Kohonen, 2001) are a well-known, unsupervised form of artificial neural networks that is broadly used in descriptive data analysis (Oja et al., 2002). The following phases are involved in the training procedure of a single SOM:

(1) Initialization of the reference vectors of neurons

(2) Searching for a best matching unit (BMU) for the first input vector based on Euclidean distance metrics

(3) Recalculating the reference vector of the BMU based on the values of the input vector

(4) Recalculating the reference vectors of the neurons located next to BMU

(5) Iteration of steps 2–4 for every input vector in sequence

(6) Fine tuning of the map by iterating steps 2–5 using a smaller learning rate

(7) Searching for the eventual BMUs for the input vectors

In other words, n-dimensional data are assigned to structural units (i.e. neurons), which are usually hexagonal and which can be arranged in a multidimensional lattice (i.e. map). Generally, a two-dimensional map is used, because it can be visualized and interpreted more easily. In the supervised training procedure, every neuron on the map adapts to two characteristic features: (1) the position on the lattice, and (2) an *n*-dimensional reference vector. The first one of these features describes the degree of similarity of the neuron to other neurons (the closer the neurons are to each other, the more similar they are), whereas the other one presents a description of the archetypal features of the original data samples assigned to this neuron. Therefore, the reference vectors can be considered, on one hand, as generalized descriptions of the original data vectors assigned to each unit, and on the other hand, as estimators of the similarity of neurons.

The drawback of the traditionally used visualization methods of SOM is that they do not offer a direct way of monitoring the process dynamics and evolution. Another visualization approach for use with SOM is the trajectory analysis, which can be used for illustrating the short- and long-term behavior of a dynamical system and the mechanisms by which the system is changing with time. A SOM trajectory demonstrates the order of BMUs in the SOM in response to a set of input vectors taken over a given period of time (Kohonen, Oja, Simula, Visa, & Kangas, 1996). Generally speaking, trajectory points are first acquired by determining the BMU of each original data sample, and these points can then be illustrated successively on the map. In practice, trajectories can thus be formed using consecutive images of operating points that can be displayed on the map after training (Kohonen, 2001).

Dynamical Monitoring Approach

The SOM-based system for adaptive emission monitoring is organized as follows:

(a) Train the SOM using recorded measurement data archives to create the background model to be used by the system.

(b) Input the limits of variables to be monitored (only when implemented for the first time or when these limits are changed).

(c) Feed input data consisting of the most recent measurements (consistent feed).

(d) Obtain output data including the process evolution (i.e. trajectories) and the most recent conditions (i.e. neurons) to be shown on the map, as well as alerts and information on possible emission problems, which are also reported and analyzed.

The main stages of the adaptive monitoring approach are illustrated in Figure 1, and a diagram describing the operation of the system in a decision-making task is presented in Figure 2. Right after feeding in the measurement data obtained at a single moment of time, including the emission components to be monitored (NO_X and SO_2 emissions), to the system, the BMU of this data row is first determined by terms of Euclidean distance. Using this information and a number of most recent BMUs, the present and former conditions can be shown as a trajectory on the SOM. Generally speaking, the application determines the prevailing condition of the process by finding a model (i.e. the reference vector of a neuron) which is the most similar to the most recent measurement.

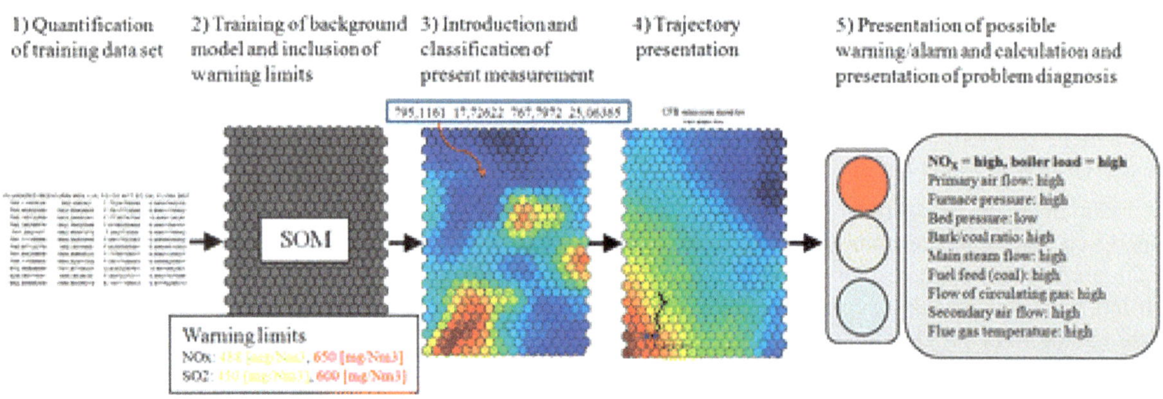

Figure 1: The stages of the dynamical monitoring approach.

After this, the recently measured emission (NO_X and SO_2) release rates are compared to the preset limits to analyze if they are exceeded. An exceeding of the red limit for an effluent component leads to a notification and an alarm sound, and in the meantime a red circle is shown on the map. In case there is an exceeding of the yellow warning level, the color tone of the circle is extracted from a color map comprising tones from yellow to red, depending on the degree of the exceeding. As a consequence, the colored circle appearing on the map illustrates, not only the problem itself but also the severity of the problem.

This step is followed by a diagnosis which eventually proposes the potential reasons for the problem situation. First, the reference vector of the present BMU is analyzed by the system to search for those variable components that have the largest absolute values. As a result, those variable components which distinguish this neuron the most from other neurons can be illustrated on the screen, and this way the possible reasons for the problem become achievable. Furthermore, the latest values of these variables are shown by the graph, which enables the comparison between the most recent measurements and the reference vector. This makes it possible to evaluate how similar the prevailing conditions are to the prototype conditions to which they have been assimilated.

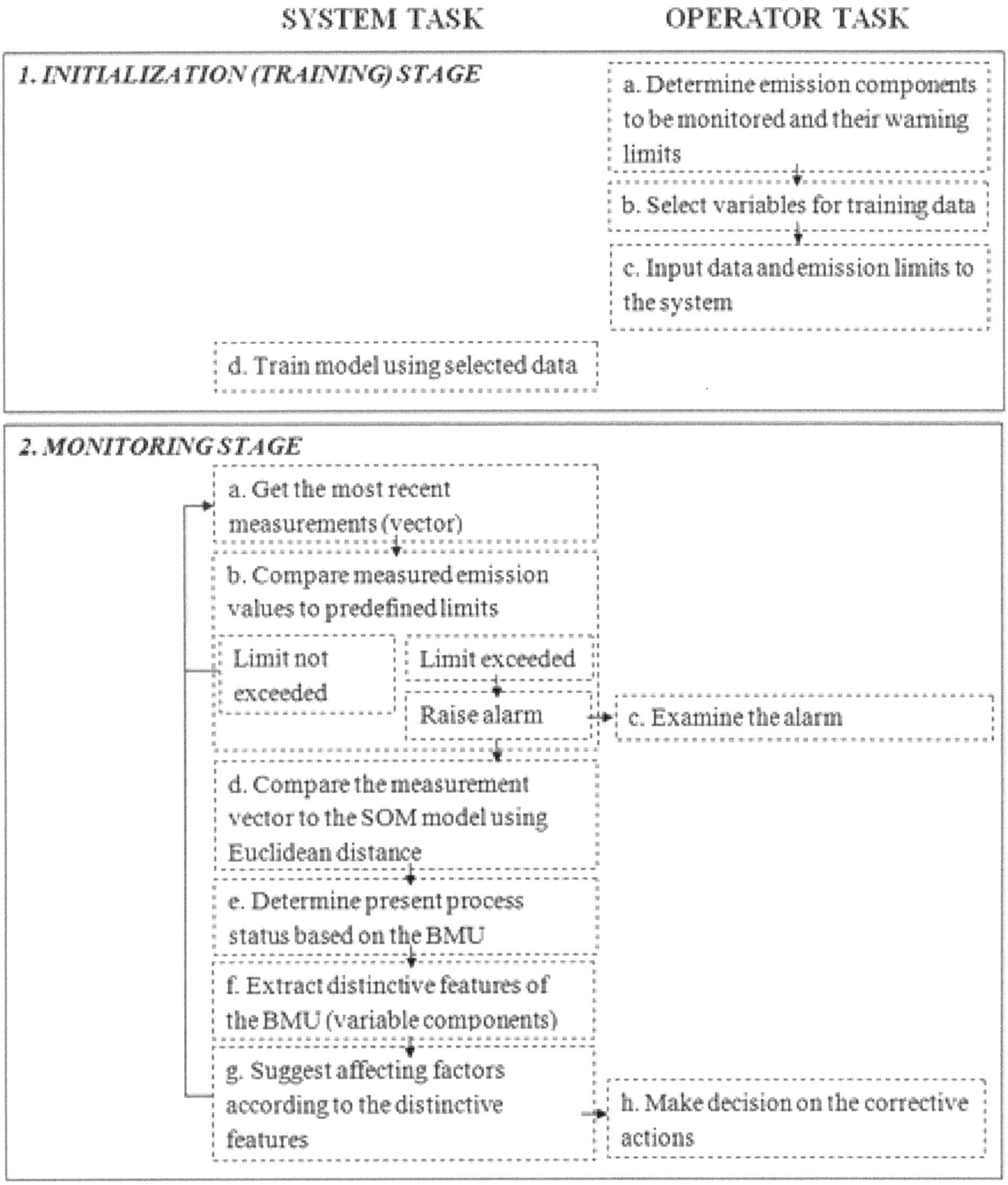

Figure 2: Illustration of the system operation in a decision-making task.

Diagnosis of High Emission Rates Using Condition Analysis

As reference vectors provide generalized descriptions of the original data vectors assigned to each unit, each neuron represents a model derived from a number of data vectors that are closely related to each other in terms of Euclidean distance. These models can be regarded as condition models illustrating the prevailing state of the process.

For example, let r_m be an n-dimensional reference vector describing the condition of a SOM unit:

$$r_m = (r_{m1}, r_{m2}, \ldots, r_{mn}), (m = 1, \ldots, M)$$

where n is the number of model variables and M refers to the total number of map neurons. Due to the nature of SOM, vectors r_1 and r_2 are quite similar as well as vectors r_{107} and r_{108}, whereas vectors r_1 and r_{107} are significantly different, as can be seen in Figure 3. For the same reason, reference vectors may also present the potential reasons for any occurred problems: if the problem is detected in a situation that most likely belongs to the neuron having the reference vector r_1, the reasons have to be different than those of a problem detected in a situation that belongs to the neuron of reference vectorr_{107}. In other words, the factors that separate these conditions from the other conditions are important.

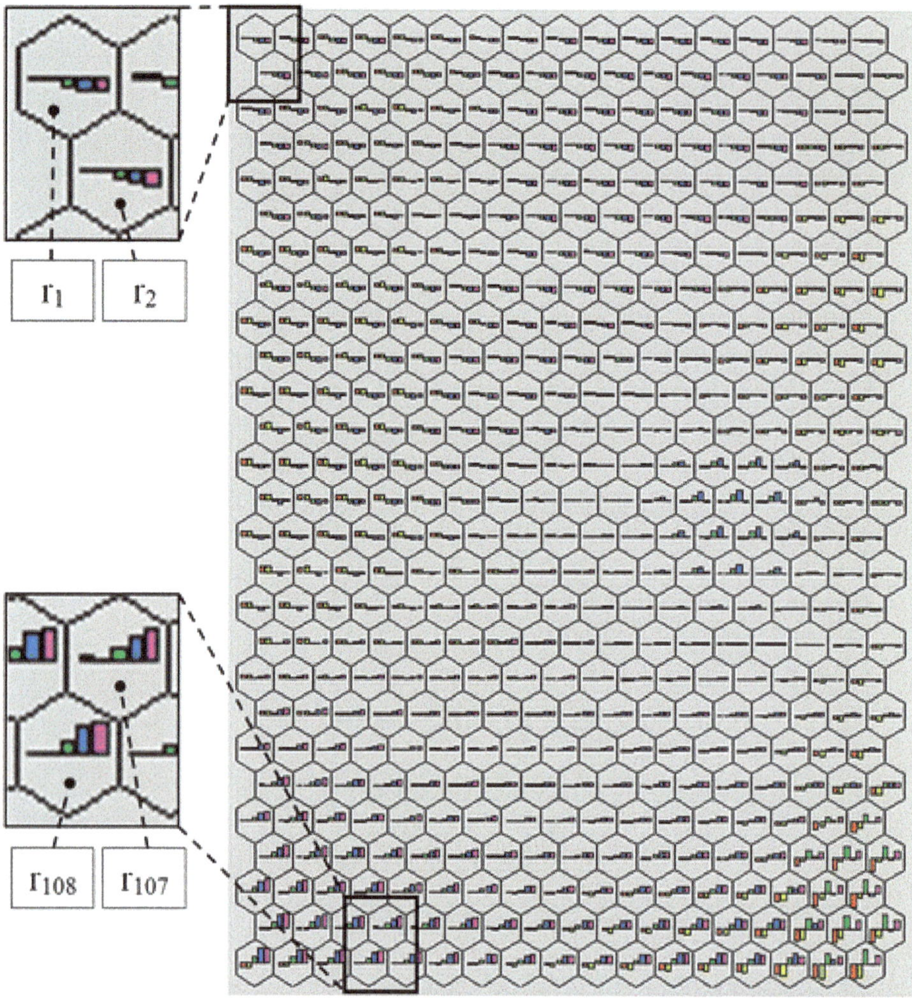

Figure 3: Example of a SOM grid having 27 × 17 neurons, and their reference vectors consisting of five variable components (illustrated by bars).

Case: Fluidized bed Process and data

FBC technology is intended primarily for producing energy from solid fuels such as coal, peat, biomass, or waste-derived fuels. CFBs, an example of which can be seen in Figure 4, are a widely used form of FBC, which are able to burn a large variety of fuels without a major reduction in performance (Koornneef, Junginger, & Faaij, 2007). This flexibility allows co-combustion of solid fuels, including fossil fuels such as coal, biomass-derived fuels such as wood and agricultural residues, peat, and even waste-derived fuels like demolition wood.

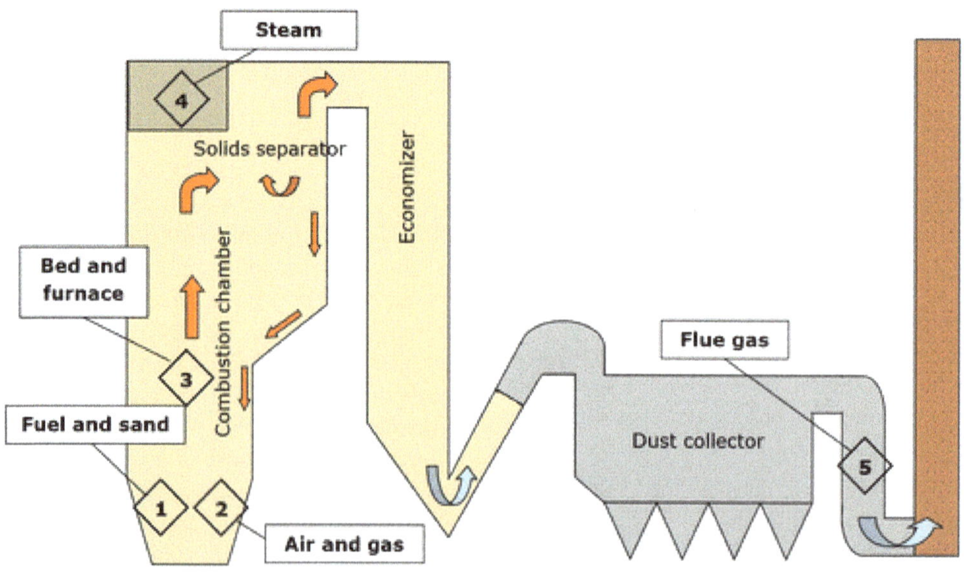

Figure 4: The CFB boiler and its main parts.

The demonstration environment involves a 150 MWth industrial CFB boiler fired by forest residue (mainly bark) as the primary fuel and coal as the supporting fuel. The total number of samples is 8,640 from a period of one month of operation (5-min sample interval). The total number of input variables is 57 (see Table 1).

Table 1: The variables of sample data

Variable	Measure-ments[1]
(1) Fuel and sand feed	
Fuel screw speed (bark)	2
Fuel screw speed (coal)	2
Bark/coal ratio	1
Total bark flow	1
Fuel feed (bark)	2
Fuel feed (coal)	2
Flow of start-up oil	1
Bark moisture	1
(2) Air and gas feed	
Primary air flow	1
Secondary air flows	4
Secondary air flow (total)	1
Flow of circulating gas	1
Primary air temperature after fan	1
Secondary air temperature after fan	1
(3) Bed and furnace conditions	
Bed pressure	3
Furnace pressure	5
Bed temperature	5
Bed temperature (avg)	1
Furnace temperature	3
(4) Steam system	
Main steam flow	1
(5) Flue gas	
O_2, flue gas	3
CO, flue gas	1
SO_2, flue gas	1
NO_x, flue gas	1
H2O, flue gas	1
Flue gas temperature	5
Cyclone temperature	2

[1]Number of measurements for each variable.

CASE DEMONSTRATIONS

Monitoring of NO_X and SO_2 in 150 MWth CFB

The emission monitor based on SOM-derived condition models is coded on the Matlab (Mathworks, Natick, USA) platform. To demonstrate the use of the system, we present two examples of problem situations on increasing emission rates. An example in which an increased NO_X concentration is detected can be seen in Figure 5. In the component planes (i.e. the colored maps), the colors of individual hexagonal neurons varying from red to blue indicate the value of the variable concerned: a red color tone on the SOM indicates that the value of the variable, in general, is on a high level in that neuron, whereas a blue color tone denotes a low value of that variable.

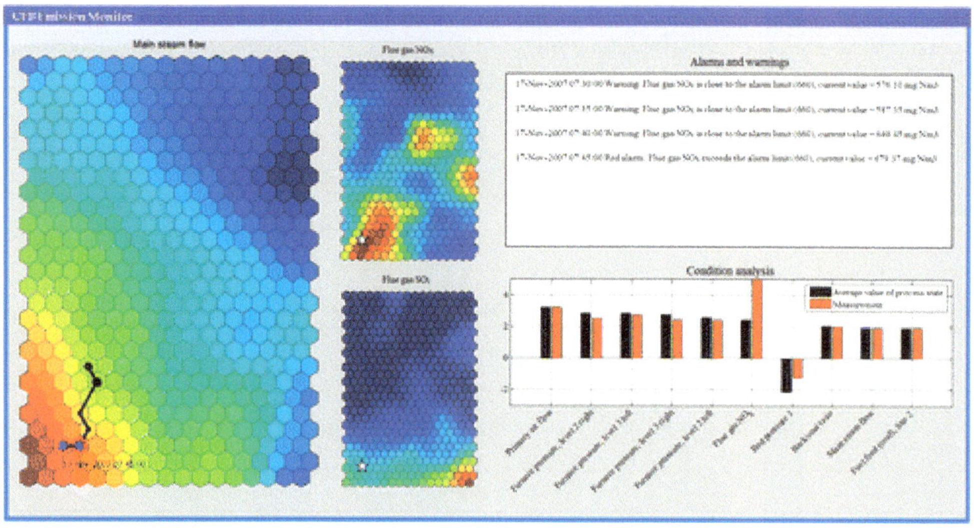

Figure 5: An example of the emission monitor showing an increased rate of NO_X **emission, and an analysis of the problem situation and prevailing process status.**

Thus, it can be seen in the large map that the data examples associated with a high level of main steam flow (i.e. the boiler load level) have been concentrated on the lower left corner of the map. The trajectory shown on the leftmost map illustrates that the process evolves from a state of medium power to that of high power, whereby also the NO_X emission rate is increased. Meanwhile, warnings are presented in the right-hand list. Finally, an alarm is given as the NO_X emission continues to increase.

A condition analysis based on the reference vectors is then performed and presented as a bar graph shown on the right in Figure 5. As can be seen in the graph, this situation is characterized by a high boiler load. Moreover, it is shown that the bark to coal ratio is high, meaning that the share of bark is high compared to the share of coal. Perhaps there is a set of biomass fuel in use that has a high content of nitrogen. In addition, the bed pressure seems to be low in this problem situation.

Another problem situation in which an increased SO_2 concentration is detected can be seen in Figure 6. As can be seen, the power of the boiler is on the low level in this case. It is also notable that in this situation, the boiler is fueled mainly by the supporting fuel (i.e. coal), which is not the preferable fuel combination based on which the boiler was designed. In summary, low boiler load and perhaps relatively unstable combustion conditions, added to the use of a sulfur-rich fuel, has led to a release of a relatively high concentration of sulfur dioxide.

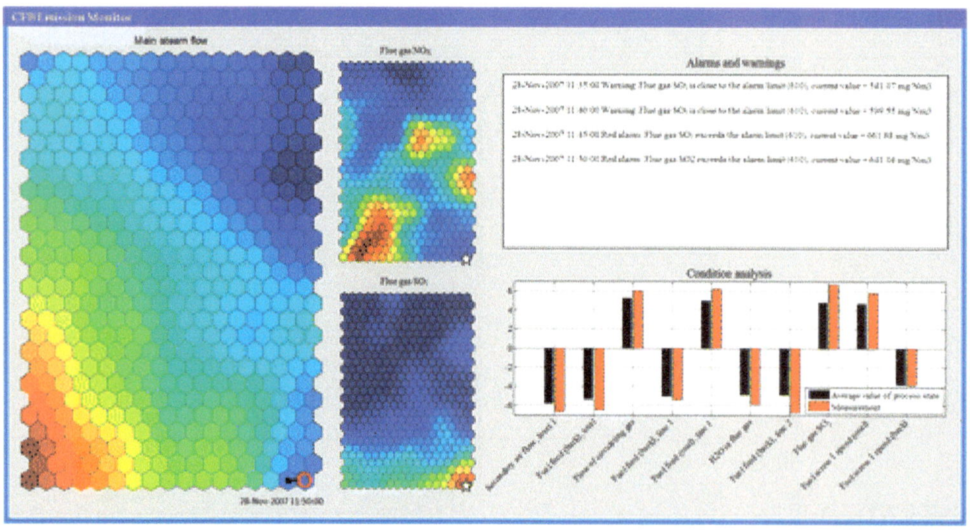

Figure 6: An example of the emission monitor showing an increased rate of SO_2 **emission, and an analysis of the problem situation and prevailing process status.**

Analysis of Problem Situations

In addition to online performance monitoring, another useful feature of the approach is the offline (post-process) analysis of emissions. The summary of problem situations in the CFB case can be seen in Figure 7. It can be seen that the problems related to a high SO_2 concentration are concentrated in two quite similar process states (i.e. neurons), illustrated on the bottom right corner of the map. The combination of a low boiler load and a high coal to bark ratio are characteristic features for these neurons.

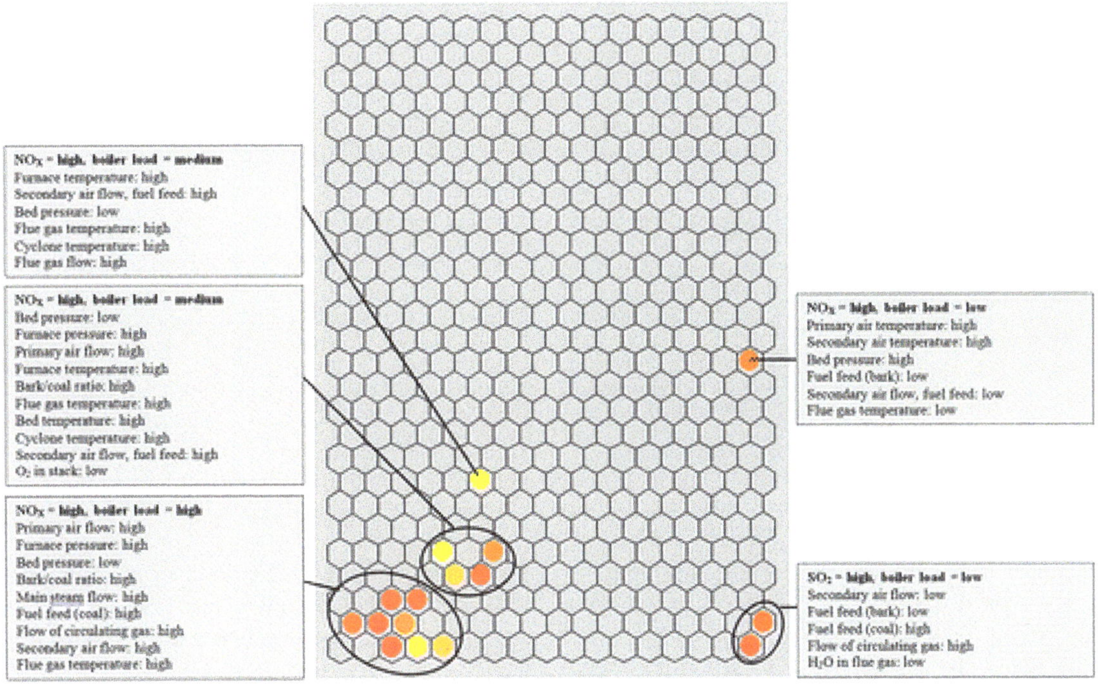

Figure 7: Summary of the detected problem situations and their diagnosis.

The situation is essentially different when it comes to problems related to a high NO_X concentration, however. First of all, the NO_X-related problems can be linked with all degrees of boiler load. Under low load conditions, the temperatures of feed air, a high bed pressure, and a high coal to bark ratio are the potential reasons for increased NO_X release rate, whereas high temperatures and a low bed pressure are characteristics for the problem situations detected under medium load conditions. The latter cases (medium power) seem to be situations in which the load is presently being reduced, and the process has not yet gained its stability. In other words, these factors indicate that there has been some instability in the process. When it comes to high boiler load, it is obvious that some of the problems originate from the fact that the process is working at full power and is therefore approaching the unacceptable emission limits. On the other hand, the low bed pressure may be a sign of a disturbance in the combustion, which has led to an exceptionally increased NO_X concentration in the flue gas.

DISCUSSION

Environmental awareness has been increased exponentially in recent decades, and it is clear that power plants will have to seek new ways of managing emissions more efficiently in the future. Measurement databases are valuable process history records, because they contain information on the past operation of the process and on the mutual interactions between different factors affecting the process performance. When it comes to emission management, these data can be exploited in several tasks, such as estimating future emission concentrations or analyzing the emissions already released. This kind of refined information also provides support for decisions aimed at reducing the emissions in the future. Methodologies based on computational intelligence, including intelligent monitoring systems and soft sensors, offer a promising alternative route to process improvement, by which a more efficient and environmentally friendly process can be achieved.

As industrial processes are getting more complex and the amount of measurements is increasing continuously, it is obvious that process monitoring based on single measurements is not the most efficient way of controlling the processes anymore. Monitoring systems having diagnosing abilities and based on multivariable measurement data can provide mutual information that provides deeper knowledge and is thus more useful, but this kind of information is accessible only by models which are able to extract the essential features from the data. On the other hand, the outputs of the multivariable process monitoring methods used traditionally are not very easy to interpret, as they lack holistic visualization abilities for presenting the extracted information such as the prevailing condition of the process and its direction. What is even more important, their capability of analyzing problem situations and their potential reasons is generally restricted, because they are designed principally to detect normal conditions.

Interpretation of results

In the present paper, an adaptive monitoring platform based on the SOM method is described, and the system is demonstrated using measurement data from an industrial CFB boiler fired by forest residue as the primary fuel and coal as the supporting fuel. The system is capable, not only of visualizing the process behavior and detecting the problem situations but also of providing analysis of these cases automatically. Based on the results, the novel method for monitoring emissions overcomes the existing ones in certain aspects that should be important in future monitoring applications. The pros and cons of the most general multivariate methods with diagnosing abilities for emission monitoring can be seen in Table 2.

Table 2: Summary of multivariate methods with diagnosing abilities for emission monitoring

Attribute	PCA	PLS	SOM
Ability to detect abnormalities	++	++	++
Ability to use multivariate data	++	++	++
Interpretation of results		+	+
Nonlinearity	*	*	++
Visualization of transitions			+
Ability to provide status information	+	+	+
Ability to diagnose problems	+	+	++
Ability to learn			++
Requirements for data quality	++	++	+
Ease of use (without a tailored solution)	+	+	
Commercial availability	++	+	

Notes: PCA = principal component analysis, PLS = partial least squares, SOM = self-organizing map.
*Nonlinear version exists also.

The chemical and physical phenomena occurring in combustion are typically dynamical, which has to be taken into account when developing monitoring systems. The present results show that an adaptive approach able to learn the changing behavior is necessary to adapt to the dynamics of the CFB combustion. It can also be concluded that this kind of dynamical evolution can be comprehended more easily by monitoring the change between different conditions, i.e. the transition from certain identified status to another one. This can be described by a model constructed using a group of key measurements and by visualizing this model in an understandable manner.

This, combined with the presented results, leads us to the most important conclusion, which is that the SOM-based adaptive monitoring approach demonstrated here offers a descriptive, visual way of modern emission monitoring and analyzing both immediate changes and long-term evolution in CFB combustion. The model behind the visual presentation is able to learn with time, which makes it capable of detecting new, previously unobserved phenomena. The system also facilitates the extraction of information from a large amount of multidimensional measurement data and can be used for comprehensive, model-based condition monitoring.

Challenges of Condition Monitoring

Because manual processing of measurement data tends to be laborious and time consuming, advanced data mining methods can provide resource savings in the monitoring, diagnosis, and even control of industrial processes. The so-called "big data" problem, i.e. the continuously increasing amount of measurement data, and increasingly complicated processes require appropriate software tools to extract the essential information. For the same reasons, intelligent software systems designed for advanced process diagnostics can prove themselves useful. The results of this study provide a framework that can be easily implemented in a software application. This kind of software provides a competent way of analyzing a large amount of process data, and the results show that it could provide a useful modeling tool for industrial applications.

However, it is doubtful that systems such as presented here can be completely autonomous in the near future, and it is presumable that human intervention is needed in certain steps of evaluation. For example, in this case the model working in the background of the system also involves the functionality of analyzing those situations which are classified as problematic. Nonetheless, it is crucial to be aware of the fact that the system makes, in principle, just a preliminary suggestion on the possible reasons for the problem events. These potentially affecting factors can be viewed on the screen and have to be reviewed by an expert who has adequate experience on the process performance and the required empirical knowledge to judge which ones of the presented variables are the most logical origins of this particular problem. This evaluation is critical, because the variables brought up by the system are simply a mathematical representation on the factors by which the prevailing process state differs most from the others.

The requirement for proper training data is another important issue when implementing a system such as presented here. It is important to bear in mind that the ability to adapt to prevailing conditions (i.e. the most recent data) is a prerequisite when applying the system in dynamical processes. Therefore, the model has to be kept up-to-date by training the SOM regularly using a representative sample data-set. On the other hand, in some cases it might be useful to preserve also some older information, and not just the most recent. For example, some phenomena affecting emission rates may happen regularly but rarely, so the system should be able to remember some older occurrences as well. Additionally, if the reason for a detected problem or a combination of reasons is not represented in the training data-set (the origins of the problem are new to the system), the system will not be able to evaluate the situation.

Recommendations and Future Aspects

Based on the results, it is possible to recommend the most beneficial purposes of use for the approach presented in this study. When it comes to process design and development towards cleaner processes, it is obvious that physical emission models will be the standard parts in the emission modeling toolbox also in the future. Therefore, the most practical use of advanced data-driven methods would be in supporting the physical emission models. A promising purpose of use for data-driven emission models is in advanced monitoring systems. The main advantages of these systems are that they facilitate the extraction of information from multidimensional measurement data, assist in interpreting the prevailing process status, and may describe its evolution in a comprehensive manner. The ability to adapt to prevailing conditions (i.e. the most recent measurement data) is a prerequisite when applying these methods in energy systems. The most practical way of using these methods would be in advanced software solutions for decision-support which are designed to diagnose and monitor industrial processes.

The most promising target of use of the presented approach is in decision support systems aimed at reducing air emissions and their management cost in energy boilers. In the future, an advanced emission monitoring system should be able to propose actions for improving the emission performance, so that the most problematic situations could be prevented by controlling the process beforehand. An intelligent monitoring system could be able to analyze the status of the process and assist the plant personnel in interpreting the prevailing risk level using the available measurements. The system could then suggest a code of practice, so that the process operator would be able to act correctly and quickly. It would also be important to know the future condition of the process and the future emission concentration, to anticipate the future status and to justify operative actions to improve and optimize the process. This could be achieved by a combination of intelligent data-based monitoring and soft sensors.

CONCLUSIONS

We have presented a new approach for advanced, model-based emission monitoring in CFBs, which is realized by combining existing methodology and multivariate data consisting of process measurements in a novel way. A real-world CFB process is presented as a case study. It can be concluded that data-driven emission models can provide functional platforms for advanced monitoring systems.

Based on the results the main advantages of the model-based monitoring system are as follows:

- Good descriptive capabilities including visualization of short- and long-term changes.
- Ability to analyze: the system is able to diagnose problem situations, recognize prevailing process status and provide decision-support for corrective actions.
- Ability to adapt: the system is able to adapt to prevailing conditions.
- Ability to learn: the system is able to learn from process history.

As the industrial monitoring platforms are moving towards more intelligent solutions which provide, not only the traditional detection of abnormal conditions but also many predictive and analytical capabilities, the systems such as presented here offer potential platforms for this purpose. As the regulations regarding the industrial emissions will be tightened in the near future, the results serve the development of advanced decision-support systems intended to reduce air emissions.

ACKNOWLEDGEMENTS

The Finnish Funding Agency for Technology and Innovation (TEKES) and the project Intelligent Software and Service Concept of the Industrial Internet (InDiGO!) are acknowledged.

REFERENCES

1. Basu, P. (1999). Combustion of coal in circulating fluidized-bed boilers: A review. *Chemical Engineering Science, 54*, 5547–5557.10.1016/S0009-2509(99)00285-7
2. Beér, J. M. (2000). Combustion technology developments in power generation in response to environmental challenges.*Progress in Energy and Combustion Science, 26*, 301–327.10.1016/S0360-1285(00)00007-1
3. Bishop, C. M., Svensén, M., & Williams, C. K. I. (1998). GTM: The generative topographic mapping. *Neural Computation, 10*, 215–234.10.1162/089976698300017953
4. Chang, C. C., & Chen, S. H. (2015). A comparative analysis on artificial neural network-based two-stage clustering. *Cogent Engineering, 2*, 995785.10.1080/23311916.2014.995785
5. Chea, R., Grenouillet, G., & Lek, S. (2016). Evidence of water quality degradation in lower Mekong basin revealed by self- organizing map. *PloS One, 11*, e0145527.10.1371/journal.pone.0145527
6. Coda Zabetta, E., Hiltunen, M., Moulton, B., & Hotta, A. (2009). Biomass and waste co combustion - European experience. In*Proceedings of Electric Power 2009 Conference*. Rosemont, IL.
7. Dittenbach, M., Merkl, D., & Rauber, A. (2000). The growing hierarchical self-organizing map. In *Proceedings of International Joint Conference on Neural Networks (IJCNN 2000)* (Vol. VI, pp. 15–19). Los Alamitos, CA: IEEE Computer Society.
8. Gorban, A. N., & Zinovyev, A. (2005). Elastic principal graphs and manifolds and their practical applications. *Computing, 75*, 359–379.10.1007/s00607-005-0122-6
9. Hill, S. C., & Douglas Smoot, L. (2000). Modeling of nitrogen oxides formation and destruction in combustion systems.*Progress in Energy and Combustion Science, 26*, 417–458.10.1016/S0360-1285(00)00011-3
10. Iliyas, S. A., Elshafei, M., Habib, M. A., & Adeniran, A. A. (2013). RBF neural network inferential sensor for process emission monitoring. *Control Engineering Practice, 21*, 962–970.10.1016/j.conengprac.2013.01.007
11. Juntunen, P., Liukkonen, M., Lehtola, M., & Hiltunen, Y. (2013a). Cluster analysis by self-organizing maps: An application to the modelling of water quality in a treatment process. *Applied Soft Computing, 13*, 3191–3196.10.1016/j.asoc.2013.01.027
12. Juntunen, P., Liukkonen, M., Lehtola, M. J., & Hiltunen, Y. (2013b). Dynamic soft sensors for detecting factors affecting turbidity in drinking water. *Journal of Hydroinformatics, 15*, 416–426.10.2166/hydro.2012.052
13. Keskitalo, J., la Cour Jansen, J., & Leiviskä, K. (2010). Calibration and validation of a modified ASM1 using long-term simulation of a full-scale pulp mill wastewater treatment plant. *Environmental Technology, 31*, 555–566.10.1080/09593330903563802
14. Kohonen, T. (2001). *Self-organizing maps* (3rd ed.). New York, NY: Springer-Verlag.10.1007/978-3-642-56927-2

15. Kohonen, T., Oja, E., Simula, O., Visa, A., & Kangas, J. (1996). Engineering applications of the self-organizing map.*Proceedings of the IEEE, 84*, 1358–1384.10.1109/5.537105

16. Koornneef, J., Junginger, M., & Faaij, A. (2007). Development of fluidized bed combustion – An overview of trends, performance and cost. *Progress in Energy and Combustion Science, 33*, 19–55.10.1016/j.pecs.2006.07.001

17. Liukkonen, M., Havia, E., Leinonen, H., & Hiltunen, Y. (2011). Quality-oriented optimization of wave soldering process by using self-organizing maps. *Applied Soft Computing, 11*, 214–220.10.1016/j.asoc.2009.11.011

18. Liukkonen, M., Heikkinen, M., Hiltunen, T., Hälikkä, E., Kuivalainen, R., & Hiltunen, Y. (2011). Artificial neural networks for analysis of process states in fluidized bed combustion. *Energy, 36*, 339–347.10.1016/j.energy.2010.10.033

19. Liukkonen, M., & Hiltunen, T. (2014). Adaptive monitoring of emissions in energy boilers using self-organizing maps: An application to a biomass-fired CFB (circulating fluidized bed). *Energy, 73*, 443–452.10.1016/j.energy.2014.06.034

20. Liukkonen, M., Hiltunen, T., Hälikkä, E., & Hiltunen, Y. (2011). Modeling of the fluidized bed combustion process and NOx emissions using self-organizing maps: An application to the diagnosis of process states. *Environmental Modelling & Software,26*, 605–614.

21. Liukkonen, M., Hälikkä, E., Hiltunen, T., & Hiltunen, Y. (2012). Dynamic soft sensors for NOx emissions in a circulating fluidized bed boiler. *Applied Energy, 97*, 483–490.10.1016/j.apenergy.2012.01.074

22. Liukkonen, M., Hälikkä, E., Hiltunen, T., & Hiltunen, Y. (2013). Adaptive soft sensor for fluidized bed quality: Applications to combustion of biomass. *Fuel Processing Technology, 105*, 46–51.10.1016/j.fuproc.2011.06.015

23. Liukkonen, M., Juntunen, P., Laakso, I., & Hiltunen, Y. (2013). A software platform for process monitoring: Applications to water treatment. *Expert Systems with Applications, 40*, 2631–2639.10.1016/j.eswa.2012.11.013

24. Liukkonen, M., Laakso, I., & Hiltunen, Y. (2013). Advanced monitoring of industrial wastewater treatment: Multivariable approach using the self-organizing map. *Environmental Modelling & Software, 48*, 193–201.

25. Lv, Y., Liu, J., Yang, T., & Zeng, D. (2013). A novel least squares support vector machine ensemble model for NOx emission prediction of a coal-fired boiler. *Energy, 55*, 319–329.10.1016/j.energy.2013.02.062

26. MacGregor, J. A., & Cinar, A. (2012). Monitoring, fault diagnosis, fault-tolerant control and optimization: Data driven methods. *Computers and Chemical Engineering, 47*, 111–120.10.1016/j.compchemeng.2012.06.017

27. Nussbaumer, T. (2003). Combustion and co-combustion of biomass: Fundamentals, technologies, and primary measures for emission reduction. *Energy & Fuels, 17*, 1510–1521.

28. Oja, M., Kaski, S., & Kohonen, T. (2002). Bibliography of self-organizing map (SOM) papers: 1998-2001 Addendum. *Neural Computing Surveys, 3*, 1–156.

29. Qin, S. J. (2012). Survey on data-driven industrial process monitoring and diagnosis. *Annual Reviews in Control, 36*, 220–234.10.1016/j.arcontrol.2012.09.004

30. Sanjay,C.,&Prithvi,C.(2014).Hybridintelligencesystems andartificialneuralnetwork(ANN)approachformodeling of surface roughness in drilling. *Cogent Engineering, 1*, 943935.10.1080/23311916.2014.943935 [

31. Shah-Hosseini, H., & Safabakhsh, R., (2000). TASOM: The time adaptive self-organizing map. In *Proceedings of International Conference on Information Technology: Coding and Computing* (pp. 422–427). Los Alamitos, CA: IEEE Computer Society.

32. Shakil, M., Elshafei, M., Habib, M. A., & Maleki, F. A. (2009). Soft Sensor for NO_X and O_2 Using Dynamic Neural Networks.*Computers and Electrical Engineering, 35*, 578–586.10.1016/j.compeleceng.2008.08.007

33. Smrekar, J., Potočnik, P., & Senegačnik, A. (2013). Multi-step-ahead prediction of NOx emissions for a coal-based boiler.*Applied Energy, 106*, 89–99.10.1016/j.apenergy.2012.10.056

34. Strebel, K., Espinosa, G., Giralt, F., Kindler, A., Rallo, R., Richter, M., & Schlink, U. (2013). Modeling airborne benzene in space and time with self-organizing maps and Bayesian techniques. *Environmental Modelling & Software, 41*, 151–162.

35. Wang, F., Huang, Q., Liu, D., Yan, J., & Cen, K. (2008). Improvement of load-following capacity based on the flame radiation intensity signal in a power plant. *Energy Fuels, 22*, 1731–1738.10.1021/ef700620n

36. Yu, H., Song, Y., Liu, R., Pan, H., Xiang, L., & Qian, F. (2014). Identifying changes in dissolved organic matter content and characteristics by fluorescence spectroscopy coupled with self-organizing map and classification and regression tree analysis during wastewater treatment. *Chemosphere, 113*, 79–86.10.1016/j.chemosphere.2014.04.020

37. Zhang, Y., Liang, X., Wang, Z., & Xu, L. (2015). A novel approach combining self-organizing map and parallel factor analysis for monitoring water quality of watersheds under non-point source pollution. *Scientific Reports, 5*, 16079.10.1038/srep16079

CHAPTER 8

MEASUREMENT-INDUCED-NONLOCALITY FOR DIRAC PARTICLES IN GARFINKLE–HOROWITZ–STROMINGER DILATION SPACE–TIME

Juan He[a], [b], Shuai Xu[c], Liu Ye[a],

[a] School of Physics & Material Science, Anhui University, Hefei, 230601, People's Republic of China

[b] School of Physics & Electronics Science, Fuyang Normal College, Fuyang, 236037, People's Republic of China

[c] Institute of Plasma Physics, Chinese Academy of Sciences, Hefei, 230031, People's Republic of China

ABSTRACT

We investigate the quantum correlation via measurement-induced-nonlocality (MIN) for Dirac particles in Garfinkle–Horowitz–Strominger (GHS) dilation space–time. It is shown that the physical accessible quantum correlation decreases as the dilation parameter increases monotonically. Unlike the case of scalar fields, the physical accessible correlation is not zero when the Hawking temperature is infinite owing to the Pauli exclusion principle and the differences between Fermi–Dirac and Bose–Einstein statistics. Meanwhile, the boundary of MIN related to Bell-violation is derived, which indicates that MIN is more general than quantum nonlocality captured by the violation of Bell-inequality. As a by-product, a tenable quantitative relation about MIN redistribution is obtained whatever the dilation parameter is. In addition, it is worth emphasizing that the underlying reason why the physical accessible correlation and mutual information decrease is that they are redistributed to the physical inaccessible regions.

INTRODUCTION

The combination of relativity theory and quantum information science has received considerable attention [1], [2], [3], [4], [5], [6], [7] and [8] in recent years, since it not only leads us to a deeper explanation of quantum mechanics but also provides a new way to comprehend the information paradox when black holes [9], [10], [11], [12] and [13] are involved. For example, Pan et al. investigated the effect of Hawking temperature on the entanglement and teleportation for the scalar field in an asymptotically flat black hole with spherical symmetry [14]. Downes et al. studied the entangling moving cavities in noninertial frames [15]. More recently, Iizuka and Kabat discussed the mutual information of two Hawking particles emitted consecutively by an evaporating black hole[7]. Particularly, it is worth noticing that since the investigation of quantum correlation in the background of a Garfinkle–Horowitz–Strominger (GHS) dilation black hole can be related to quantum information theory, string theory and loop quantum gravity[16] and [17], it is considered to be a better understanding of black holes in the perspective of quantum mechanics.

On the other hand, nonlocality is a fundamental feature of quantum states not available in the classical world, and has many intriguing and subtle manifestations. In general, most

Citation: Juan He, Shuai Xu, Liu Ye, Measurement-induced-nonlocality for Dirac particles in Garfinkle–Horowitz–Strominger dilation space–time, Physics Letters B, Volume 756, 10 May 2016, Pages 278-282. DOI: 10.1016/j.physletb.2016.02.073

researchers explore the quantum nonlocality by means of Bell-inequality which is satisfied by any local hidden variable theory. Recently, Luo and Fu [18] introduced a geometric measure of nonlocality, named measurement-induced-nonlocality (MIN), which is based on the consideration that measurements in quantum mechanics usually cause disturbance. MIN is a more general kind of correlation in some sense and is defined via maximizing the Hilbert–Schmidt distance between the pre- and post-measurement states under certain constraint. Enlightened by that, in this article, we investigate quantum correlation for Dirac particles in the background of a GHS dilation black hole. Our attention is focused on finding some significant rules about MIN and quantum mutual information redistribution. The problem is illustrated in a situation that two observers, Alice and Bob, share a Werner state at the same initial point in flat Minkowski space–time. At one moment, Alice remains at the asymptotically flat region but Bob freely falls toward a dilation black hole and locates near the event horizon. We explore quantum correlation in detail and find some interesting properties in this background.

The organization of this paper is as follows. In section 2, we deduce the vacuumstructure for Dirac fields in GHS dilation space–time. In section 3, MIN and mutual information for Dirac particles in the background of a GHS dilation black hole are discussed in detail. Finally a conclusion is given in the section 4.

THEORETICAL DERIVATION OF THE VACUUM STRUCTURE FOR DIRAC FIELDS IN GHS DILATION SPACE–TIME

In this section, we deduce the vacuum structure for Dirac particles in the background of a GHS dilation black hole by using the similar research method of Schwarzschild space–time [19]. The spherically symmetric line element of the GHS black hole can be written as[20] and [21]

$$ds^2 = -\left(\frac{r-2M}{r-2D}\right)dt^2 + \left(\frac{r-2M}{r-2D}\right)^{-1} dr^2 + r(r-2D)(d\theta^2 + \sin^2\theta d\phi^2), \quad (1)$$

where M and D are parameters related to the mass of black hole and dilation field, consider G, c, ħ and κ_B as unity in this paper for simplicity. The relationship amongM, the charge Q and D is given by $D=Q^2/2M$.

In the curved space–time, Dirac equation is described as [22]

$$[\gamma^a e_a^\mu(\partial_\mu + \Gamma_\mu)]\psi = 0, \quad (2)$$

where γ^a is the Dirac matrices, e_a^μ corresponds to the inverse of the tetrad e_μ^a and Γ_μ represents the spin connection coefficient. Introducing a tortoise coordinate

$$r_* = r + 2(M-D)\ln[(r-2M)/(2M-2D)], \quad (3)$$

defining the retarded time u and advanced time v

$$u = t - r_* \qquad v = t + r_*. \quad (4)$$

Then, by solving the Dirac equation in the GHS dilation space–time, we can obtain the positive (fermions) frequency outgoing solutions outside and inside regions of the event horizon $\chi_{\vec{k}}^{I+} = \Re e^{-i\omega_i u}$ and $\chi_{\vec{k}}^{II+} = \Re e^{i\omega_i u}$, where $\Re$ is a 4-component Dirac spinor and ω_i represents the monochromatic frequency of the Dirac field. Obviously, $\chi_{\vec{k}}^{I\pm}$ and $\chi_{\vec{k}}^{II\pm}$ form a complete orthogonal family. So, we can expand the Dirac field χ_{out} as

$$\chi_{out} = \sum_{i,\xi}\int dk[a_{\vec{k}}^{\xi}\chi_{\vec{k}}^{\xi+} + b_{\vec{k}}^{\xi*}\chi_{\vec{k}}^{\xi-}], \quad (5)$$

where $\xi=(I,II)$, $a_{\vec{k}}^{I}$ and $b_{\vec{k}}^{I*}$ correspond to the fermion annihilation and antifermion creation operators acting on the state of the exterior region, and $a_{\vec{k}}^{II}$ and $b_{\vec{k}}^{II*}$ are the fermion annihilation and antifermion creation operators acting on the state of the interior region, respectively.

On the other hand, the generalized light-like Kruskal coordinates U and V for the GHS dilation space–time [23] is introduced

$$u = -(4M-4D)\ln[-U/(4M-4D)], \quad (6)$$
$$v = (4M-4D)\ln[V/(4M-4D)], \quad \text{if } r > r_+;$$
$$u = -(4M-4D)\ln[U/(4M-4D)],$$
$$v = (4M-4D)\ln[V/(4M-4D)], \quad \text{if } r < r_+.$$

Then, utilizing the relationship between Kruskal coordinates and black hole coordinates, we can obtain a complete basis for positive energy modes which are analytic for all real U and V by making an analytic continuation for $\chi_{\vec{k}}^{I\pm}$ and $\chi_{\vec{k}}^{II\pm}$ according to the suggestion of Damour–Ruffini

$$\delta_{\vec{k}}^{I+} = e^{2(M-D)\pi\omega_i}\chi_{\vec{k}}^{I+} + e^{-2(M-D)\pi\omega_i}\chi_{-\vec{k}}^{II-},$$
$$\delta_{\vec{k}}^{II+} = e^{-2(M-D)\pi\omega_i}\chi_{-\vec{k}}^{I-} + e^{2(M-D)\pi\omega_i}\chi_{\vec{k}}^{II+}. \quad (7)$$

Considering them as new basis, we can expand the Dirac fields in the Kruskal space–time

$$\chi_{out} = \sum_{i,\xi} d\vec{k}[2\cosh(4(M-D)\pi\omega_i)]^{-1/2}[\ell_{\vec{k}}^{\xi}\delta_{\vec{k}}^{\xi+} + \eta_{\vec{k}}^{\xi*}\delta_{\vec{k}}^{\xi-}], \quad (8)$$

where $\ell_{\vec{k}}^{\xi}$ and $\eta_{\vec{k}}^{\xi*}$ represent the annihilation and creation operators acting on the Kruskal vacuum. Eqs. (5) and (8) correspond to the decomposition of the Dirac fields in GHS dilation and Kruskal modes respectively, then the appropriate Bogoliubov transformations [24] between the creation and annihilation operators in the GHS dilation and Kruskal coordinates can be obtained. Considering the orthonormality of modes, each annihilation operator $\ell_{\vec{k}}^{I}$ can be obtained as a combination of dilation particle operators of only one frequency ω_i [25] and [26]

$$\ell_{\vec{k}}^{I} = [e^{-8(M-D)\pi\omega_i} + 1]^{-1/2}a_{\vec{k}}^{I} - [e^{8(M-D)\pi\omega_i} + 1]^{-1/2}b_{\vec{k}}^{II*}. \quad (9)$$

Since the GHS dilation space–time can be divided to the physical inaccessible and accessible region, the mode of ground state in a GHS dilation black hole coordinate corresponds to a two-mode squeezed state in Kruskal coordinate. After properly normalizing the state vector, the vacuum state of the Kruskal particle for mode $\vec{k}$ can be expressed as

$$|0_{\vec{k}}\rangle_K^+ = [e^{-8(M-D)\pi\omega_i} + 1]^{-1/2}|0_{\vec{k}}\rangle_I^+|0_{-\vec{k}}\rangle_{II}^- + [e^{8(M-D)\pi\omega_i} + 1]^{-1/2}|1_{\vec{k}}\rangle_I^+|1_{-\vec{k}}\rangle_{II}^-, \quad (10)$$

where $\{|n_{\vec{k}}\rangle_I^+\}$ and $\{|n_{-\vec{k}}\rangle_{II}^-\}$ correspond to the orthonormal bases for the outside and inside regions of the event horizon respectively, the superscript on the kets $\{+,-\}$ indicate the particle and antiparticle vacua. In the same way, the only excited state can be described as

$$|1_{\vec{k}}\rangle_K^+ = |1_{\vec{k}}\rangle_I^+ |0_{-\vec{k}}\rangle_{II}^-, \tag{11}$$

Here and after, we consider $\{|n_{\vec{k}}\rangle_I^+\}$ and $\{|n_{-\vec{k}}\rangle_{II}^-\}$ as $\{|n\rangle_I\}$ and $\{|n\rangle_{II}\}$ respectively for simplicity.

MIN FOR DIRAC PARTICLES IN THE BACKGROUND OF A GHS DILATION BLACK HOLE

In order to discuss quantum correlation, we assume Alice and Bob share a generically quantum state in flat Minkowski space–time. Alice has a detector only detects mode $|n\rangle_A$, and Bob has another detector sensitive only to the mode $|n\rangle_B$. At one moment, Alice remains at the asymptotically flat region but Bob freely falls toward a GHS dilation black hole and then hovers near the event horizon. In essence, due to the Hawking effect, Bob's detector will detect a thermal Fermi–Dirac distribution of particles as he traverses the Kruskal vacuum. That is, the mode $|n\rangle_B$ must be specified in the coordinates of the GHS dilation black hole in order to describe what Bob sees in this curved space–time. The Hawking temperature felt by Bob can be expressed as $T=1/[8\pi(M-D)]$. We choose Werner state [27] as the initial state

$$\rho_{AB} = \frac{1-\eta}{4} I + \eta |\varphi^-\rangle\langle\varphi^-|, \tag{12}$$

where $|\varphi^-\rangle = (|01\rangle - |10\rangle)/\sqrt{2}$ is a maximally entangled state and the parameter $\eta \in [0,1]$. Note that Werner state is an entangled state in the case of $\eta > \frac{1}{3}$, while is a disentangled state for $\eta \le \frac{1}{3}$.

Meanwhile, we can rewrite Eq. (12) in terms of Minkowski modes for Alice and Kruskal modes for Bob based on Eqs. (10) and (11). The part of physical accessible comes from A and B_I since the exterior region is causally disconnected from the interior region of the black hole. Taking trace over the state of the interior region, a density matrix can be obtained

$$\begin{aligned}
\rho_{AB_I} &= Tr_{B_{II}}(\rho_{AB_I B_{II}}) \\
&= (1-\eta)\left[e^{-8(M-D)\pi\omega}+1\right]^{-1}/4|00\rangle\langle 00| \\
&\quad + \left[(1+\eta)+\left[e^{8(M-D)\pi\omega}+1\right]^{-1}(1-\eta)\right]/4|01\rangle\langle 01| \\
&\quad + (1+\eta)\left[e^{-8(M-D)\pi\omega}+1\right]^{-1}/4|10\rangle\langle 10| \\
&\quad + \left[(1-\eta)+\left[e^{8(M-D)\pi\omega}+1\right]^{-1}(1+\eta)\right]/4|11\rangle\langle 11| \\
&\quad - \eta\left[e^{-8(M-D)\pi\omega}+1\right]^{-\frac{1}{2}}/2\left(|01\rangle\langle 10|+|10\rangle\langle 01|\right).
\end{aligned} \tag{13}$$

On the other hand, MIN can be utilized to capture quantum correlation. Consider a bipartite quantum state ρ shared by two parts A and B with respective Hilbert spaces H^A and H^B. MIN is defined as follows according to Luo and Fu [18]

$$MIN(\rho) = Max_{\Pi^A} \|\rho - \Pi^A(\rho)\|^2 \tag{14}$$

where the maximum is taken over the local von Neumann measurements $\Pi^A = \{\Pi_l^A\}(l=1,2)$ that do not disturb ρ^A in the sense that $\sum_l \Pi_l^A \rho^A \Pi_l^A = \rho^A$. $\|\cdot\|^2$ represents the Hilbert–Schmidt norm with $\|Y\|^2 = \mathrm{tr}\,Y^+Y$.

For convenience, we introduce the arbitrary two-qubit states in the Bloch decomposition as

$$\rho = \frac{1}{4}\left[I\otimes I + \sum_i x_i\sigma_i \otimes I + \sum_i y_i \otimes I\sigma_i + \sum_{i,j} t_{ij}\sigma_i\otimes\sigma_j\right] \tag{15}$$

where $\sigma_{i,j}(i,j=1,2,3)$ are the Pauli matrices, I is the 2×2 identity matrix, the Bloch vectors $\{x_1,x_2,x_3\}$ and $\{y_1,y_2,y_3\}$ are $x_i=tr[(\sigma_i\otimes I)\rho^X]$ and $y_i=tr[(I\otimes\sigma_i)\rho^X]$, the correlation tensor t_{ij} is $t_{ij}=tr[(\sigma_i\otimes\sigma_j)\rho^X]$.

The density matrix defined in Eq. (13) is a two-qubit X-shaped real state. A general real X-state as:

$$\rho^X = \begin{pmatrix} \rho_{11} & 0 & 0 & \rho_{14} \\ 0 & \rho_{22} & \rho_{23} & 0 \\ 0 & \rho_{23} & \rho_{33} & 0 \\ \rho_{14} & 0 & 0 & \rho_{44} \end{pmatrix}, \tag{16}$$

where $\rho_{ij}(i,j=1,2,3,4)$ are all real parameters. The characterizing parameters of X-state in Bloch decomposition can be expressed as [28]

$$\begin{aligned}
x_3 &= \mathrm{tr}(\sigma_z^A\rho^X) = \rho_{11}+\rho_{22}-\rho_{33}-\rho_{44}, \\
y_3 &= \mathrm{tr}(\sigma_z^B\rho^X) = \rho_{11}-\rho_{22}+\rho_{33}-\rho_{44}, \\
t_{11} &= \mathrm{tr}(\sigma_x^A\sigma_x^B\rho^X) = 2\rho_{14}+2\rho_{23}, \\
t_{22} &= \mathrm{tr}(\sigma_y^A\sigma_y^B\rho^X) = -2\rho_{14}+2\rho_{23}, \\
t_{33} &= \mathrm{tr}(\sigma_z^A\sigma_z^B\rho^X) = \rho_{11}-\rho_{22}-\rho_{33}+\rho_{44},
\end{aligned} \tag{17}$$

$x_1=x_2=y_1=y_2=t_{12}=t_{21}=t_{13}=t_{23}=t_{31}=t_{32}=0.$

According to the Theorem 3 in Ref. [18], MIN can be acquired for two-qubit X states

$$MIN_A(\rho^X) = \begin{cases} \frac{1}{4}(t_{11}^2+t_{22}^2) & \text{if } x\neq 0 \\ \frac{1}{4}(t_{11}^2+t_{22}^2+t_{33}^2-\delta_{\min}) & \text{if } x=0 \end{cases} \tag{18}$$

where $\delta_{\min} = \min\{t_{11}^2, t_{22}^2, t_{33}^2\}$.

Then, through mathematical calculations, it is easy to obtain the MIN of physical accessible state

$$MIN(\rho_{AB_I}) = \frac{\eta^2}{2(1+e^{-8(M-D)\pi\omega})}. \tag{19}$$

From Eq. (19), we know that the quantum correlation not only depends on the initial state, but also depends on the mass of black hole, dilation parameter, the frequency of the field, and so on. It is worth emphasizing that MIN of physical accessible is equal to that of initial states in the limit of $(M-D)\to+\infty$, corresponding to the situation that thermal Fermi–Dirac distribution of particles detected by Bob's detector which is nonexistent. Besides, the physically accessible MIN vs D are plotted in Fig. 1 which shows how the dilation parameter would affect the properties of quantum correlation. It is not difficult to find that the physical accessible correlation decreases as the dilation parameter D growing monotonically. This indicates that the effect of black hole's dilation makes the difference between pre- and post-measurement states become smaller, namely, the disturbance caused by local measurements tend to feeble. What is interesting that, for the scalar field, the quantum correlation is exactly zero when the Hawking temperature is infinite $(M-D\to 0)$, however, for the case of Dirac field, the quantum correlation does not go to zero any more. The intrinsic reasons for this phenomenon are the differences between Fermi–Dirac and Bose–Einstein statistics and the principle of Pauli exclusion. For a system composed of Fermi particles, the quantum states cannot be infinitely excited.

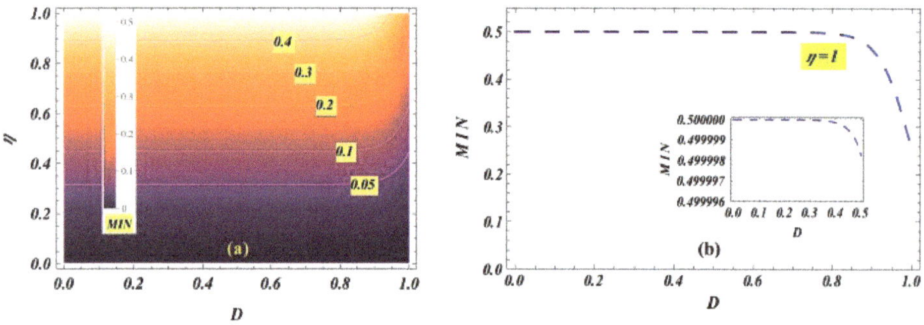

Figure 1: (Color online.) (a) The contour plot of physically accessible MIN versus dilation parameter D and state parameter η with the fixed ω and M. (b) Physically accessible MIN as a function of dilation parameter D for the case of η = 1 with the fixed ω and M.

Since MIN can be introduced to indicate some properties about the nonlocality, it is necessary to discuss the relation between MIN and the violation of Bell-inequality. As an index for quantum non-locality, Bell–Clauser–Horne–Shimony–Holt (Bell-CHSH) inequality plays most significant role. The so called CHSH inequality can be expressed as $|\langle B_{CHSH}\rangle| = |tr(\rho B_{CHS}H)| \leq 2$, where $B_{CHSH} = \mathbf{a}\cdot\boldsymbol{\sigma}\otimes(\mathbf{b}+\mathbf{b}')\cdot\boldsymbol{\sigma}+\mathbf{a}'\cdot\boldsymbol{\sigma}\otimes(\mathbf{b}-\mathbf{b}')\cdot\boldsymbol{\sigma}$, **a**, **a**′, **b**, **b**′ are the three-dimensional unit vector and $\boldsymbol{\sigma} = \sigma_x\mathbf{i}+\sigma_y\mathbf{j}+\sigma_z\mathbf{k}$.

According to the Horodecki criterion [29], for X states, utilizing the Bloch decomposition, the maximum violation of CHSH inequality can be simplified to

$$B_{CHSH}^{Max}(\rho^X) = \mathrm{Max}_{\mathbf{a},\mathbf{a}',\mathbf{b},\mathbf{b}'}\, tr(\rho^X B_{CHSH}) = 2\sqrt{\sum_{i=1}^{3} t_{ii}^2 - \delta_{\min}}. \tag{20}$$

Unexpectedly, we acquire an interesting result, namely,

$$B_{CHSH}^{Max}(\rho^X) = 4\sqrt{\mathrm{MIN}(\rho^X)}. \tag{21}$$

Thus, when $\mathrm{MIN}(\rho AB_I) > 0.25$, the Bell-CHSH inequality is violated, it means that these exists nonlocal quantum correlation. Of course, if $\mathrm{MIN}(\rho AB_I) \leq 0.25$, it informs us that these have quantum correlation but obey the Bell inequality. That is, we cannot be sure whether quantum correlation exists or not when the Bell inequality is not violated. Consequently, MIN can reveal more general properties about quantum system than Bell inequality.

An open question can be put forward: whether the lost correlation is destroyed or transferred to anywhere? To solve this perplexing problem, we calculate the MIN of other possible bipartite states which are physically inaccessible. By tracing over the exterior region mode B_I and mode A respectively, the corresponding density matrix can be obtained

$$\begin{aligned}\rho_{AB_{II}} = &\left[(1-\eta)\left[e^{-8(M-D)\pi\omega}+1\right]^{-1}+(1+\eta)\right]/4|00\rangle\langle00| \\ &+\left[e^{8(M-D)\pi\omega}+1\right]^{-1}(1-\eta)/4|01\rangle\langle01| \\ &+\left[e^{8(M-D)\pi\omega}+1\right]^{-1}(1+\eta)/4|11\rangle\langle11| \\ &+\left[(1+\eta)\left[e^{-8(M-D)\pi\omega}+1\right]^{-1}+(1-\eta)\right]/4|10\rangle\langle10| \\ &-\eta\left[e^{8(M-D)\pi\omega}+1\right]^{-1/2}/2\left(|00\rangle\langle11|+|11\rangle\langle00|\right)\end{aligned} \tag{22}$$

and

$$\begin{aligned}\rho_{B_I B_{II}} = &\left[e^{-8(M-D)\pi\omega}+1\right]^{-1}/2|00\rangle\langle00| \\ &+\left[e^{8(M-D)\pi\omega}+1\right]^{-1}/2|11\rangle\langle11|+1/2|10\rangle\langle10| \\ &-\left[e^{-8(M-D)\pi\omega}+1\right]^{-1/2} \\ &\times\left[e^{8(M-D)\pi\omega}+1\right]^{-1/2}/2\left(|00\rangle\langle11|+|11\rangle\langle00|\right).\end{aligned} \tag{23}$$

Similarly, MIN of the physical inaccessible states can be acquired

$$\mathrm{MIN}(\rho_{AB_{II}}) = \frac{\eta^2}{2(1+e^{8(M-D)\pi\omega})}, \tag{24}$$

$$\mathrm{MIN}(\rho_{B_I B_{II}}) = \frac{1}{4[1+\cosh[8(M-D)\pi\omega]]} \tag{25}$$

Notably, the quantum properties of $\rho_{B_I B_{II}}$ have nothing to do with parameter η. The reason is that the initial subsystem B of Werner state, which can be obtained by tracing over subsystem A, is already irrelevant to parameter η. Meanwhile, from the Fig. 2which plots all the bipartite MIN as a function of dilation parameter D, the physical accessible quantum correlation degrades while the inaccessible correlations increase with the dilation parameter growing. Owing to monogamy of quantum correlations, the disappeared quantum correlation is redistributed to the physical inaccessible region, and a quantitative inequality is always tenable whatever the dilation parameter is, that is

$$\mathrm{MIN}(\rho_{AB_I}) + \mathrm{MIN}(\rho_{AB_{II}}) \le \frac{1}{2}S(\rho_A) \tag{26}$$

where $S(\rho)=-\mathrm{tr}(\rho\log_2\rho)$ is the von Neumann entropy and $\rho_A=\mathrm{tr}_B(\rho AB)$ denotes the reduced density operator of subsystem A . Besides, it is interesting that in the limit of $M-D\to 0$, corresponding to the case of infinite Hawking temperature, the quantum correlations of ρAB_I and $\rho ABII$ is identical.

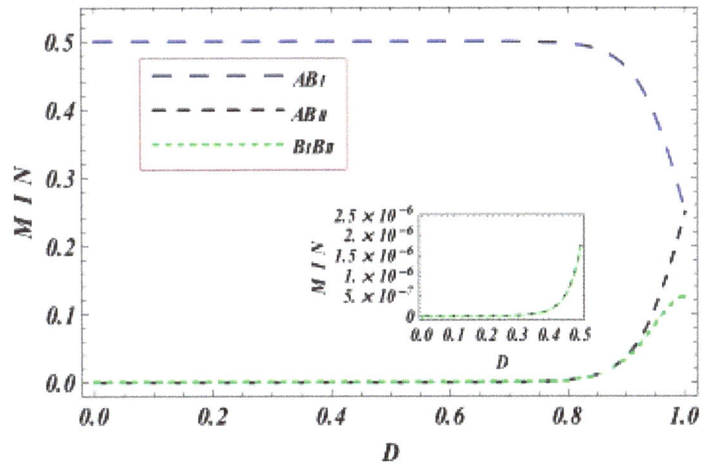

Figure 2: (Color online.) All the bipartite MIN as a function of dilation parameter D with the fixed η and M.

Additionally, to better understand the information redistribution, we also investigate the all bipartite mutual information of the given system. For a bipartite system ρAB, the quantum mutual information can be defined as [29]

$$I(\rho AB)=S(\rho_A)+S(\rho_B)-S(\rho AB) \tag{27}$$

Through a series of calculation, we plot mutual information of all the bipartite states as a function of dilation parameter D in Fig. 3. It is not difficult to find that, just like the trends of quantum correlation, the mutual information of subsystem ρAB_I degrades while the others increase with the dilation parameter growing. If we accept the mutual information is a good measure for total correlations, the difference between mutual information and MIN can be regarded as the classical correlation. By the further analysis, without surprising, we assure that the transfer is also occurred for classical correlation. As a consequence, a significant conclusion can be reached: the lost quantum correlation, classical correlation and mutual information from mode A and mode B_I is allocated to inaccessible systems, especially between mode A and mode B_I.

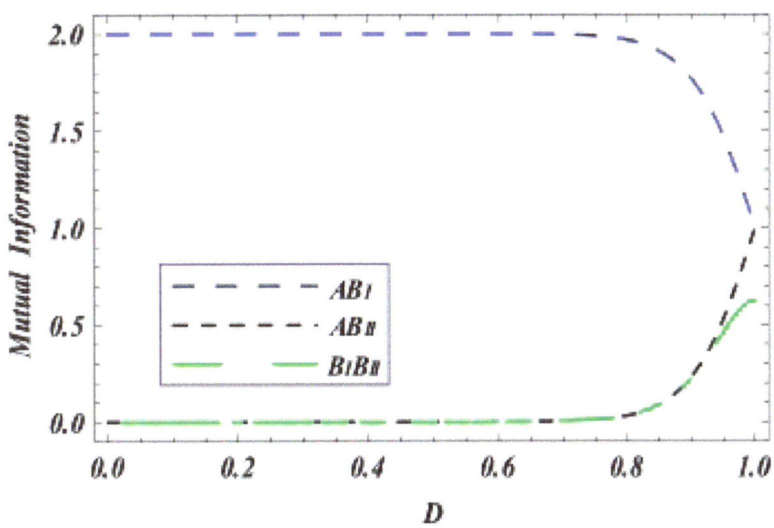

Figure 3: (Color online.) All the bipartite mutual information as a function of dilation parameter D with the fixed η andM.

CONCLUSIONS

In this article, we analytically investigate the quantum correlation between two modes of a Dirac field as detected by Alice who stays stationary at an asymptotically flat region and Bob who locates near the event horizon in GHS dilation space–time. We choose Werner state as the initial state, which contains entangled state and unentangled state. The result indicates that the physical accessible correlation decreases as the dilation parameter increasing monotonically. Due to the Pauli Exclusion Principle and the differences between Fermi–Dirac and Bose–Einstein statistics, the physical accessible correlation is not zero when the Hawking temperature is infinite, which is different from the scalar field. Beyond this, the reason why the physical accessible correlation (including quantum and classical correlation) and mutual information are reduced is that they are redistributed to the physical inaccessible regions. In addition, the boundary of MIN related to Bell-violation is derived, which manifests that MIN is more general than quantum nonlocality captured by the violation of Bell-inequality. Meanwhile, we provide a quantitative inequality about the redistribution of MIN and note that the inequality is always established whatever the dilation parameter is.

ACKNOWLEDGEMENTS

This work was supported by the National Science Foundation of China under Grants No.61275119 and No. 11575001, the Doctoral Foundation of the Ministry of Education of China under Grant No. 20103401110003, the Natural Science Research Project of Education Department of Anhui Province of China under Grants No. KJ2013A205 and No.KJ2016A547, the Doctoral Foundation of Fuyang Normal College under Grant No.FYNU1602, and also by the Personal Development Foundation of Anhui Province(2008Z018).

REFERENCES

1. . Fuentes-Schuller, R.B. Mann, Phys. Rev. Lett. 95 (2005) 120404.
2. G. Adesso, I. Fuentes-Schuller, M. Ericsson, Phys. Rev. A 76 (2007) 062112.
3. M. Aspachs, G. Adesso, I. Fuentes, Phys. Rev. Lett. 105 (2010) 151301.
4. I. Lamata, M.A. Martin-Deigado, E. Solano, Phys. Rev. Lett. 97 (2006) 250502.
5. J. Wang, Q.Y. Pan, J.L. Jing, Phys. Lett. B 692 (2010) 202.
6. J. León, E. Martín-Martínez, Phys. Rev. A 80 (2009) 012314.
7. N. Iizuka, D. Kabat, Phys. Rev. D 88 (2013) 084010.
8. J. Wang, Q. Pan, S. Chen, J. Jing, Phys. Lett. B 677 (2009) 186.
9. L. Bombelli, R.K. Koul, J. Lee, R. Sorkin, Phys. Rev. D 34 (1986) 373.
10. S.W. Hawking, Nature 248 (1974) 30; S.W. Hawking, Phys. Rev. D 14 (1976) 2460.
11. Y. Du, Y. Tian, Phys. Lett. B 739 (2014) 250–255.
12. A. Farahi, L.A.P. Zayas, Phys. Lett. B 734 (2014) 31–35.
13. H. Terashima, Phys. Rev. D 61 (2000) 104016.
14. Q. Pan, J. Jing, Phys. Rev. D 78 (2008) 065015.
15. T.G. Downes, I. Fuentes, T.C. Ralph, Phys. Rev. Lett. 106 (2011) 210502.
16. O. Dreyer, Phys. Rev. Lett. 90 (2003) 081301.
17. S. Chen, J. Jing, Class. Quantum Gravity 22 (2005) 533.
18. S. Luo, S. Fu, Phys. Rev. Lett. 106 (2011) 120401.
19. J. Wang, Q. Pan, J. Jing, Ann. Phys. 325 (2010) 1190.
20. D. Garfinkle, G.T. Horowitz, A. Strominger, Phys. Rev. D 43 (1991) 3140.
21. A. Gareia, D. Galtsov, O. Kechkin, Phys. Rev. Lett. 74 (1995) 1276.
22. D.R. Brill, J.A. Wheeler, Rev. Mod. Phys. 29 (1957) 465.
23. T. Damoar, R. Ruffini, Phys. Rev. D 14 (1976) 332.
24. S.M. Barnett, P.M. Radmore, Methods in Theoretical Quantum Optics, Oxford University Press, New York, 1997, pp. 67–80.
25. E. Martín-Martínez, L.J. Garay, Juan León, Phys. Rev. D 82 (2010) 064006.
26. D.E. Bruschi, J. Louko, E. Martín-Martínez, A. Dragan, I. Fuentes, Phys. Rev. A 82 (2010) 042332.
27. R.F. Werner, Phys. Rev. A 40 (1989) 4277.
28. A.R.P. Rau, J. Phys. A 42 (2009) 412002.
29. R.S. Ingarden, A. Kossakowski, M. Ohya, Information Dynamics and Open Systems – Classical and Quantum Approach, Kluwer Academic Publishers, Dordrecht, 1997.

CHAPTER 9 DIRECT MEASUREMENT OF AXIAL MOMENTUM IMPARTED BY AN ELECTROTHERMAL RADIOFREQUENCY PLASMA MICRO-THRUSTER

Christine Charles [1], Roderick W.Boswell [1], Andrew Bish [1], Vadim Khayms[2] and Edwin F.Scholz [3]

[1]Space Plasma, Power and Propulsion Laboratory, Research School of Physics and Engineering, The Australian National University, Canberra, ACT, Australia

[2]Lockheed Martin Space Systems Company, Sunnyvale, CA, USA

[3]Lockheed Martin Space Systems Company, Littleton, CO, USA

Gas flow heating using radio frequency plasmas offers the possibility of depositing power in the center of the flow rather than on the outside, as is the case with electro-thermal systems where thermal wall losses lower efficiency. Improved systems for space propulsion are one possible application and we have tested a prototype micro-thruster on a thrust balance in vacuum. For these initial tests, a fixed component radio frequency matching network weighing 90 g was closely attached to the thruster in vacuum with the frequency agile radio frequency generator power being delivered via a 50 ohm cable. Without accounting for system losses (estimated at around 50%), for a few 10 s of Watts from the radio frequency generator the specific impulse was tripled to ~48 s and the thrust tripled from 0.8 to 2.4 milli-Newtons.

INTRODUCTION

Over the past decade, neutral-gas heating has been experimentally studied both in low-pressure high-power (with and without an applied magnetic field) radiofrequency (rf) discharges [1] and in high-pressure low-power rf discharges [2]. When neutral pressure varies along the axis or across a radius, ionization and transport become coupled and this greatly affects the plasma steady-state [3]. It has been analytically shown [3, 4] that the transport process varies with collisionality (effectively operating pressure and discharge geometry) and power, and also with neutral gas heating [5], thereby yielding very different radial and axial profiles of neutral and charged particle densities and temperatures inside the plasma cavity. The plasma parameters will also vary with the rf power coupling mechanism [6], i.e., the rf electrode configuration and operating conditions. When these rf discharges are considered for use as thrusters, it is necessary to have a detailed understanding of how axial momentum is imparted in the presence of a plasma flow and plasma expansion in vacuum.

Electric thrusters such as ion gridded thrusters, Hall Effect thrusters, arcjets and resisto-jets are routinely used on satellites [7, 8]. Their usefulness is maximized compared to cold gas and chemical thrusters when their specific impulse exceeds a few hundred seconds but their miniaturization often leads to decreased performances while remaining costly (i.e., gridded engines and Hall effect thrusters require an additional electron neutralizer, usually a Hollow Cathode which needs a certain period of time to achieve optimum operating characteristics). Low-cost low-volume and low-weight thrusters using "green" and safe propellants suitable for use on micro-satellites such as "CubeSats" (based on the 10 cm by 10 cm by 10 cm modular unit) are not readily available. Electrothermal thrusters such

Citation: Charles C, Boswell RW, Bish A, Khayms V and Scholz EF (2016) Direct Measurement of Axial Momentum Imparted by an Electrothermal Radiofrequency Plasma Micro-Thruster. Front. Phys. 4:19. doi: 10.3389/fphy.2016.00019

as resisto-jets and arcjets aim at heating the propellant and are often in part chemical. For example, arcjets commonly operate with hydrazine or ammonia. Some recent arcjets are essentially cathode spot vacuum arc discharges and erode the cathode producing a stream of metallic blobs and ions. Pulsed Plasma Thrusters also operate by eroding the electrodes and can be scaled down. To date it is extremely difficult to obtain actual thrust balance measurements of microthrusters in general. The recently developed *Pocket Rocket* concept is essentially an instant "on" radio frequency plasma heated gas thruster [2].

Fruchtman [3] has reported on the effect of the energy deposited in the neutral gas by a collisional plasma and this has been recently confirmed experimentally using volume-average ro-vibrational spectroscopic measurements of neutral gas temperatures in a small-diameter collisional radiofrequency plasma jet known as *Pocket Rocket* [2, 9]. Neutral gas heating in *Pocket Rocket* mostly results from ion-neutral collisions in the bulk plasma (charge-exchange and elastic) and from heating at the plasma cavity radial walls [6, 10]. Recent 1D3v Particle In Cell simulations of this plasma jet [11] have provided details on the neutral and charged particles density profiles along the expansion which closely relate to the "open system" analytical analysis by Fruchtman [3], where a net mass flow is considered. Direct measurements of the axial momentum generated by a xenon rf plasma jet expanding in vacuum are reported as a function of two parameters; rf power and gas flow. In essence it is a proof of concept that this type of plasma device generates thrust rather than a test of a space qualified ready system. Future refinements to the rf and gas delivery systems are expected to improve the overall efficiency of the system by a factor of 2–3.

EXPERIMENTAL SET UP

Experiments are carried out in the 1 m-diameter 2 m-long *Wombat* vacuum chamber [12] recently upgraded with a new vacuum pumping system (scroll, turbomolecular, and cryogenic pumps) and a suite of access and viewing ports (Figure 1). The small diameter *Pocket Rocket* plasma thruster called (*MiniPR*), is attached to a thrust balance comprising a grounded open cube frame (0.3 × 0.3 × 0.3 m^3) suspended from four "chimneys" (Figure 1). A laser-displacement sensor system [13, 14] positioned near the top of the thrust balance mounting is used to measure the displacement of the thrust balance and hence the axial force imparted by the plasma heated gas flow. The base pressure is measured to be about 7 × 10^{-7} Torr using an ion gauge and a baratron gauge located 2 m downstream of the thrust balance.

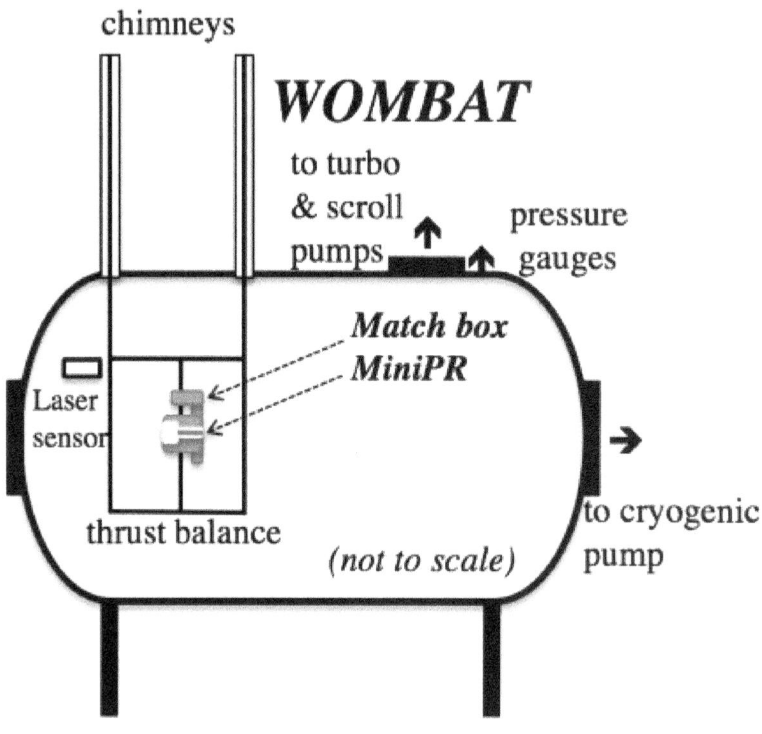

Figure 1: Schematic of the *Wombat* **vacuum chamber showing the** *MiniPR* electrothermal radiofrequency plasma thruster body with *Match box* **mounted on the thrust balance**.

Calibration of the thrust balance is carried out in air using precise weights attached to a wire hanging from a pulley (shown in Charles et al. [15]) and yields a typical force-displacement calibration factor of about 0.1 mN/μm. The procedure consisted in recording the balance displacement in μm in air using a motor to drive the pulley and sequentially apply an increasing gravitational force (generated by zero up to 10 weights of 2.1 g each) to yield the calibration factor. The balance would be returned to its resting position of zero displacement (no weight) prior to pump down. Following cold gas and plasma thrust measurements under vacuum, *Wombat* would be pressurized and the calibration process repeated in air to check the initial calibration factor. The discrepancy was <2.5%.

A cartoon of *MiniPR* is shown in Figure 2 and is fully described in Charles et al. [15, 16]. Briefly, it comprises an

18 mm long 1.5 mm inner diameter ceramic alumina tube plasma cavity having a central annular 5 mm wide rf copper electrode and two grounded 3 mm-wide aluminum electrodes (forming the grounded 8.5 cm diameter aluminum housing), each separated by two macor rings, one 3 mm wide (gas inlet side) and the other 4 mm wide (plasma exhaust side). A 40 mm-diameter 24 mm-long aluminum cavity which acts as gas plenum is terminated by a 26 mm-diameter, 12 mm-deep viewing window (not shown on Figure 2 for clarity).

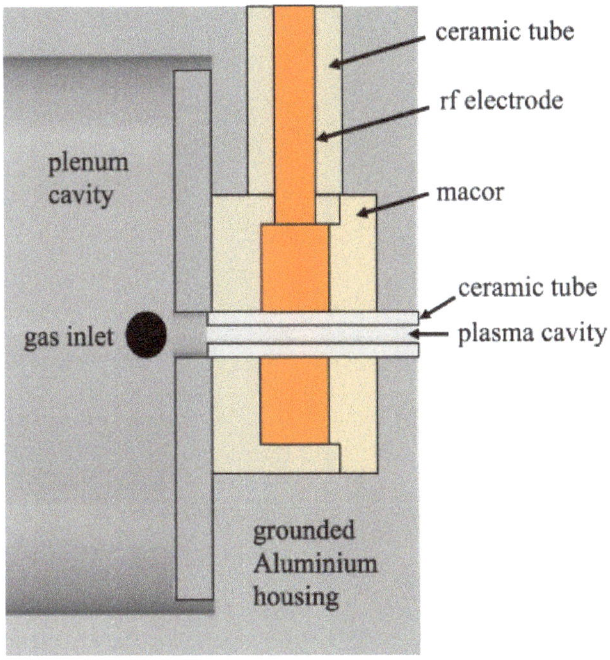

Figure 2: Cartoon showing *MiniPR* **main components**.

Xenon propellant gas is introduced via the plenum, flows along the alumina tube, and expands into vacuum. The complete gas line consists of a flow controller placed outside the vacuum system, a gas line feed-through mounted on *Wombat* and a 4 m long 1 mm inner diameter teflon tube for direct gas injection into the plenum cavity (Figure 2). To minimize vibration noise only the turbomolecular/scroll pumping system is used.

MiniPR's central rf annular electrode is attached to a variable frequency solid state impedance matching network [16] in a grounded aluminum box ("*Match box*" on Figure 1) via a 20 cm long semi-rigid *RG401* coaxial cable. A frequency adjustable MKS generator is used to send power into the match using a 1 m-long coaxial cable inside vacuum and a 4 m-long cable outside vacuum. With the present configuration there is significant power loss (30–50%) in the 12 turns 16 mm-diam 35 mm-long inductance of the solid state matching network [16]. The operating resonant frequency is initially determined for each rf power (typically around 13.8 MHz) by carrying out a number of short plasma shots of a few seconds in duration. This frequency is subsequently used to generate the plasma and obtain direct thrust measurement. Both the rf cable and gas lines inside vacuum are anchored at one position at the top of the thrust balance mounting to minimize mechanical resistance and maintain adequate sensitivity of the thrust balance. This challenging configuration is very distinct from previous thrust measurements using higher-power lower-pressure inductive rf thrusters (in which axial momentum is primarily imparted by accelerated ions rather than neutrals) which did not require mechanical contact between the gas injector and the plasma cavity or between the rf antenna and the plasma cavity [13,14].

MOMENTUM RESULTS AND DISCUSSION

Experimental Thrust Components

The two parameters of interest are the rf power and the xenon gas flow rate. Measurements of the axial momentum imparted by *MiniPR* are obtained as follow: the manual and automatic gas valves outside vacuum are both opened to inject a constant xenon flow which initially generates substantial oscillations of the balance before a "baseline" position is reached after about one minute. A series of 20 s long plasma shots are subsequently carried out to measure the thrust component imparted by the plasma for varying rf power and a constant flow of 52.5 sccm (5.16 mg.s^{-1}) or for varying xenon flow (the latter is obtained by a fine control of the flow controller which changes the baseline) and a constant rf power of 32 W. The measured thrust is the total thrust F_{total} and since the ionization degree in the discharge is low (no neutral depletion with the plasma "on") F_{total} can be written as

$$F_{total} = F_{plasma} + F_{cold\ gas} \tag{1}$$

where F_{plasma} is the axial force generated by the plasma and $F_{cold\ gas}$ is the axial force generated by the cold gas flow. After completion of the plasma "shots" the manual shut off gas valve is closed leaving a displacement which corresponds to the $F_{cold\ gas}$ or "no plasma" thrust component (fortunately, closing the gas valve generates less oscillations than opening the gas valve). For the present study the thrust gain is defined as

$$G = \frac{F_{plasma}}{F_{cold\ gas}} \tag{2}$$

Although the cold gas force would be reduced by the formation of the plasma in highly ionized plasmas (i.e., at high rf power), here the ionization rate is <1% [6] and it is assumed that gas heating does not significantly alters the chocked flow approximation discussed below.

Effect of Radiofrequency Power

F_{total} and $F_{cold\ gas}$ are shown on Figure 3 vs. increasing rf power and constant gas flow of 52.5 sccm (5.16 mg.s^{-1}), giving an operating plenum pressure of 7.7 Torr (1026 Pa). Since the flow is constant, $F_{cold\ gas}$ is constant and measured to be about 0.8 ± 0.25 mN. Four data sets (Series 1 to 4) were sequentially taken over the course of three consecutive days (without breaking vacuum) yielding an experimental error of ±0.25 mN. Series 1 on day 1 (red filled circles), series 2 on day 2 (open triangles), and series 4 on day 3 (green crosses) were carried out using continuous plasma excitation. Since a rf power larger than about 20 W was required for plasma ignition, series 3 (blue filled squares) was carried out on day 2 using pulsed plasma excitation with a period of 1 ms and a 50 percent duty cycle yielding time-averaged values of 12 and 16.7 W, respectively. Other points in series 3 were obtained using increasing duty cycle values of 20, 40, 60, 80, and 100% to yield increasing time-averaged power values of 10–50 W.

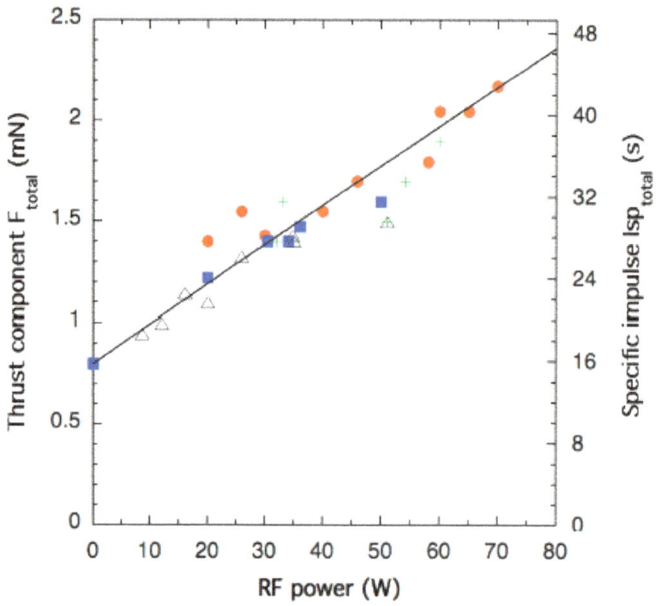

Figure 3: F_{total} thrust component and $I_{sptotal}$ specific impulse measured as a function of rf power for four data sets: series 1 (filled red circles), series 2 (open triangles), and series 4 (green crosses) respectively taken on days 1, 2, and 3 for continuous plasma excitation and series 3 (filled blue squares) taken on day 2 for pulsed plasma excitation. The xenon gas flow is constant at 52.5 sccm. The error bar on thrust components is ±0.25 mN. A solid black line is added as a visual guide.

Neglecting the neutral gas pressure term, the axial momentum generated by the cold gas can be approximated for an isentropic choked flow regime [17] by the momentum term

$$F_{cold\ gas} = \dot{m}c_s \tag{3}$$

where $\dot{m}$ is the xenon mass flow rate and $c_s = \sqrt{\frac{\gamma_{Xe}RT}{m_{Xe}}}$ is the xenon gas sound speed (T is the gas temperature, R = 287 J.kg^{-1}K^{-1} is the universal gas constant, γ_{Xe} is the specific heat capacity for xenon and m_{Xe} is the xenon molar mass). At T~300 K, c_s is about 170 m.s^{-1} which yields $F_{cold\ gas}$ of 0.87 mN very close to the experimental value of 0.8 ± 0.25 mN. As a result of the measurement procedure and the absence of a gas flow valve directly located at the plenum, it is expected that the measured value of $F_{cold\ gas}$ may be somewhat below the true cold gas thrust due to the residual gas trapped in the small diameter 4 m long teflon tubing gas line which requires hours to be effectively pumped out. This uncertainty may be combined with other effects related to the geometric details of the orifice/nozzle, its sonic surface, and the "free jet" expansion [18, 19].

The results on Figure 3 show that F_{total} and hence F_{plasma} increases linearly with rf power and the thrust gain G (defined in Equation 2) is about 1 for a rf power of 40 W. Any gain in thrust is attributed to an increase in the gas exhaust velocity due to neutral gas heating by the plasma. Figure 3 also shows the result of the "effective" or total specific impulse defined as

$$I_{sp\ total} = \frac{F_{total}}{\dot{m}g_0} = \frac{(G+1)F_{cold\ gas}}{\dot{m}g_0} \sim (G+1)I_{sp\ cold\ gas} \tag{4}$$

where g_0 is the standard gravitational acceleration constant (9.81 m.s^{-2}) and

$$I_{sp\ cold\ gas} = \frac{c_s}{g_0} \text{ and } I_{sp\ plasma} = \frac{Gc_s}{g_0} \tag{5}$$

are the "cold gas" specific impulse and "plasma" specific impulse, respectively ($I_{sptotal} = I_{spplasma} + I_{spcold gas}$). From Equations 5, $I_{spcold\ gas}$ is 15.8 s and $I_{spplasma}$ increases linearly with rf power (Figure 3). This linear variation of F_{plasma} and $I_{spplasma}$ with rf power is consistent with a similar variation of plasma density vs. power measured in a previous prototype in the vicinity of the rf electrode for constant argon gas flow [6].

Effect of Gas Flow

Figure 4 shows thrust and specific impulse measurements obtained for a constant rf power of 32 W and varying xenon gas flow rate from 16.4 to 52.5 sccm (1.61–5.16 mg.s^{-1}), giving a plenum pressure increasing from 3.2 Torr (426 Pa) to 7.7 Torr (1026 Pa). Visual inspection of the discharge through windows on *Wombat* shows that the most luminous part of the discharge is located in the plasma cavity only (not in the plenum) and that a noticeable "sphere-shaped" plasma plume extending a few mm out is only seen for flows exceeding 30 sccm. From Equation (5), $I_{spcold\ gas}$ is constant at 15.8 s in this case but $F_{cold\ gas}$ increases linearly with flow (Equation 3): the open circles on Figure 4 are direct measurements of $F_{cold\ gas}$ while the open triangles are calculated using Equation (3) (isentropic chocked flow approximation with 1 sccm of Xenon corresponding to 0.0983 mg.s^{-1} as detailed in Goebel and Katz [7], p. 464). For the lowest gas flow rate of 16.4 sccm (1.61 mg.s^{-1}), F_{total} is 0.75 mN, i.e., three times greater than the cold gas thrust value $F_{cold\ gas}$ of 0.25 mN. Hence for this condition the thrust gain for the plasma is G = 2 and from Equation (4), $I_{sptotal}$ is 48 s, i.e., three times that the $I_{spcold\ gas}$ value. With the plasma "on" at a fixed forward rf power, the effect of increasing the xenon gas flow leads to a decrease of $I_{sptotal}$ and an increase of F_{total} (Figure 4). At a flow rate of 41 sccm (4.03 mg.s^{-1}), F_{total} is 1.4 mN (G = 1) and remains constant thereafter.

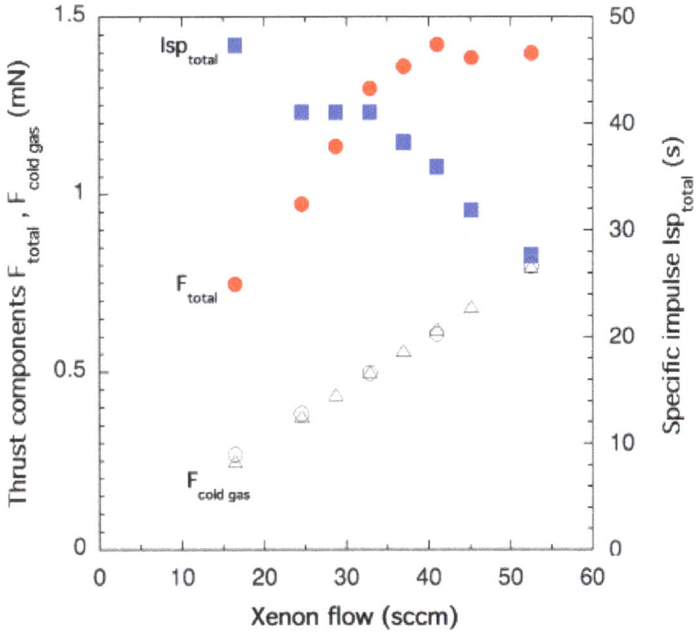

Figure 4: Total F_{total} (filled red circles) and cold gas $F_{cold\,gas}$ (open circles) thrust components and total specific impulse $I_{sptotal}$ (filled blue squares) measured vs. xenon gas flow rate for constant rf power of 32 W and a driving frequency of 13.78 MHz; cold gas specific impulse $I_{spcold\,gas}$ is constant at ~16 s; $F_{cold\,gas}$ (open triangles) is calculated using Equation (3). The error bar on thrust components is ± 0.25 mN.

In a typical DC arcjet [20] or cascaded arc [21], the electron density is very high (~10^{15}cm^{-3}) with the discharge near Local Thermal Equilibrium (neutral gas temperature approaching electron temperature) and the energy transfer is dominated by elastic electron-neutral collisions. In such high electron density mode typical of an arc or a filament there are often issues with stability of the arc and electrode erosion but interesting thruster characteristics have been reported for non-reactive gases such as helium [22]. Although *Mini P R*operating at higher powers and higher pressures may well exhibit a filamentary type of discharge, the currently investigated plasma coupling mode corresponds to an electron density in the 10^{12}cm^{-3} range for a pressure of a few Torr for which propellant heating mostly results from ion-neutral collisions (elastic and charge exchange) in the volume and neutral collisions with the plasma cavity wall (resisto-jet effect) which is bombarded by the ion flux especially in the rf electrode region where a self-bias is present. Using a total ion-neutral collisions cross-section of about $\sigma_{Xe^{+}-Xe} \sim 10^{-18}\mathrm{m}^2$ for xenon [23], the corresponding mean free path $\lambda_i = \frac{1}{n_g \sigma_{Xe^{+}-Xe}}$ at a mid cavity pressure of 3.85 Torr (513 Pa), half of the plenum pressure for 52.5 sccm gas flow rate, and a neutral gas density of $n_g \sim 1.36 \times 10^{23}\ \mathrm{m}^{-3}$ is 7.35 microns yielding 200 collisions across a plasma cavity diameter of 1.5 mm (about two thirds of which would be ion-neutral charge exchange collisions).

The observed decrease of the I_{sp} with increasing flow (Figure 4) is most likely due to the plasma power being shared among an increasing number of gas atoms and a consequent decrease in the upstream temperature and sound speed. The saturation of the thrust with increasing flow is not presently understood and is being investigated. It is likely that the increasing collisionality affects the radial and axial form of the electric fields that heat the ions which subsequently heat the neutrals (the low ionization degree gives rise to weak electric fields within the plasma plume). Flow rates above 40 sccm leads to higher rates for elastic and charge exchange collisions which could result in a poorer collimation of charged particles inside the choked flow and therefore in lower momentum along the axial direction. Higher mass flow rates of xenon also represents higher pressures in the plenum, where a secondary plasma could be triggered thereby distributing the applied rf power to the plasma plume, decreasing the charge production and leading to lower thrust.

Since Figure 3 corresponds to a flow of 52.5 sccm (maximum investigated flow in Figure 4), and $I_{sptotal}$ increases by a factor of two when the flow is decreased to 16.4 sccm (Figure 4), a linear extrapolation suggests that a specific impulse of about 100 s could be obtained for a power of 70 W. An increase of power efficiency by a factor of two to four can be expected by optimizing this type of plasma jet for space use, including investigation of the role of the plasma cavity geometry, the type of material use for the rf electrode and plasma cavity, and optimizing the rf match configuration. For a thrust gain G of about 2 (16.4 sccm on Figure 4), the corresponding gas velocity is about three times the sound speed (510 m.s^{-1}) which yields an estimated gas temperature of about 1600 K in good agreement with typical results recently obtained by ro-vibrational spectroscopy in argon in a similar geometry [15].

The results indicate that the choked flow approximation is appropriate. The sonic surface at the exit of the cavity can be determined by computer fluid simulation as shown by Charles et al. [15] and it is this surface averaged velocity which should determine the cold gas flow under choked flow approximation. Downstream of this sonic surface, the gas will expand and accelerate. The latter is still under investigation: there is an axial pressure gradient and the approximation looses its validity downstream of the plasma cavity.

The Hollow Cathode Thruster which has been flown by *Surrey Satellite Technology Ltd* on *TechDemoSat-1* has similar performance: tests with xenon performed on the ground indicated that the Hollow Cathode Thruster would provide 85 s of specific impulse, 1.5 mN of thrust. It required 50–60 W during heating and 40–60 W during discharge at nominal conditions [24].

CONCLUSIONS

In summary, direct measurements of plasma and cold gas axial thrust components generated by a xenon rf plasma jet have been performed. The cold gas thrust and specific impulse are in agreement with calculations made assuming an isentropic choked flow regime. The momentum gain from the plasma increases linearly with rf power and the I_{sp} increases with decreasing gas flow. At low flow rates the thrust and I_{sp} are consistent with an upstream gas temperature of 1600 K reported previously for a similar system operating outside vacuum [15]. The tests with varying rf power showed a possible advantage of *MiniPR*: when pulsed, the thrust was that expected for a continuous plasma. Consequently, it would be possible to use this thruster with a switch on time in the microsecond range, and combined with gas puffing would allow controlled thrusts down into the micro-Newton range.

AUTHOR CONTRIBUTIONS

All authors contributed as a team effort. The experiments were carried out at the Australian National University by the authors affiliated to that institution.

CONFLICT OF INTEREST STATEMENT

The authors declare that the research was conducted in the absence of any commercial or financial relationships that could be construed as a potential conflict of interest.

ACKNOWLEDGMENTS

This research was partially funded by the Australian Space Research Program [*Wombat* upgrade as part of the *Australian Plasma Thruster* (APT) project] of the Department of Industry, Innovation, Science, Research and Tertiary Education (Australian Government), the Australian Research Council Discovery Project (*Pocket Rocket* thruster research as part of the DP140100571 project), and Lockheed Martin US (*Pocket Rocket* thruster Research and Development contract).

REFERENCES

1. 1. Aanesland A, Liard L, Leray G, Jolly J, Chabert P. Direct measurements of neutral density depletion by two-photon absorption laser- induced fluorescence spectroscopy. *Appl Phys Lett.*(2007) **91**:121502. doi: 10.1063/1.2786601
2. 2. Greig A, Charles C, Hawkins R, Boswell R. Direct measurement of neutral gas heating in a radio-frequency electrothermal plasma micro-thruster. *Appl Phys Lett.* (2013) **103**:074101. doi: 10.1063/1.4818657
3. 3. Fruchtman A. Energizing and depletion of neutrals by a collisional plasma. *Plasma Sources Sci Technol.* (2008) **17**:024016. doi: 10.1088/0963-0252/17/2/024016
4. 4. Fruchtman A, Makrinich G, Chabert P, Rax JM. Enhanced plasma transport due to neutral depletion. *Phys Rev Lett.* (2005) **95**:115002. doi: 10.1103/PhysRevLett.95.115002
5. 5. Liard L, Raimbault J-L, Rax J-M, Chabert P. Plasma transport under neutral gas depletion conditions. *Phys Plasmas* (2007) **40**:5192. doi: 10.1088/0022-3727/40/17/026
6. 6. Charles C, Boswell R. Measurement and modeling of a radiofrequency micro-thruster. *Plasma Sources Sci Technol.* (2012) **21**:022002. doi: 10.1088/0963-0252/21/2/022002
7. 7. Goebel DM, Katz I. *Fundamentals of Electric Propulsion.* Hoboken, NJ: Wiley (2008).
8. 8. Charles C. Plasmas for spacecraft propulsion. *J Phys D Appl Phys.* (2009) **42**:163001. doi: 10.1088/0022-3727/42/16/163001
9. 9. Boswell RW. *Plasma Micro-Thruster.* International Patent WO 2012151639 A1. Madison, CT: IFI CLAIMS Patent Services.
10. 10. Greig A, Charles C, Paulin N, Boswell R. Volume and surface propellant heating in an electrothermal radio-frequency plasma micro-thruster. *Appl Phys Lett.* (2014) **105**:054102. doi: 10.1063/1.4892656
11. 11. Charles C, Hawkins R, Boswell R. Particle in cell simulation of a radiofrequency plasma jet expanding in vacuum. *Appl Phys Lett.* (2015) **106**:093502. doi: 10.1063/1.4914109
12. 12. Charles C, Boswell RW, Bish A. Variable frequency matching to a radiofrequency source immersed in vacuum. *J Phys D Appl Phys.* (2013) **46**:365203. doi: 10.1088/0022-3727/46/36/365203
13. 13. Lafleur T, Takahashi K, Charles C, Boswell RW. Direct thrust measurements and modelling of a radio-frequency expanding plasma thruster. *Appl Phys Lett.* (2011) **18**:080701. doi: 10.1063/1.3610570
14. 14. Takahashi K, Lafleur T, Charles C, Alexander P, Boswell RW. Electron diamagnetic effect on axial force in an expanding plasma: experiments and theory. *Phys Rev Lett.* (2011) **107**:235001. doi: 10.1103/PhysRevLett.107.235001
15. 15. Charles C, Bish A, Boswell RW, Dedrick J, Greig A, Hawkins R, et al. A short review of experimental and computational diagnostics for radiofrequency plasma micro-thrusters. *Plasma Chem Plasma Process.* (2016) **36**:29–44. doi: 10.1007/s11090-015-9654-5
16. 16. Charles C, Boswell RW, Bish A. Low-weight fixed ceramic capacitor impedance matching system for an electrothermal plasma microthruster. *J Propul Power* (2014) **30**:1117. doi: 10.2514/1.B35119
17. 17. Wilcox DC. *Basic Fluid Mechanics, 3rd Edn.* San Diego, CA: DCW Industries (2007).
18. 18. Murphy HR, Miller DR. Effects of nozzle geometry on kinetics in free-jet expansion. *J Phys Chem.* (1984) **88**:4474.
19. 19. Arakoni RA, Ewing JJ, Kushner MJ. Microdischarges for use as microthrusters: modelling and scaling. *J Phys D Appl Phys.* (2008) **41**:105208. doi: 10.1088/0022-3727/41/10/105208

20. 20. Zube DM, Myers RM. Thermal nonequilibrium in a low-power arcjet nozzle. *J Propul Power*(1993) **9**:545.

21. 21. Van de Sanden MCM, de Regt JM, Schram DC. Recombination of argon in an expanding plasma jet. *Phys Rev E* (1993) **47**:2792.

22. 22. Willmes GF, Burton RL. Low-Power helium pulsed arcjet. *J Propul Power* (1999) **15**:440.

23. 23. Piscitelli D, Phelps AV, de Urquijo J, Basurto E, Pitchford LC. Ion mobilities in Xe/Ne and other rare-gas mixtures. *Phys Rev E* (2003) **68**:046408. doi: 10.1103/PhysRevE.68.046408

24. 24. Lamprou D, Lappas VJ, Shimizu T, Gibbon D, Perren M. Hollow cathode thruster design and development for small satellites. In: *IEPC-2011-151 Proceedings, The 32nd International Electrical Propulsion Conference*. Wiesbaden (2011).

CHAPTER 10

INCONCLUSIVE QUANTUM MEASUREMENTS AND DECISIONS UNDER UNCERTAINTY

Vyacheslav I. Yukalov [1, 2] and Didier Sornette [1, 3]

[1]Department of Management, Technology and Economics, ETH Zürich, Zürich, Switzerland

[2]Bogolubov Laboratory of Theoretical Physics, Joint Institute for Nuclear Research, Dubna, Russia

[3]Swiss Finance Institute, University of Geneva, Geneva, Switzerland

INTRODUCTION

The standard theory of quantum measurements [1] is based on the projection operator measure corresponding to operationally testable events. Simple measurements really have to be operationally testable in order to possess physical meaning. However, if a measurement is composite, consisting of several parts, the intermediate stages do not have to necessarily be operationally testable, but can be inconclusive.

As a typical example, we can recall the known double-slit experiment, when particles pass through a screen with two slits and then are registered by particle detectors some distance away from the screen. This experiment can be treated as a composite event consisting of two parts, one being the passage through one of the slits and second, registration by detectors. The registration of a particle by a detector is an operationally testable event, since the particle is either detected or not, with the result being evident for the observer. But the passage of the particle through one of the slits is not directly observed, and the experimentalist does not know which of the slits the particle has passed through. In that sense, the passage of the particle through a slit is an inconclusive event. The existence of this inconclusive event, occurring at the intermediate stage of the experiment, is intimately associated with an interference effect. Otherwise, if the experimentalist would precisely determine the slit through which the particle has passed, the interference pattern registered by the particle detectors would be destroyed. The existence of interference is precisely due to the presence of the inconclusive event that happened at the intermediate stage.

The occurrence of inconclusive events in decision making is even more frequent and important. Practically any decision, before it is explicitly formulated, passes through a stage of deliberation and hesitation accompanying the choice. That is, any decision is actually a composite event consisting of an intermediate stage of deliberation and of the final stage of taking a decision. The final stage of decision making is equivalent to an operationally testable event in quantum measurements. While the intermediate stage of deliberation is analogous to an inconclusive event.

The analogy between the theory of quantum measurements and decision theory has been mentioned by von Neumann [1]. Following this analogy, Quantum Decision Theory (QDT) has been advanced [2–7], with the mathematical structure that is applicable to both decision making as well as to quantum measurements. The generality of our framework, being equally suitable for quantum measurements and decision making, is its principal difference from all other attempts that employ

Citation: Yukalov VI and Sornette D (2016) Inconclusive Quantum Measurements and Decisions under Uncertainty. Front. Phys. 4:12. doi: 10.3389/fphy.2016.00012

quantum techniques in psychological sciences. An extensive literature on various quantum models in psychology and cognitive science can be found in the books [8–11] and review articles [12–15].

Any approach, applying quantum techniques to decision theory, is naturally based on the notion of probability. This is because quantum theory is intrinsically probabilistic. Respectively, the intrinsically probabilistic nature of QDT is what makes it principally different from stochastic decision theory, where the choice is treated as always being deterministic, while in the process of choosing the decision maker acts with errors [16–20]. Such stochastic decision theories can be termed as "deterministic theories embedded into an environment with stochastic noise." The standard way of using a stochastic approach is to assume a probability distribution over the values characterizing the errors made by the subjects in the process of decision making. Then the parameters entering the distribution are fitted to a posteriori empirical data by maximizing the log-likelihood function. Such a procedure allows one to better fit the given set of data to the assumed basic deterministic decision theory, in particular due to the introduction of additional fitting parameters. However, it does not essentially change the structure of the underlying deterministic theory, although improving it slightly. And, being descriptive, the classical stochastic approach does not provide quantitative predictions.

Contrary to classical stochastic theory, in the quantum approach, we do not assume that the choice of a decision maker is deterministic, with just some weak disturbance by errors. Following the general quantum interpretation, we consider the choice process, including deliberations, hesitations, and subconscious estimation of competing outcomes, as intrinsically random. The probabilistic decision, in the quantum case, is not just a stochastic decoration of a deterministic process, but it is an unavoidable random part of any choice. The existence of the hidden, often irrational subconscious feelings and deliberations, results in the appearance of quantum interference and entanglement. The difference between classical stochastic decision theory and QDT is similar to the difference between classical statistical physics and quantum theory. In the former, all processes are assumed to be deterministic, with statistics coming into play because of measurement uncertainties, such as no precise knowledge of initial conditions and the impossibility of measuring exactly the locations and velocities of all particles. In contrast, quantum theory is principally probabilistic, which becomes especially important for composite measurements.

A detailed mathematical theory of quantum measurements in the case of composite events has been developed in our previous papers [21–23]. In the present paper, we concentrate our attention on composite measurements including intermediate inconclusive events and on the application of this notion for characterizing decision making under risk and uncertainty. The importance of composite events, including intermediate inconclusive events, in decision theory makes it necessary to pay a special attention to the correct mathematical formulation of such events and to the description of their properties allowing for the quantitative evaluation of the corresponding quantum probabilities. We show that, despite uncertainty accompanying inconclusive events, it is possible to give quantitative evaluations for quantum probabilities in decision theory, based on non-informative priors. Considering, as an illustration, the decoy effect, we demonstrate that even the simple non-informative priors provide predictions in very good agreement with experimental data.

COMPOSITE QUANTUM MEASUREMENTS AND EVENTS

In this section, we give a brief summary of the general scheme for defining quantum probabilities for composite events. As we have stressed above, in our approach, the mathematics is the same for describing either quantum measurements or decision making. Therefore, when referring to an event, we can keep in mind either a fact of measurement or a decision action.

Let A_n be a conclusive operationally testable event labeled by an index n. And let $B = \{B_\alpha\}$ be a set of inconclusive events labeled by α. Defining the space of events as a Hilbert space HH, we associate with an event A_n a state $|n\rangle$ in this Hilbert space and an event operator $\hat{P}_n$,

$$A_n \to |n\rangle \to \hat{P}_n = |n\rangle\langle n| \, . \tag{1}$$

The event operator for an operationally testable event is a projector.

The set of inconclusive events B generates in the Hilbert space HH the state $|B\rangle$ and the event operator $\hat{P}_n$,

$$B \to |B\rangle \to \hat{P}_B = |B\rangle\langle B| \, , \tag{2}$$

where the state reads

$$|B\rangle = \sum_\alpha b_\alpha |\alpha\rangle \, , \tag{3}$$

with coefficients b_α being random complex numbers. The event operator for an inconclusive event is not necessarily a projector, but a member of a positive operator-valued measure [7, 21–23].

The space of events, in the quantum approach, is the Hilbert space

$$\mathcal{H} = \mathcal{H}_A \bigotimes \mathcal{H}_B \tag{4}$$

that is a tensor product of the spaces

$$\mathcal{H}_A = \text{span}\{|n\rangle\} \, , \qquad \mathcal{H}_B = \text{span}\{|\alpha\rangle\} \, .$$

Each decision maker is characterized by an operator $\hat{\rho}$ that can be termed the strategic state of a decision maker, which, in quantum theory, corresponds to a statistical operator. The pair $\{\mathcal{H}, \hat{\rho}\}$, in physics, is named a statistical ensemble, and in decision theory, it is a decision ensemble.

A composite event is called a prospect and is denoted as

$$\pi_n = A_n \bigotimes B \, . \tag{5}$$

A prospect π_n generates a state $|\pi_n\rangle$ in the Hilbert space of events H and a prospect operator P^(πn)P^(πn),

$$\pi_n \to |\pi_n\rangle \to \hat{P}(\pi_n) = |\pi_n\rangle\langle \pi_n| \, , \tag{6}$$

with the prospect state

$$|\pi_n\rangle = |n\rangle \bigotimes |B\rangle = \sum_\alpha b_\alpha |n\alpha\rangle \, . \tag{7}$$

The prospect operator is a member of a positive operator-valued measure, which implies that these operators satisfy the resolution of unity [21, 23]. Since they contain random quantities b_α, the corresponding random resolution has to be understood not as a direct equality between numbers, but, e.g., as the equality in mean [24].

The prospect probability is

$$p(\pi_n) = \mathrm{Tr}\, \hat{\rho}\hat{P}(\pi_n) \,, \tag{8}$$

with the trace over the space HH. To form a probability measure, the prospect probabilities are to be normalized:

$$\sum_n p(\pi_n) = 1 \,, \qquad 0 \leq p(\pi_n) \leq 1 \,. \tag{9}$$

Taking explicitly the trace in expression (Equation 8) and separating diagonal and off-diagonal terms, we see that the prospect probability

$$p(\pi_n) = f(\pi_n) + q(\pi_n) \tag{10}$$

is represented as a sum of a positive-definite term

$$f(\pi_n) = \sum_\alpha |b_\alpha|^2 \langle n\alpha|\hat{\rho}|n\alpha\rangle \tag{11}$$

and a sign-undefined term

$$q(\pi_n) = \sum_{\alpha\neq\beta} b_\alpha^* b_\alpha \langle n\alpha|\hat{\rho}|n\beta\rangle \,. \tag{12}$$

The appearance of the sign-undefined term is a purely quantum effect responsible, in quantum measurements, for interference patterns. The attenuation of this quantum term is called decoherence. In quantum theory, decoherence can be due to external as well as to internal perturbations and the influence of measurements [25–27]. And in QDT, decoherence can occur due to the accumulation of information [28].

The disappearance of the quantum term implies the transition to classical theory. This is formulated as the *quantum-classical correspondence principle*[29], which in our case reads as

$$p(\pi_n) \to f(\pi_n) \,, \qquad q(\pi_n) \to 0 \,. \tag{13}$$

This principle tells us that the term $f(\pi_n)$ plays the role of classical probability, hence is to be normalized:

$$\sum_n f(\pi_n) = 1 \,, \qquad 0 \leq f(\pi_n) \leq 1 \,. \tag{14}$$

When decisions concern a choice between lotteries, the classical term $f(\pi_n)$ has to be defined according to classical decision theory based either on expected utility theory or on some non-expected value functional. This suggests to call $f(\pi_n)$ a *utility factor*, since it is defined on rational grounds and reflects the utility of a choice. The quantum term is caused by the interference and entanglement effects in quantum theory, which correspond, in decision making, to irrational effects describing the attractiveness of choice. Therefore, it can be called the attraction factor. From Equations (9) and (14), it follows the *alternation law*

$$\sum_n q(\pi_n) = 0 \,, \qquad -1 \leq q(\pi_n) \leq 1 \,. \tag{15}$$

Note that, in quantum theory, the definition of the composite event in the form of prospect (Equation 5) is valid for any type of operators, since they are defined on different spaces. No problem with non-commutativity of operators defined on a common Hilbert space arises. This way makes it possible to introduce joint quantum probabilities for several measurements [21, 23]. Contrary to this, considering operators on the same Hilbert space does not allow one to define joint probabilities for non-commuting operators. Sometimes, one treats the Lüders probability of consecutive measurements as a conditional probability. This, however, is not justified from the theoretical point of view [21, 23, 30] and also contradicts experimental data [31, 32]. But defining the quantum joint probability according to expression (Equation 8) contains no contradictions.

In the present section, the general scheme of QDT is presented. Being limited by the length of this paper, we cannot go into all mathematical details that have been thoroughly described in our previous publications. However, we would like to stress that for the purpose of practical applications, it is not necessary to study all these mathematical peculiarities, but it is sufficient to employ the final formulas following the prescribed rules. One can use the formulated rules as given prescriptions, without studying their justification. The main formulas of this section, which are necessary for the following application, are: the form of the quantum probability (Equation 10), the normalization conditions (Equations 9 and 14), and the alternation law (Equation 15). More details required for practical application will be given in the sections below.

NON-INFORMATIVE PRIOR FOR UTILITY FACTORS

To make the above scheme applicable to decision theory, it is necessary to specify how one should quantify the values of the utility factors and attraction factors. Here we show how these values can be defined as non-informative priors.

Let us consider a set of lotteries $L_n = \{x_i, p_n(x_i) : i = 1, 2, \ldots, N_n\}$, enumerated by the index $n = 1, 2, \ldots, N_L$, with payoffs x_i and their probabilities $p_n(x_i)$. The related expected utilities $U(L_n) = \sum_i u(x_i)p_n(x_i)$ can be defined according to the expected utility theory [33]. For the present consideration, the utility functions $u(x)$ do not need to be specified. For instance, they can be taken as linear functions, since this choice has the advantage of making the utility factors independent from the units measuring the payoffs.

In QDT, the act of choosing a lottery L_n, denoted as A_n, together with the accompanying set of inconclusive events B, including the decision-maker hesitations [6, 30], compose the prospect (Equation 5). Depending on whether the expected utilities are positive on negative, there can be two cases.

If the expected utilities of the considered set of lotteries are all positive (non-negative), such that

$$U(L_n) \geq 0 \qquad (n = 1, 2, \ldots, N_L) \,, \tag{16}$$

then it is reasonable to require that zero utility corresponds to zero utility factor:

$$f(\pi_n) \to 0\,, \qquad U(L_n) \to 0\,. \tag{17}$$

The case where the utility factor is simply proportional to the related expected utility trivially obeys this condition (Equation 17). Taking into account the normalization condition (Equation 14) gives the utility factor

$$f(\pi_n) = \frac{U(L_n)}{\sum_n U(L_n)}\,. \tag{18}$$

When the expected utilities are negative, which happens in the domain of losses, such that

$$U(L_n) < 0 \qquad (n = 1, 2, \ldots, N_L)\,, \tag{19}$$

the required condition is that an infinite loss corresponds to zero utility factor:

$$f(\pi_n) \to 0\,, \qquad |U(L_n)| \to \infty\,. \tag{20}$$

The simplest way to satisfy this condition (Equation 20) is that the utility factor is inversely proportional to the related expected utility. Taking into account the normalization condition, we get

$$f(\pi_n) = \frac{|U(L_n)|^{-1}}{\sum_n |U(L_n)|^{-1}}\,. \tag{21}$$

The utility-factor forms (Equations 18 and 21) coincide with the choice probabilities in the Luce choice axiom [34]. It is possible to show that generalized forms for the utility factors can be derived by maximizing a conditional Shannon entropy or from the principle of minimal information [12, 35, 36].

In the case of positive expected utilities, we consider the information functional, taking into account the normalization condition (Equation 14) and the expected log-likelihood Λ. This functional reads as

$$I[\, f(\pi_n)\,] = \sum_n f(\pi_n) \ln f(\pi_n) + \lambda \left[\sum_n f(\pi_n) - 1 \right] + \alpha \left[\sum_n f(\pi_n)\Lambda(\pi_n) - \Lambda \right]\,, \tag{22}$$

where

$$\Lambda(\pi_n) = -\ln U(L_n)\,, \qquad U(L_n) \geq 0\,.$$

Minimizing functional (Equation 22) results in the utility factor

$$f(\pi_n) = \frac{U^\alpha(L_n)}{\sum_n U^\alpha(L_n)} \qquad (\alpha > 0)\,, \tag{23}$$

in which the positive sign of α is prescribed by the condition that the larger utility implies the larger factor.

In the case of negative expected utilities, the information functional takes the form

$$I[\, f(\pi_n)\,] = \sum_n f(\pi_n) \ln f(\pi_n) + \lambda \left[\sum_n f(\pi_n) - 1 \right] + \gamma \left[\Lambda - \sum_n f(\pi_n)\Lambda(\pi_n) \right]\,, \tag{24}$$

where

$$\Lambda(\pi_n) = -\ln |U(L_n)|\,, \qquad U(L_n) < 0\,.$$

Then its minimization yields the utility factor

$$f(\pi_n) = \frac{|U(L_n)|^{-\gamma}}{\sum_n |U(L_n)|^{-\gamma}} \qquad (\gamma > 0)\,, \tag{25}$$

with the positive sign of γ prescribed by the requirement that the larger cost implies the smaller factor.

The utility factors (Equations 23 and 25) are the examples of power-law distributions that are known in many applications [35–37].

NON-INFORMATIVE PRIOR FOR ATTRACTION FACTORS

Although the attraction factor characterizes irrational features of decision making, it can be estimated by invoking non-informative prior assumptions. An important consequence of the latter is the *quarter law* derived earlier [4,5, 12]. Here we first give the new, probably the simplest, derivation of the quarter law and, second, we show how this law can be used for estimating the attraction factors in the case of an arbitrary number of prospects.

Let us consider the sum

$$\frac{1}{N_L} \sum_{n=1}^{N_L} |q(\pi_n)| = \int_0^1 \varphi(x) x \, dx \tag{26}$$

of the attraction factor moduli, where

$$\varphi(x) \equiv \frac{1}{N_L} \sum_{n=1}^{N_L} [\, \delta(x - q(\pi_n)) + \delta(x + q(\pi_n))\,] \tag{27}$$

plays the role of the attraction-factor distribution. The latter is normalized as

$$\int_{-1}^1 \varphi(x) \, dx = 1\,, \tag{28}$$

since the attraction factors, in view of condition (Equation 15), vary in the interval [−1, 1]. If $q(\pi_n)$ does not equal zero, then normalization (Equation 28) is evident. And when $q(\pi_n) = 0$, then one should use the identity

$$\int_0^1 \delta(x) \, dx = \frac{1}{2}$$

for the semi-integral of the Dirac function.

The use of a non-informative prior implies that the values of the attraction factor are not known. A full ignorance is captured by a uniform distribution, which, according to normalization (Equation 28), gives

$$\varphi(x) = \frac{1}{2}\,. \tag{29}$$

In that case, integral (Equation 26) results in the *quarter law*

$$\frac{1}{N_L} \sum_{n=1}^{N_L} |q(\pi_n)| = \frac{1}{4} . \quad (30)$$

If the prospect lattice LL = $\{\pi_n\}$ consists of N_L prospects, we can always arrange the corresponding attraction factors in the ascending order, such that

$$q(\pi_n) > q(\pi_{n+1}) \qquad (n = 1, 2, \ldots, N_L - 1) . \quad (31)$$

We denote the largest attraction factor as

$$q_{max} \equiv q(\pi_1) > 0 . \quad (32)$$

Given the unknown values of the attraction factors, the non-informative prior assumes that they are uniformly distributed and at the same time they must obey the ordering constraint (Equation 31). Then, the joint cumulative distribution of the attraction factors is given by

$$\Pr[q(\pi_1) < \eta_1, q(\pi_2) < \eta_2, \ldots, q(\pi_{N_L}) < \eta_{N_L} | \eta_1 \le \eta_2 \le \ldots \le \eta_{N_L}] = \int_0^{\eta_1} dx_1 \int_{x_1}^{\eta_2} dx_2 \ldots \int_{x_{N_L-1}}^{\eta_{N_L}} dx_{N_L} , \quad (33)$$

where the series $\eta_1 \le \eta_2 \le \ldots \le \eta_{NL}$ of inequalities ensure the ordering. It is then straightforward to show that the average values of the $q(\pi_n)$ are equidistant, i.e., the difference between any two neighboring factors is on average

$$\Delta \equiv \langle q(\pi_n)\rangle - \langle q(\pi_{n+1})\rangle = const \qquad (\text{independent of } n) . \quad (34)$$

Taking their average values as determining their typical values, we omit the symbol $\langle . \rangle$ representing the average operator and use Equation (34) to represent the n-th attraction factor as

$$q(\pi_n) = q_{max} - (n-1)\Delta . \quad (35)$$

With notations (Equations 32 and 34), the alternation condition (Equation 15) yields

$$q_{max} = \frac{N_L - 1}{2} \Delta . \quad (36)$$

And the quarter law (Equation 30) leads to the gap

$$\Delta = \frac{N_L}{2\sum_n |N_L + 1 - 2n|} . \quad (37)$$

If N_L is even, then

$$\sum_{n=1}^{N_L} |N_L + 1 - 2n| = \frac{N_L^2}{2} \qquad (N_L \; even) ,$$

while when N_L is odd, then

$$\sum_{n=1}^{N_L} |N_L + 1 - 2n| = \frac{N_L^2 - 1}{2} \qquad (N_L \; odd) .$$

This allows us to represent gap (Equation 37) as

$$\Delta = \begin{cases} \frac{1}{N_L} & (N_L \; even) \\ \frac{N_L}{N_L^2 - 1} & (N_L \; odd) . \end{cases} \quad (38)$$

And for the largest attraction factor, we find

$$q_{max} = \begin{cases} \frac{N_L - 1}{2N_L} & (N_L \; even) \\ \frac{N_L}{2(N_L + 1)} & (N_L \; odd) . \end{cases} \quad (39)$$

The above expressions make it possible to evaluate, on the basis of the non-informative prior, the whole set

$$Q_{N_L} \equiv \{q(\pi_n) : \; n = 1, 2, \ldots, N_L\} \quad (40)$$

of the attraction factors:

$$q(\pi_n) = \begin{cases} \frac{1}{2N_L}(N_L - 2n + 1) & (N_L \; even) \\ \frac{N_L}{2(N_L^2 - 1)}(N_L - 2n + 1) & (N_L \; odd) . \end{cases} \quad (41)$$

For example, in the case of two prospects, we have

$$\Delta = \frac{1}{2} , \qquad q_{max} = \frac{1}{4} \qquad (N_L = 2) ,$$

which yields the attraction set

$$Q_2 = \left\{ \frac{1}{4} , \; -\frac{1}{4} \right\} .$$

For three prospects, we get

$$\Delta = \frac{3}{8} , \qquad q_{max} = \frac{3}{8} \qquad (N_L = 3) ,$$

hence

$$Q_3 = \left\{ \frac{3}{8} , \; 0 , \; -\frac{3}{8} \right\} .$$

Similarly, for four prospects, we find

$$\Delta = \frac{1}{4} , \qquad q_{max} = \frac{3}{8} \qquad (N_L = 4) ,$$

with the attraction set

$$Q_4 = \left\{ \frac{3}{8} , \; \frac{1}{8} , \; -\frac{1}{8} , \; -\frac{3}{8} \right\} .$$

When there are five prospects, then

$$\Delta = \frac{5}{24} , \qquad q_{max} = \frac{5}{12} \qquad (N_L = 5) ,$$

from where

$$Q_5 = \left\{ \frac{5}{12} , \; \frac{5}{24} , \; 0 , \; -\frac{5}{24} , \; -\frac{5}{12} \right\} .$$

Thus, we can evaluate the attraction factors for any number of prospects, obtaining a kind of a quantized attraction set. In the case of an asymptotically large number N_L of prospects, we have

$$\Delta \simeq \frac{1}{N_L} , \qquad q_{max} \simeq \frac{1}{2} \qquad (N_L \gg 1) , \tag{42}$$

and

$$q(\pi_n) \simeq \frac{1}{2} - \frac{2n-1}{2N_L} . \tag{43}$$

The non-informative priors can be employed for predicting the results of decision making. This makes the principal difference compared with the introduction into expected utility of adjustment parameters that are fitted*post-hoc* to the given experimental data [38].

QUANTITATIVE EXPLANATION OF DECOY EFFECT

We now show how the non-informative priors of the attraction factors can be employed to explain the decoy effect and for quantitative prediction in decision-making. Throughout this section, we denote, for simplicity, the objects of choice, say *A*, as well as the act of choosing an object *A*, by the same letter *A*. As has been emphasized above, the act of choice under uncertainty is a composite prospect. But, again for simplicity, we employ the same letter for denoting the action *A* and the related prospect (Equation 5).

The decoy effect was first studied by Huber et al. [39], who called it the effect of asymmetrically dominated alternatives. Later this effect has been confirmed in a number of experimental investigations [40–43]. The meaning of the decoy effect can be illustrated by the following example. Suppose a buyer is choosing between two objects, *A* and *B*. The object *A* is of better quality, but of higher price, while the object *B* is of slightly lower quality, while less expensive. As far as the functional properties of both objects are not drastically different, but *B* is cheaper, the majority of buyers value the object *B* higher. At this moment, the salesperson mentions that there is a third object *C*, which is of about the same quality as *A*, but of a much higher price than *A*. This causes the buyer to reconsider the choice between the objects *A* and *B*, while the object *C*, having the same quality as *A* but being much more expensive, is of no interest. Choosing now between *A* and *B*, the majority of buyers prefer the higher quality but more expensive object *A*. The object *C*, being not a choice alternative, plays the role of a decoy. Experimental studies confirmed the decoy effect for a variety of objects: cars, microwave ovens, shoes, computers, bicycles, beer, apartments, mouthwash, etc. The same decoy effect also exists in the choice of human mates distinguished by attractiveness and sense of humor [44]. It is common as well for animals, for instance, in the choice of female frogs of males with different attraction calls characterized either by low-frequency and longer duration or by faster call rates [45].

The decoy effect contradicts the regularity axiom in decision making telling that if *B* is preferred to *A* in the absence of *C*, then this preference has to remain in the presence of *C*.

In the frame of QDT, the decoy effect is explained as follows. Assume buyers consider an object *A*, which is of higher quality but more expensive, and an object *B*, which is of moderate quality but cheaper. Suppose the buyers have evaluated these objects *A* and *B*, which implies that the initial values of the objects are described by the utility factors *f*(*A*) and *f*(*B*). In experiments, the latter correspond to the fractions of buyers evaluating higher the related object. When the decoy *C*, of high quality but essentially more expensive, is presented, it attracts the attention of buyers to the quality characteristic. The role of the decoy is well understood as attracting the attention of buyers to a particular feature of the considered objects, because of which the decoy effect is sometimes named the attraction effect [40]. In the present case, the decoy attracts the buyer attention to quality. The attraction, induced by the decoy, is described by the attraction factors *q*(*A*) and *q*(*B*). Hence the probabilities of the related choices are now

$$p(A) = f(A) + q(A) , \qquad p(B) = f(B) + q(B) .$$

Since the quality feature becomes more attractive, $q(A) > q(B)$. According to the non-informative prior, we can estimate the attraction factors as $q(A) = 1/4$ and $q(B) = -1/4$.

To be more precise, let us take numerical values from the experiment of Ariely and Wallsten [43], where the objects under sale are microwave ovens. The evaluation without a decoy results in $f(A) = 0.4$ and $f(B) = 0.6$. In the presence of the decoy, we predict that the choice probabilities can be evaluated as

$$p(A) = f(A) + 0.25 , \qquad p(B) = f(B) - 0.25 .$$

This gives $p(A) = 0.65$ and $p(B) = 0.35$. The experimental values for the choice between *A* and *B*, in the presence of but excluding *C*, correspond to the fractions $p_{exp}(A) = 0.61$ and $p_{exp}(B) = 0.39$, which is close to the predicted probabilities.

Another example can be taken from the studies of the frog mate choice [45], where frog males have attraction calls differing in either low-frequency sound or call rate. The males with lower frequency calls are denoted as *A*, while those with high call rate, as *B*. In an experiment with 80 frog females, without a decoy, it was found that females evaluate higher the fastest call rate, so that$f(A) = 0.35$ and $f(B) = 0.65$. In the presence of an inferior decoy, attracting attention to the low-frequency characteristic, the non-informative prior predicts the probabilities

$$p(A) = 0.35 + 0.25 = 0.6 , \qquad p(B) = 0.65 - 0.25 = 0.4 .$$

The empirically observed fractions are found to be $p_{exp}(A) = 0.6$ and $p_{exp}(B) = 0.4$, in remarquable agreement with our predictions.

To make it clear how the decoy effect fits the title of the paper "Inconclusive quantum measurements and decisions under uncertainty," it is worth extending the comments that have been mentioned in the Introduction.

Our principal point of view is that decision making, generally, almost always deals with composite events, since any choice is accompanied by subconscious feelings and irrational biases. The latter are often difficult to formalize and, even more, their weights usually are not known and are practically unmeasurable. This is why these subconscious irrational factors can be treated as what is called inconclusive events. When choosing between several possibilities, say A_n, one actually considers composite prospects, as defined in Equation (5). And the composite nature of choices requires the use of quantum techniques, as has been explained in our previous paper [30]. Otherwise, the probabilities of simple events could be characterized by classical theory. It is the composite nature of the considered prospects that yields the appearance of the quantum term $q(\pi_n)$ related to interference and coherence effects. In that way, the choice between the objects in the decoy effect is also a composite prospect, being composed of the choice as such and accompanying subconscious feelings forming an inconclusive set. This is why the use of QDT here is necessary and why it gives so good results.

It is admissible to give a schematic picture of the choice in the decoy effect by analogy with the double-slit experiment in physics, which is mentioned in the Introduction. Thus, making a concrete selection of either an object A or B is the analog of the registration of the particle by a detector. But before such a selection is done, there exists the uncertainty of deciding which of the object features are actually more important. These not precisely defined acts of hesitation play the role of the slits, with the uncertainty associated with which of them the particle has passed through. When it is known which of the slits the particle has passed through, then the interference effects in physics disappear. Similarly, in decision theory, if the values of each object are clearly defined, there are no hesitations, no interference, and the selection can be based on classical rules. Such an objective evaluation in the decoy effect happens in the absence of any decoy, when a decision maker rationally evaluates the features of the given objects, say quality and price. The appearance of a decoy induces hesitations concerning which of the features are actually more important. These hesitations before the choice are the analogs of the uncertainty of which slits will be visited by the traveling particle. The uncertainty results in the interference and the arising quantum term, whether in the registration of a particle or in the final choice of a decision maker.

DISCUSSION

We have presented a mathematical formulation for the concept of inconclusive quantum measurements and events. This type of measurements in physics happens at intermediate stages of composite measuring procedures, while the final measurement stage is operationally testable. In decision making, inconclusive events correspond to the intermediate stage of deliberations. Invoking non-informative priors, it is possible to estimate the prospect probabilities, thus, predicting the results of decision making.

GENERALLY, invoking more information on the properties of the attraction factor, it is possible to define its form more accurately than the value given by non-informative prior. For example, from condition (Equation 9) it follows that

$$-f(\pi_n) \leq q(\pi_n) \leq 1 - f(\pi_n) .$$

Hence, for a positive $q(\pi_n)$, we have

$$0 \leq q(\pi_n) \leq 1 - f(\pi_n) .$$

While for a negative $q(\pi_n)$, we get

$$-f(\pi_n) \leq q(\pi_n) \leq 0 .$$

Therefore, the attraction factor has to satisfy the limits

$$q(\pi_n) \rightarrow +0 , \qquad f(\pi_n) \rightarrow 1 ,$$

$$q(\pi_n) \rightarrow -0 , \qquad f(\pi_n) \rightarrow 0 .$$

This suggests that the absolute value of the attraction factor can be modeled by an expression proportional to

$$q(\pi_n) \propto f^{\mu}(\pi_n)[1 - f(\pi_n)]^{\nu} ,$$

with μ and ν being positive parameters and the sign defined by the ambiguity and risk aversion principle [4–6, 12]. More detailed study of such a form will be given in a separate paper.

But it turns out that even the simple non-informative prior provides us a rather good estimate allowing for quantitative predictions in decision making. And we have illustrated the approach by the decoy effect for which the non-informative priors yield quantitative predictions in very good agreement with empirical data.

In this paper, decision making by separate subjects is considered. We think that the theory can be generalized by considering societies of decision makers. The exchange of information in a society should certainly influence the decisions of the society members. To develop a theory of many agents, it is necessary to generalize the apporach by treating a dynamical model of agents exchanging information. Then, we think, it would be feasible to describe the behavior of the agents operating in a financial market and taking decisions about buying or selling shares in the presence of information asymmetry. And it would be possible to explain the known stylized facts in financial markets, such as, for example, the fat tails of return distributions, and volatility clustering, as well as transient bubbles and crashes, which are connected with herding behavior. Some first results in that direction are reported in our previous papers [6, 28], where the role of additional information, received by decision makers, is analyzed and it is shown that the amount of the additional information essentially influences the value of the quantum term. Further work on the generalization of the approach toward a dynamical theory of decision-maker societies is in progress.

AUTHOR CONTRIBUTIONS

All authors listed, have made substantial, direct and intellectual contribution to the work, and approved it for publication.

CONFLICT OF INTEREST STATEMENT

The authors declare that the research was conducted in the absence of any commercial or financial relationships that could be construed as a potential conflict of interest.

ACKNOWLEDGMENTS

Financial support from the ETH Zürich Risk Center is appreciated. We are greateful for discussions to E.P. Yukalova.

REFERENCES

1. von Neumann J. *Mathematical Foundations of Quantum Mechanics*. Princeton, NJ: Princeton University (1955).
2. Yukalov VI, Sornette D. Quantum decision theory as quantum theory of measurement. *Phys Lett A* (2008) 372:6867–71. doi: 10.1016/j.physleta.2008.09.053
3. Yukalov VI, Sornette D. Physics of risk and uncertainty in quantum decision making. *Eur Phys J B* (2009a) 71:533–48. doi: 10.1140/epjb/e2009-00245-9
4. Yukalov VI, Sornette D. Mathematical structure of quantum decision theory. *Adv Complex Syst.*(2010) 13:659–96. doi: 10.1142/S0219525910002803
5. Yukalov VI, Sornette D. Decision theory with prospect interference and entanglement. *Theor Decis.* (2011) 70:283–328. doi: 10.1007/s11238-010-9202-y
6. Yukalov VI, Sornette D. Manipulating decision making of typical agents. *IEEE Trans Syst Man Cybern Syst.* (2014a) 44:1155–68. doi: 10.1109/TSMC.2014.2314283
7. Yukalov VI, Sornette D. Positive operator-valued measures in quantum decision theory. *Lect Notes Comput Sci.* (2015a) 8951:146–61. doi: 10.1007/978-3-319-15931-7_12
8. Khrennikov A. *Ubiquitous Quantum Structure*. Berlin: Springer (2010).
9. Busemeyer JR, Bruza P. *Quantum Models of Cognition and Decision*. Cambridge: Cambridge University (2012).
10. Bagarello F. *Quantum Dynamics for Classical Systems*. Hoboken, NJ: Wiley (2013).
11. Haven E, Khrennikov A. *Quantum Social Science*. Cambridge: Cambridge University (2013).
12. Yukalov VI, Sornette D. Processing information in quantum decision theory. *Entropy* (2009b)11:1073–120. doi: 10.3390/e11041073
13. Sornette D. Physics and financial economics (1776–2014): puzzles, Ising and agent-based models. *Rep Prog Phys.* (2014) 77:062001. doi: 10.1088/0034-4885/77/6/062001
14. Ashtiani M, Azgomi MA. A survey of quantum-like approaches to decision making and cognition. *Math Soc Sci.* (2015) 75:49–50. doi: 10.1016/j.mathsocsci.2015.02.004
15. Haven E, Khrennikov A. Quantum probability and mathematical modelling of decision making.*Philos Trans R Soc A* (2016) 374:20150105. doi: 10.1098/rsta.2015.0105
16. Hey JD. Experimental investigations of errors in decision making under risk. *Eur Econ Rev.*(1995) 39:633–40. doi: 10.1016/0014-2921(09)40007-4
17. Loomes G, Sugden R. Testing different stochastic specifications of risky choice. *Economica*(1998) 65:581–98. doi: 10.1111/1468-0335.00147
18. Carbone E, Hey JD. Which error story is best? *J Risk Uncert.* (2000) 202:161–76. doi: 10.1023/A:1007829024107
19. Hey JD. Why we should not be silent about noise. *Exp Econ.* (2005) 8:325–45. doi: 10.1007/s10683-005-5373-8
20. Conte A, Hey JD, Moffatt PG. Mixture models of choice under risk. *J Econometr.* (2011)162:79–88. doi: 10.1016/j.jeconom.2009.10.011
21. Yukalov VI, Sornette D. Quantum probabilities of composite events in quantum measurements with multimode states. *Laser Phys.* (2013) 23:105502. doi: 10.1088/1054-660X/23/10/105502
22. Yukalov VI, Sornette D. Quantum theory of measurements as quantum decision theory. *J Phys Conf Ser.* (2015b) 594:012048. doi: 10.1088/1742-6596/594/1/012048
23. Yukalov VI, Sornette D. Quantum probability and quantum decision making. *Philos Trans R Soc A* (2016) 374:20150100. doi: 10.1098/rsta.2015.0100
24. Kallenberg O. *Foundations of Modern Probability*. Berlin: Springer (2001).
25. Yukalov VI. Equilibration and thermalization in finite quantum systems. *Laser Phys Lett.*(2011) 8:485–507. doi: 10.1002/lapl.201110002
26. Yukalov VI. Equilibration of quasi-isolated quantum systems. *Phys Lett A* (2012a) 376:550–4. doi: 10.1016/j.physleta.2011.11.015
27. Yukalov VI. Decoherence and equilibration under nondestructive measurements. *Ann Phys.*(2012b) 327:253–63. doi: 10.1016/j.aop.2011.09.009
28. Yukalov VI, Sornette D. Role of information in decision making of social agents. *Int J Inf Technol Dec Mak.* (2015c) 14:1129–66. doi: 10.1142/S0219622014500564
29. Bohr N. *Collected Works. The Correspondence Principle*, Vol. 3. Amsterdam: North-Holland (1976).
30. Yukalov VI, Sornette D. Conditions for quantum interference in cognitive sciences. *Top Cogn Sci.* (2014b) 6:79–90. doi: 10.1111/tops.12065
31. Boyer-Kassem T, Duchên S, Guerci E. Testing quantum-like models of judgment for question order effects. arXiv:1501.04901 (2015a).
32. Boyer-Kassem T, Duchên S, Guerci E. *Quantum-Like Models Cannot Account for the Conjunction Fallacy*. Working Paper, Tilburg University, Tilburg (2015b).
33. von Neumann J, Morgenstern O. *Theory of Games and Economic Behavior*. Princeton, NJ: Princeton University (1953).
34. Luce RD. *Individual Choice Behavior: A Theoretical Analysis*. New York, NY: Wiley (1959).
35. Frank SA. The common patterns of nature. *J Evol Biol.* (2009) 22:1563–85. doi: 10.1111/j.1420-9101.2009.01775.x
36. Batty M. Space, scale, and scaling in entropy maximizing. *Geogr Anal.* (2010) 42:395–421. doi: 10.1111/j.1538-4632.2010.00800.x
37. Saichev A, Malevergne Y, Sornette D. *Theory of Zipf's Law and Beyond*. Heidelberg: Springer (2010).
38. Sinisalchi M. Vector expected utility and attitudes toward variation. *Econometrica* (2009)77:801–55. doi: 10.3982/ECTA7564
39. Huber J, Payne JW, Puto C. Adding asymmetrically dominated alternatives: violations of regularity and

similarity hypothesis. *J Consum Res.* (1982) 9:90–8. doi: 10.1086/208899

40. Simonson I. Choice based on reasons: the case of attraction and compromise effects. *J Consum Res.* (1989) 16:158–74. doi: 10.1086/209205
41. Wedell DH. Distinguishing among models of contextuality induced preference reversals. *Learn Memory Cogn.* (1991) 17:767–78. doi: 10.1037/0278-7393.17.4.767
42. Tversky A, Simonson I. Context-dependent preferences. *Manag Sci.* (1993) 39:1179–89. doi: 10.1287/mnsc.39.10.1179
43. Ariely D, Wallsten TS. Seeking subjective dominance in multidimensional space: an explanation of the asymmetric dominance effect. *Org Behav Human Decis Process.* (1995) 63:223–32. doi: 10.1006/obhd.1995.1075
44. Bateson M, Healy SD. Comparative evaluation and its implications for mate choice. *Trends Ecol Evol.* (2006) 20:659–64. doi: 10.1016/j.tree.2005.08.013
45. Lea AM, Ryan MJ. Irrationality in mate choice revealed by tungara frogs. *Science* (2015)349:964–6. doi: 10.1126/science.aab2012

CHAPTER 11

STATE OF THE ART OF TACTILE MICRO COORDINATE METROLOGY

Rudolf Thalmann, Felix Meli and Alain Küng

Federal Institute of Metrology METAS, 3003 Bern-Wabern, Switzerland

ABSTRACT

Micro parts are increasingly found in a number of industrial products. They often have complex geometrical features in the millimeter to micrometer range which are not accessible or difficult to measure by conventional coordinate measuring machines or by optical microscopy techniques. In the last years, several concepts of tactile micro coordinate measuring machines have been developed in research laboratories and were partly commercialized by industry. The major challenges were related to the development of innovative micro probes, to the requirements for traceability and to the performance assessment at reduced measurement uncertainty. This paper presents a review on state of the art developments of micro coordinate measuring machines and 3D micro probes in the last 20 years, as far as these were qualified in a comparable way, with a special emphasis on research conducted by the Federal Institute of Metrology METAS in this field. It outlines the accuracy limitations for the probe head including the probing element and for the geometrical errors of the machine axes. Finally, the achieved performances are summarized and the challenges for further research are addressed.

INTRODUCTION

Coordinate measuring machines (CMMs) have become versatile and widespread metrology tools to perform complex dimensional measurement tasks in an efficient way. There is an ongoing trend to miniaturization in mechanical and optical production technology, leading to industrial products with generally larger functionality in a smaller volume and with less energy consumption. Small components with often complex geometrical features are found, e.g., in medical products such as ear implants or hearing aids, in gears of micro motors, in small freeform lenses of mobile phones, in injection systems for the automotive industry, or in the telecom sector for fiber optic or next generation radio frequency technology components. Hence, there is a new demand for highly accurate dimensional measurements on micro parts having geometrical features in the meso scale,*i.e.*, in the sub-millimeter to micrometer range. Such micro parts are often too complex and too large for optical microscopy techniques and their tiny structures are hardly accessible by means of conventional tactile coordinate measuring machines.

The major limiting factors of conventional CMMs for the measurement of small object features are the size of the

Citation: Rudolf Thalmann, Felix Meli and Alain Küng, State of the Art of Tactile Micro Coordinate Metrology, doi:10.3390/app6050150

probing element, the contact force as well as the CMM stage accuracy. Whereas modern multi-sensor CMMs with optical probes, mostly based on imaging systems, are well suited for the measurement of small features, only tactile probes have the fundamental properties needed for high precision full 3D capability with well established traceability. The paper discusses the key elements of tactile micro coordinate metrology, *i.e.*, the probe head for probing objects with small styli and weak forces, the precision stages with an Abbe error free metrology system, as well as the system calibration and performance verification.

Several review papers have been published in the field of micro coordinate metrology, in particular on probing systems [1], on aspects of tactile probing [2] and more recently on existing performance verification infrastructure of micro CMMs [3]. The purpose of this paper is to present a review on state of the art machine and probe developments seeking ultimate accuracy for three-dimensional measurements of micro parts, with a special emphasis on those instruments, where comparable qualification and performance assessment results are available. The research conducted on the micro CMM at METAS [4] is presented in more detail, since this machine has been qualified extensively using methods relying as closely as possible on existing written standards.

PROBE DEVELOPMENTS

When measuring small features using a tactile probe, the influence of the probe-surface interaction gets more and more critical with decreasing size of the probing element. For smaller probe sphere radii, the contact area gets smaller and therefore effects of elastic or even plastic deformation of the surface due to static or dynamic measurement forces become very critical. The effects of surface deformation by over-travel forces due to the probe stiffness and of impact forces due to the probe mass, dynamic excitations due to vibrations, surface forces as well as tip rotations due to stylus bending or rotation has been extensively studied by Bos [2]. Conventional probe systems of CMMs did not fulfil the stringent requirements regarding low stiffness and low dynamic mass nor the accuracy requirements set by ultra precision CMMs, therefore new systems had to be developed. A review over a wide range of probing systems has been given by Weckenmann *et al.* [1].

One the first micro probes has been developed jointly by PTB and Werth in 1998 [5] and soon after implemented on commercial multi-sensor CMMs. It is based on a glass fiber with a quasi spherical melted end whose position is measured with an opto-electronic system. Tip diameters down to 20 μm can be realized. Whereas the first probes had only 2D capability, Werth has developed in the mean time a fiber probe with full 3D capability [6]. The probe is available on Werth multi-sensor CMMs and has proven its usefulness in many applications, in particular for the measurement of small holes. Probing errors of a few hundred nanometers are specified, limited mainly by the relatively large form deviations of the melted glass sphere tips.

PTB developed a probe based on a micro-fabricated silicon membrane with piezo-resistive sensors [7], first published in 1999. The system measures two angular and one translational movement of the probing element. The stiffness in horizontal direction depends on the shaft length and is usually not the same as in the vertical direction. The probe design has been commercialized in the meantime and is available on the F25 micro CMM from Zeiss. The manufacturer specifies a probing error of 250 nm [8].

Approximately at the same time, NPL has also developed a micro-fabricated probe based on three angled flexible elements with capacitive sensors [9]. This original design has been commercialized as the Triskelion probe by IBS [10]. The manufacturer indicates the 3D measurement uncertainty of tip deflection to be below 20 nm. More recently NPL developed a novel vibrating micro-probe, based on the Triskelion design with six piezoelectric sensors and actuators, two on each flexure [11]. The design is optimized to feature high isotropy and very low contact forces in the μN range, thus reducing effects of surface interaction forces [2]. Experimental results to qualify the probe are not yet published.

A similar probe design using three angled flexible elements, but based on a monolithic silicon chip with piezo-resistive sensors and suitable for mass production, has been developed by TUE [12] and first published in 2001. Meanwhile, this is commercialized as the Gannen probe by XPRESS [2,13]. It is specified with a combined 3D probe uncertainty of 45 nm.

Another micro probe which is commercially available was developed by the CMM manufacturer Mitutoyo. The UMAP probe, implemented on the UMAP vision system since 2002 [14], is based on an ultrasonic micro-vibration sensor and available with stylus tip diameters between 15 μm and 300 μm. The accuracy of the probe itself is not specified.

In order to overcome the trade-off between the stiffness requirements of the probing system at contact and during approach, the University of Nottingham is actually developing a probing system for micro CMMs with variable stiffness [15]. While still under development, first promising experiments resulted in an expanded uncertainty of the probing system of about 60 nm. Issues of drift and stylus tip displacement while switching the stiffness yet need to be addressed.

Further probe developments still in the research phase are conducted in Asia: The Hefei University of Technology has been working on several concepts, all based on glass or tungsten styli attached to a floating plate, the latter being suspended on wires or leaf springs [16,17,18]. The deflection is measured with laser focus sensors or with a laser interferometer combined with an autocollimator based optical sensor for the angular motion. Their latest development includes fiber Bragg grating sensors to measure the deflection of the three arc-shaped cantilevers holding the stylus [19]. The University of Kitakyushu is working on a vibrating fiber probing system [20]. It consists of an extremely fine optical fiber with a glass tip diameter of 5 μm, but it is limited to probing in *x*/*y*-direction only.

METAS has developed, in collaboration with the Swiss Federal Institute of Technology EPFL and the industrial partner Mecartex, a 3D probe manufactured out of a single piece of aluminum by conventional milling and electro-discharge machining [4,21]. Its filigree parallel kinematic structure (Figure 1) with 60 μm thick flexure hinges minimizes the moving mass and leaves the probing sphere exactly three translational degrees of freedom. It has a stiffness of 20 N/m and exhibits perfectly isotropic probing forces below 0.5 mN at 20 μm deflection. The probe head allows for exchangeable styli with probe sphere diameters ranging from 0.1 mm to 1 mm. The probe deflection, *i.e.*, the *x/y/z* components of the translational motion, is measured by three inductive sensors. The tilted orientation of the probe head coordinate system (Figure 1) greatly facilitates the access to the probe head for exchangeable styli and also implicates, that all three axes are affected by the gravitational force in an identical way, allowing for isotropic probing forces even when changing the stylus weight.

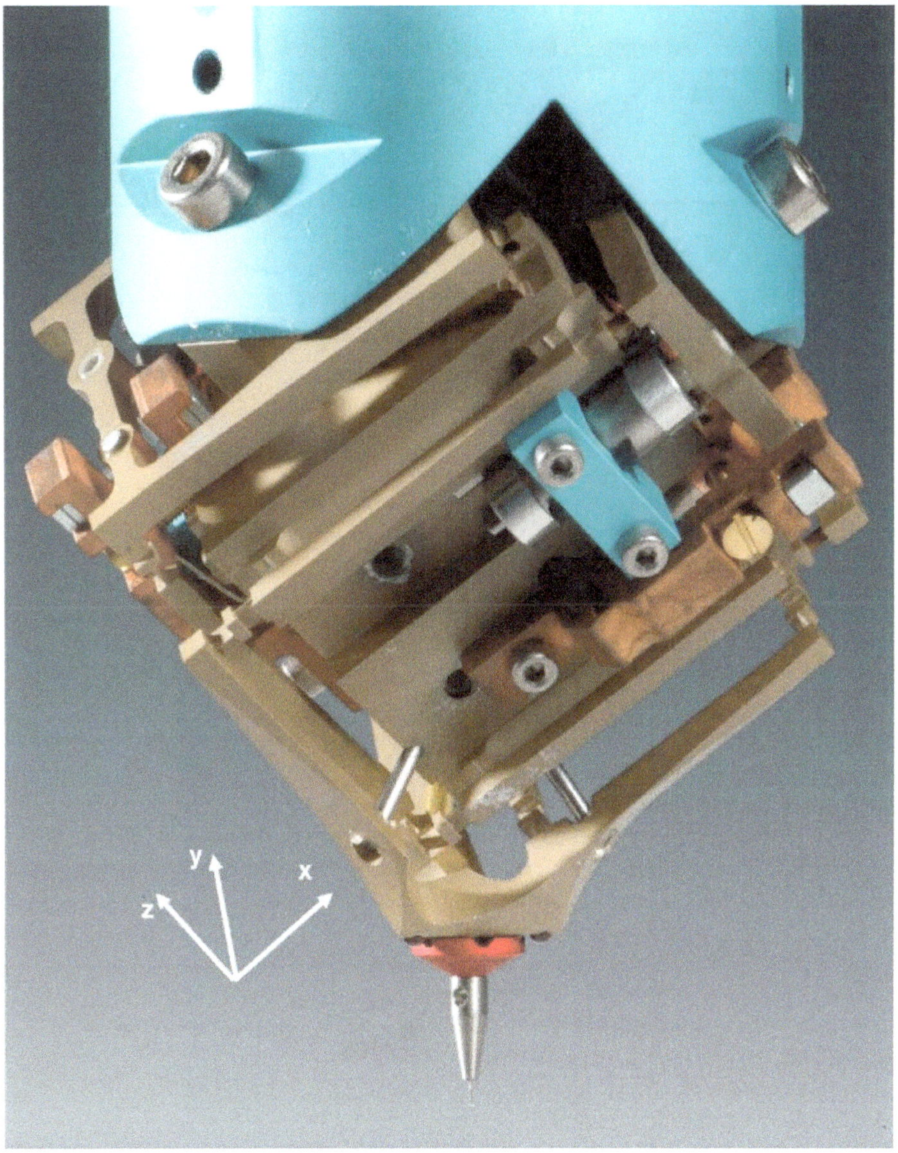

Figure 1: METAS tactile probe for micro parts (housing removed).

STAGE AND METROLOGY SYSTEMS DESIGN

The second critical element of an accurate micro CMM is the 3D translation stage with the metrology system. In order to achieve smallest uncertainties, the Abbe principle must be strictly fulfilled, *i.e.*, the reference scale must be in line with the measured length and thus always pointing towards the probe element for all positions of the stage, which imposes severe design restrictions. For classical CMMs with large measurement volumes, this cannot be realized due to limited space and disproportional cost.

Two major classes of metrology systems can be distinguished, *i.e.*, laser interferometer and incremental scale based systems. Abbe free stages with interferometers use a moving frame with mirrors onto which usually the workpiece is placed, as it is shown for the two-dimensional case inFigure 2 (left). Plane mirror interferometers are used to measure the displacement of the mirror frame with respect to the metrology frame with the probe. To build an Abbe free measurement stage using incremental scales is somewhat more complicated because the scales need to be connected with both frames and therefore require additional bearings. Figure 2 (right) shows a possible solution in 2D, as proposed by Vermeulen *et al.* [22]. Whereas measurement stages with incremental scales are more robust for industrial environment and potentially cheaper, they have the disadvantage, that the additional bearing between the two frames is part of the measurement loop and thus induces an additional uncertainty contribution.

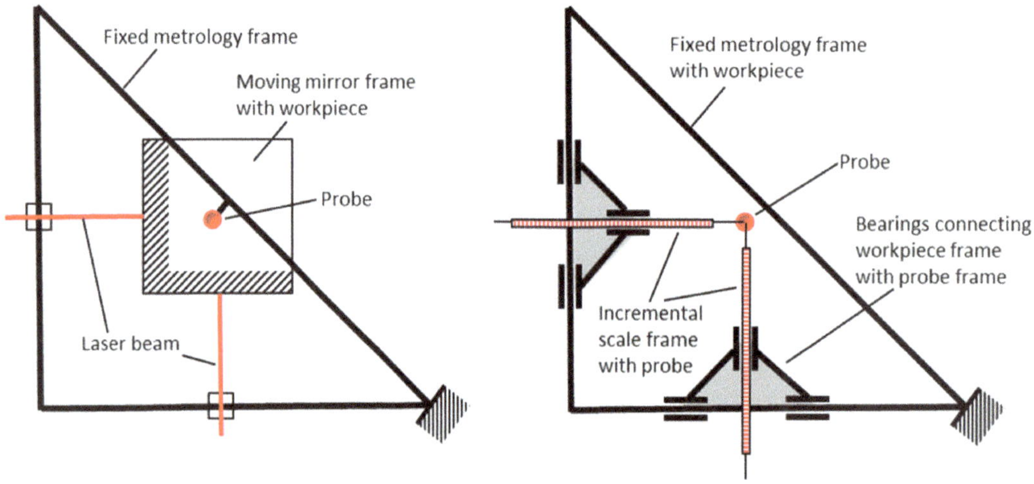

Figure 2: Principle of measurement stages (2D case for simplification of drawing). (left) Laser interferometer scales and moving workpiece stage with reference mirror frame; (right) Incremental scales and moving probe stage with additional bearings between frames.

Incremental scale based measurement stages have been proposed by TUE [22], which was later on commercialized by Zeiss for the micro coordinate machine F25 [8] (the instrument F25 is no longer part of the product portfolio of Zeiss). In this design, the Abbe principle is respected for the*x*/*y*-stage, while for a *z*-displacement of the probe out of the *x*/*y*-plane an Abbe offset must be taken into account. The F25 provides a measurement volume of 100 mm × 100 mm × 100 mm. A similar design, but with all three axes in Abbe, has been realized by TUE [23] for the TriNano micro CMM and is commercialized by XPRESS [13].

One of the first Abbe free interferometric stages for coordinate metrology was realized at NPL: a conventional CMM was used for the 3D movement of a cube corner supporting the mirrors for reflecting the interferometer beams [9]. TU Illmenau designed an ultra-precise 3D stage with Zerodur® (Schott AG, Jena, Germany) frame and a multi-axis interferometer including angular motion (yaw) measurement for their nano-positioning and nano-measuring machine NMM [24], which is commercialized by SIOS [25]. The Hefei University of Technology built two prototype micro CMMs, both with roller bearing stages driven by piezo motors and laser interferometers for stage position sensing [26,27].

An original ultra precision stage development at Philips CFT [28] has been installed at METAS and was the basis for a new development by IBS [29] leading to the Isara 400 machine [30]. The ultraprecision stage of the METAS micro CMM [4,28] has vacuum preloaded air bearings. The stages are driven by Lorenz actuators and their motion is measured and controlled by three plane mirror heterodyne interferometers. The original V-configuration of the guideways with wedges as shown in Figure 3 makes the stage very compact and stiff. The working volume of the stage is 90 mm × 90 mm × 38 mm. According to Figure 2 (left) the probe head remains in a fixed position connected to the metrology frame while the stage is moving the workpiece during measurements. The workpiece is located in a Zerodur cube corner with three perpendicular flat mirrors forming the reference coordinate system. All three interferometer beams measuring the workpiece displacement are pointing to the center of the probing sphere, thus limiting the residual Abbe offset to the probe radius.

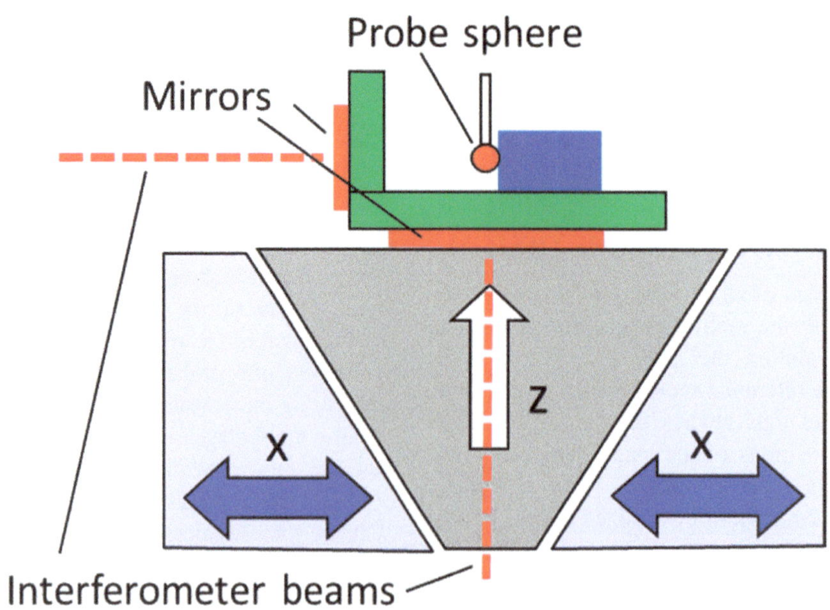

Figure 3: Principle of the Abbe free CMM stage at METAS. *y*-axis perpendicular to drawing plane.

SYSTEM CALIBRATION AND PERFORMANCE VERIFICATION

Probe Calibration

Single Point Probing

The single point probing behavior of a CMM probe can be characterized by the P_{FTU} value (single stylus form error) as specified in the standard ISO 10360-5 [31], based on the peak-to-valley deviation of 25 probing points on a sphere arranged according to Figure 4 (left). For two of the micro CMMs presented above, corresponding results are available. The P_{FTU} value for the Zeiss F25 probe was found from the average of 14 measurements to be 189 nm with 60 nm standard deviation and all values but one complying with the manufacturer specification of 250 nm [8]. A repeatable pattern was found in these measurements, most probably due to the anisotropic behavior of the silicon chip membrane as suggested by Bergmans *et al.* [8].

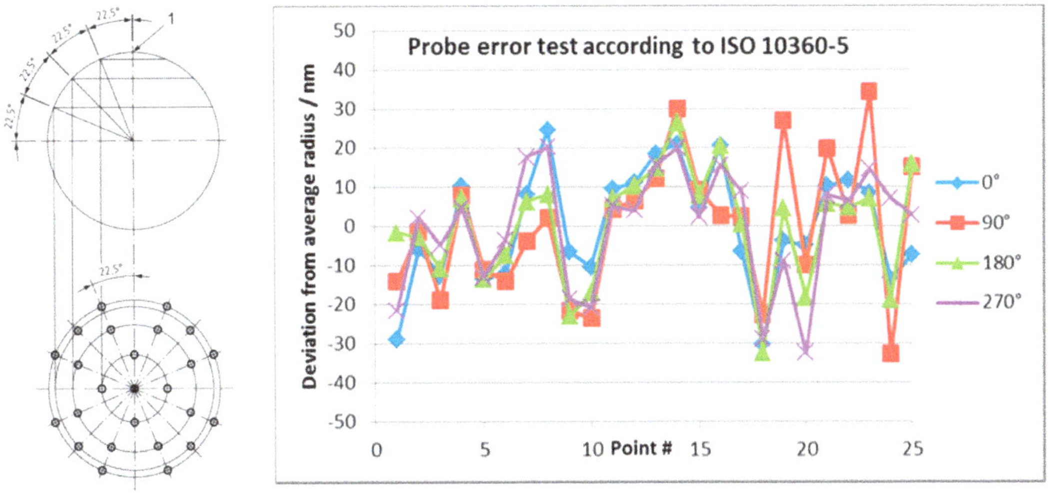

Figure 4: Determination of the single point probing error according to ISO 10360-5 on a 1 mm Si_3N_4 sphere: (left) Arrangement of probing points on the reference sphere; (right) Deviation from average radius for four measurements with sphere rotated by 90°.

METAS has carried out the same standardized test [31] for single point probing error on a 1 mm diameter sphere [32]. The test was performed four times with the sphere rotated by 90° between each test. The sphere radius was found to be between 0.500202 nm and 0.500211 nm with a standard deviation of 4 nm. The test resulted in P_{FTU} values between 53 nm and 67 nm with an average value of 58 nm. The measurements in Figure 4 (right) show a systematic behavior probably due to an uncompensated form deviation of the probing sphere. The Si_3N_4 test sphere [33] had an estimated form deviation of less than 15 nm.

Scanning

The standard ISO 10360-4 [34] describes the scanning probing error T_{ij} which is the range of radial distances obtained from four circular scans on a sphere: one full circle in an equatorial plane, a second full circle in a plane parallel to the first, a third half circle through the pole, and a fourth half circle in a plane perpendicular to the third at a distance of about 1/2 of the radius from the pole (Figure 5, left).

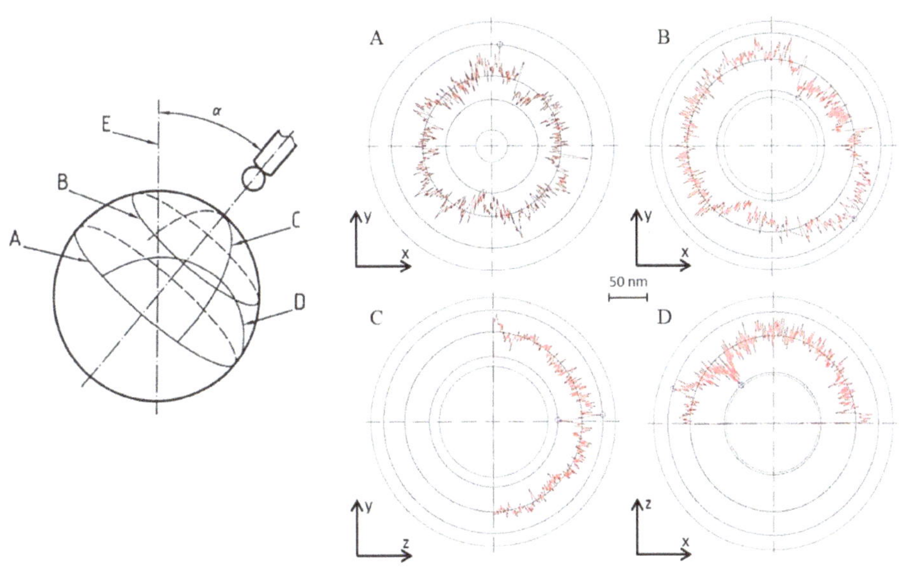

Figure 5: Determination of the single point probing error according to ISO 10360-4 on a 1 mm ruby sphere: (left) Arrangement of scanning profile planes A, B, C and D; (right) Measured profiles A to D resulting in a scanning probing error T_{ij} = 87 nm.

This test was performed with the METAS probe on a 1 mm ruby sphere with a point density of 300 pts/mm [35]. The LS diameter fitted through the scan profiles resulted in 1.000806 mm, compared to the value of 1.000804 mm found by the independent probe sphere diameter determination as shown in Section 4.1.3. For T_{ij} a value of 87 nm was obtained. The corresponding values for the four individual scans were 51 nm, 75 nm, 67 nm and 78 nm, respectively. It must be noted that these values comprise the form deviation of the reference sphere (53 nm), the uncompensated form deviation of the probing sphere and surface roughness contributions of both spheres, as no filtering was applied.

Probe Sphere Diameter and Form Correction

On a conventional CMM the absolute diameter of the probe sphere is calibrated on a reference sphere. At the same time the anisotropic behavior of the probe head including stylus bending is corrected. This requires an independent calibration of the diameter of the reference sphere and assumes the roundness deviation of both, the reference sphere and the probe sphere, to be much smaller than the overall probing error, which is generally the case. For high precision micro coordinate metrology, the requirements for the reference sphere are much higher and cannot be fulfilled by independent calibrations. Therefore, an error separation method was developed involving three nominally identical spheres compared against each other in various configurations, using one of the spheres both as the probe sphere and the reference sphere, respectively [36]. In this way, both the absolute diameter and the sphericity deviation of three 1 mm ruby spheres could be mapped over their entire accessible surface. This method finally relies directly on the traceability of the CMM stage interferometer and on the repeatability of the probe being usually below 4 nm (standard deviation). Figure 6 shows a sphericity map of one of the measured spheres with a form deviation of 33 nm.

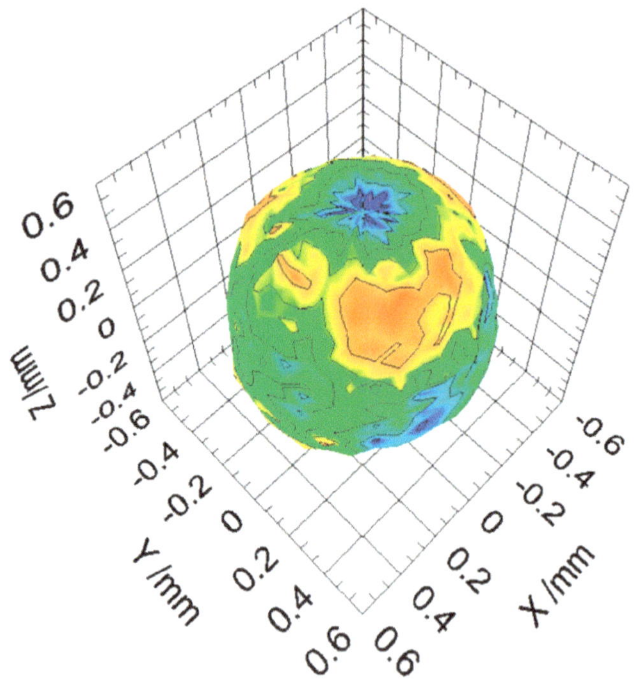

Figure 6: Sphericity map of a 1 mm ruby sphere with 33 nm form deviation. The 3D plot shows results obtained by probing the surface with a 10° resolution in longitude and a 5° resolution in latitude.

Stage Calibration and Error Correction

The way of calibrating the measurement stage depends much on its design (Section 3). Measurement stages for micro CMMs with incremental scales are calibrated and corrected similarly to conventional machines, *i.e.*, the position accuracy is measured with a laser interferometer and the geometrical errors are usually determined with the help of artefacts. For the F25 machine some measurement results were published by PTB [37]. They carried out straightness measurements on a Zerodur gauge block assembly along the horizontal *y*-axis over 43 mm, resulting in a straightness deviation of about 30 nm with a reproducibility that would allow for a further correction down to a few nanometers residual error. A similar measurement along the vertical *z*-axis over a length of 1.75 mm (limited by the length of the stylus shaft) resulted in a straightness deviation of less than 10 nm.

The calibration of interferometer based stages includes the calibration of the laser wavelength (given by the optical frequency and the refractive index determined by the compensation unit), and—more critically—the flatness and squareness of the faces of the reference mirror cube. For the flatness calibration of the reference mirrors, high precision reference artefacts are used. IBS reported on flatness measurements using a Zerodur block with a metallic coating allowing for a capacitive sensor to be used instead of the tactile probe [29] und thus improving resolution and noise due to surface roughness and probe repeatability. The squareness calibration was made using the same artefact by applying a reversal technique.

The calibration and correction of the Zerodur reference mirror of the METAS micro CMM has been made in a similar way as described above [35]. For this, a Zerodur block of 100 mm × 60 mm × 40 mm with form deviations <10 nm was used. The gold coated block was scanned by a capacitance sensor. Where possible, reversal techniques were applied to separate the errors. The resulting errors were then mapped in correction files. The calibration of the squareness in *x*/*y*-plane was made with the help of a ball plate. The squareness of the *z*-axis with respect to horizontal axes was calibrated with the help of a Ø 44 mm tungsten carbide sphere, by scanning the sphere in two half-circles in the *x*/*z*- and the *y*/*z*-planes and adjusting the corresponding orthogonality angles as to obtain minimum residuals from a circle. Although somewhat less sensitive than checking the orthogonality by, e.g., measuring a ball bar in diagonal directions, the method using a sphere is more efficient and provides an independent performance check.

Performance Tests

Performance tests of micro CMMs can be made in the same way as for conventional machines following the procedure of the acceptance and reverification test ISO 10360-2 [38] using specially designed small artefacts. Since the target uncertainties are in the nanometer range, the stability of these artefacts is a particular challenge. The quantity to be determined in the above cited ISO test isE_0, the error of indication of a calibrated test length.

Ball bars of different lengths ranging from 20 mm to 100 mm were manufactured using Zerodur bars and ruby spheres (Figure 7). The results shown in Figure 7 assess the E_0 value of the machine in the whole volume to be 27 nm + 0.2 ×

$10^{-6} \times L$ [35]. Since the test requires the ball bar to be measured along the diagonals of the volume, one can compute the squareness between the three machine axes. It is an accurate method since the longest diagonal length can be measured.

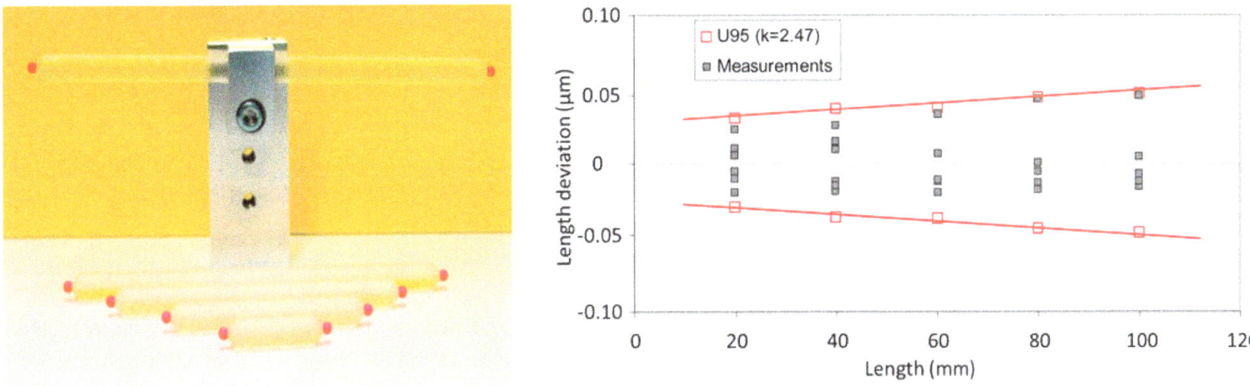

Figure 7: (left) Micro ball bars used for performance verification; (right) Length deviations measured with 5 different ball bars each oriented in different axial and diagonal directions. Red squares denote the symmetric 95% confidence intervals, red lines the fitted 95% confidence length measurement error $E_0 = 27\ \text{nm} + 0.2 \times 10^{-6} \times L$.

Ball plates are another suitable artefact to test the performance of a CMM. The calibration method requires measurements in 4 positions (two from each side of the plate) in order to be able to separate the errors of the CMM from the deviations of the ball plate nominal positions. The ball plate used in the test hereafter was manufactured using 25 ruby spheres, Ø 3 mm, pressed into the holes of an 85 mm × 85 mm invar frame (Figure 8).

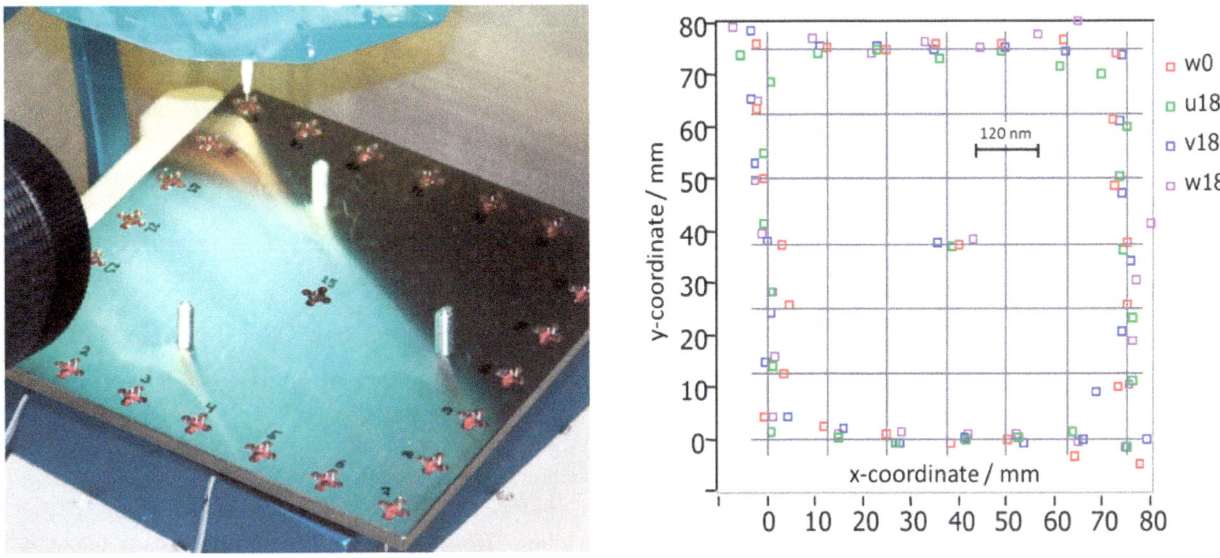

Figure 8: (left) Micro ball plate used for performance verification; (right) Deviations measured for 25 ball center positions in four different orientations of the ball plates. The evaluation of all relative ball distances results in a 95% confidence length measurement error of $E_0 = 26\ \text{nm} + 0.23 \times 10^{-6} \times L$.

Apart from the squareness error between the axes, as it is also obtained from ball bar measurements described above, the method also provides information about other error contributions, such as axes straightness or angular distortions. As shown in Figure 8 the measurement results on the METAS micro CMM exhibit a residual squareness error of (0.17 ± 0.05)". The contributions from other error sources, which are smaller than 40 nm, are largely dominated by temperature drifts. Evaluating the 300 relative distances between all measured ball center positions in a similar way as in Figure 7 (right) leads to a 95% confidence length measurement error of $E_0 = 26\ \text{nm} + 0.23 \times 10^{-6} \times L$, a value very close to the one obtained using the ball bars.

PTB and METAS carried out a bilateral comparison on a 90 mm × 90 mm Zerodur plate with hemispheres [39]. The average difference of the hemisphere *x*- and *y*-coordinates was 12 nm and 7 nm respectively, with a maximum value of 28 nm. It must be noted that, since an error separation calibration procedure was applied, these measurements do not represent the deviations of the machine, but only the capability and comparability of calibrating a ball plate standard.

CHALLENGES FOR FURTHER DEVELOPMENT

To extend the use of tactile micro coordinate metrology for a large variety of industrial applications, a series of challenges need to be addressed.

Accessibility of 3D Features

Quite often, the geometrical features of a micro-parts are not easily accessible with a spherical probe on a straight stylus, but limited by the probe diameter for small holes, by the stylus length for high aspect features such as deep holes, or by the direction of the probe stylus. Therefore, smaller probe elements with smaller and better known form deviation, longer styli to access deep holes and smaller probing forces to allow for longer styli need to be developed. To achieve a better accessibility of normally hidden features, probes with multiple, *i.e.*, star-shaped styli were built [40], their application turned out to be delicate for standard use, however. Another approach is to implement additional rotational axes on the micro CMM [41], obviously at the cost of accuracy due to uncorrected guide and positional errors of the rotation axes.

Probe/Surface Interaction

The smaller the workpiece features, the probe sphere diameter and the targeted measurement uncertainty are, the more important becomes the interaction of the probe with the surface. Not only effects of surface indentation due to measurement forces [2], but also stylus wear and deposition of wear debris on the probe sphere become relevant [42]. These effects can be significantly reduced by selection of appropriate materials for the probe sphere, *e.g.*, diamond coated spheres [43].

For very weak probing forces, effects of attractive forces between stylus tip and surface get relevant and may cause sticking effects [2,11]. These problems may be overcome by novel concepts such as vibrating probes [11,14,20].

Task Specific Uncertainty

Performance verification tests as described in Section 4 may result in standardized length measurement errors and qualify the probe errors, but do provide only some basic contributions to the uncertainty of a specific geometrical feature to be determined. Length measurement errors are certainly not adequate for the uncertainty of derived geometrical elements, such as a local radius, a form deviation or an angle. Methods for the estimation of reliable, task specific measurement uncertainties still need to be developed further and are not yet mature for day to day applications. A promising approach is the *Virtual CMM* based on numerical simulation [44,45], which has been successfully implemented on the METAS micro CMM [46], but which still requires a considerable effort in modelling and computing. Further research is needed for refining and implementing the error model adapted to the particular characteristics of a micro CMM and integrating this directly into a high level software, allowing to achieve estimations of uncertainty not only off-line, but with much higher computing speed almost in real-time during the measurement process.

Application in Industry

For a wider use of tactile micro coordinate metrology in industry, several issues need to be further addressed. The need for fast and efficient measurements contradicts to the requirements of slow approaching speed to achieve small contact forces and thus asks for improvements of the dynamic properties of machines and probes. Another problem is the fragility of probing systems and styli, which easily break due to overtravel or uncontrolled manipulation. Systems with the possibility of exchangeable styli obviously help to save costs of replacing the entire probe system. Finally, industry is asking for comparability of measurement results and of system specifications. A written standard for performance verification dedicated to micro CMMs, or preferably an amendment of an existing standard is clearly needed [3].

STATE OF THE ART AND CONCLUSIONS

As outlined earlier there are actually a few types of machines for high precision tactile micro coordinate measurements available, which have full 3D measurement capability and are operational with a high-level software platform to provide industrial measurement services: the well established Zeiss F25 [8], from which about 25 machines were sold and are operating in research labs and institutes worldwide, but is unfortunately no longer commercially available; two more recent developments, the Isara 400 [30] and the TriNano [13,23] are both in principle commercially available, but the market uptake did not really happen so far; the micro CMM at METAS [4] is a prototype development, which is fully operational since 2006 and almost daily used for providing calibration and measurement services to industry. Also to be mentioned is the commercially available SIOS nanopositioning and nanomeasuring machine [24,25], which can be configured with a suitable probe to perform 3D coordinate measurements. Furthermore, micro CMMs were developed by Hefei University of Technology [26,27], complemented by tactile micro probe developments [16,17,18,19] to achieve 3D measurement capability. These latter laboratory instruments, however, are still in the development phase and have not proven yet their full measurement capability following standardized procedures as outlined in Section 4. Not discussed further in this context are multi-sensor CMMs, such as [6,14], which have accuracies closer to conventional CMMs.

State of the art probes of above mentioned instruments do achieve a repeatability of a few nanometers, but the best values reported for the 3D probing error according to standardized specification tests [31,34] are around 50 nm, with similar performance for single point probing and scanning. The achievable probing errors are limited mainly by the form deviation and the roughness of the probing sphere, and the overall stability of the instrument during the test procedure. State of the art CMM stages were demonstrated to have maximum errors of indicated length according to a standardized specification test [38] in the entire volume and in arbitrary direction of about 60 nm, limited by the uncorrected flatness deviation of the reference mirrors, residual squareness errors and the errors of the probe, the latter being part of the mentioned specification test.

In conclusion, in spite of the big research and development effort spent worldwide for tactile micro coordinate metrology, there are still very few results available for a comparable performance verification according to standardized procedures. The reason for this is partly that many instruments and probe systems are still in the development phase, but there is also a lack of generally accepted and widely available material standards and written standards dedicated to the particular characteristics of micro CMMs.

ACKNOWLEDGMENTS

This work was funded by the R & D-Program at METAS and the European Metrology Research Program EMRP.

AUTHOR CONTRIBUTIONS

M.F., A.K. and R.T. conceived and designed the experiments; A.K. performed the experiments and analyzed the data; R.T. wrote the paper.

CONFLICTS OF INTEREST

The authors declare no conflict of interest.

REFERENCES

1. Weckenmann, A.; Estler, T.; Peggs, G.; McMurtry, D. Probing systems in dimensional metrology. *Ann. CIRP* 2004, *53*, 657–684.
2. Bos, E.J.C. Aspects of tactile probing on the micro scale. *Precis. Eng.* 2011, *35*, 228–240.
3. Claverley, J.D.; Leach, R.K. A review of the existing performance verification infrastructure for micro-CMMs. *Precis. Eng.* 2014, *39*, 1–15.
4. Küng, A.; Meli, F.; Thalmann, R. Ultraprecision micro-CMM using a low force 3D touch probe. *Meas. Sci. Technol.* 2007, *18*, 319–327.
5. Guilun, J.; Schwenke, H.; Trapet, E. Opto-tactile sensor for measuring small structures on coordinate measuring machines. *Proc. ASPE* 1998, *18*, 25–28.
6. Werth 3D Fibre Probe. Available online: http://www.werth.de/index.php?id=262&L=1(accessed on 26 April 2016).
7. Buetefisch, S.; Dai, G.; Danzebrink, H.-U.; Koenders, L.; Solzbacher, F.; Orthner, M.P. Novel design for an ultra high precision 3D micro probe for CMM applications. *Procedia Eng.* 2010,*5*, 705–712.
8. Bergmans, R.H.; Nieuwenkamp, H.J.; van Veghel, M.G.A. Probing behavior of a microCMM. In Proceedings of the 11th Euspen International Conference, Como, Italy, 23–26 May 2011; Volume 1, pp. 104–107.
9. Peggs, G.N.; Lewis, A.; Oldfield, S. Design of a compact high-accuracy CMM. *Ann. CIRP*1999, *48*, 417–420.
10. IBS Precision Engineering. Available online: http://www.ibspe.com/category/isara-400-3d-cmm/triskelion-touch-probe.htm (accessed on 26 April 2016).
11. Claverley, J.K.; Leach, R.K. Development of a three-dimensional vibrating tactile probe for miniature CCMs. *Precis. Eng.* 2013, *37*, 491–499.
12. Haitjema, H.; Pril, W.O.; Schellekens, P.H.J. Development of a silicon-based nanoprobe system for 3-D measurements. *Ann. CIRP* 2001, *50*, 365–368.
13. XPRESS Precision Engineering B.V. Available online: http://www.xpresspe.com/probe2.php(accessed on 26 April 2016).
14. Mitutoyo. Available online: http://www.mitutoyo.co.jp/eng/new/news/2002/02_13.html(accessed on 26 April 2016).
15. Alblalaihid, K.; Kinnell, P.; Lawes, S.; Desgaches, D.; Leach, R. Performance Assessment of a New Variable Stiffness Probing System for Micro-CMMs. *Sensors* 2016, *16*, 492. [PubMed]
16. Fan, K.C.; Cheng, F.; Wang, W.; Chen, Y.; Lin, J.Y. A scanning contact probe for a micro-coordinate measuring machine (CMM). *Meas. Sci. Technol.* 2010, *21*, 054002.
17. Li, R.J.; Fan, K.C.; Miao, J.W.; Huang, Q.X.; Tao, S.; Gong, E.R. An analogue contact probe using a compact 3D optical sensor for micro/nano coordinate measuring machines. *Meas. Sci. Technol.* 2014, *25*, 094008.
18. Li, R.J.; Fan, K.C.; Huang, Q.X.; Zhou, H.; Gong, E.R.; Xiang, M. A long-stroke 3D contact scanning probe for micro/nano coordinate measuring machine. *Precis. Eng.* 2016, *43*, 220–229.
19. Liu, F.F.; Chen, L.J.; Wang, J.F.; Xia, H.J.; Li, R.J.; Yu, L.D.; Fei, Y. Modeling and prototyping of a fiber Bragg grating-based dynamic micro-coordinate measuring machine probe. *Meas. Sci. Technol.* 2016, *27*, 025016.
20. Murakami, H.; Katsuki, A.; Sajima, T.; Suematsu, T. Study of a vibrating fiber probing system for 3-D micro-structures: Performance improvement. *Meas. Sci. Technol.* 2014, *25*, 094010.
21. Meli, F.; Fracheboud, M.; Bottinelli, S.; Bieri, M.; Thalmann, R.; Breguet, J.-M.; Clavel, R. High precision, low force 3D touch probe for measurements on small objects. In Proceedings of the Euspen International Topical Conference, Aachen, Germany, May 2003; pp. 411–414.
22. Vermeulen, M.M.P.A.; Rosielle, P.C.J.N.; Schellekens, P.H.J. Design of a high-precision 3D-coordinate measuring machine. *Ann. CIRP* 1998, *47*, 447–450.
23. Van Riel, M.; Moers, T. Nanometer uncertainty for a micro price. *Mikroniek* 2010, *50*, 13–17.
24. Jäger, G.; Manske, E.; Hausotte, T.; Buchner, H.-J. The metrological basis and operation of nanopositioning and nanomeasuring machine. *Tech. Mess.* 2009, *76*, 227–234.
25. SIOS Meßtechnik GmbH. Available online: http://www.sios-de.com/?page_id=635 (accessed on 26 April 2016).
26. Fan, K.C.; Fei, J.T.; Yu, X.F.; Chen, Y.J.; Wang, W.L.; Chen, F.; Liu, S.L. Development of a low-cost micro-CMM for 3D micro/nano measurements. *Meas. Sci. Technol.* 2006, *17*, 524–532.
27. Huang, Q.X.; Wu, K.; Wang, C.; Li, R.J.; Fan, K.C.; Fei, J.T. Development of an Abbe Error Free Micro Coordinate Measuring Machine. *Appl. Sci.* 2016, *6*, 97.
28. Ruijl, T. Ultra Precision Coordinate Measuring Machine; Design, Calibration and Error Compensation. Ph.D. Thesis, Technical University of Delft, Delft, The Netherlands, February 2001.
29. Widdershoven, I.; Spaan, H.A.M. Calibration of the ISARA 400 ultra-precision CMM. In Proceedings of the First International Workshop on Advances in IT—Service Process Engineering, Guadeloupe, France, 23–28 February 2011.
30. IBS Precision Engineering. Available online: http://www.ibspe.com/category/isara-400-3d-cmm.htm (accessed on 26 April 2016).
31. ISO 10360-5. *Geometrical Product Specifications (GPS)—Acceptance and Reverification Tests for*

Coordinate Measuring Machines (CMM)—Part 5: CMMs Using Single and Multiple Stylus Contacting Probing Systems; International Organization for Standardization: Geneva, Switzerland, 2010.

32. Küng, A. *Towards Truly 3D Metrology for Advanced Micro-Parts*. EURAMET Project 1088. 2012. Available online: http://www.euramet.org/technical-committees/length/tc-l-projects/(accessed on 26 April 2016).
33. Saphirwerk. Available online: http://www.saphirwerk.com/ (accessed on 26 April 2016).
34. ISO 10360-4. *Geometrical Product Specifications (GPS)—Acceptance and Reverification Tests for Coordinate Measuring Machines (CMM)—Part 4: CMMs Used in Scanning Measurement Mode*; International Organization for Standardization: Geneva, Switzerland, 2000.
35. Küng, A.; Meli, F. Comparison of three independent calibration methods applied to an ultra-precision μ-CMM. In Proceedings of the 7th Euspen International Conference, Cranfield, UK, 20–24 May 2007; Volume 1, pp. 230–233.
36. Küng, A.; Meli, F. Self calibration method for 3D roundness of spheres using an ultraprecision coordinate measuring machine. In Proceedings of the 5th Euspen International Conference, Montpellier, France, May 2005; pp. 193–196.
37. Neugebauer, M. Precision size and form measurements with a micro-CMM F25. In Proceedings of the IXth International Scientific Conference Coordinate Measuring Technique, Ustroń, Poland, 14–16 April 2010.
38. ISO 10360-2. *Geometrical Product Specifications (GPS)—Acceptance and Reverification Tests for Coordinate Measuring Machines (CMM)—Part 2: CMMs Used for Measuring Linear Dimensions*; International Organization for Standardization: Geneva, Switzerland, 2009.
39. Neugebauer, M. Bilateral Comparison on Micro-CMM Artefacts between PTB and METAS; EURAMET Project 1105 Final Report 2011. Available online: http://www.euramet.org/get/?tx_stag_base%5Bfile5D=3457&tx_stag_base%5Baction%5D=downloadRaw&tx_stag_base%5Bcontroller%5D=Base (accessed on 26 April 2016).
40. Küng, A.; Meli, F. Versatile probes for the METAS 3D Micro-CMM. In Proceedings of the 8th Euspen International Conference, Zürich, Switzerland, 18–22 May 2008; Volume 1, pp. 269–272.
41. Küng, A.; Meli, F.; Nicolet, A. A 5 degrees of freedom μCMM. In Proceedings of the 14th Euspen International Conference, Dubrovnik, Croatia, 2–6 June 2014; Volume 1, pp. 269–272.
42. Küng, A.; Nicolet, A.; Meli, F. Study of sapphire probe tip wear when scanning on different materials. *Meas. Sci. Technol.* 2012, *23*, 094016.
43. Küng, A.; Nicolet, A.; Meli, F. Study of wear of diamond-coated probe tips when scanning on different materials. *Meas. Sci. Technol.* 2015, *29*, 084005.
44. ISO/TS 15530-4. *Geometrical Product Specifications (GPS)—Coordinate Measuring Machines (CMM) Technique for Determining the Uncertainty of Measurement—Part 4: Evaluating Task-specific Measurement Uncertainty Using Simulation*; International Organization for Standardization: Geneva, Switzerland, 2008.
45. Härtig, F.; Trapet, E.; Wäldele, F.; Wiegand, U. Traceability of coordinate measurements according to the virtual CMM concept. In Proceedings of the 5th IMEKO TC-14 Symposium on Dimensional Metrology in Production and Quality Control, Zaragoza, Spain, 25–27 October 1995; pp. 245–254.
46. Küng, A.; Meli, F.; Nicolet, A.; Thalmann, R. Application of a virtual coordinate measuring machine for measurement uncertainty estimation of aspherical lens parameters. *Meas. Sci. Technol.* 2014, *25*, 094011.

CHAPTER 12 SIMULATION MODEL FOR CORRECTION AND MODELING OF PROBE HEAD ERRORS IN FIVE-AXIS COORDINATE SYSTEMS

Adam Gąska, Piotr Gąska and Maciej Gruza

Laboratory of Coordinate Metrology, Cracow University of Technology, Kraków 31-155, Poland

ABSTRACT

Simulative methods are nowadays frequently used in metrology for the simulation of measurement uncertainty and the prediction of errors that may occur during measurements. In coordinate metrology, such methods are primarily used with the typical three-axis Coordinate Measuring Machines (CMMs), and lately, also with mobile measuring systems. However, no similar simulative models have been developed for five-axis systems in spite of their growing popularity in recent years. This paper presents the numerical model of probe head errors for probe heads that are used in five-axis coordinate systems. The model is based on measurements of material standards (standard ring) and the use of the Monte Carlo method combined with select interpolation methods. The developed model may be used in conjunction with one of the known models of CMM kinematic errors to form a virtual model of a five-axis coordinate system. In addition, the developed methodology allows for the correction of identified probe head errors, thus improving measurement accuracy. Subsequent verification tests prove the correct functioning of the presented model.

INTRODUCTION

The development of production engineering and the demand for high quality manufacturing brings about new challenges in the coordinate metrology, which serves as the key tool for quality assurance systems in many branches of industry (automotive, aerospace, machine, etc.) [1,2,3]. With the increase in manufacturing accuracy, the measurements and methods used for assessing measurement uncertainty have to be more precise. The time spent on performing measurements and evaluations also plays a crucial role, as it has a fairly significant impact on the total costs of manufacturing. Therefore, new solutions are being introduced in regards to both the construction of measuring systems, and the methods for the assessment of coordinate measurement accuracy [4,5].

Simulative models constitute one of the newest developments in accuracy estimation. In practice, such models are based on developing virtual models of measuring systems. Their main advantage is the possibility to obtain measurement uncertainty following only a single measurement, whereas the common methods of uncertainty estimation (such as the calibrated workpiece method [6,7], or multiple measurement strategy [8,9]) would require multiple measurements. In

Citation: Adam Gąska, Piotr Gąska and Maciej Gruza, Simulation Model for Correction and Modeling of Probe Head Errors in Five-Axis Coordinate Systems, doi:10.3390/app6050144

some instances, simulative models may also be used for error compensation, and consequently, for improving measurement accuracy.

The virtual Coordinate Measuring Machine (CMM) model was first developed in the 1990s at Physikalisch-Technische Bundesanstalt in Germany [10,11]. The original model was based on the determination of all possible error causes in the entire measuring volume of the CMM, and the evaluation of their impact on measurement results. Thanks to this, the mathematical model of ideal measurement was supplemented by the influence of possible errors that may occur during measurements, and used for performing multiple simulations of the considered measurement. Measurement uncertainty was then determined based on the variability of simulated results and simple statistical operations.

So far, several models have been developed for typical three-axis CMMs. All of those models utilize a similar working principle as the PTB model, and were presented in detail in [12,13,14,15,16,17]. Lately, a virtual model of measurements performed on the Articulated Arm CMM was also developed [18]. Further works focused on the construction of a virtual model for the Laser Tracker system are currently in progress both in Germany and the UK.

In the last years, it has been possible to observe an increase in the popularity of the so-called five-axis coordinate systems. The five-axis system is created by adding two rotations of a probe head to the standard movements of the machine. Thanks to this solution, it is possible to speed up the measuring process, especially for solids of revolution such as rings, cylinders, spheres, etc., without accuracy loss. Workpieces of this type may be measured using mainly movements of the probe head kinematic pairs. Unlike the three-axis CMM, there is no need to move the whole body of the machine during measurements of all points. The reduction in measurement time is clearly visible, and because of that, the number of installations or retrofits for five-axis systems is becoming a fairly common trend in the industry. Nevertheless, hardly any literature can be found on this subject, and few scientists pay sufficient attention to this type of system. For this reason, the authors decided to investigate the field of errors of probe heads used in five-axis systems, and to attempt to develop a model of this type.

The main sources of errors occurring during coordinate measurements performed on a CMM are kinematic system errors and probe head errors [19,20,21,22]. These types of errors are included in all virtual models of CMMs. The models of probe head errors are usually based on error identification with the use of material standards, such as spherical standards and standard rings. Some currently used probe head error simulation models are described below.

A probe head error model utilizing an artificial neural network was developed at the Laboratory of Coordinate Metrology (LCM) of the Cracow University of Technology as part of a Virtual CMM based on assumptions of the matrix method [16]. The model utilizes the Probing Error Function (PEF) described in [13,16]. Probing Error (PE) depends on a number of factors, among them: deformation of the measuring ball during contact, deflection of the stylus under the force of the measurement, probe pretravel, the form error of the measuring ball, non-linearity and differentiation of the transducer characteristics in different directions of deflection, i.e., the directions of probe head movements (u, v, w), reaction to uneven load of the probe head, etc. The PE can be defined as a sum of a number of contributing factors. The value of PE changes significantly for different probing directions, so the PEF may be obtained in relation to the probe deflection angle (α). Values of PEF can be determined through measurements of calibrated material standards with known form deviations. The value of the PE at a certain measured point, measured at a specified angle, may be expressed as a radial error obtained during point measurement [16].

The model is divided into two modules: one responsible for systematic error simulation and the second for random error modeling. The ring gauge of a 25 mm diameter has to be measured 32 times, in 64 evenly distributed points, to gather data for model preparation. It was assumed that the 64 measured points are sufficient to model the probe head functioning for any probing direction. On the basis of obtained results, the mean value of the PEF should be attributed to each measured point. These values are treated as a systematic part of the probing system error. The obtained data is used as a learning data set for the constructed artificial neural network, which links the probe head deflection angle and systematic error. The network consists of three layers, and utilizes back propagation and different activation functions (Figure 1). The module was prepared both for inner dimensions—using the gauge ring as a reference—and outer dimensions. In the case of the latter, the entire procedure should be repeated, except that instead of a ring, the spherical standard of a 25 mm diameter should be measured at the reference sphere equator.

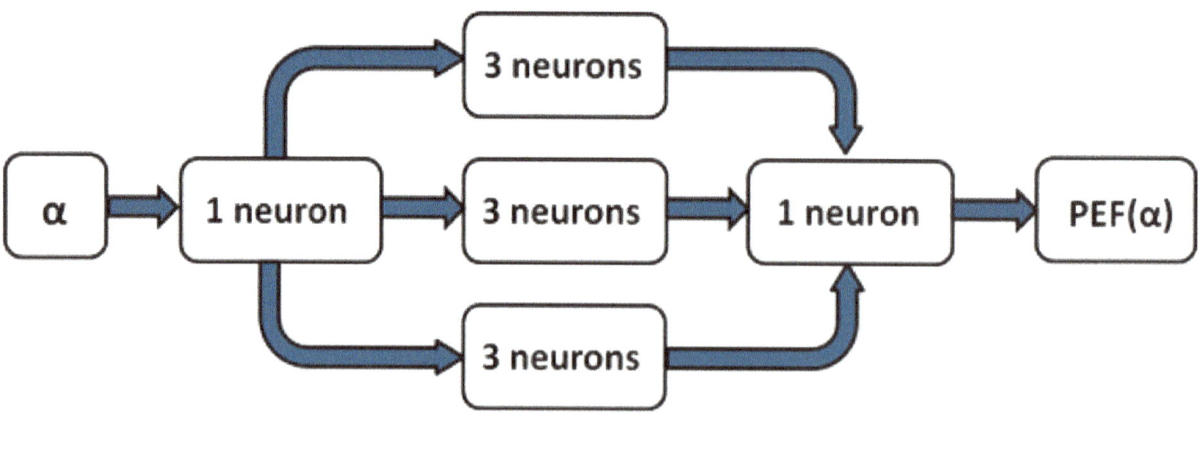

Figure 1: Construction of the neural network used for probe error modeling. α: probe deflection angle; PEF: Probe Error Function.

The second module is responsible for a simulation of probe head random errors. The chi-squared test should be used to check the distribution of results obtained during the gauge ring measurement. Tests performed previously indicated that the resulting distributions are mostly similar to a normal distribution. In this case, the random error of the probe head connected with a certain probing direction could be modeled using the standard deviation obtained for the considered point. In order to interpolate the values of standard deviation between measured points, the artificial neural network with identical construction to the one in the previous case was used. Finally, a normal distribution generator based on the Monte Carlo method may be used in order to simulate the random error. More details regarding this model may be found in [16].

Another probe head error model developed at LCM is based on the Monte Carlo method. The main idea of the model is the same as in the previously discussed model, namely to link probe head errors with the probing direction; however, this model was adapted to describe the probe spatial approach to the measured surface. In such a situation at least two angles defining tip deflection should be used to sufficiently describe the probe head behavior. Assuming that deflection can be described similarly to the definition of coordinates in a spherical coordinate system, the final probe head error function (PEF) can be written in the following form in Equation (1) [13]:

$$PEF = (\alpha, \beta, PE) \quad (1)$$

The standard used during the model preparation stage is the calibration sphere which fulfills the requirements formulated in [13]. The sphere is measured in evenly distributed points on the upper hemisphere of the standard. These points define the nodes of the reference grid of the model, in which probe head errors are determined empirically through measurements. The arrangement of the points can be described using angles (α, β). The sphere is measured at least 10 times, and on the basis of the acquired results the mean value of the PEF at each point is calculated, as well as standard deviation associated with it. Then, a t-distribution described by $\bar{x}$, s, v parameters is assigned to each grid node, where $\bar{x}$ is the mean value of the distribution, equal to the mean value of the PEF, s is the dispersion equal to the calculated standard deviation, and v is the number of degrees of freedom equal to the number of measurements minus one. When the direction is different than the one for which errors were determined experimentally, the bilinear interpolation adapted for the spherical system must be used in order to determine the values describing the probability distribution.

The value of the PE for the chosen probing direction is simulated using the Monte Carlo method. Since the PE can be treated as a vector with the same direction as a probing direction, it is possible to calculate its effect on the coordinates of measured points. The probing errors along axes x, y, z of a machine's datum system (PE_x, PE_y, PE_z) can be expressed as follows:

$$PE_x = PEF \times nx \quad (2)$$

$$PE_y = PEF \times ny \quad (3)$$

$$PE_z = PEF \times nz \quad (4)$$

where:

nx, ny, nz—approach direction cosines

PEF—value of probe head error function

The experiment described in [23] was conducted to check the number of points that should be included in the reference grid. As a result, it was found that the model based on 46 points accurately describes the probe errors. More details regarding this model can be found in [13].

The modules presented above served as a starting point for adapting the probe error function to describe the field of errors of the probe heads utilized on five-axis coordinate systems.

DEVELOPED SIMULATIVE MODEL AND STEPS TO ITS IMPLEMENTATION

This section describes the concept of the developed simulative model, and the experiments that were designed and performed in order to implement this model and verify its reliability.

Description of Simulative Model

The developed simulative model is based on the concept of PEF described in Section 1. In this case, the number of input quantities that has to be specified in order to obtain the probe error (PE) is equal to three. The PEF may be written as presented in Equation (5):

$$PEF = (A, B, \alpha, PE) \quad (5)$$

where:

A—rotation angle along the horizontal axis of the probe head,

B—rotation angle along the vertical axis of the probe head,

α—angle in which the touch-trigger module is working,

PE—probe error given as a result of PEF usage for considered angles.

The A and B angles were presented in Figure 2. They have their zero positions when the probe is oriented vertically. The α angle is dependent on the A and B angles. It is defined on the plane perpendicular to the probe when it is oriented using Aand B angular positions, and has its zero indication in the direction in which the probe rotates along the A axis in the positive direction. The angle increases in the counterclockwise direction.

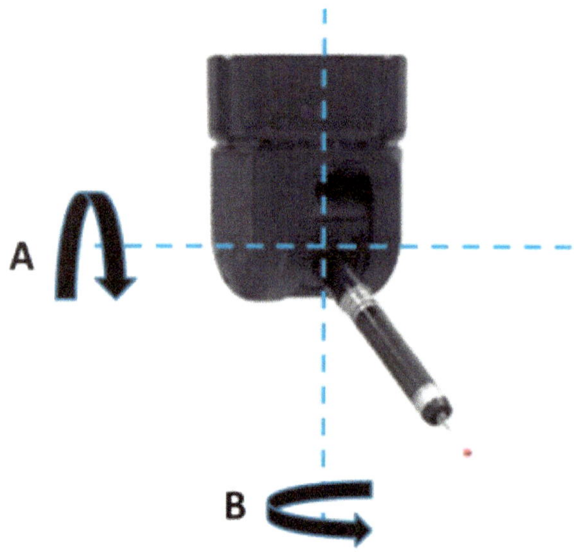

Figure 2: Articulated probe head used in five-axis coordinate systems with revolution axes marked. A: rotation angle along the horizontal axis of the probe head; B: rotation angle along the vertical axis of the probe head.

Depending on those three input parameters, the model provides the output value, which is the simulated value of the PE. The simulation is based on the Monte Carlo method and trilinear interpolation adapted for usage in polar systems. It uses the values of PE identified experimentally for selected A, B and α angles (experiments aimed at determining these values are presented in Section 2.2). The points determined using these angles for the chosen stylus used during measurements may be treated as nodes of the reference point grid. Let A_s, B_s and α_s denote the values of angles for which the simulation of errors has to be performed. A_{s-1}, B_{s-1} and α_{s-1} are the values of angles for the nearest node with angles lower than A_s, B_s and α_s, respectively, and A_{s+1}, B_{s+1} and α_{s+1} are the values of angles for the nearest node with corresponding angles higher than A_s,B_s and α_s. The values of PEs in nodes surrounding the point defined using A_s, B_s and α_s have to be used. In order to simulate the PE value for the point defined using A_s, B_s and α_s, the values of PEs in nodes surrounding this point have to be simulated using the Monte Carlo method. For all of the simulations presented here, the Monte Carlo method uses the scaled and shifted t-distributions with parameters ($\bar{x}$, s, v), where $\bar{x}$ denotes the mean radial PE, s is the standard deviation associated with $\bar{x}$ and v is the number of degrees of freedom. The parameters of these distributions are determined using the experiment presented in Section 2.2. Hence, the simulation should be performed for nodes (A_{s-1}, B_{s-1}, α_{s-1}), (A_{s-1}, B_{s-1}, α_{s+1}), (A_{s-1}, B_{s+1}, α_{s-1}), (A_{s-1}, B_{s+1}, α_{s+1}), (A_{s+1}, B_{s-1}, α_{s-1}), (A_{s+1}, B_{s-1}, α_{s+1}), (A_{s+1}, B_{s+1}, α_{s-1}) and (A_{s+1}, B_{s+1}, α_{s+1}). Then, a trilinear interpolation according to Formula (6) should be performed in order to obtain the PE value for the simulated point.

$$PEF(A_s, B_s, \alpha_s) = ((A_{s+1} - A_s)/(A_{s+1} - A_{s-1}) \times ((B_{s+1} - B_s)/(B_{s+1} - B_{s-1}) \times P2 + (B_s - B_{s-1})/(B_{s+1} - B_{s-1}) \times P4)) + ((A_s - A_{s-1})/(A_{s+1} - A_{s-1}) \times ((B_{s+1} - B_s)/(B_{s+1} - B_{s-1}) \times P1 + (B_s - B_{s-1})/(B_{s+1} - B_{s-1}) \times P3)) \quad (6)$$

where:

$$P1 = (((\alpha_{s+1} - \alpha_s)/(\alpha_{s+1} - \alpha_{s-1})) \times PEF(A_{s+1}, B_{s-1}, \alpha_{s-1})) + (((\alpha_s - \alpha_{s-1})/(\alpha_{s+1} - \alpha_{s-1})) \times PEF(A_{s+1}, B_{s-1}, \alpha_{s+1}))$$

$$P2 = (((\alpha_{s+1} - \alpha_s)/(\alpha_{s+1} - \alpha_{s-1})) \times PEF(A_{s-1}, B_{s-1}, \alpha_{s-1})) + (((\alpha_s - \alpha_{s-1})/(\alpha_{s+1} - \alpha_{s-1})) \times PEF(A_{s-1}, B_{s-1}, \alpha_{s+1}))$$

$$P3 = (((\alpha_{s+1} - \alpha_s)/(\alpha_{s+1} - \alpha_{s-1})) \times PEF(A_{s+1}, B_{s+1}, \alpha_{s-1})) + (((\alpha_s - \alpha_{s-1})/(\alpha_{s+1} - \alpha_{s-1})) \times PEF(A_{s+1}, B_{s+1}, \alpha_{s+1}))$$

$$P4 = (((\alpha_{s+1} - \alpha_s)/(\alpha_{s+1} - \alpha_{s-1})) \times PEF(A_{s-1}, B_{s+1}, \alpha_{s-1})) + (((\alpha_s - \alpha_{s-1})/(\alpha_{s+1} - \alpha_{s-1})) \times PEF(A_{s-1}, B_{s+1}, \alpha_{s+1}))$$

The presented model may serve two primary purposes. The first involves its use as part of virtual model of a five-axis coordinate system. In this application, it should be connected with one of the modules responsible for the simulation of the CMM's kinematic system errors (presented in [10,12,13,16]). For all of the points considered in a simulated measurement, the kinematic errors should first be simulated, and then the probe head errors should be simulated using the presented model. A single simulation of the probe head error for the considered point is done using the methodology presented above. For proper functioning of the virtual model, the simulation should be repeated n times, where n is the number of Monte Carlo trials set by the user or manufacturer of the virtual CMM software.

The second possible usage of the developed model is to utilize it for simulation and correction of probe head errors during the measurements performed in five-axis mode. A detailed description of this process is presented in Section 2.4.

Implementation Measurements

In order to gather the input data set for model, the reference ring was measured with different angular orientations of the probe head. The reference rings used during experiments were attached to a solid block which was installed in a swivel and tilting vise. A ring with a 20 mm diameter was chosen for the measurements, while other rings were used during datum definition and in further experiments (Section 3.3). In each ring arrangement, the standard was measured 15 times in 64 evenly distributed measured points. For the datum system definition, the CMM works in standard three-axis mode while all reference ring measurements were done using only rotational moves of the probe head. Therefore, the possible influences of machine kinematics are minimized. The positions were selected in a manner that allows covering the majority of the probe head working range. The orientation of the reference ring was changed together with the angular orientation of the probe. The utilized vise allows us to rotate the installed object around two perpendicular axes of revolution, and thus it is possible to set the ring so its axis will be oriented parallel to the stylus in A, B orientation. The authors decided to base the model on measurements carried out in 24 positions defined using B and A probe angles. Angle B changes at 60° in the range between −120° and 180°, while angle A changes at 30° in the range between 0° and 90°. Measurements for one of the described positions are presented in Figure 3.

Figure 3: Measurements of standard ring in one of the selected positions.

Verification Measurements

After the preparation measurements, the ring was measured in further positions to create the verifying data set. The 84 additional positions were selected with B changing at 20° between (−160°; 180°), and A changing at 15° between (0°, 90°). The same measuring procedure was used. The results obtained during these measurements are used to check the difference between the simulated values and the empirically gathered data. The authors assumed that the PE value may be considered as simulated correctly if it differs from the corresponding value obtained experimentally not more than ±3 * standard deviation assigned to the considered point.

Correction of Probe Head Errors Using the Developed Model

To perform a simulation and correction of probe head errors during the measurements performed in five-axis mode, theA, B and α angles of all points that were measured have to be recorded. Then the simulations according to the presented methodology have to be performed n times for all of these points. The mean from n simulated values is taken as the probe head error (PE) for the considered point (depending on α). To perform the correction of this error, the PE_x, PE_y and PE_zcomponents have to be calculated using Equations (2)–(4). Then, the values of these components should be subtracted from the actual values of point coordinates x, y, z, giving the corrected point coordinates x_{corr}, y_{corr} and z_{corr} from Equations (7)–(9).

$$x_{corr} = x - PE_x \tag{7}$$

$$y_{corr} = y - PE_y \tag{8}$$

$$z_{corr} = z - PE_z \tag{9}$$

All measured features and relations should be calculated once again using the corrected point coordinates x_{corr}, y_{corr} and z_{corr}. In the presented research, the process of correction of the probe head errors was assisted by a script written in the Python programming language cooperating with macro prepared in Modus software. The raw measurement data including point coordinates, approach vectors and stylus orientation angles was sent to the Python script, which performed the correction of probe head errors and sent back corrected point coordinates to the metrological software, in which the calculation of measured features was done once again.

RESULTS

Results presented in this section are the results of experiments described in Section 2. All experiments were performed at the Laboratory of Coordinate Metrology on a Zeiss WMM 850S machine (Carl Zeiss, Jena, Germany) equipped with a Renishaw PH20 probe head and a TP20 standard force module and a stylus with a 4 mm ball and a length of 10 mm. The machine is placed in an air-conditioned room with thermal stability at the level of 20 ± 0.5 °C. Modus metrological software was used on the machine.

Results of Identification of Errors

Figure 4 presents the results for a B angle equal to 120° and A changing in the range 0–90°.

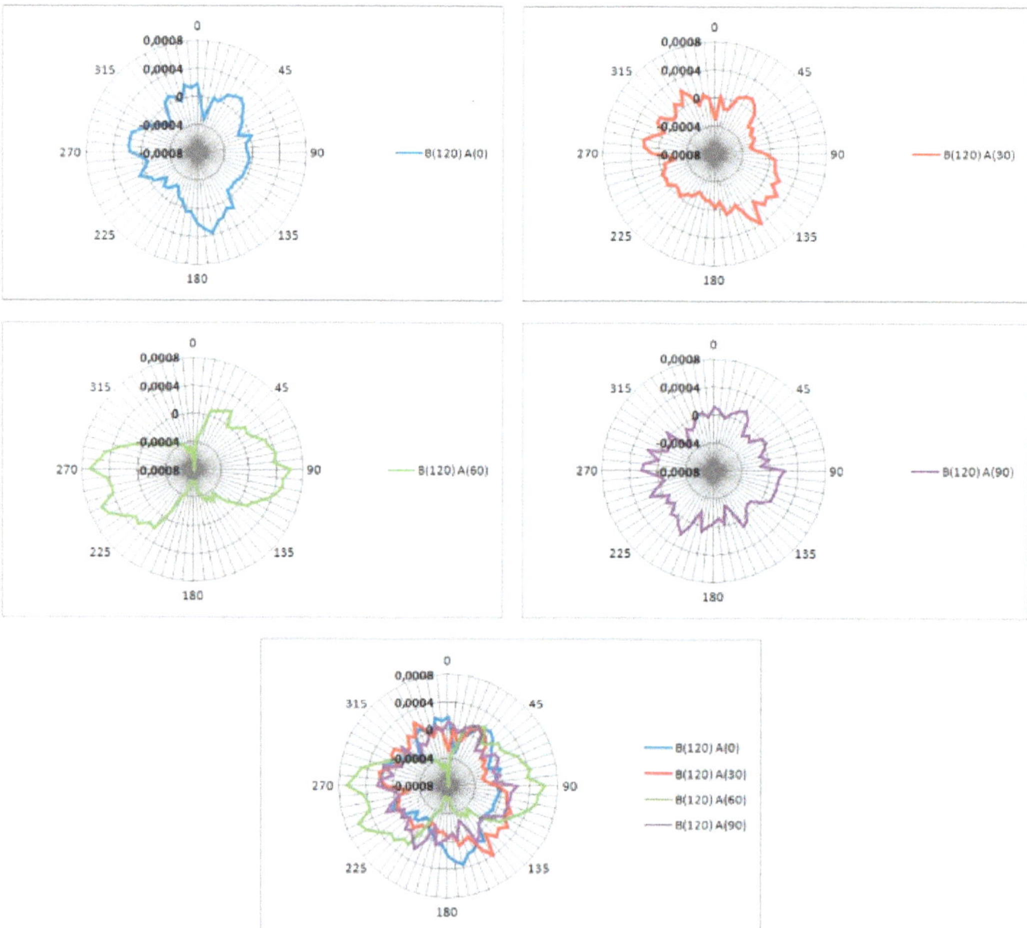

Figure 4: Results of identification of errors for a B angle equal to 120° and A changing in the range 0–90°. All angles given in degrees and errors in mm.

The results of probe head error identification are used as an input data for the presented simulative model. They are given in tabular form. The example of such a table is presented in Table 1.

Table 1: Example of input data for simulative model for A = 30° and B = −120°. A, B, α given in degrees; $\bar{x}$, s in mm.

A	B	α	$\bar{x}$	s	A	B	α	$\bar{x}$	s
30	−120	0.000	−0.00072	0.00048	30	−120	180.000	−0.00017	0.00016
30	−120	5.625	0.00007	0.00019	30	−120	185.625	−0.00034	0.00019
30	−120	11.250	0.00007	0.00020	30	−120	191.250	−0.00024	0.00020
30	−120	16.875	0.00015	0.00019	30	−120	196.875	−0.00034	0.00017
30	−120	22.500	0.00022	0.00013	30	−120	202.500	−0.00010	0.00019
30	−120	28.125	0.00029	0.00016	30	−120	208.125	0.00005	0.00029
30	−120	33.750	0.00015	0.00016	30	−120	213.750	0.00016	0.00027
30	−120	39.375	0.00010	0.00017	30	−120	219.375	0.00032	0.00019
30	−120	45.000	0.00020	0.00023	30	−120	225.000	0.00018	0.00019
30	−120	50.625	0.00019	0.00016	30	−120	230.625	0.00010	0.00012
30	−120	56.250	0.00029	0.00017	30	−120	236.250	0.00012	0.00018
30	−120	61.875	0.00009	0.00021	30	−120	241.875	0.00039	0.00022
30	−120	67.500	0.00010	0.00018	30	−120	247.500	0.00035	0.00020
30	−120	73.125	0.00012	0.00020	30	−120	253.125	0.00026	0.00016
30	−120	78.750	0.00012	0.00016	30	−120	258.750	0.00036	0.00015
30	−120	84.375	0.00016	0.00014	30	−120	264.375	0.00027	0.00019
30	−120	90.000	0.00014	0.00011	30	−120	270.000	0.00009	0.00014
30	−120	95.625	0.00004	0.00010	30	−120	275.625	0.00012	0.00020
30	−120	101.250	−0.00001	0.00023	30	−120	281.250	−0.00003	0.00018
30	−120	106.875	0.00000	0.00012	30	−120	286.875	0.00003	0.00018
30	−120	112.500	−0.00003	0.00015	30	−120	292.500	−0.00001	0.00024
30	−120	118.125	−0.00007	0.00013	30	−120	298.125	−0.00004	0.00016
30	−120	123.750	0.00019	0.00011	30	−120	303.750	−0.00037	0.00024
30	−120	129.375	0.00004	0.00017	30	−120	309.375	−0.00020	0.00018
30	−120	135.000	−0.00004	0.00017	30	−120	315.000	−0.00016	0.00014
30	−120	140.625	−0.00030	0.00019	30	−120	320.625	−0.00006	0.00018
30	−120	146.250	−0.00021	0.00016	30	−120	326.250	−0.00007	0.00013
30	−120	151.875	−0.00023	0.00016	30	−120	331.875	−0.00010	0.00014
30	−120	157.500	−0.00019	0.00017	30	−120	337.500	−0.00032	0.00013
30	−120	163.125	−0.00014	0.00021	30	−120	343.125	−0.00033	0.00015
30	−120	168.750	−0.00009	0.00021	30	−120	348.750	−0.00027	0.00015
30	−120	174.375	−0.00011	0.00011	30	−120	354.375	−0.00021	0.00016

The mean absolute radial error for all positions of the standard ring was equal to 0.0002 mm. The mean value of the standard deviation was equal to 0.0002 mm. The rest of the results of the error identification may be found in thesupplementary materials added to this paper (Table S1).

The authors also repeated measurements of the ring in select positions to determine whether the probe error distributions retain their character. These measurements were carried out after the verification measurements to assess how the considerable exploitation of the probe head (the whole measuring procedure for one position consists of 960 measured points) affects the obtained results. Figure 5 presents the results of verification measurement for position B = 0°, A = 45° and the results obtained for repeated measurements in the same position. The measurements were repeated after measurements of 46 other positions. The results obtained for the checked positions show satisfactory consistency; therefore, it is justified to base the model on a data set obtained using the experiments described in Section 2.2.

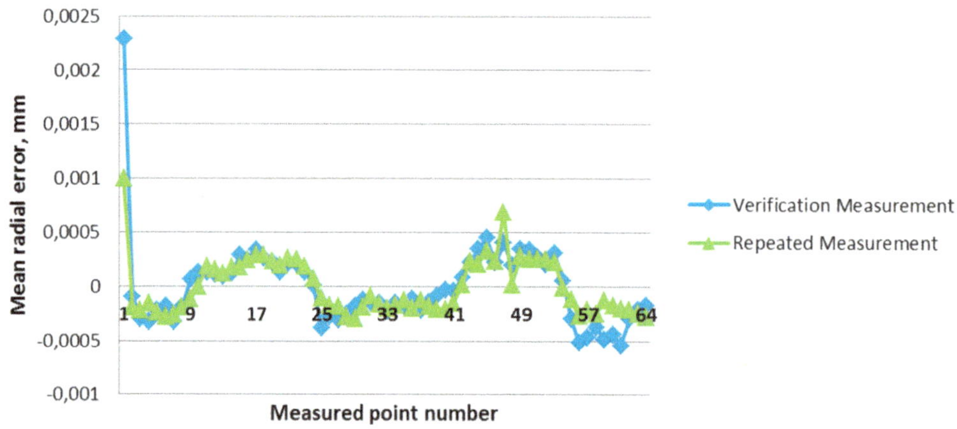

Figure 5: Results of mean Probing Error (PE) values obtained for the chosen position during verification measurements and measurements repeated after a relatively long time of probe head functioning.

Results of Model Verification

The verification of the model was done according to methodology presented in Section 2.3. The verification was done for 84 positions of the standard ring. Figure 6 and Figure 7 present the comparison between the results of the experimental probe head error identification and the simulation using the developed model for A = 75° and B = 80° (Figure 5) and A = −100° and B = 30° (Figure 6).

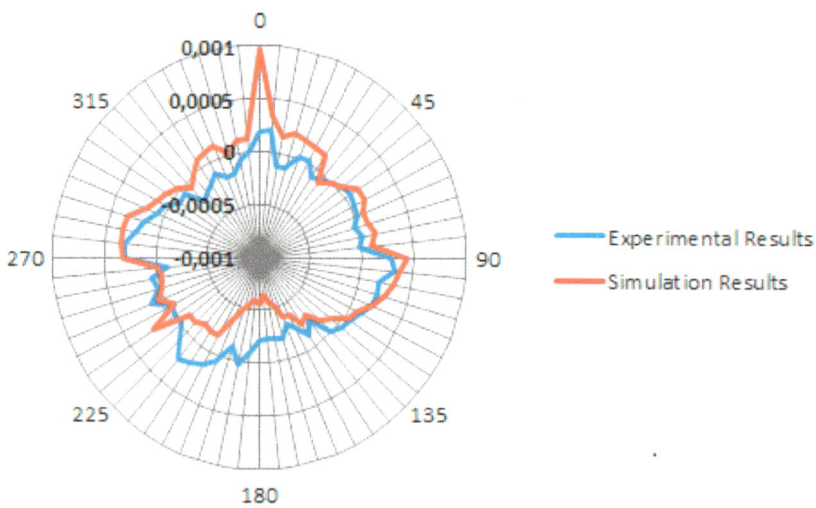

Figure 6: Comparison between results of experimental probe head errors identification and simulation using developed model for A = 75 and B = 80.

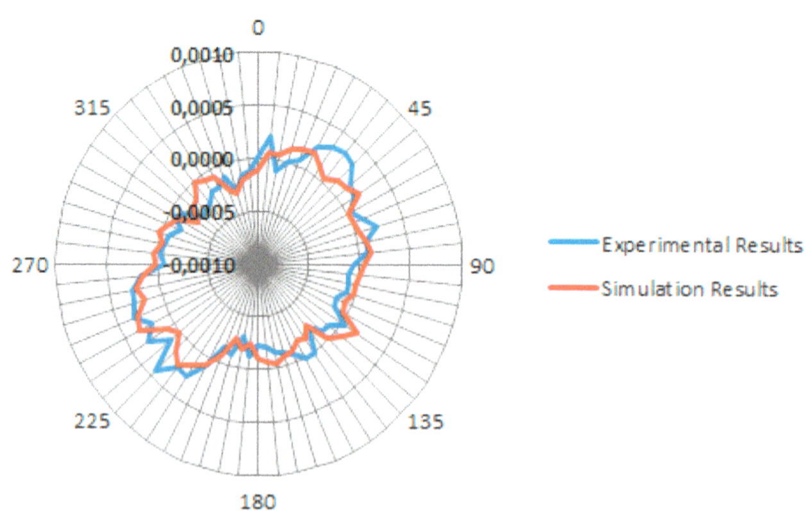

Figure 7: Comparison between results of experimental probe head errors identification and simulation using developed model for A = −100 and B = 30.

For the first position the mean absolute difference between the real values of the probe head errors (determined experimentally) and the simulated ones was equal to 0.0002 mm and for the second position 0.0001 mm. The authors assumed that the PE value may be considered as simulated correctly if it differs from the corresponding value obtained experimentally not more than ±3 × standard deviation assigned to the considered point. Out of 64 simulated values of PE for each position, 61 (95.31%) were properly simulated for the first position and 64 (100%) for the second position (according to assumption presented in Section 2.3).

Example of Probe Head Error Correction

The possibility of correcting probe head errors was presented on the example of measurements of the standard ring with a diameter of 34 mm. The values of diameter and roundness deviation were determined during the measurements. The ring was mounted in the plane perpendicular to the probe oriented using A = 43° and B = 91°. Table 2 presents the results obtained during the real measurements without compensation of the probe head errors and after the compensation done using the methodology presented in Section 2.4. Results were compared with the data given on the calibration certificate of the ring.

Table 2: Results of roundness deviation measurements (given in mm) for standard ring with and without probe head errors correction.

Number of Points Used	Without Correction	With Correction	Calibration Certificate
64	0.0013	0.0010	0.0004
16	0.0012	0.0008	0.0004
8	0.0010	0.0005	0.0004

DISCUSSION

The primary conclusion of the presented experiments is that the developed simulation model allows for faithful reproduction of the error field for probe heads used in five-axis coordinate systems. In the case of the two presented positions used during verification measurements, the percentage of properly simulated error values was 95.31% and 100%, respectively. Such high numbers prove the effectiveness of the presented model and suggest its possible application as a module for probe head error simulation in virtual models of five-axis CMMs.

The developed model may also be helpful during the correction of probe head errors. As was shown in Table 2, the precision of the roundness deviation measurements has noticeably improved following the implementation of the correction using the methodology presented in Section 2.4. It should be noted that the effectiveness of probe head error correction is more pronounced in the case of a smaller number of measuring points used for the calculation of geometrical features. In this paper, example measurements were carried out on a circle, but the authors' previous research suggests similar relationships in the case of other features. This phenomenon is connected with the averaging tendency of the least squares method, usually utilized for calculating features.

The results presented in Figure 5 indicate good stability of probe head error characteristics. It can be concluded from this that irrespective of the usage of the developed model, the initial measurements aimed at gathering input data for the model would not have to be repeated very frequently. The ideal interval for the re-identification of probe head errors should be investigated for different probe head types, but the authors suppose that it could be expressed in months.

As for future research directions, the most important goal is to develop a fully functional virtual model of a five-axis coordinate system. The research presented in this paper marks the first step in this direction. The simulative model discussed here should be connected to a model responsible for simulating kinematic system errors of the machine. Such a combined virtual model would reduce the time spent on the determination of the uncertainty of measurements performed on the five-axis coordinate system. Seeing as numerous systems of this type are presently used in industrial conditions, this would be likely to significantly reduce costs associated with quality control, and as a natural consequence, the total production costs would also drop down.

The model presented in this paper was checked for the measurements of regular rotary features for which the probes such as the one discussed here are usually applied in practice. In case of features such as cylinders, circles, spheres, cones, etc., the measurements may be easily performed using only the movements of the probe head, which significantly reduces their total time (in relation to the standard three-axis CMM). The measurements of more complicated shapes are also possible but the reduction of the measurement duration is not that meaningful as in the case of regular features. So it may be concluded that the model works properly for the majority of practical usages of the discussed probes. However, in the future, it will be advisable to also include in it the possibility of simulating the measurements of more complicated geometries, when the stylus has to move on complex trajectories and the α angle cannot be easily defined on the plane perpendicular to the probe when it is oriented using A and B angular positions.

The next important possibility of improving the presented model is to reduce the number of orientations and points used during identification measurements. It should be possible to minimize their number in a way that simultaneously reduces the time needed for model implementation, and provides faithful reproduction of a real field of probe head errors.

SUPPLEMENTARY MATERIALS

The following are available online at www.mdpi.com/2076-3417/6/5/144/s1, Table S1: Complete results of probe head errors identification.

ACKNOWLEDGMENTS

Reported research was realized as part of a project financed by National Science Centre, Poland, grant No. 2015/17/D/ST8/01280. Costs to publish in open access were also paid from this source of funding.

AUTHOR CONTRIBUTIONS

Adam Gąska and Piotr Gąska conceived and designed the simulative model and the experiments; Piotr Gąska and Maciej Gruza performed the experiments; Piotr Gąska prepared Python scripts and Modus macros; Piotr Gąska and Adam Gąska analyzed the data; Adam Gąska and Piotr Gąska wrote the paper.

CONFLICTS OF INTEREST

The authors declare no conflict of interest.

REFERENCES

1. Kunzmann, H.; Pfeifer, T.; Schmitt, R.; Schwenke, H.; Weckenmann, A. Productive metrology-adding value to manufacture. CIRP Ann. Manuf. Technol. 2005, 54, 691–704.
2. Caja, J.; Gomez, E.; Maresca, P. Optical measuring equipments. Part I: Calibration model and uncertainty estimation.Precis. Eng. 2015, 40, 298–304.
3. Asano, Y.; Furutani, R.; Ozaki, M. Verification of interim check method of CMM. Int. J. Autom. Technol. 2011, 5, 115–119.
4. Cuesta, E.; Alvarez, B.; Sanchez-Lasheras, F.; Gonzalez-Madruga, D. A statistical approach to prediction of the CMM drift behaviour using a calibrated mechanical artefact. Metrol. Meas. Syst. 2015, 22, 417–428.
5. Takamasu, K.; Takahashi, S.; Abbe, M.; Furutani, R. Uncertainty estimation for coordinate metrology with effects of calibration and form deviation in strategy of measurement. Meas. Sci. Technol. 2008, 19.
6. International Organization for Standardization. ISO 15530-3:2011—Geometrical Product Specifications (GPS)—Coordinate Measuring Machines (CMM): Technique for Determining the Uncertainty of Measurement—Part 3: Use of Calibrated Workpieces or Measurement Standards; ISO: Geneva, Switzerland, 2011.
7. Weckenmann, A.; Lorz, J. Monitoring coordinate measuring machines by calibrated parts. J. Phys. Conf. Ser. 2005, 13, 190–193.
8. Sato, O.; Osawa, S.; Kondo, Y.; Komori, M.; Takatsuji, T. Calibration and uncertainty evaluation of single pitch deviation by multiple-measurement technique. Precis. Eng. 2010, 34, 156–163.
9. Osawa, S.; Busch, K.; Franke, M.; Schwenke, H. Multiple orientation technique for the calibration of cylindrical workpieces on CMMs. Precis. Eng. 2005, 29, 56–64.
10. Trapet, E.; Franke, M.; Hartig, F.; Schwenke, H.; Wadele, F.; Cox, M.; Forbers, A.; Delbressine, F.; Schnellkens, P.; Trenk, M.; et al. Traceability of Coordinate Measuring Machines According to the Method of the Virtual Measuring Machine; PTB: Braunschweig, Germany, 1999.
11. Schwenke, H.; Siebert, B.R.L.; Wäldele, F.; Kunzmann, H. Assessment of uncertainties in dimensional metrology by Monte Carlo simulation: Proposal of a modular and visual software. CIRP Ann. Manuf. Technol. 2000, 49, 395–398.
12. Wilhelm, R.G.; Hocken, R.; Schwenke, H. Task specific uncertainty in coordinate measurement. CIRP Ann. Manuf. Technol. 2001, 50, 553–563.
13. Sładek, J.; Gąska, A. Evaluation of coordinate measurement uncertainty with use of virtual machine model based on Monte Carlo method. Measurement 2012, 45, 1564–1575.
14. Aggogeri, F.; Barbato, G.; Barini, E.M.; Genta, G.; Levi, R. Measurement uncertainty assessment of coordinate measuring machines by simulation and planned experimentation. CIRP J. Manuf. Sci. Technol. 2011, 4, 51–56.
15. Giusca, C.L.; Leach, R.K.; Forbes, A.B. A virtual machine-based uncertainty evaluation for a traceable areal surface texture measuring instrument. Measurement 2011, 44, 988–993.
16. Sładek, J. Coordinate Metrology: Accuracy of Systems and Measurements; Springer: Berlin, Germany, 2016.
17. Maresca, P.; Gomez, E.; Caja, J.; Barajas, C. Virtual environment for the simulation of a horizontal coordinate measuring machine in the teaching of dimensional metrology. Procedia Eng. 2013, 63, 234–242.
18. Ostrowska, K.; Gąska, A.; Sładek, J. Determining the uncertainty of measurement with the use of a Virtual Coordinate Measuring Arm. Int. J. Adv. Manuf. Technol. 2014, 71, 529–537.
19. Cuesta, E.; Talenti, A.; Patino, H.; Gonzalez-Madruga, D.; Martínez-Pellitero, S. Sensor prototype to evaluate the contact force in measuring with coordinate measuring arms. Sensors 2015, 15, 13242–13257.
20. Nafi, A.; Mayer, J.R.R.; Wozniak, A. Novel CMM-based implementation of the multi-step method for the separation of machine and probe errors. Precis. Eng. 2011, 35, 318–328.
21. Gąska, A.; Krawczyk, M.; Kupiec, R.; Ostrowska, K.; Gąska, P.; Sładek, J. Modeling of the residual kinematic errors of coordinate measuring machines using Laser Tracer system. Int. J. Adv. Manuf. Technol. 2014, 73, 497–507.
22. Dobosz, M.; Woźniak, A. CMM touch trigger probes testing using a reference axis. Precis. Eng. 2005, 29, 281–289.
23. Gąska, A.; Harmatys, W.; Gąska, P.; Gruza, M.; Mathia, T.; Sładek, J. Optimization of probe head errors model used in Virtual CMM systems. In Proceedings of the 11th International Conference on Laser Metrology and Machine Performance (LAMDAMAP) 2015, Huddersfield, UK, 17–18 March 2015; Blunt, L., Hansen, H.N., Eds.; euspen: Cranfield, UK, 2015.

CHAPTER 13

A SYNCHROPHASOR BASED OPTIMAL VOLTAGE CONTROL SCHEME WITH SUCCESSIVE VOLTAGE STABILITY MARGIN IMPROVEMENT

Heng-Yi Su [1], Yi-Chung Chen [2] and Yu-Liang Hsu [3]

[1]Department of Electrical Engineering, Feng Chia University (FCU), No.100, Wenhwa Road, Seatwen, Taichung 40724, Taiwan

[2]Department of Information Engineering and Computer Science, Feng Chia University (FCU), No.100, Wenhwa Road, Seatwen, Taichung 40724, Taiwan

[3]Department of Automatic Control Engineering, Feng Chia University (FCU), No.100, Wenhwa Road, Seatwen, Taichung 40724, Taiwan

ABSTRACT

This paper proposes an optimal control scheme based on a synchronized phasor (synchrophasor) for power system secondary voltage control. The framework covers voltage stability monitoring and control. Specifically, a voltage stability margin estimation algorithm is developed and built in the newly designed adaptive secondary voltage control (ASVC) method to achieve more reliable and efficient voltage regulation in power systems. This new approach is applied to improve voltage profile across the entire power grid by an optimized plan for VAR (reactive power) sources allocation; therefore, voltage stability margin of a power system can be increased to reduce the risk of voltage collapse. An extensive simulation study on the IEEE 30-bus test system is carried out to demonstrate the feasibility and effectiveness of the proposed scheme.

INTRODUCTION

Voltage instability has been regarded as one of the primary threats to the security of modern power network operation during the past few decades. Power system disturbances such as a continuous load increase and/or a major change in network topology can result in voltage collapse. The voltage collapse problem, which is characterized by the loss of voltage magnitude at certain locations of the power grid, has caused several severe blackout events worldwide [1,2].

A number of planning and operation technologies have been proposed to mitigate the risk of voltage collapse [3]. Among these technologies, phasor measurement unit (PMU) based schemes to secure power systems have become one of the enabling techniques which are under active investigations. Indeed, the use of PMUs in modern power systems becomes popular [4,5].

In order to provide a better voltage support in transmission networks, the coordinated voltage control has been developed. It is organized as a hierarchical structure with three levels: the primary, secondary, and tertiary voltage control. Significant attention has been given to the study of the secondary level, which is an automatic regulation of voltage and reactive power for power systems [6,7,8,9,10,11,12,13,14,15,16,17,18,19]. The task of the secondary voltage control (SVC) is to regulate system voltage profile and to protect against potential voltage instability scenario at an early stage. The conventional

Citation: Heng-Yi Su, Yi-Chung Chen and Yu-Liang Hsu, A Synchrophasor Based Optimal Voltage Control Scheme with Successive Voltage Stability Margin Improvement, doi:10.3390/app6010014

approaches use only the voltage information at the observed buses (the so-called pilot nodes) as a triggering signal, but just monitoring voltage magnitudes alone may give the wrong indication in static voltage stability study [20]. In other words, voltage stability issues cannot be fully prevented by the traditional SVC methods; therefore, an additional accepted measure of static voltage stability is required for a more reliable SVC scheme.

The voltage stability margin (VSM), which is defined to be the distance between the given operating point and the voltage-collapse point, can serve as the static voltage stability indicator. A wide variety of approaches have been proposed for static VSM evaluation [21,22,23,24,25,26], wherein the measurement-based methods [25,26] address such problem by using the impedance-matching concept, *i.e.*, at the maximum loading condition, load impedance is equal to Thevenin equivalent impedance in magnitude. Due to the elegance and simplicity of the measurement-based method, it becomes an attractive alternative in static voltage stability monitoring. In this paper, a VSM index computed from synchrophasor data is utilized.

This paper is concerned with designing an optimal voltage control scheme using synchrophasor measurements. In the proposed technique, the voltage magnitude and static voltage stability margin are considered as two key criteria for activating the proposed control strategy; therefore, a more efficient secondary voltage control can be accomplished. The rest of the paper is organized as follows: In Section 2, the proposed scheme based on the PMU technique is presented. Section 3describes the fundamental theories and mathematical principles of the proposed scheme. Numerical simulations and test results are given and discussed in Section 4. Section 5 concludes the paper.

OVERVIEW OF THE PROPOSED SCHEME

The overall architecture of the proposed scheme is shown in Figure 1. The functions and principles of the scheme are briefly described in this section.

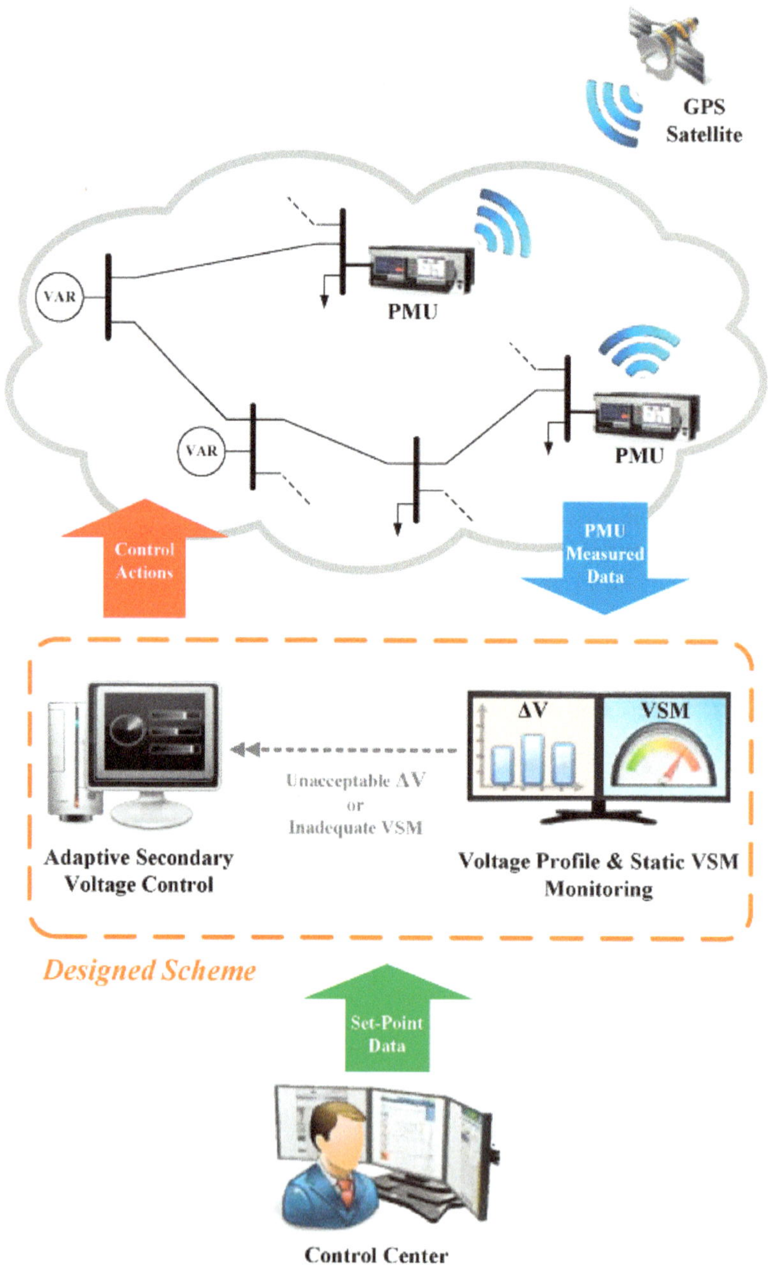

Figure 1: Schematic view of the proposed scheme.

The designed scheme is illustrated with the part enclosed by the dashed line shown in Figure 1. Indeed, it consists of two primary functions: (i) voltage profile and static VSM monitoring function, which monitors not only voltage magnitudes but static VSM as well using synchrophasor data from PMUs; and (ii) adaptive secondary voltage control (ASVC) function which provides appropriate control actions to the controllers. The proposed scheme aims to regulate voltage profiles and to enlarge static voltage stability margin by means of a proper adjustment of VAR sources when the power grid has poor voltage profile and/or shows the risk to voltage collapse.

The flowchart of the proposed scheme is illustrated in Figure 2 and is described in the following:

Step 1: In this study, we assume that PMUs are installed at the selected pilot buses. Using the precise timing signal provided by GPS as the common time base for PMUs, both magnitude and phase angle of voltage and current signals at different PMU locations can be measured, at exactly the same time instant from all observable system buses. Once the PMU measurements with time tags are entered into the designed scheme, these coherent and real-time measured quantities are then used to process the tasks of the proposed scheme.

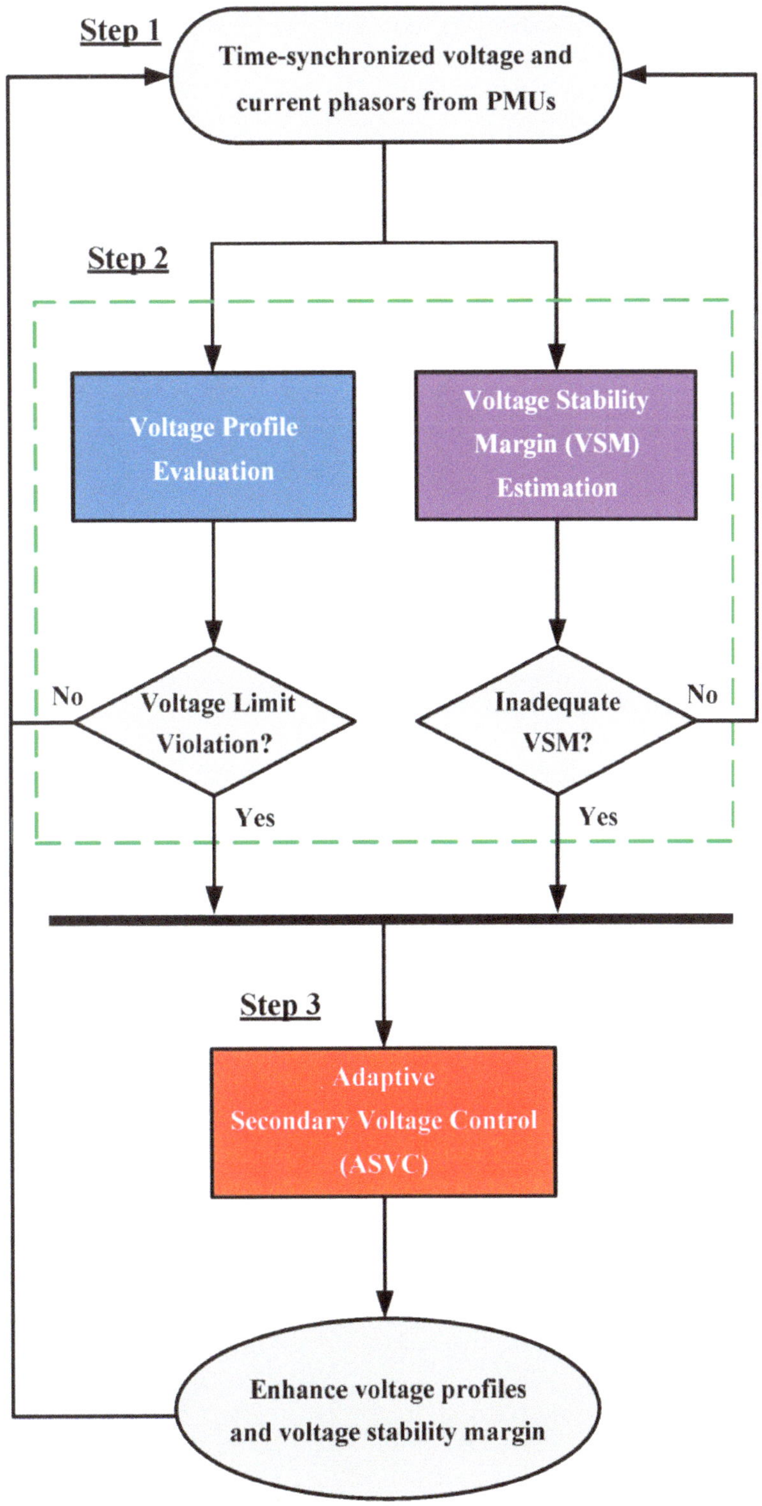

Figure 2: Flowchart of the proposed scheme.

Step 2: At this stage, the proposed scheme performs voltage profile evaluation and static voltage stability margin estimation simultaneously.

(1)The voltage magnitudes, which are extracted from the PMUs in step 1, are used to evaluate the voltage magnitudes of the monitored buses. As long as the measured voltages are not within the specified limit, voltage violation is confirmed. Then, the scheme will issue a trigger signal to the ASVC function. Otherwise, the scheme will go back to Step 1.

(2)As mentioned before, depending on bus voltages for the activation of control actions is not a very reliable strategy to be adopted. Another important index, which measures the proximity of an operating point to voltage instability, is the voltage stability margin (VSM). Therefore, the scheme needs to further check to see whether the current power system state is operated within a sufficient safety margin or not. If the computed value VSM is less than the predetermined security margin, inadequate VSM is identified. This will initiate the ASVC action accordingly.

Step 3: When critical voltage level or inadequate VSM is detected, the ASVC method will be activated automatically. The proposed control strategy can provide appropriate control actions to improve voltage profiles as well as voltage stability margin of the power system. That is, the proposed scheme is capable of preventing the system from possible voltage collapse.

The following section shows the detailed principles and methods used in the proposed scheme.

PRINCIPLES OF THE SCHEME

The proposed scheme is intended to automatically improve power system voltage profiles and static voltage stability margin simultaneously. The basic principles of the scheme will be derived in the following subsections.

Adaptive Secondary Voltage Control (ASVC) Function

First, consider that the approximate model of the small disturbance voltage-var control is represented by

$$\begin{bmatrix} \Delta Q_G \\ \Delta Q_L \end{bmatrix} = \begin{bmatrix} B_{GG} & B_{GL} \\ B_{LG} & B_{LL} \end{bmatrix} \begin{bmatrix} \Delta |V_G| \\ \Delta |V_L| \end{bmatrix} \tag{1}$$

where ΔQ and $\Delta|V|$ are the reactive power and voltage magnitude change vectors; B stands for the system susceptance matrix; and the subscripts L and G denote the load and the voltage-controlled buses, respectively. In the above matrix equation, load voltage changes $\Delta|V|$ can be expressed as

$$\Delta |V_L| = J_1 q - J_2 u \tag{2}$$

where

$$J_1 = B_{LL}^{-1}$$
$$J_2 = J_1 B_{LG} \tag{3}$$
$$q = \Delta Q_L$$
$$u = \Delta |V_G|. \tag{4}$$

Notice that q and u are considered as reactive power load disturbances and control variables, respectively.

In this study, the feedback control law is used to develop the secondary voltage controller. The control vector is determined by using the worst-case design, the technique which addresses the problem by applying the minimization of the maximum load voltage deviation as an objective function for the optimal control model. Mathematically, the problem is to search an optimal set of control solutions such that the l_∞ norm of load voltage changes is minimized.

Suppose that PMUs are placed at the chosen pilot nodes; the voltage deviations at the observed buses can be elaborated by

$$\Delta |V_P| = |V_P^*| - |V_P| \tag{5}$$

Where $|V_P^*|$ and $|V_P|$ denote the pilot-node set-point voltages determined by tertiary level and pilot-node measured voltages obtained from installed PMUs, respectively. In the proposed secondary voltage control, both voltage profile and static voltage stability margin are adopted to serve as trigger signals, as shown in Figure 3. It utilizes pilot-bus voltage changes $\Delta|Vp|$ as input signals, and generates control actions u as output signals.

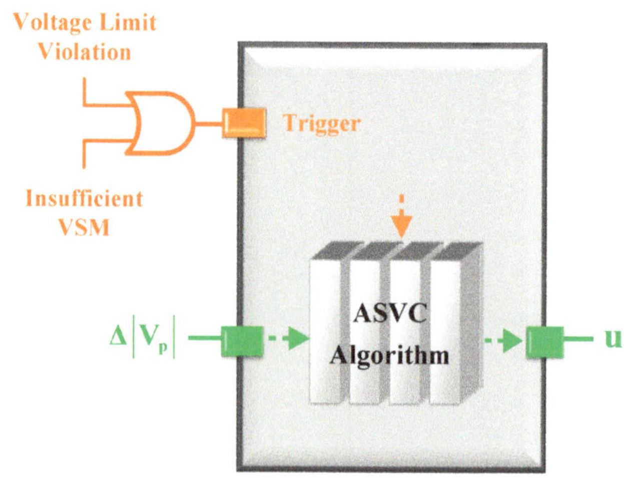

Figure 3: Block diagram of the proposed ASVC method. Once voltage limit violation or insufficient VSM is identified, the ASVC algorithm will be triggered automatically.

Based on different numbers of pilot buses being used, the following conditions are considered.

(1) All Load Buses as Pilot Buses: This means that voltage measurements are available at all load buses. Under this circumstance, the first term J1q in Equation (2) can be replaced by $\Delta|Vp|$; therefore, the problem formulation of the voltage control strategy is stated as

$$\min_u \| \Delta |V_P| - J_2 u \|_\infty$$
$$\textit{subject to } |V_G^{min}| \leq |V_G| \leq |V_G^{max}| \tag{6}$$

Where $|V_G^{min}|$ and $|V_G^{max}|$ are the lower and upper bounds of generator voltage magnitudes, respectively.

(2) Partial Load Buses as Pilot Buses: This means that voltage measurements are only available at the given

pilot buses, and thus $\Delta|V_p|$ only has some elements of the vector J1q under this situation. Indeed, $\Delta|V_p|$ can be related to q by the equation $J_{pq}=\Delta|V_p|$, in which J_p is the matrix with the rows of J1 corresponding to the pilot points selected.

Since there are fewer measurements than variables to be estimated, the equation $J_{pq}=\Delta|V_p|$ is underdetermined. In the proposed method, however, any reactive power disturbance variable vector q that satisfies $Jpq=\Delta|V_p|$ can be utilized to determine feasible control signals that minimize $\| \Delta|VL| \|_\infty$. The least-norm technique [27], which is the most commonly used for solving an underdetermined set of linear equations, is applied to approximate q. In this case, the problem is to findq that satisfies $J_{pq}=\Delta|V_p|$ and minimizes ‖q‖2. Thus, the estimate of q is given by

$$\mathbf{q}^* = J_p^T \left(J_p J_p^T\right)^{-1} \Delta \left|\mathbf{V_p}\right| \tag{7}$$

With the optimal q^*, the proposed control strategy can be formulated in the following search problem:

$$\min_{\mathbf{u}} \| J_1\mathbf{q}^* - J_2\mathbf{u} \|_\infty \quad \textit{subject to } |\mathbf{V}_\mathrm{G}^{\min}| \le |\mathbf{V}_\mathrm{G}| \le |\mathbf{V}_\mathrm{G}^{\max}|. \tag{8}$$

The constrained optimization problems in Equations (6) and (8) can be reformulated as linear programming problems [27] and be solved by using a linear programing solver such as *linprog* function in the MATLAB optimization toolbox [28]. The optimal control actions, which are obtained from Equations (6) or (8), ensure that the resulting worst load voltage change can precisely remain within the range of the predefined constraint regardless of any unexpected load disturbances acting on the system.

Voltage Stability Margin Estimation Function

A common static voltage stability index is expressed by the voltage stability margin which shows how far the system is away from a possible instability event. In order to rapidly assess static voltage stability of a power grid, the local measurement-based methods have been presented in the works [25,26]. The key idea of these techniques is provided here.

To start with, consider a load at bus *i* connected to a complex power system, which can be simplified to a single-machine-infinite-bus system by an estimated Thevenin equivalent network as shown in Figure 4, where E_i^{th}, Z_i^{th}, and Z_i^L correspond to Thevenin equivalent voltage, Thevenin equivalent impedance, and load impedance in phasor representation at bus *i*, respectively. When the power transmitted is maximum at bus *i* $|Z_i^L|$ is identical to $|Z_i^{th}|$. Based on the impedance match theory, tracking $|Z_i^L|$ and $|Z_i^{th}|$ plays an important role in real-time voltage instability detection.

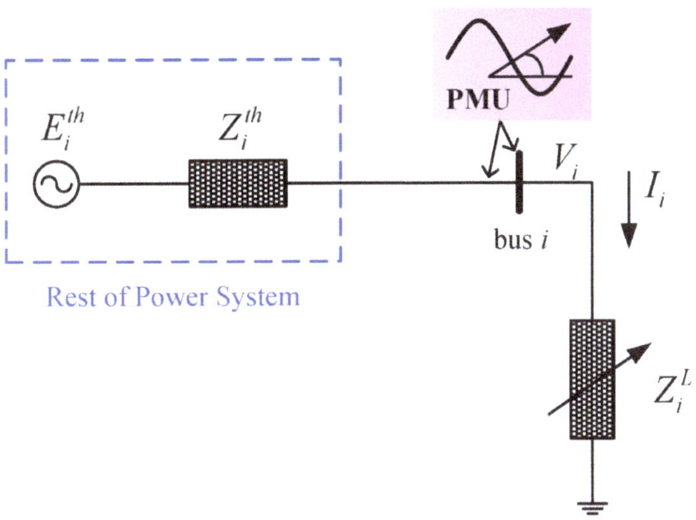

Figure 4: Thevenin equivalent network at bus *i* with PMU installed.

Using the approach above, the impedance based VSM index of bus *i* is defined as

$$VSM_i = \frac{|Z_i^M|}{|Z_i^L|} = 1 - \frac{|Z_i^{th}|}{|Z_i^L|} \tag{9}$$

Where $|Z_i^L|$ and $|Z_i^{th}|$ can be obtained by using the local measured voltage and current phasors which are available from the installed PMU [25,26]. The static voltage stability margin for the entire power system is defined to be

$$VSM = \min\{VSM_i \mid i = 1, 2, \ldots, p\} \tag{10}$$

where *p* represents the set of numbers of pilot buses. Note that the value of VSM is between 0 and 1. At the voltage-collapse point, the VSM is equal to 0. Although the approach proposed in [25] was employed to estimate VSM, it is noteworthy that any other measurement-based method of VSM analysis can be applied for this purpose, for example, the one proposed in [26].

The index VSM incorporated with the voltage magnitudes at the monitored buses will be used as the triggering signals for activating the proposed ASVC approach. Figure 5 illustrates the proposed control strategy, where ε1 and ε2 are the threshold levels for voltage change and security margin, respectively. In Figure 5, if the power system is operated at the critical state with poor voltage profile or insufficient VSM, then the ASVC function will be automatically initiated to steer the power system away from the point which is prone to voltage collapse.

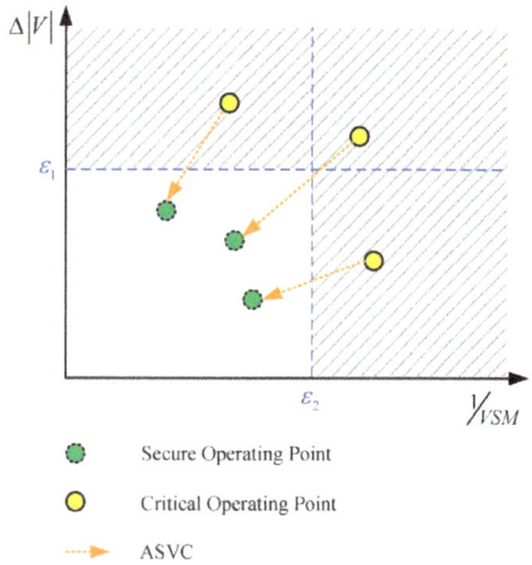

Figure 5: Illustration of the proposed control scheme.

SIMULATION RESULTS

This section presents numerical examples of the developed control scheme, *i.e.*, simultaneous system voltage profiles improvement as well as static voltage stability margin enlargement, on a sample power network. The simulation program is coded using MATLAB® and implemented on a personal computer which has a CPU with Intel® Core™2 Duo 2.66 GHz and a memory with 4 GB.

The IEEE 30-bus system is used as an example to show the effectiveness of the proposed scheme. This sample system consists of 41 transmission lines, 6 VAR sources, and 24 loads. The system data including line parameters and bus data are given in [29]. For all examples, the permissible value adopted for voltage magnitude is set at $0.95 \le |V| \le 1.05$ p.u., and the security margin for the test system is set as VSM≥0.3 p.u.. These are the threshold levels for triggering the proposed control algorithm. The voltage profile improvement index x^{rms} used in this research is expressed by the root mean square value of voltage changes at all load buses, *i.e.*,

$$x^{rms} = \left(\frac{1}{m}\sum_{j=1}^{m} \|x_j\|_2^2\right)^{1/2} \tag{11}$$

where m is the number of load buses in the test system, and xj for each j=1, 2, … , m denotes the resulting voltage change at each of the load buses. In this test system, m is set to be 24. Note that the smaller the value of xrms, the better the voltage profile will be.

In order to verify the performance of the presented methodology to power system secondary voltage control, we have studied a lot of experiments. These simulation cases include different load levels, different load patterns, various pilot-bus selections, and various branch outage contingencies. Among those investigated cases, some typical test results for the scenarios given in Table 1 are briefly summarized in the following.

Table 1: Case studies for the IEEE 30-bus system ("#" Indicates" number" & "/" indicates "None").

Case	PMU Locations at	Load Change Pattern	Line Outage Contingency
I	#30	Single bus	/
II	#3, #12	All buses	/
III	#24, #29	Several buses	Line #27–#28
IV	All	Several buses	/

CASE I

In the first case, bus load change pattern is considered. The load at bus #30 is gradually increased by 15% of its initial load level from the time instant 1 min to 2.5 min of the simulation.

In this test, the voltage magnitude |V30|=0.9388 p.u. of pilot bus #30 drops below the constraint of 0.95 p.u. at t=2.5min. This low voltage violation activates the proposed ASVC immediately, and Figure 6a illustrates the voltage trace at pilot bus #30 during the simulation test of case I. When the proposed scheme is used to carry out the secondary voltage control, voltage violation will be eliminated effectively. In addition, the value of VSM is increased from 0.51 to 0.72 p.u., as shown inFigure 6b. This demonstrates that larger VSM can be achieved simultaneously with improvement of overall network voltage.

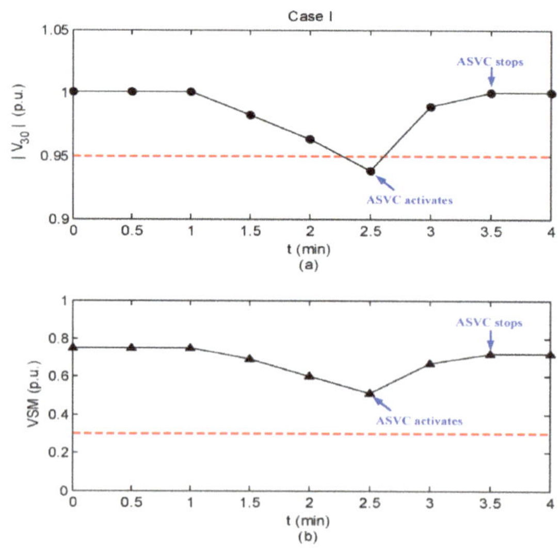

Figure 6: Simulation result of case I. (a) The trace of |V30|; (b) The trace of VSM.

Case II

In case II, a new operating condition and new load change pattern are investigated. From the time instant 0.5 min to 2 min of the simulation, all the loads in the IEEE 30-bus system are increased by 20% based on the initial load levels.

During the load increase, the voltage magnitudes at the monitored buses (bus #3 and #12 in this case) are within the pre-determined voltage limits, but static voltage stability margin of the power system is less than the threshold level of 0.3 p.u. at t=2 min. This means that the system has a great potential of voltage collapse without any remedial control. Under such situation, however, no control actions will be initiated by the traditional SVC methods which consider only bus voltages as trigger signal. Instead, the proposed scheme employs not only voltage magnitudes but also static voltage stability margin as criteria; therefore, inadequate VSM (VSM=0.23<0.3 p.u.) is identified.

After applying the proposed ASVC, all the voltage magnitudes are maintained within the range of predefined constraint and the static voltage stability margin is enlarged as well. The simulation result is illustrated in Figure 7, showing that the VSM is significantly increased from 0.23 to 0.57 p.u.. The enlargement of the static voltage stability margin is a result of the improvement in the system voltage profile.

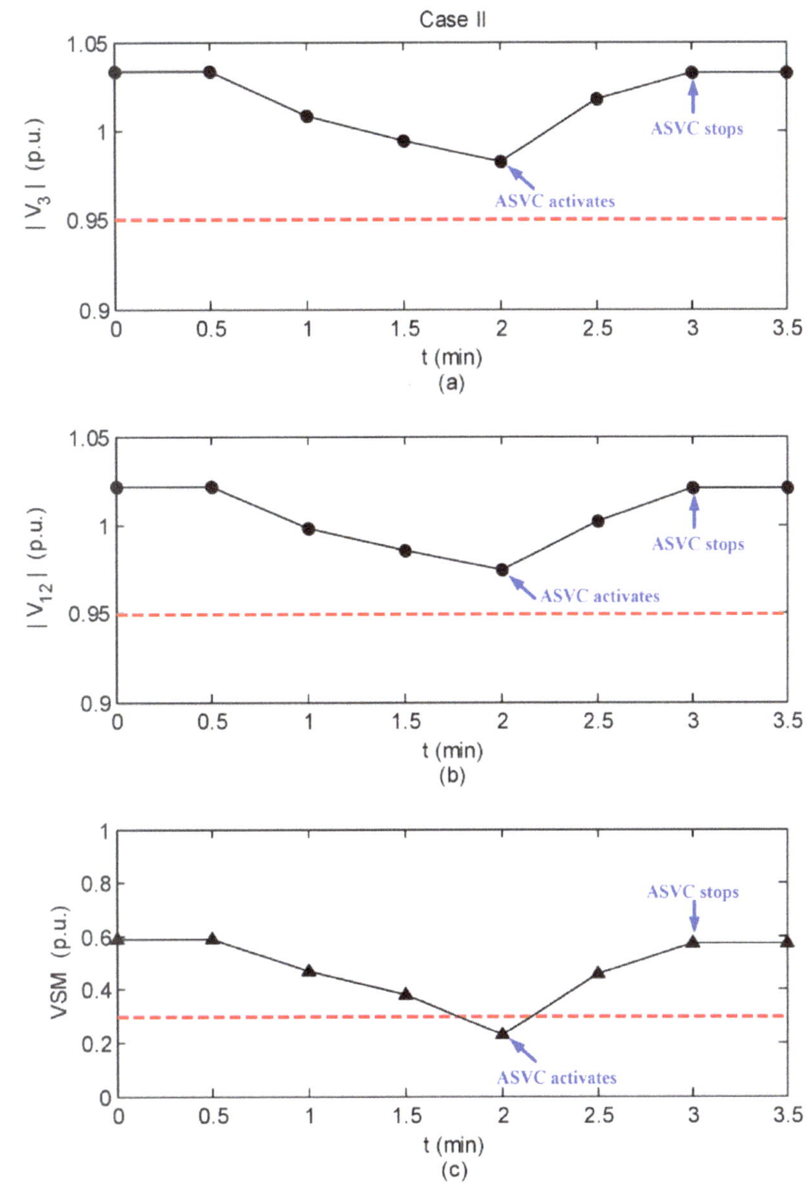

Figure 7: Simulation result of case II. (a) The trace of |V3|; (b) The trace of |V12|; (c) The trace of VSM.

Case III

This example is to demonstrate the performance of the proposed scheme with respect to branch outage contingency following several loads increase condition. In this case, the loads at bus #12, #24, and #29 are increased by 20% from the time instant 0.5 min to 2 min, and the transmission line connected between bus #27 and #28 is tripped in a contingency at 2 min.

As can be expected, the system has poor voltage profile and unsecure margin due to the disturbances of both load increase and line outage. Figure 8 shows the simulation result of case III. One can observe that the proposed scheme works satisfactorily in this case.

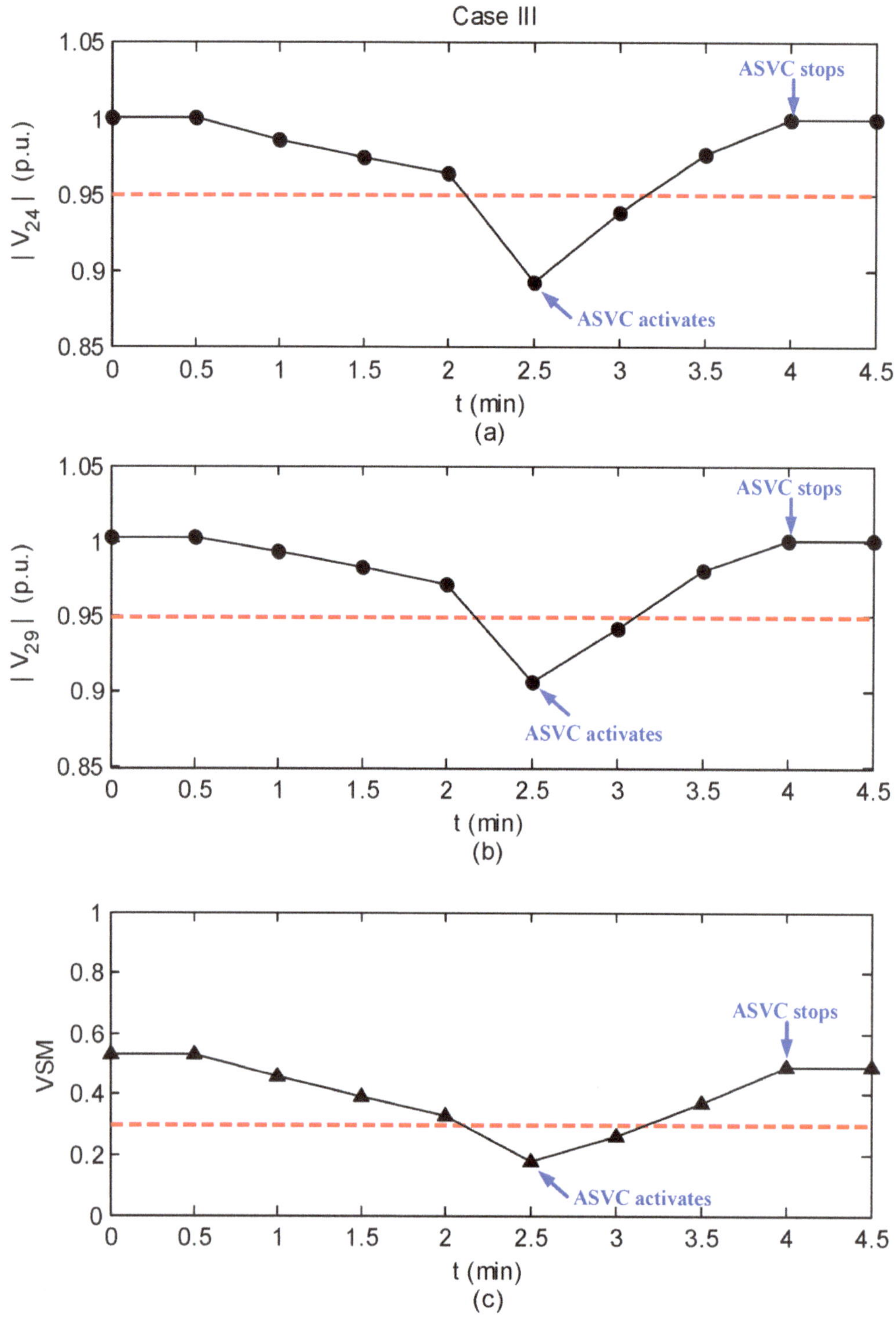

Figure 8: Simulation result of case III. (a) The trace of |V24|; (b) The trace of |V29|; (c) The trace of VSM.

Case IV

In the fourth case, we assume that PMUs are installed at all load buses. In addition, several loads change is considered. The loads at bus #3, #16, and #20 are gradually increased by 15% from the time instant 0.5 min to 1.5 min of the simulation.

Figure 9 shows the traces of the selected pilot buses (bus #23 and #26 in this case) and *VSM* during the simulation of case IV. Since insufficient *VSM* (VSM=0.26<0.3) is detected at t=1.5 min, the control strategy will be activated accordingly. Thus, the voltage levels and voltage stability margin are considerably enhanced after ASVC. This can be regarded clearly inFigure 9.

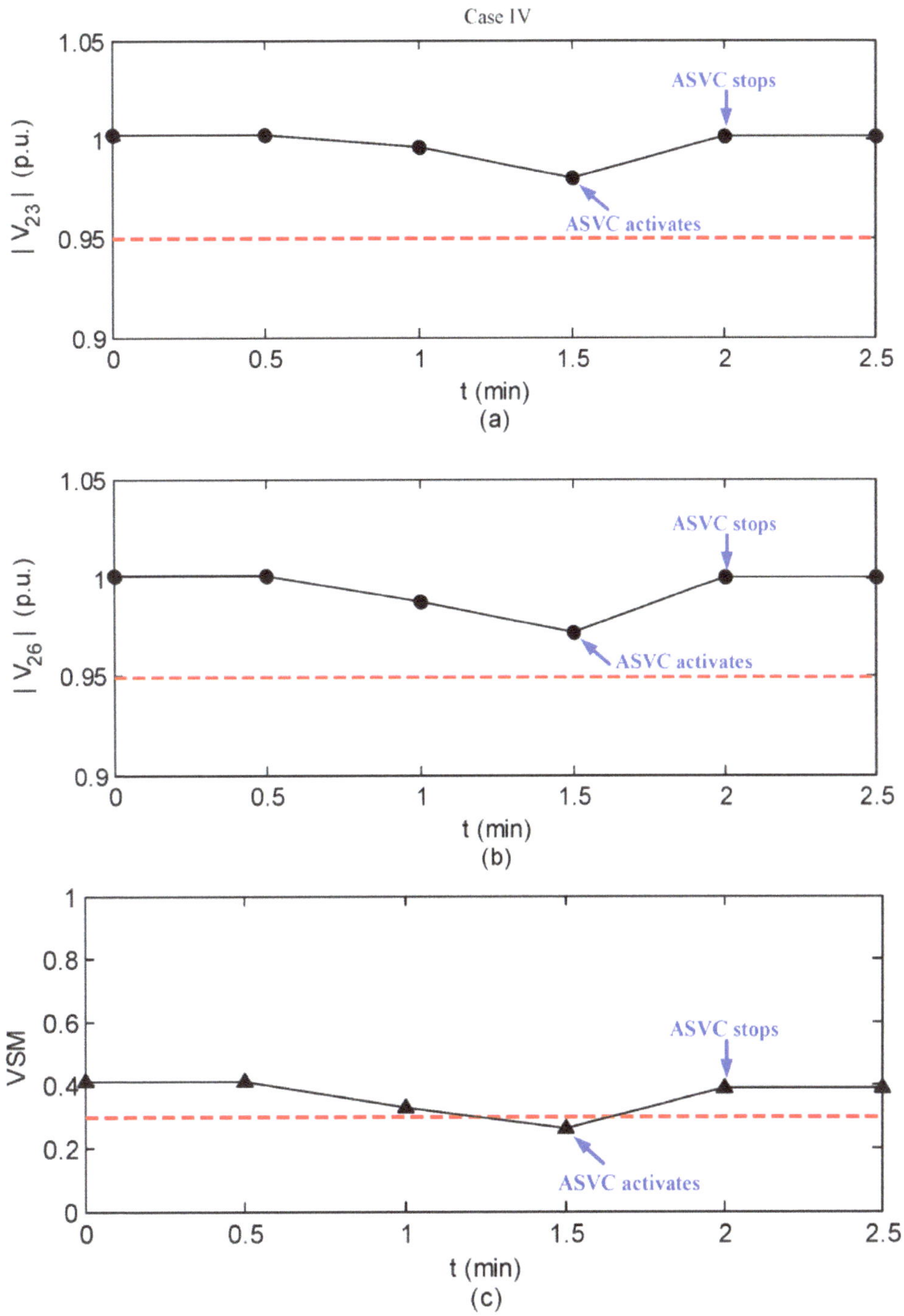

Figure 9: Simulation result of case IV. (a) The trace of |V23|; (b) The trace of |V26|; (c) The trace of VSM.

Table 2 summarizes the test results for the above three cases. An inspection from the shown table indicates that both voltage profile and static voltage stability margin are greatly improved after applying the proposed control scheme.

Table 2: Performance evaluation for the proposed scheme on the IEEE 30-bus system.

Case	x^{rms} (p.u.)		VSM (p.u.)	
	Before	After	Before	After
I	0.0601	0.0355	0.51	0.72
II	0.0348	0.0205	0.23	0.57
III	0.0803	0.0411	0.18	0.49
IV	0.0416	0.0194	0.26	0.39

CONCLUSIONS

A synchrophasor based optimal voltage control scheme, which considers both voltage profile and static voltage stability margin, is developed in order to achieve secure grid operations. A detailed derivation of the principles used in the scheme is presented with illustrated figures. An extensive simulation studies on the IEEE 30-bus test system is carried out to demonstrate the feasibility and effectiveness of the proposed scheme.

ACKNOWLEDGMENTS

This work was supported by the Ministry of Science and Technology of Taiwan, ROC., under Contracts MOST 104-2221-E-035-041 and MOST 104-2119-M-035-002.

AUTHOR CONTRIBUTIONS

Heng-Yi Su proposed the approach and wrote the manuscript with input from Yi-Chung Chen and Yi-Liang Hsu.

CONFLICTS OF INTEREST

The authors declare no conflict of interest.

REFERENCES

1. Kundur, P. *Power System Stability and Control*; McGraw-Hill: New York, NY, USA, 1994.
2. Kundur, P.; Paserba, J.; Ajjarapu, V.; Andersson, G.; Bose, A.; Canizares, C.; Hatziargyriou, N.; Hill, D.; Stankovic, A.; Taylor, C.; *et al.* Definition and classification of power system stability IEEE/CIGRE joint task force on stability terms and definitions. *IEEE Trans. Power Syst.* 2004, *19*, 1387–1401.
3. Morison, G.K.; Gao, B.; Kundur, P. Voltage stability analysis using static and dynamic approaches. *IEEE Trans. Power Syst.* 1993, *8*, 1159–1171.
4. Phadke, A.G.; Thorp, J.S. *Synchronized Phasor Measurements and Their Applications*; Springer-Verlag: New York, NY, USA, 2008.
5. De La Ree, J.; Centeno, V.; Thorp, J.S.; Phadke, A.G. Synchronized phasor measurement applications in power systems. *IEEE Trans. Smart Grid.* 2010, *1*, 20–27.
6. Thorp, J.S.; Ilic-Spong, M.; Varghese, M. An optimal secondary voltage-var control technique. *Automatica* 1986, *22*, 217–222.
7. Ilic, M.; Christensen, J.; Eichorn, K.L. Secondary voltage control using pilot point information. *IEEE Trans. Power Syst.* 1988, *3*, 660–668.
8. Stankovic, A.; Ilic, M.; Maratukulam, D. Recent results in secondary voltage control of power systems. *IEEE Trans. Power Syst.* 1991, *6*, 94–101.
9. Zobian, A.; Ilic, M.D. A steady state voltage monitoring and control algorithm using localized least square minimization of load voltage deviations. *IEEE Trans. Power Syst.* 1996, *11*, 929–938.
10. Wang, H.; Li, H.; Chen, H. Coordinated secondary voltage control to eliminate voltage violations in power system contingencies. *IEEE Trans. Power Syst.* 2003, *18*, 588–595.
11. Berizzi, A.; Marannino, P.; Merlo, M.; Pozzi, M.; Zanellini, F. Steady-state and dynamic approaches for the evaluation of loadability margin in the presence of secondary voltage regulation. *IEEE Trans. Power Syst.* 2004, *9*, 1048–1057.
12. Paul, J.P.; Leost, J.T.; Tesseron, J.M. Survey of the secondary voltage control in France: Present realization and investigations. *IEEE Trans. Power Syst.* 1987, *2*, 505–511.
13. Lagonotte, P.; Sabonnadiere, J.C.; Leost, J.Y.; Paul, J.P. Structural analysis of the electrical system: Application to the secondary voltage control in France. *IEEE Trans. Power Syst.* 1989, *4*, 1477–1484.
14. Corsi, S.; Marannino, P.; Losignore, N.; Moreschini, G.; Piccini, G. Coordination between the reactive power scheduling function and the hierarchical voltage control of the EHV ENEL system. *IEEE Trans. Power Syst.* 1995, *10*, 686–694.
15. Sancha, J.L.; Fernandez, J.L.; Cortes, A.; Abarca, J.T. Secondary voltage control: Analysis, solutions, simulation results for the Spanish transmission system. *IEEE Trans. Power Syst.* 1996, *11*, 630–638.
16. Corsi, S.; Pozzi, M.; Sabelli, C.; Serrani, A. The coordinated automatic voltage control of the Italian transmission grid-Part I: Reasons of the choice and overview of the consolidated hierarchical system. *IEEE Trans. Power Syst.* 2004, *19*, 1723–1732.
17. Corsi, S.; Pozzi, M.; Sforna, M.; Dell'Olio, G. The coordinated automatic voltage control of the Italian transmission grid-part II: Control apparatuses and field performance of the consolidated hierarchical system. *IEEE Trans. Power Syst.* 2004, *19*, 1733–1741.
18. Conejo, A.; Aguilar, M.J. Secondary voltage control: Nonlinear selection of pilot buses, design of an optimal control law, and simulation results. *IEE Proc. Gen. Trans. Distrib.* 1998, *145*, 77–81.
19. Yu, C.; Yoon, Y.T.; Ilic, M.D.; Catelli, A. On-line voltage regulation: The case of New England. *IEEE Trans. Power Syst.* 1995, *10*, 631–638.
20. Begovic, M.; Fulton, D.; Gonzalez, M.R.; Goossens, J.; Haas, R.W.; Henville, C.F.; Michel, G.L.; Postforoosh, J.; Williams, J.B.; Zimmerman, K. Summary of system protection and voltage stability. *IEEE Trans. Power Syst.* 2004, *19*, 1387–1401.
21. Flatabo, N.; Ognedal, R.; Carlsen, T. Voltage stability condition in a power transmission system calculated by sensitivity methods. *IEEE Trans. Power Syst.* 1990, *5*, 1286–1293.
22. Flatabo, N.; Fosso, O.; Ognedal, R.; Carlsen, T. A method for calculation of margins to voltage instability applied on the Norwegian system for maintaining required security level. *IEEE Trans. Power Syst.* 1993, *8*, 920–928.
23. Lof, P.A.; Semed, T.; Andersson, G.; Hill, D.J. Fast calculation of a voltage stability index. *IEEE Trans. Power Syst.* 1992,*7*, 54–64.

24. Lof, P.A.; Andersson, G.; Hill, D.J. Voltage stability indices for stressed power systems. *IEEE Trans. Power Syst.* 1993, *8*, 326–335.
25. Vu, K.; Begovic, M.M.; Novosel, D.; Saha, M.M. Use of local measurements to estimate voltage-stability margin. *IEEE Trans. Power Syst.* 1999, *14*, 1029–1035.
26. Smon, I.; Verbic, G.; Gubina, F. Local voltage-stability index using Tllegen's theorem. *IEEE Trans. Power Syst.* 2006, *21*, 1267–1275.
27. Boyd, S.; Vandenberghe, L. *Convex Optimization*; Cambridge University Press: Cambridge, UK, 2004.
28. MathWorks, Inc. MATLAB Optimization Toolbox. Available online: http://www.mathworks.com/products/optimization/ (accessed on 4 March 2005).
29. University of Washington College of Engineering. Power Systems Test Case Archive. Available online: http://www.ee.washington.edu/re-serach/pstcal/ (accessed on 17 September 2005).

CHAPTER 14

A MICRO-COORDINATE MEASUREMENT MACHINE (CMM) FOR LARGE-SCALE DIMENSIONAL MEASUREMENT OF MICRO-SLITS

So Ito [1], Hirotaka Kikuchi [1], Yuanliu Chen [1], Yuki Shimizu [1], Wei Gao [1], Kazuhiko Takahashi [2], Toshihiko Kanayama [2], Kunmei Arakawa [2] and Atsushi Hayashi [2]

[1]Department of Finemechanics, Tohoku University, Sendai 980-8579, Japan

[2]Engineering Department, MMC RYOTEC Corporation, Gifu 503-2301, Japan

ABSTRACT

This paper presents a micro-coordinate measuring machine (micro-CMM) for large-scale dimensional measurement of a micro-slit on a precision die coater by using a shear-mode micro-probe. A glass micro sphere with a nominal diameter of 52.3 μm was attached on one end of a tapered glass capillary tube as a probe tip ball. The micro-slit width of a slot die coater with a nominal slit width of 85 μm was measured by the micro-CMM. The probe tip was placed in the slit for the measurement. The effective working length of the probe was confirmed experimentally to be at least 1 mm. In order to measure the gap width uniformity over the entire slot die length of 200 mm, an air-bearing linear slide with a travelling stroke of 300 mm was employed in the micro-CMM to position the probe along the length direction of the slot die. The angular alignment error and the motion error of the air-bearing linear slide as well as those of the stages for positioning the probe along the direction perpendicular to the length direction of the slot die were investigated for evaluation of the expanded uncertainty of gap width measurement.

INTRODUCTION

Precisely machined micro-features have been utilized widely with the development of micro-fabrication technologies such as micro-cutting [1,2], micro- electrical discharge machining (EDM) [3,4], micro- and nano-lithography [5,6], nanoimprint [7] and micro-molding [8]. In particular, more precision micro-features with high aspect ratios represented by deep micro-holes and micro-slits are utilized for precision tools and measuring instruments that require sub-micrometric or nanometric dimensional accuracies. The requirements for dimensional measurements of precision micro-features with high aspect ratios are thus increasing. The micro-slit of a slot die coater is a good example of such precision micro-features. The slot die is a precision tool for coating of functional liquid materials on the surface of a substrate such as a glass plate or a film. On the top of the slot die coater is a micro-slit from which the liquid material is extruded. The micro-slit has a high aspect ratio with a gap width in the order of 100 μm and a depth of more than 10 mm. Since the gap width of the micro-slit as well as its uniformity along the length direction of the

Citation: So Ito , Hirotaka Kikuchi , Yuanliu Chen , Yuki Shimizu , Wei Gao, Kazuhiko Takahashi, Toshihiko Kanayama, Kunmei Arakawa and Atsushi Hayashi, A Micro-Coordinate Measurement Machine (CMM) for Large-Scale Dimensional Measurement of Micro-Slits, doi:10.3390/app6050156

slot die coater is a key parameter for determining the coating quality, it is necessary to measure the micro-slit gap width and its uniformity accurately.

For dimensional measurement of micro-features with high-aspect ratios, the micro-coordinate measurement machine (micro-CMM) with a micro-probing system is one of the prospective measuring instruments because the inside of a micrometric feature such as a micro-hole or a micro-slit can be detected by the micro-probe with a micro-tip ball. On the other hand, the accuracy and reliability of the micro-probing system are determined by the positioning accuracy of the probe qualification of the probe tip, measurement strategies, and so on. The plastic deformation caused by the contact force between the probe and the measured surface is also a main error source, which requires the probing system to have a high sensitivity for detecting the probe–surface contact so that a small contact force can be achieved for reduction of the plastic deformation [9].

The contact detection is typically carried out by detecting the elastic deformation of the springs for supporting the stylus of a micro-probe. Different kinds of sensors have been integrated into the micro-probe for this purpose. For example, three integrated strain gauges have been employed to detect the probe–surface contact [10]. Optical systems have also been integrated into the micro-probes for contact detection. Cui et al. have proposed a three-dimensional fiber probe for the measurement of micro parts. [11]. Muralikrishnan et al. have proposed a fiber probe with an ellipsoidal tip for three-dimensional measurement on micro-features [12]. Fan et al. have developed low-cost and high precision micro-CMM by utilizing the focusing probe [13]. Schwenke et al. have developed an opto-tactile sensor for the contact detection of a micro-probe in two and three directions at Physikalisch-Technische Bundesanstalt (PTB) in Germany [14], where the micro-probe has a tip ball of 60 μm in diameter. The probe displacement can be detected by using an optical beam focused on the probe tip ball. Takaya et al. have proposed a nano-probe system based on the laser-trapping technique [15,16]. A microsphere with a diameter of 8 μm, which is held by the focused laser beam, is used as the probe tip ball with a piconewton measuring force. Such optical-based detection methods are more sensitive than the traditional strain gauges. However, the effective working distance of the probe is confined by the numerical aperture of the focusing lens and it is difficult to measure the inside of a high-aspect ratio micro-feature.

Detection of electric contact has been applied for micro-probing systems for highly sensitive probe-surface contact detection. Masuzawa et al. have introduced a micro-probe, in which the probe-surface contact is detected based on the electrical contact. Since the a micro-probe tip is attached on the end of a long and sharp stylus shaft, the probe can be used to measure the inside of micro-holes [17,18]. The micro-probe tip has a pyramidal edge and it can detect the contact in one direction. As for the multi-axis detection of contact, Weckenmann et al. have developed a probing technique based on the method of scanning tunneling microscopy (STM) [19], in which a spherical probe tip is employed. However, the applications of such methods are only effective for measurement of conductive materials.

Improvement of sensitivity for contact detection can also be achieved by applying mechanical vibrations to the micro-probe. Bauza et al. have proposed a vibrating shank for measurement of the inside of a micro-hole [20,21]. A fiber with a diameter of 7 μm is vibrated in one direction by a crystal resonator, which is used as a virtual micro-probe for contact detection in one axis. The effective diameter of the vibrating fiber probe is approximately 30 μm. With respect to a probe with multi-axis detection, Claverley et al. have developed a three-dimensional vibrating probe composed of PZT sensors and actuators [22,23]. The probe tip of the three-dimensional vibrating probe has a spherical tungsten tip with a diameter of approximately 70 μm. Murakami et al. have presented an optical fiber probe which has a glass tip ball with a diameter of 5–50 μm [24,25]. Since the vibrating probe allows the use of a probe stylus with a lower stiffness, it results in a lower probe–surface contact force. In addition, it is an effective means for the vibrating probe tip to reduce the influence of adhesion force of the measured surface against the probe. However, the measurement is influenced by the water layers on both the measured surface and the probe surface [9,22,25]. In responding to this problem, the authors have developed a shear-mode micro-probe, which can reduce the influence of the water layers through detecting the interaction force between the probe and the measured surface over the water layers [26]. It has been demonstrated that the shear-mode detection probe can achieve nanometric resolution and probing repeatability.

The motivation of this research is to apply the shear-mode micro-probe for measurement of the gap width of the micro-slit on a precision slot die coater. The precision slot die is composed of two precision flat plates [27] made of stainless steel with a length of 200 mm. The two precision flat plates are separated with a gap with a nominal gap width of 85 μm to form the micro-slit on the coater edge. It is required to make gap width measurement of the micro-slit with the expanded uncertainty of less than 100 nm [28]. Traditionally, mechanical feeler gauges have been used for the measurement of micro-slit gap width. However, the mechanical feeler gauge often induces damages on the edge areas of the micro-slit due to the large contact force. Although a CCD camera based optical system [28] has been proposed for non-contact measurement of the micro-slit gap width, it is difficult for the optical system to achieve sub-micrometer resolutions because of the diffraction limit of the optical system. As for the detection of single edge position, it can be achieved by using the optical system. However, with regard to the precision measurement of the slit gap width, it is necessary to detect the positions of two edges quantitatively. Therefore, the resolution in lateral direction is one of the important requirement for the gap width measurement. In addition, it cannot be used for measuring the gap width inside the micro-slit. Recently, a non-contact measurement system with a higher resolution by using a capacitive displacement sensor has been developed [27]. However, the measurable minimum gap width of the electric gap sensor is limited to 150 μm, due to the sensor thickness. Compared with the measurement techniques mentioned above, a micro-CMM is more effective for such a measurement in terms of the measurable minimum gap width and measurement resolution, as well as the flexibility and accessibility to the inside of the micro-slit.

This paper presents a micro-CMM specifically designed for the slit width measurement of the precision die coater. In addition to the newly developed shear-mode micro-probe, an air-bearing linear slide with a travelling stroke of 300 mm was employed in the micro-CMM so that there is a uniformity of the gap width over the entire length of the micro-slit of 200 mm. This is a significant difference from an existing micro-CMM that can only cover a measurement length from several millimetres to several centimetres. The calibration of the conventional micro-CMMs is carried out by using the calibration artifact with a length of 20 mm at the maximum [29]. Therefore, a longer travelling range of the probe positioning system is essential for the gap width measurement of the slot die. The effective working length of the shear-mode micro-probe is designed to be 1 mm, which is longer than that of the existing micro-probes [25,30], so that the inside of the micro-slit can be accessed by the probe tip while avoiding the influence of the chamfered edges of the precision parts forming the micro-slit. In order to shorten the measurement time, a measurement strategy of gap width measurement based on two probing points is also introduced. Next, a description of the experimental setup for the micro-CMM, experiments and uncertainty analysis of the measurement results are presented.

EXPERIMENTAL SETUP FOR THE MICRO-CMM

Figure 1a shows the photographs of the shear-mode micro-probe. The micro-stylus is composed of a glass micro-stylus shaft and a glass micro-sphere. The stylus shaft is made of a capillary glass tube which is thermally pulled by using a commercially available glass pipette puller. A micro-glass sphere with a nominal diameter of 52.6 ± 3.2 μm (Certified size standard Cat No. 9050, Duke Scientific Corporation, California, CA, USA) [31] is attached on one end of the glass micro-stylus by thermosetting resin. The diameter of the commercially available glass micro-sphere was selected based on the nominal gap width of the precision slot die. Figure 1b shows the schematic of a shear-mode micro-probe. The micro-stylus is then attached on a tuning fork quartz crystal resonator (TF-QCR). A PZT transducer is employed to vibrate the assembly, which is composed of the TF-QCR, the micro-stylus, and the micro-sphere at the resonance frequency of the TF-QCR. The variation of the vibration frequency is detected by the TF-QCR based on the piezoelectric effect of the quartz crystal. The vibration of the TF-QCR is detected and amplified by current-to-voltage (I-V) converter. The output of the I-V converter is then utilized to keep the vibration frequency constant by using a self-excitation circuit that consists of a phase locked loop circuit (PLL). It should be noted that the actual resonance frequency of the TF-QCR used in the assembly is confirmed to be approximately 30 kHz, which is smaller than that of the single TF-QCR element of 32.768 kHz, due to the influence of the masses of the micro-stylus and the micro-sphere attached on the TF-QCR. The vibration direction is set in parallel with the measuring surface.

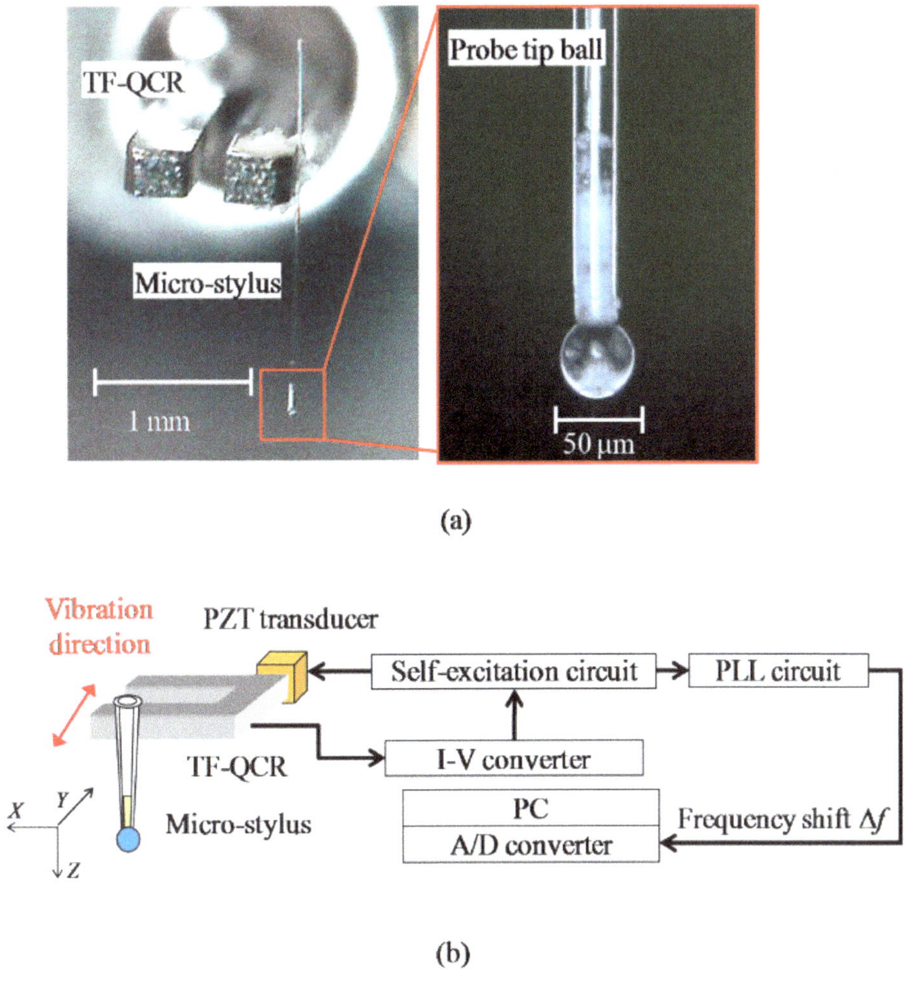

Figure 1: Configuration of shear-mode micro-probe. (a) Photograph of shear-mode micro-probe; (b) Schematic diagram of shear-mode micro-probe.

To determine the trigger point for the probing of the micro-probe to the measured surface, the interaction force of the water layer on the measured surface against the probe is detected through detecting the change of the probe vibration with the TC-QCR [26]. When the tip ball of the vibrating probe contacts with the water layer of the measured surface, the frequency of the probe vibration is shifted due to the adhesion force of the water layer. The output of the TF-QCR is input to a phase locked loop (PLL) circuit for detection of the amount of frequency shift. The PLL circuit output is then employed as the trigger signal for the probing of the micro-probe to the measured surface. The PLL output signal is fed into analog-to-digital (A/D) converter of a personal computer (PC). A certain frequency shift amount is set to be a threshold of the trigger signal.

Figure 2 shows a schematic diagram of the micro-CMM by using the shear-mode micro-probe. The probe is mounted on the PZT-driven linear stages (P-621.2CL and P-621.1ZCL, PI), which can be moved in the X- and Z-directions independently, for fine positioning of the probe. The stroke and resolution of the PZT stages are 100 μm and 1 nm, respectively. The positioning accuracy in the X- and Z-directions are ±2 nm and ±1 nm, respectively. Since the gap width of the slot die is calculated based on the X-directional probe displacement, a commercial laser interferometer (RLE10, RENISHAW) with a resolution of 0.39 nm is employed for the measurement of the X-directional probe displacement. A plane mirror is mounted on the opposite side of the probe on the X-direction PZT stage as a moving mirror for the laser interferometer. In order to reduce the influence of the Abbe error, the optical axis of the laser interferometer is aligned with the center of the probe tip. The PZT stages are mounted on the DC servo motor driven positioning stages (M-111.1DG, PI) for coarse positioning of the probe in the X- and Z-directions. The stroke and resolution of the servo motor coarse stages are 25 mm and 6.9 nm, respectively. The positioning accuracy resolution of the servo motor stages are 0.1 μm. For positioning the probe in the Y-direction, an air bearing linear slide (TAAT30SL-19+C, NTN) with a travel stroke of 300 mm is employed. The resolution and positioning accuracy of the Y-directional linear slide are 20 nm and ±0.1 μm, respectively. The precision slot die with a length of 200 mm is placed on the moving table of the Y-directional linear slide for measurement.

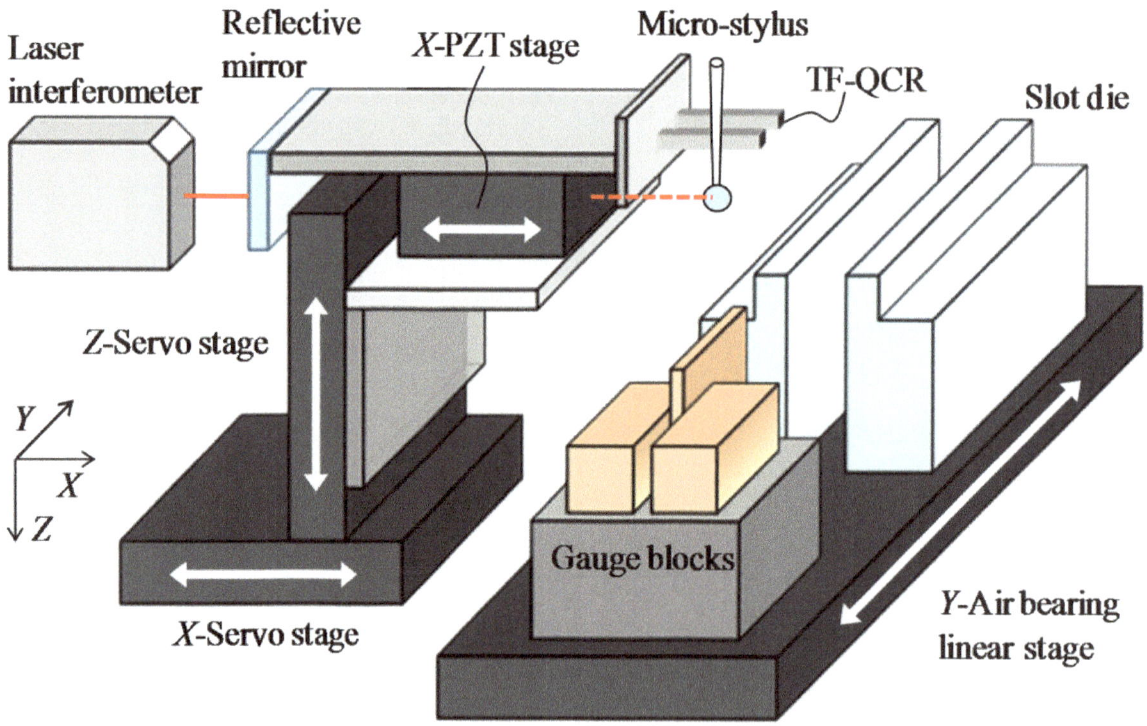

Figure 2: Experimental configuration of the micro-CMM (micro-coordinate measurement machine) by using a shear-mode micro-probe.

When a typical CMM is employed for measuring the gap distance between two parallel surfaces, at least three points on one surface are probed first to define the surface. A point on the other surface is then probed for obtaining the coordinates of the point. The gap distance between the two surfaces can thus be accurately obtained from the coordinates of the point on the second surface with respect to the defined first surface even if the normal of the surfaces are not parallel/vertical to the coordinate axes of the CMM. However, it is necessary to conduct the probing at least three points on the first surface, which is a time-consuming process. In order to shorten the measurement time, the gap width of the precision slot die is measured in the micro-CMM, as shown in Figure 3, by probing only one point on each of the surfaces forming the micro-slit under the condition that the axes of the X-directional positioning stages are carefully aligned to be parallel to the normal of the slit surfaces and perpendicular to the Y-axis of the linear slide.

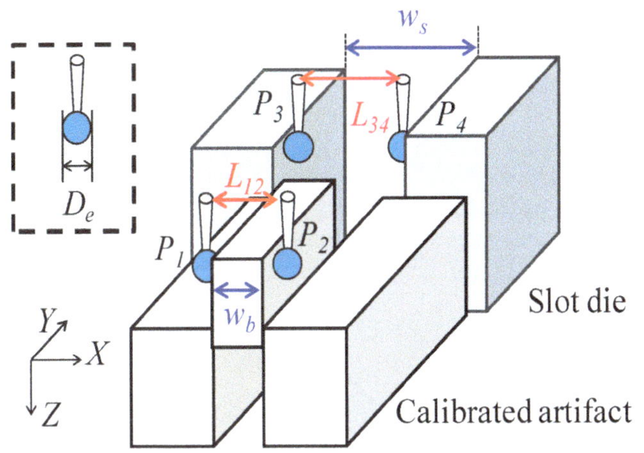

Figure 3: Schematic diagram of on-line qualification and length measurement of gap width.

Consequently, the gap width of the slot die w_s can be calculated by the following equation.

$$w_s=L_{34}+D_e+2s_{ts} \quad (1)$$

where L_{34} is the distance between the trigger positions P_3 and P_4 along the X-direction. D_e is the effective diameter of the probe tip ball. s_{ts} is the thickness of the water layer on the slot die at the trigger point. The effective diameter D_e is influenced by the water layer on the measured surface, the stiffness of the stylus shaft, and so on. The probing can be carried out by the shear-mode micro-probe through detection of the interaction between the water layer on the measured surface and that on the surface of the probe tip ball. Similar to the authors' previous research [32], on-line qualification of the effective diameter D_e is carried out soon before the gap width measurement to reduce the combined standard uncertainty caused by the variations in the water layer thickness of the water layer, which is dependent on measurement conditions such as temperature, humidity, and surface materials. Firstly, the effective diameter of the probe tip is estimated by using a calibrated artifact. A grade K gauge block (Mitsutoyo) with a thickness of 100 μm and a tolerance of ±10 nm is used as the calibrated artifact. The tolerance of the gauge block was determined based on the calibration certification. Both the slot die and the gauge block are made of stainless steel. The effective diameter D_e can be estimated by:

$$D_e=L1_2-w_b-2s_{tb} \quad (2)$$

where w_b is the thickness of the calibrated artifact, L_{12} is the distance between the trigger points P_1 and P_2 along the X-direction, and s_{tb} is the thickness of the water layer on the calibrated artifact.

The stroke of the PZT stage used in the micro-CMM is 100 μm. It is not enough to move the probe across the calibrated artifact for probing the two sides of the artifact. The X-servo motor stage is therefore used to move the probe to the opposite side of the artifact for the probing at P_2 after the probing at P_1. In order to prevent the motion error induced by the Z-servo motor stage, it is desired to move the probe in the X-direction without lifting the probe up in the Z-direction. For this purpose, the artifact is moved by the Y-slide table in the negative Y-direction until the moving path of the probe along theX-direction is not blocked by the artifact. After the probe is moved across the artifact by the X-servo motor stage, the artifact is moved back to its original position by the Y-slide table in the positive Y-direction. Finally, the probing at P_2 is carried out by using the X-directional PZT stage. As a result, Equation (1) can be rewritten as follows.

$$w_s=L_{34}+D_e+2s_{ts}=L_{34}+L_{12}-w_b-2(s_{tb}-s_{ts}) \quad (3)$$

Since s_{tb} is almost the same as s_{ts}, the influence of the water layer thickness can be cancelled. Figure 4 shows the vibration frequency shifts when probing the slot die and the gauge block, respectively. The vertical axis in Figure 4 indicates the frequency shift Δf and the horizontal axis indicates the probe displacement in the X-direction. When the frequency shift is extended to the trigger threshold, the probe position in the X-direction is measured by the laser interferometer to determine the trigger point of the probing. The probe tip is then retracted back from the measured surface immediately. As shown inFigure 4, if the frequency shift Δf of 0.1 Hz is set to be the trigger threshold for the probing, the difference of the probe displacements for the two results in Figure 4 is less than 2 nm. Consequently, the gap width of the slot die can be measured accurately by the on-line qualification of the effective diameter of the tip ball.

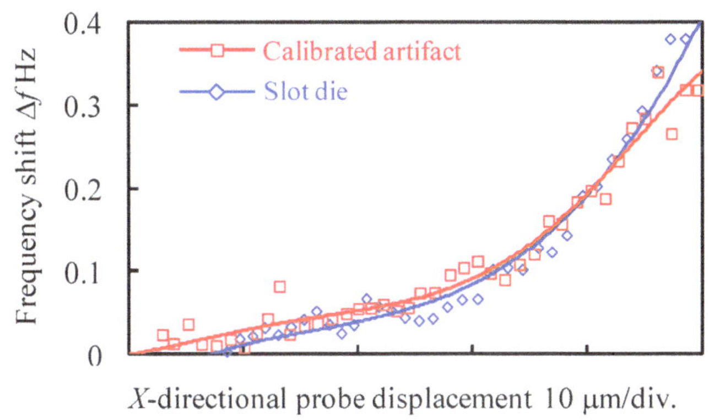

Figure 4: Vibration frequency shifts when probing the calibrated artifact and the slot die.

After the on-line qualification procedure, the probe tip is positioned inside of the micro-slit of the slot die for the gap width measurement based on Equation (1). The gap width uniformity can also be obtained by measuring the gap widths at different Y-directional positions through moving the slot die with the Y-directional linear slide.

EXPERIMENTAL RESULT AND DISCUSSION

Evaluation of Alignment Errors of the Measurement System

As described above, the gap width measurement is carried out in the developed micro-CMM by probing one point on each of the surfaces of the micro-slit. The reliability of the measurement results is influenced by the angular alignment errors and motion errors of the moving stages. Figure 5 shows a schematic of the angular alignment errors of the moving stages. The moving axis of the air bearing linear slide is defined as the Y-axis. The X-axis and Z-axis are shown in Figure 5. The length direction of the micro-slit of the slot die and the surface of the gauge block are arranged in parallel with the Y-axis. The angular alignment errors of the slot die and the calibrations around the Z-axis are indicated as θ_{gb} and θ_{sd}, respectively.θ_{gb} and θ_{sd} are calculated based on the X-directional deviation Δd_x obtained by moving the Y-directional linear slide. Δd_x is measured by using an optical fiber displacement sensor with a resolution of 0.49 nm. Therefore, θ_{gb} and θ_{sd} can be calculated by using the following equations.

$$\theta_{gb} = \tan\left(\frac{\Delta d_{x\_gb}}{L_{gb}}\right) \tag{4}$$

$$\theta_{sd} = \tan\left(\frac{\Delta d_{x\_sd}}{L_{sd}}\right) \tag{5}$$

where Δd_{x_gb} and Δd_{x_sd} are the X-directional displacements of the gauge block and the slot die due to the tilt around theZ-axis of the slot die, respectively. L_{gb} and L_{sd} are the Y-directional displacements caused by the linear slide, respectively. $L_{gb}$and L_{sd} are set to be 180 mm and 20 mm, respectively.

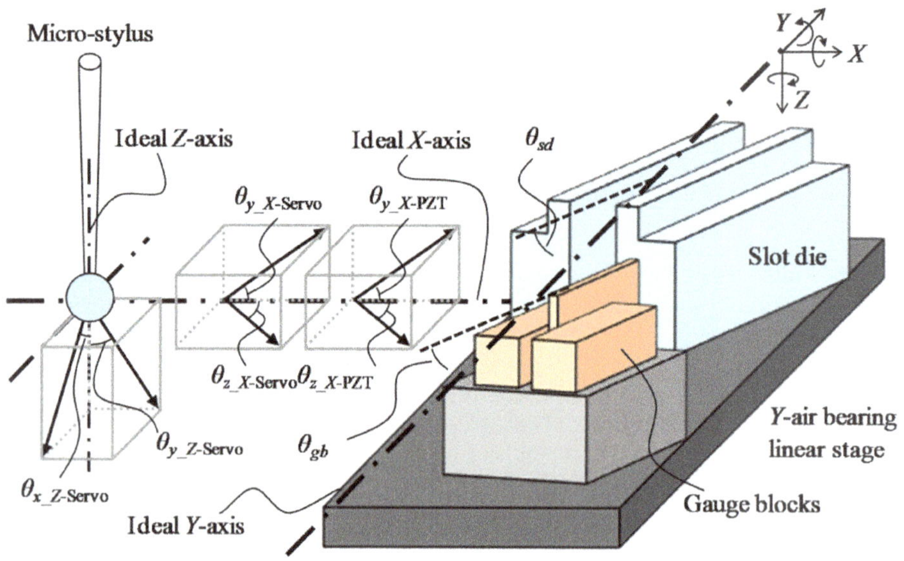

Figure 5: Schematic of the angular alignment errors of the moving axes of the measurement system.

The alignment errors of the X-directional moving axis of the PZT stage around the Y-axis and the Z-axis are indicated as$\theta_{y_X\text{-}PZT}$ and $\theta_{y\_X\text{-}PZT}$, respectively. Figure 6a shows a schematic diagram of the evaluated angular alignment errors of theX-directional stages. A right-angle prism mirror is placed on the table of the Y-linear slide for measurement of the deviations in the Y- and Z-directions with the optical fiber displacement sensor. The Y-Z plane of the prism mirror is aligned in parallel with the moving axis of the Y-linear slide. Based on Equations (4) and (5), $\theta_{y\_X\text{-}PZT}$ and $\theta_{y\_X\text{-}PZT}$ can be calculated by moving the X-directional PZT stage. Similarly, for the X-direction servo motor stages, the alignment errors around the Y-axis and Z-axis are defined as $\theta_{\_X\text{-}Servo}$ and $\theta_{z\_X\text{-}Servo}$, respectively.

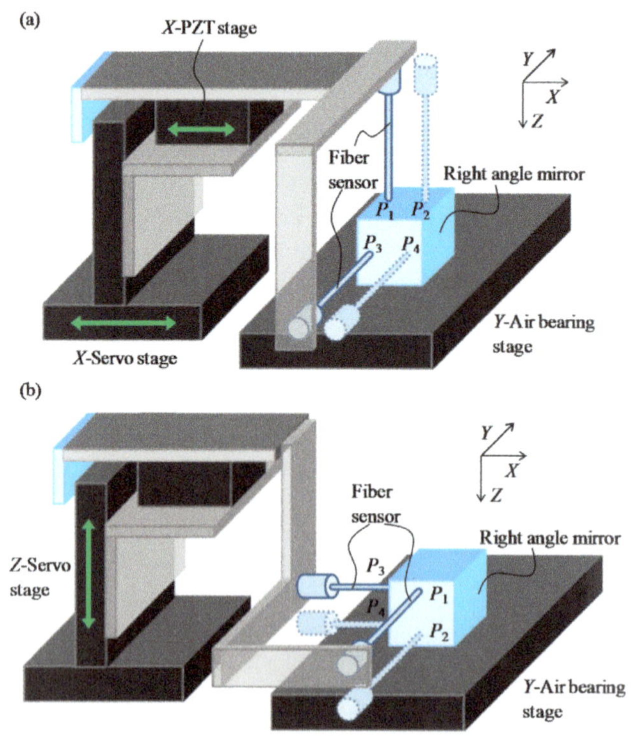

Figure 6. Schematic for evaluation of the angular alignment errors: (a) X-directional stages; (b) Z-directional stage.

The Z-directional servo motor stage is used to adjust the measurement position inside the micro-slit along the Z-direction. The alignment errors of the Z-directional servo motor stage around the X-axis and the Y-axis are indicated as $\theta_{x\_Z\text{-}Servo}$ and $\theta_{y\_Z\text{-}Servo}$, respectively. As shown in Figure 6b, $\theta_{x\_Z\text{-}Servo}$ and $\theta_{y\_Z\text{-}Servo}$ can be evaluated by using the right-angle prism mirror and the optical fiber displacement sensor.

Figure 7 shows a schematic of the tilt errors of the reflective mirror with respect to the optical axis of the laser interferometer. Since the gap width of the micro-slit is obtained from the outputs of the laser interferometer, the alignment error of the laser interferometer is also an uncertainty source for the gap width measurement. When the reflective mirror of the interferometer, which is mounted on the opposite side of the micro-probe, is not perpendicular to the X-axis, a cosine error will occur, as shown in Figure 7. θ_{y_mirror} and θ_{z_mirror} are defined as the tilt errors around the Y- and Z-axes with respect to the optical axis, respectively. Figure 8 shows a schematic of the cosine error between the laser interferometer andX-directional moving axis. L is the distance measured by the laser interferometer. L' is the actual displacement of the probe tip along the X-direction. L'/L can be expressed by the following equation.

$$\frac{L'}{L} = \frac{1}{\cos\theta_{y\_mirror}\cos\theta_{z\_mirror}} \tag{6}$$

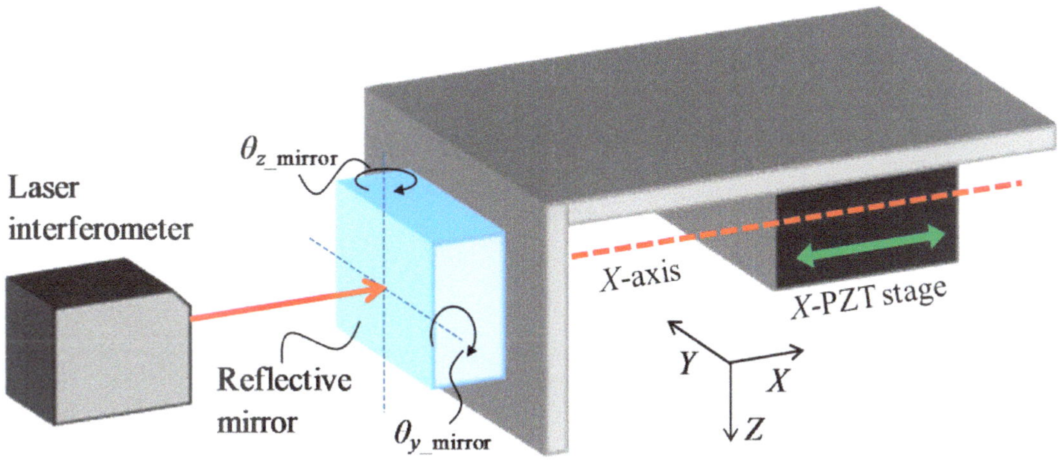

Figure 7: Tilt errors of the reflective mirror with respect to the optical axis of the laser interferometer.

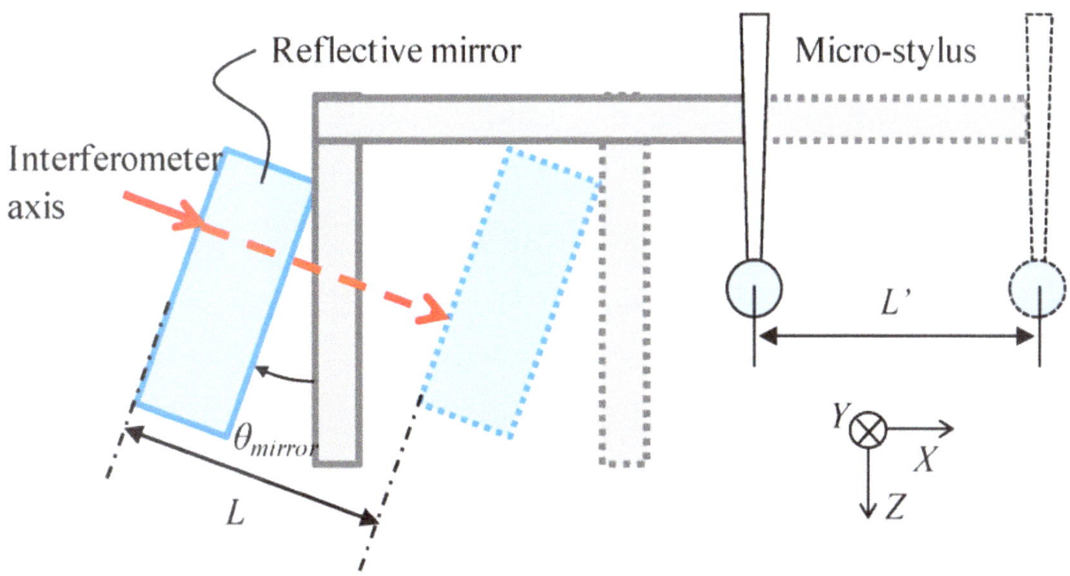

Figure 8: Schematic of the cosine error between the laser interferometer and the X-directional moving axis.

In order to evaluate the alignment errors of the reflective mirror with respect to the laser interferometer, the tilt errors of the reflective mirror were measured by using the optical fiber displacement probe. Figure 9 shows a schematic diagram for evaluation of the alignment errors of the reflective mirror. One plane of the right-angle prism mirror is arranged perpendicular to the X-axis of the PZT stage. The difference of the tilting angles between the prism mirror and the reflective mirror is measured. Table 1 shows a summary of the evaluated angular alignment errors of the developed micro-CMM.

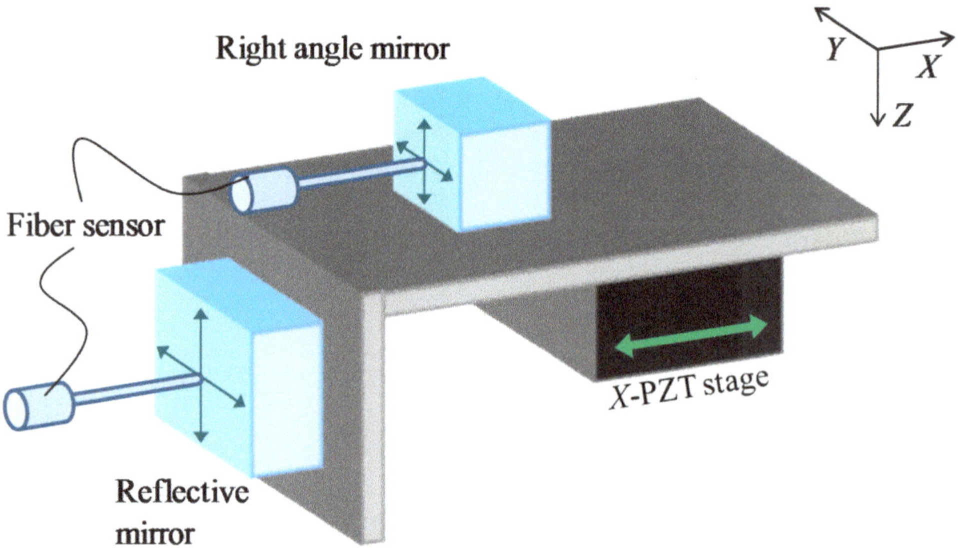

Figure 9: Schematic of angular alignment errors evaluation of the reflective mirror of the laser interferometer.

Table 1: Summary of the angular alignment errors of the positioning stages.

$\theta_{y\_X\text{-}PZT}$	$\theta_{z\_X\text{-}PZT}$	$\theta_{y\_X\text{-}Servo}$	$\theta_{z\_X\text{-}Servo}$	$\theta_{x\_Z\text{-}Servo}$
7.5 mrad	2.8 mrad	14.0 mrad	3.6 mrad	5.7 mrad
$\theta_{y\_Z\text{-}Servo}$	$\theta_{Slotdie}$	$\theta_{Gauge\text{-}block}$	θ_{y_mirror}	θ_{z_mirror}
14.1 mrad	0.03 mrad	0.07 mrad	6.8 mrad	0.34 mrad

Based on the evaluated alignment errors of the positioning stages, the geometric measurement errors are investigated.Figure 10 shows a geometrical model in the X-Y plane during the probing on the calibrated artifact. L_{gb} is the distance between the probing points on the calibrated artifact. h is the distance between the moving axes of the X-directional PZT stage on the both sides of the gauge block. D_s is the X-directional displacement of the servo motor stage. The thickness of the calibrated artifact w_b can be expressed by the following equation.

$$w_b = \left(L_{gb} - L'\right) \cos\left(\theta_{z\_X-PZT} - \theta_{gb}\right) \quad (7)$$

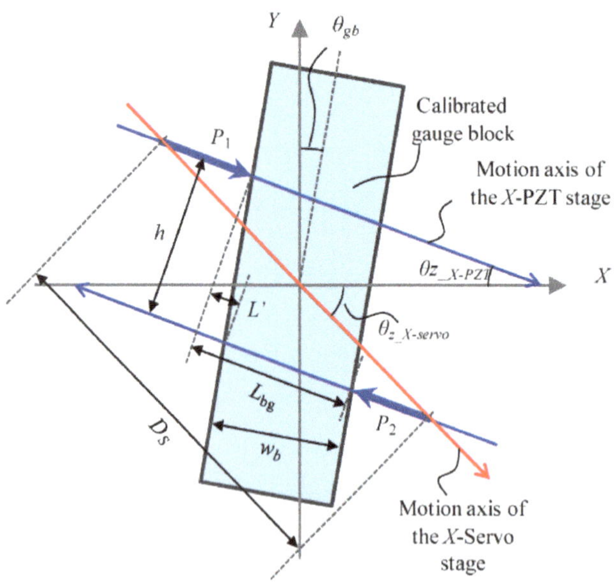

Figure 10. Geometrical model of the alignment error of the probing points during the qualification procedure.

L' can be defined as follows.

$$L' = h \tan\left(\theta_{z\_X-PZT} - \theta_{gb}\right) \quad (8)$$

h can be calculated by

$$h = D_s \sin(\theta_{z\_X-Servo} - \theta_{z\_X-PZT}) \tag{9}$$

Therefore, the measurement error Δw_b of the calibrated artifact width, which is also extended in the Y-Z plane, can be calculated by the following equation.

$$\begin{aligned} \Delta w_b &= \left(L_{gb} - D_s\right) - w_b \\ &= \left(\frac{w_b + D_s \tan(\theta_{y\_X-servo} - \theta_{y\_X-PZT})\tan(\theta_{y\_X-PZT} - \theta_{gb})}{\cos(\theta_{z\_X-PZT} - \theta_{gb})} - w_b\right) \\ &+ \left(\frac{w_b + D_s \tan(\theta_{z\_X-servo} - \theta_{z\_X-PZT})\tan(\theta_{z\_X-PZT} - \theta_{gb})}{\cos(\theta_{z\_X-PZT} - \theta_{gb})} - w_b\right) \end{aligned} \tag{10}$$

Figure 11 shows a geometrical model in the X-Y plane. The measurement error Δw_s of the slit gap width can be calculated by the following equation.

$$\Delta w_s = \frac{w_s}{\cos\left(\theta_{y\_X-PZT} - \theta_{y\_sd}\right)\cos(\theta_{z\_X-PZT} - \cos\theta_{z\_sd})} - w_s \tag{11}$$

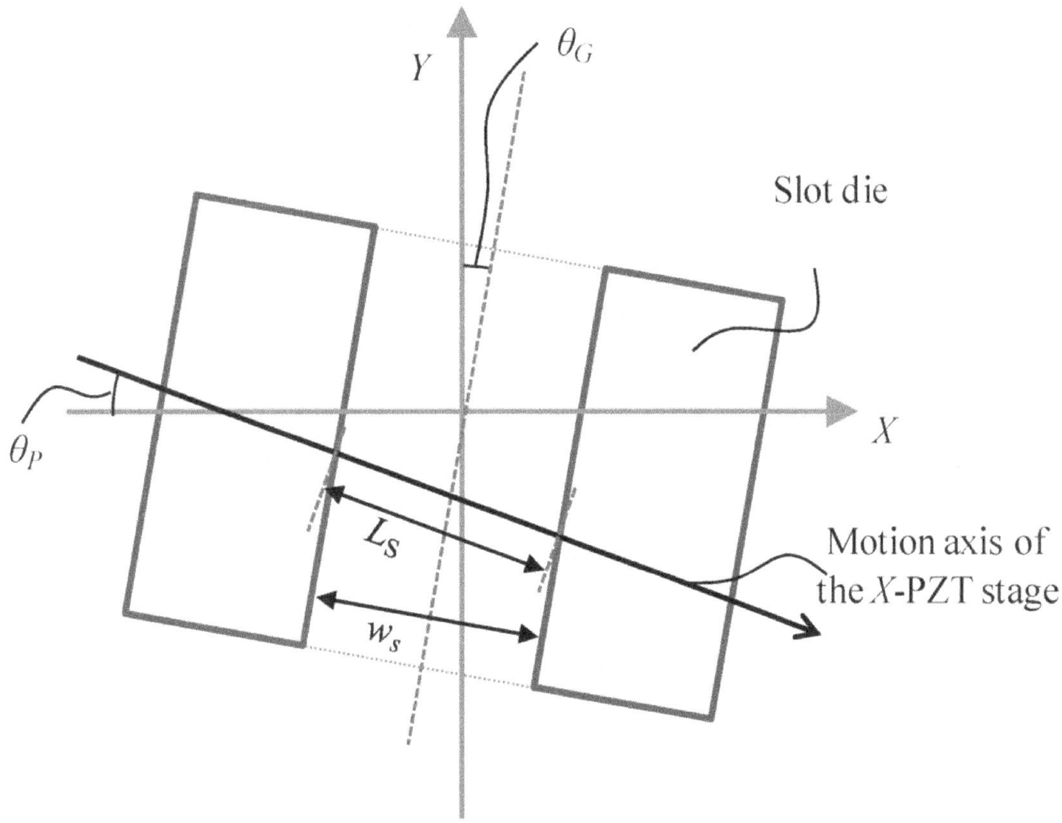

Figure 11: Geometrical model of the alignment error of the probing points during the gap width measurement.

Since the gap width measurement is carried out by the probe displacement with the laser interferometer in the X-direction, the Abbe errors for the laser interferometer have to be considered for the evaluation of the combined standard uncertainty. When the center of the probe tip is not located on the optical axis of the laser interferometer, an Abbe error component will be caused in the measurement result of the probe displacement due to the tilt errors of the X-directional stages. Figure 12 shows the Abbe error between the probe tip and the laser interferometer. As shown in Figure 12, a_1 indicates the X-directional distance between the probe tip and the reflective mirror. a_2 is the Z-directional difference between the center of the probe tip and that of the laser spot on the reflective mirror. θ_{Abbe} is the tilt error of the X-directional stages. The Abbe errors L_{y_Abbe} and L_{z_Abbe} can be expressed as follows.

$$L_{y\_Abbe} = \left(a_1 \tan\frac{\theta_{y\_X-PZT}}{2} + a_2\right)\tan\theta_{y\_X-PZT} + \left(a_1 \tan\frac{\theta_{y\_X-Servo}}{2} + a_2\right)\tan\theta_{y\_X-Servo} \tag{12}$$

$$L_{z\_Abbe} = \left(a_1 \tan\frac{\theta_{z\_X-PZT}}{2} + a_2\right)\tan\theta_{z\_X-PZT} + \left(a_1 \tan\frac{\theta_{z\_X-Servo}}{2} + a_2\right)\tan\theta_{z\_X-Servo} \tag{13}$$

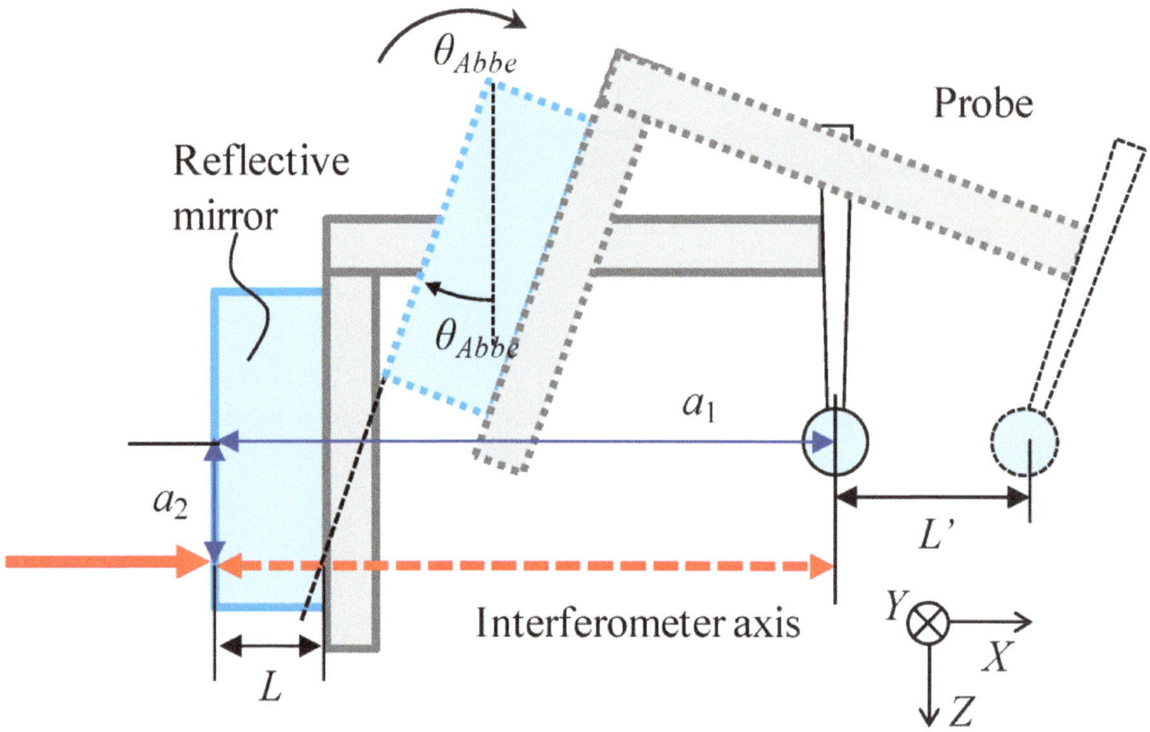

Figure 12. Schematic of Abbe error caused by the tilting motion of X-directional moving stage.

Consequently, the combined Abbe error can be expressed in the following equation.

$$L_{Abbe\_error} = \sqrt{L_{y\_Abbe}{}^2 + L_{z\_Abbe}{}^2} \tag{14}$$

The measurement error for the gap width measurement caused by the tilt errors of the positioning stages is then evaluated. When the probing points on the slot die surface are changed by moving the slot die with the Y-linear slide, errors will be introduced in the measurement result of the gap width by the tilt error about the Y-axis (Rolling) $\theta_{slide\_rolling}$ and the tilt error about the X-axis (Yawing) θ_{slide_yawing}. The gap width measurement error Δw_{y_slide} caused by the tilt errors of theY-linear slide can be indicated as follows.

$$\Delta w_{y\_slide} = \frac{w_s}{\cos\left(\theta_{slide\_rolling}\right)\cos\left(\theta_{slide\_yawing}\right)} - w_s \tag{15}$$

When the probing points are changed by moving the slot die with the Z-directional servo motor stage, errors will be introduced to the measurement result of the gap width by the tilt error $\theta_{z\text{-}servo\_rolling}$ about the Y-axis (Rolling) and the tilt error $\theta_{z\text{-}servo\_pitching}$ about the X-axis (Yawing). The gap width measurement error $\Delta w_{z\text{-}servo}$ caused by the tilt errors of the Y-linear slide can be indicated as follows.

$$\Delta w_{z-servo} = \frac{w_s}{\cos\left(\theta_{z-servo\_rolling}\right)\cos\left(\theta_{z-servo\_pitching}\right)} - w_s \tag{16}$$

Table 2 shows a summary of the tilt errors of the positioning stages. The tilt errors are measured by a laser autocollimator with a resolution of 1.21 μrad.

Table 2: Summary of tilting errors of the positioning stages.

Stage	Axis	Stroke	Tilting Error
Y-linear slide	Rolling Yawing	300 mm	8.10 mrad 12.31 mrad
Z-servo motor	Rolling Pitching	1 mm	22.59 mrad 73.40 mrad

The measurement uncertainties for on-line qualification and gap width measurement can then be evaluated based on Equations (4)–(15) as well as Table 1 and Table 2.

Experiments of On-Line Qualification and Gap Width Measurement

The micro-slit on the precision slot die for measurement has a nominal gap width of 85 ± 5 μm over a length of 200 mm. To avoid the influence of the chamfered edges of the precision flat parts forming the micro-slit, it is necessary for the probe tip to be arranged inside of the micro-slit for measurement of the gap width. In the developed micro-CMM, the probing position along the Z-direction is determined by utilizing the measurement capability of the shear-mode micro-probe in both the X- and Z-directions [26]. Figure 13 shows the detection strategy and the detection results of the chamfered edges of the two precision flat plates, respectively. It can be seen that the inner surfaces of the micro-slit can be identified from the measurement result shown in Figure 13b. The Z-directional height of the chamfered edges of the precision flat plates was estimated to be approximately 90 μm and the probing position along the Z-direction was selected to be lower than areas of the chamfered edges for the following experiments of on-line qualification and gap width measurement.

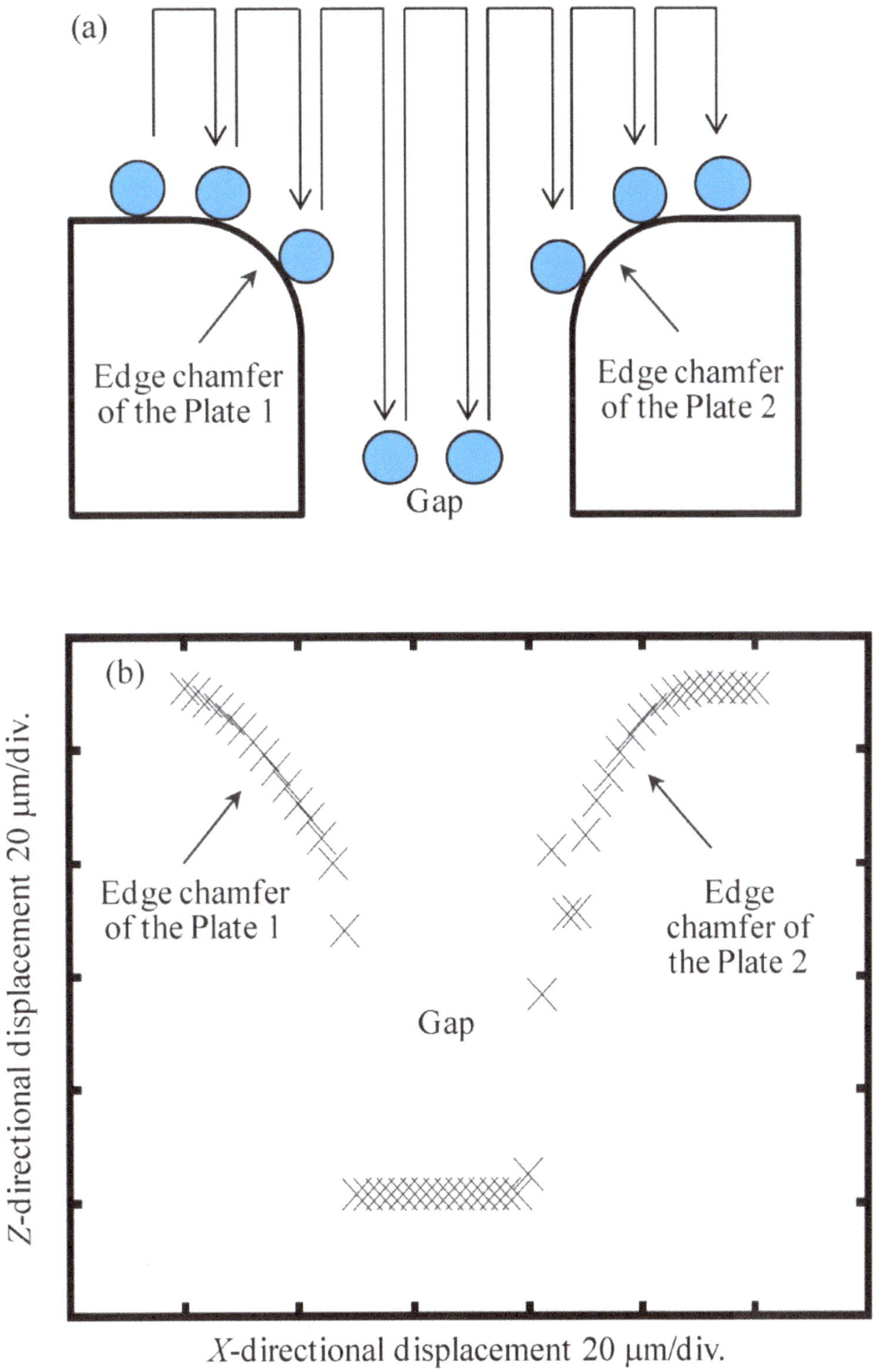

Figure 13: Chamfered edge detection by using the micro-probe: (a) Detection strategy of the chamfered edge; (b) Measurement result of the slit position.

On-line qualification of the effective diameter of the probe tip ball was then conducted by using the calibrated artifact. As a set of the on-line qualification operation, the probe was moved to the side of P_2 by the X-directional servo motor stage for another 5 times of probing at P_2 after 5 times of probing at P_1. It took approximately 50 s for the set of the operation. The same set of the operation was repeated 5 times to evaluate the repeatability of probing at the trigger points. Figure 14 shows the results of the on-line qualification. The X-directional positions of the trigger points at P_1 and P_2, which is a contact points between the probe tip ball and the inside surface of the slot die, are shown in Figure 14a,b, respectively. The vertical axes shown in Figure 14a,b show the X-directional probing positions calculated by the output of the laser interferometer. The mean values of the X-directional positions of the trigger points at P_1 and P_2 over the 5 sets of on-line qualification operations were 5.625 μm and 159.344 μm, respectively. Due to the thermal drift between the laser interferometer and the reflective mirror mounted on the X-PZT stage, the mean values at P_1 and P_2 were changed almost linearly due to the influence of thermal drifts. Figure 14c shows the value of L_{12} calculated by taking the difference of the corresponding results of Figure 14a,b, in which the thermal drift error components were removed. The mean and the standard deviation of L_{12}over the 5 sets of on-line qualification were 153.781 μm and 11.5 nm, respectively. Consequently, the effective diameter of the probe tip ball D_e was estimated to be 53.781 μm.

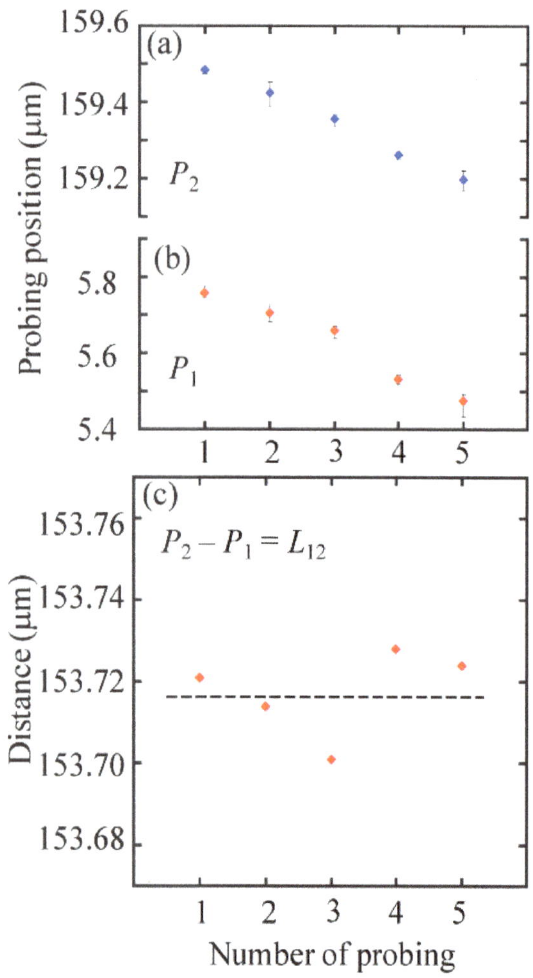

Figure 14: Experimental results of on-line qualification procedure: (a) Probing results at P_2; (b) Probing results at P_1; (c) Measurement results of L_{12}.

After the on-line qualification, the probing position was located to the inside of the micro-slit by moving the slot die with the Y-linear slide for the gap width measurement. At each Y-directional measurement position, after the probing at P_3, the probe was moved by the X-directional PZT stage toward the opposite surface of the micro-slit for the probing at P_4. This operation was repeated 5 times. The results are shown in Figure 15. Figure 15a,b show the probing positions at P_3 and P_4, which indicate the X-directional probe contact positions at the inner surface of the slot die. The mean values at P_3 and P_4 are −13.392 μm and 13.865 μm, respectively. Figure 15c shows the values of L_{34} calculated based on the results in Figure 15a,b. The mean and the standard deviation of L_{34} were estimated to be 27.243 μm and 14.8 nm, respectively. According to Equation (3), the gap width of the slot die micro-slit was estimated to be 80.962 μm.

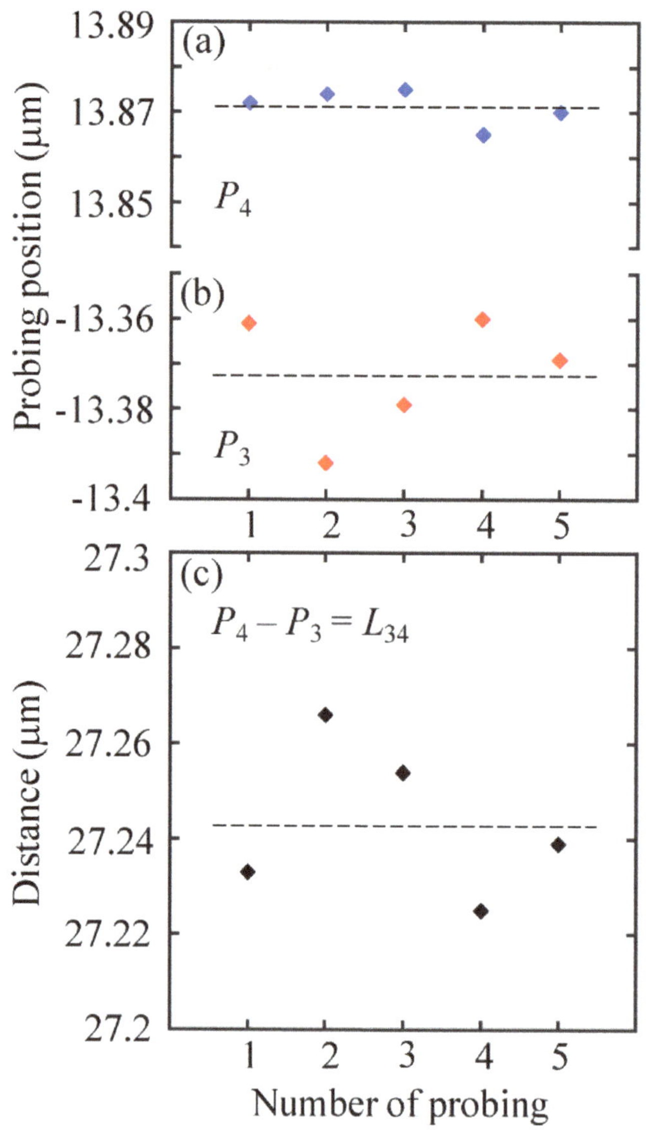

Figure 15: Experimental results of gap width measurement: (a) Probing results at P_4; (b) Probing results at P_3; (c) Measurement results of L_{34}.

In order to evaluate the accuracy of the gap width measurement result, uncertainty analysis was carried out for each of the terms in Equation (3) based on GUM (ISO Guide to the Expression of Uncertainty in Measurement) [33]. Table 3 shows a summary of the uncertainty budget for the gap width measurement. u_{ws}, which was the combined standard uncertainty of L_{34}, is calculated based on Figure 11, Equations (11) and (14). The repeatability of probing was one of the main uncertainty sources for u_{ws}. The total time for the gap width measurement was approximately 90 s, during which the temperature was measured to be 22.677 °C ± 0.011 °C. Consequently, u_{ws} is estimated to be 10.5 nm, which corresponds to the combined standard uncertainty for the measurement of L_{34}.

Table 3: Uncertainty budget (unit: nm).

Uncertainty sources	Symbol	Value	Coverage Factor	Standard Uncertainty
Uncertainty in w_S	u_{ws}	-	-	10.5
Cosine error by the alignment of the gauge block and the probing axis	$u_{cos\_slotdie}$	2.6	$\sqrt{3}$	1.7
Cosine error by the alignment of the interferometer axis and the probing axis	u_{cos_laser}	0.6	$\sqrt{3}$	0.3
Abbe error of the X-PZT stage	u_{pzt_abbe}	5.27	$\sqrt{3}$	3.0
Resolution of the interferometer	$u_{laser\_resolution}$	0.79	$\sqrt{3}$	0.5
Linearity error of the interferometer	$u_{laser\_linearity}$	5.0	$\sqrt{3}$	2.9
Thermal drift by temperature change	$u_{Thermal\_drift}$	7.0	$\sqrt{3}$	4.0
Repeatability of L_{34}	u_{rep_L34}	14.8	$\sqrt{3}$	8.5
Uncertainty in D_e	u_{De}	-	-	26.4
Cosine error by the alignment of the gauge block and the probing axis	u_{cos_gauge}	11.0	$\sqrt{3}$	6.4
Cosine error by the alignment of the interferometer axis and the probing axis	u_{cos_laser}	3.58	$\sqrt{3}$	2.1
Abbe error of the X-PZT stage	u_{pzt_abbe}	5.27	$\sqrt{3}$	3.0
Abbe error of the X-servo stage	u_{servo_abbe}	41.5	$\sqrt{3}$	24.0
Resolution of the interferometer	$u_{laser\_resolution}$	0.79	$\sqrt{3}$	0.5
Linearity error of the interferometer	$u_{laser\_linearity}$	5.0	$\sqrt{3}$	2.9
Thermal drift by temperature change	$u_{Thermal\_drift}$	7.0	$\sqrt{3}$	4.0
Repeatability of L_{12}	u_{rep_L12}	11.5	$\sqrt{3}$	6.6
Uncertainty in w_b	u_{wb}	-	-	17.3
Length tolerance(Calibrated gauge block)	$u_{tol\_calibrate}$	20.0	$\sqrt{3}$	17.3
Uncertainty in Ds	u_s	-	-	1.2
Uncertainty due to water layer	u_{water}	2.0	$\sqrt{3}$	1.2
Expanded uncertainty (with a coverage factor k = 2)	U	-	-	66.6

u_{De}, which was the combined standard uncertainty of the on-line qualification of the effective diameter of the probe tip ball, was estimated based on Figure 10, Equations (10) and (14). As shown in Table 3, the cosine error caused by the misalignment of the gauge block and the probing axis was a relatively large uncertainty source because the X-directional servo stage was moved during the on-line qualification process. The total time for the on-line qualification of the effective diameter was approximately 250 s, during which the temperature was measured to be 22.677 °C ± 0.008 °C. As a result, u_{De} was estimated to be 26.4 nm. u_{wb} was the length tolerance of the calibrated gauge block. According to the calibration certification of the gauge block used in the micro-CMM, u_{wb} was estimated to be 17.3 nm. $u_{\Delta s}$ was the uncertainty source introduced by the water layer on the measured surface. It was estimated to be 1.2 nm, based on Figure 4. Consequently, the expanded uncertainty U of the gap width measurement was estimated to be 66.6 nm (k = 2), which was smaller than the allowed maximum expanded uncertainty of 100 nm.

By changing the probing position along the Y- and Z-directions, the gap width uniformity was measured. Figure 16 shows the gap widths at measurement positions with an equal interval of 10 mm along the Y-direction. The measurement depth was set to be 50 μm. The probing was repeated 3 times at each measurement position. The dot plot in the figure shows the mean value of the repeated probing results and the error bar represents the corresponding standard deviation at each measurement position. The maximum and minimum values of the measured gap widths along the Y-direction were 78.883 μm and 81.980 μm, respectively. The total measurement time was approximately 1800 s. Since the two precision flat plates of the slot die were fastened by the screws at the center and the two ends of the slot die, the gap width tended to be smaller at the positions of the screws, which was corresponding to the measurement results in Figure 16. The maximum and minimum standard deviations of the measured gap widths were 27 nm and 4 nm, respectively. The employment of the long stroke air-bearing linear slide made it possible for measurement of the gap width uniformity over the entire slot die length of 200 mm, which was a significant improvement on a conventional micro-CMM.

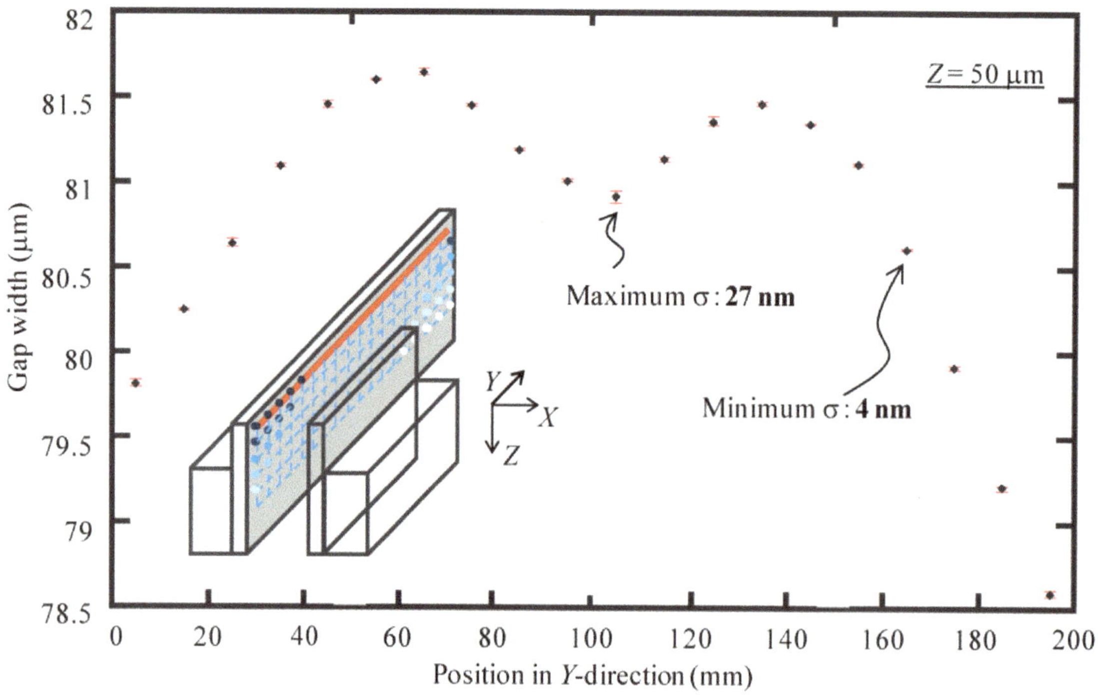

Figure 16: Measurement results of gap uniformity of slot die along the Y-direction.

Figure 17 shows the measured gap widths at the measurement positions with an equal interval of 100 μm along the Z-direction. The measurements were carried out at Y = 100 mm. The probing was repeated 5 times at each measurement position. The dot plot in the figure shows the mean value of the repeated probing results and the error bar represents the corresponding standard deviation at each measurement position. The maximum and minimum standard deviations of the measured gap widths were 27 nm and 4 nm, respectively. The total measurement time was approximately 990 s. The gap widths along the Z-direction were almost uniform. It can be seen from the figure that the gap width was successfully measured at Z = 1000 μm, indicating the effective working distance of the developed micro-CMM along the Z-direction.

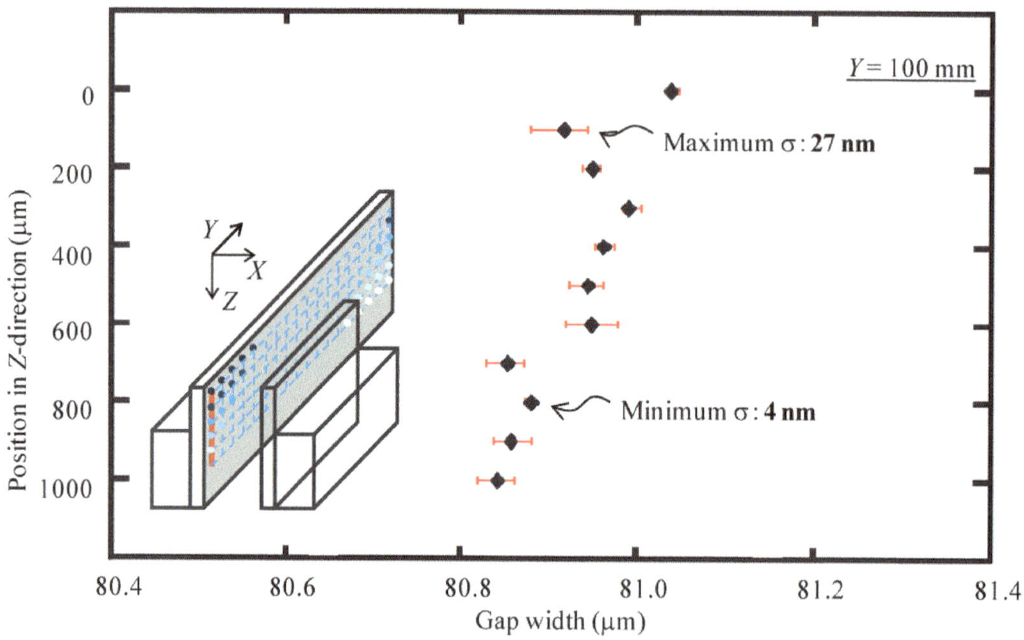

Figure 17: Measurement results of gap uniformity of slot die along the Z-direction.

The gap widths along different Y-directional lines were measured by changing the measurement depths along the Z-direction. The interval between the measurement points along the Y-direction was 10 mm and that along the Z-direction was 100 μm. The total number of probing points was 220. During the measurement, the temperature was measured to be 24.1 ± 0.1 °C. The total measurement time was approximately 200 min. As can be seen in Figure 18, the gap widths along differentY-lines were quite similar to each other. The measured gap widths at the same Y-position but different Z-positions were employed to demonstrate the gap widths at different Z-lines. The results are shown in Figure 19. It can be seen that the gap width slightly decreased with the increase in the measurement depth for all the Z-lines.

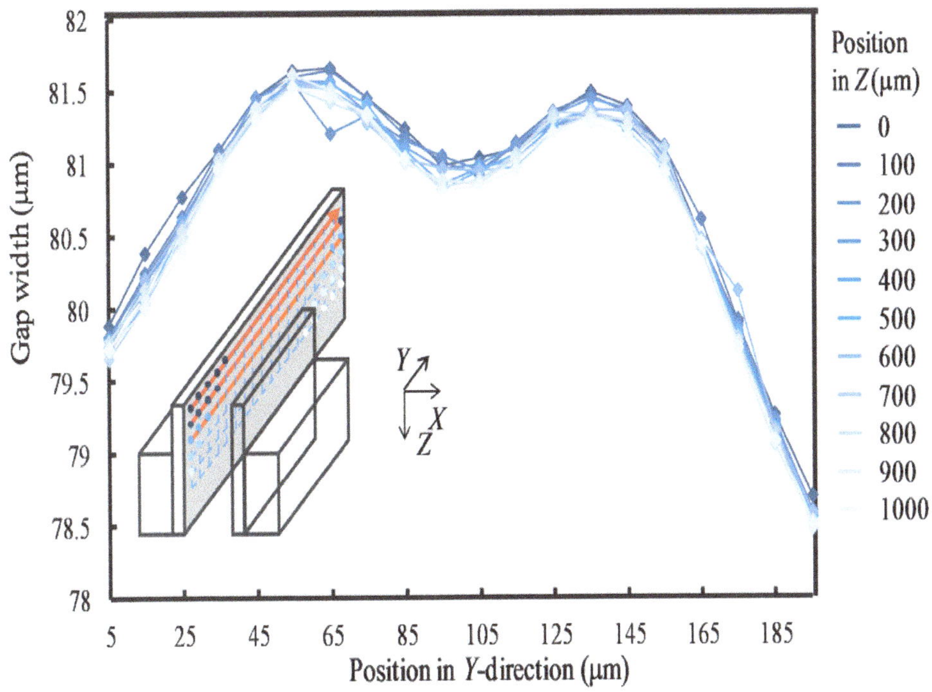

Figure 18: Gap width uniformity over the entire slot die length (Y-direction).

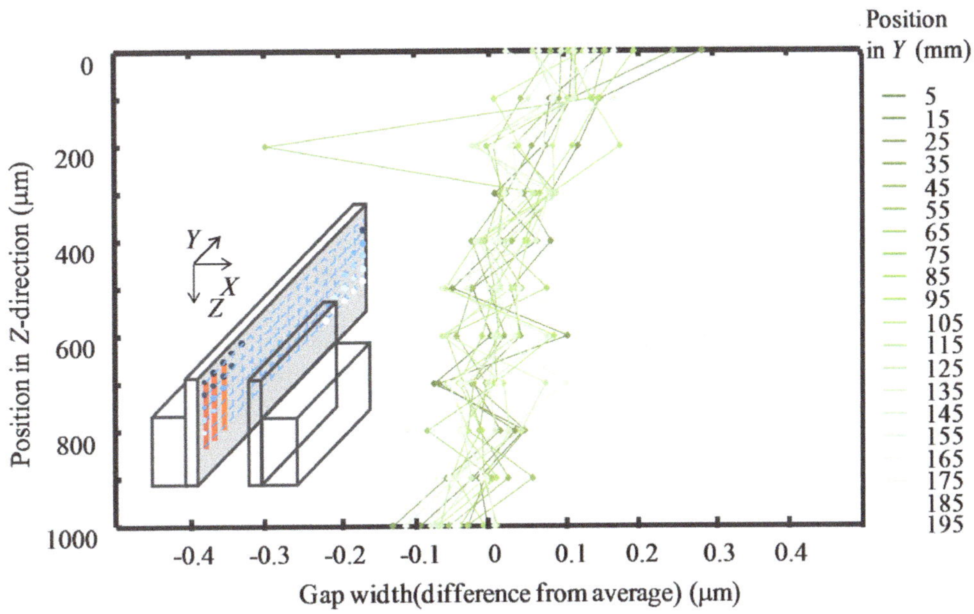

Figure 19: Gap width uniformity over the entire slot die length (Z-direction).

CONCLUSIONS

Dimensional measurement of micro-slit gap width of a precision slot die coater with a length of 200 mm and a nominal width of 85 μm has been carried out by using a specially designed and constructed micro-CMM. A shear-mode micro-probing system was employed in the micro-CMM. A calibrated artifact, which was a calibrated gauge block with a thickness of 100 μm, was placed on the same stage table as a Y-directional linear slide for the on-line qualification of the effective diameter of the probe tip ball prior to the gap width measurement of the slot die. The proposed micro-CMM, which combined the micro-probing system and a laser interferometer, was able to achieve nanometric resolution and repeatability of length measurement. The effective diameter D_e of the probe tip ball was evaluated to be 53.781 μm, and the expanded uncertainty in D_e was estimated to be 52.8 nm (k = 2). After the qualification procedure, the gap width measurement of the slot die was carried out immediately. As a result, the expanded uncertainty of the gap width measurement was estimated to be 66.6 nm (k = 2), which can satisfy the required measurement accuracy. The use of an air-bearing linear slide with a stroke of 300 mm has made it possible for the micro-CMM to cover the entire length of the precision slot die. Furthermore, a micro-stylus with an effective working length of larger than 1 mm, which was composed of a capillary glass shaft and a micro-glass sphere, has been employed to measure the gap width from the inside of the micro-slit of the precision slot die.

On the other hand, the shear-mode micro-probe was utilized as a touch-trigger probe in the developed micro-CMM and the total measurement time for the uniformity evaluation of the gap width was approximately 200 min. Reduction of the measurement time will be an important future work.

ACKNOWLEDGMENTS

This research was supported by KAKENHI. The authors would like to acknowledge the Ministry of Education, Culture, Sports, Science and Technology (MEXT) and Japan Society for the Promotion of Science (JSPS).

AUTHOR CONTRIBUTIONS

S.I. and W.G. conceived the experiments and wrote the paper; H.K. designed the experimental apparatus and performed the experiment; Y.C and Y.S. analyzed the data. K.T., T.K., K.A. and A.H. contributed to prepare slot die sample.

CONFLICTS OF INTEREST

The authors declare no conflict of interest.

REFERENCES

1. Chen, Y.L.; Gao, W.; Ju, B.F.; Shimizu, Y.; Ito, S. A measurement method of cutting tool position for relay fabrication of microstructured surface. Meas. Sci. Technol. 2014, 25, 064018.
2. Aziz, M.; Ohnishi, O.; Onikura, H. Novel micro deep drilling using micro long flat drill with ultrasonic vibration.Precis. Eng. 2012, 36, 168–174.
3. Hao, T.; Yong, L.; Long, Z.; Baoquan, L. Mechanism design and process control of micro EDM for drilling spray holes of diesel injector nozzles. Precis. Eng. 2013, 37, 213–221.
4. Goda, J.; Mitsui, K. Development of an integrated apparatus of micro-EDM and micro-CMM. Measurement 2013, 46, 552–562.
5. Vala, M.; Homola, J. Flexible method based on four-beam interference lithography for fabrication of large areas of perfectly periodic plasmonic arrays. Opt. Express 2014, 22, 18779–18789.
6. Meijer, T.; Beardmore, J.P.; Fabrie, C.G.C.H.M.; Lieshout, J.P.; Notermans, R.P.M.J.W.; Sang, R.T. Structure formation in atom lithography using geometric collimation. Appl. Phys. B 2011, 105, 703–713.
7. Chou, S.Y.; Krauss, P.R.; Renstrom, P.J. Nanoimprint lithography. J. Vacuum Sci. Technol. B 1996, 14, 4129–4133.
8. Heckele, M.; Schomburg, W.K. Review on micro molding of thermoplastic polymers. J. Micromech. Microeng. 2004, 14, R1–R14.
9. Bos, E.J.C. Aspects of tactile probing on the micro scale. Precis. Eng. 2011, 35, 228–240. [
10. Haitjema, H.; Pri, W.O.; Schellekens, P.H.J. Development of a Silicon-based Nanoprobe System for 3-D Measurements.CIRP Ann. Manuf. Technol. 2001, 50, 365–368.
11. Cui, J.; Li, J.; Feng, K.; Tan, T. Three-dimensional fiber probe based on orthogonal micro focal-length collimation for the measurement of micro parts. Opt. Express 2015, 23, 26386–26398.
12. Muralikrishnan, V.; Stone, J.; Shakarji, C.; Stoup, J. Performing three-dimensional measurements on micro-scale features using a flexible coordinate measuring machine fiber probe with ellipsoidal tip. Meas. Sci. Technol. 2012, 23, 025002.
13. Fan1, K.C.; Fei, Y.T.; Yu, X.F.; Chen, Y.J.; Wang, W.L.; Chen, F.; Liu, Y.S. Development of a low-cost micro-CMM for 3D micro/nano measurements. Meas. Sci. Technol. 2006, 17, 524–532.
14. Schwenke, H.; Waldele, F.; Weiskirch, C.; Kunzmann, H. Opto-tactile Sensor for 2D and 3D Measurement of Small Structures on Coordinate Measuring Machines. CIRP Ann. Manuf. Technol. 2001, 50, 361–364.
15. Michihata, M.; Hayashi, T.; Adachi, A.; Takaya, Y. Measurement of Probe-stylus Sphere Diameter for Micro-CMM Based on Spectral Fingerprint of Whispering Gallery Modes. CIRP Ann. Manuf. Technol. 2014, 63, 469–472.
16. Takaya, Y.; Shimizu, H.; Takahashi, S.; Miyoshi, T. Fundamental study on the new probe technique for the nano-CMM based on the laser trapping and Mirau interferometer. Measurement 1999, 25, 9–18.
17. Masuzawa, T.; Harnasaki, Y.; Fujino, M. Vibroscanning Method for Nondestructive easurement of Small Holes. CIRP Ann. Manuf. Technol. 1993, 42, 589–592.
18. Kim, B.J.; Masuzawa, T.; Bourouina, T. The vibroscanning method for the measurement of micro-hole profiles. Meas. Sci. Technol. 1999, 10, 697–705.
19. Weckenmann, A.; Hoffmann, J.; Schuler, A. Development of a tunnelling current sensor for a long-range nano-positioning device. Meas. Sci. Technol. 2008, 19, 064002.
20. Bauza, M.B.; Hocken, R.J.; Smith, S.T.; Woody, S.C. Development of a virtual probe tip with an application to high aspect ratio microscale features. Rev. Sci. Instrum. 2015, 76, 095112.
21. Bauza, M.B.; Woody, S.C.; Woody, B.A.; Smith, S.T. Surface profilometry of high aspect ratio features. Wear 2011, 271, 519–522.
22. Claverley, J.D.; Leach, R.K. Development of a three-dimensional vibrating tactile probe for miniature CMMs. Precis. Eng. 2013, 37, 491–499.
23. Leach, R.K.; Claverley, J.; Giusca, C.; Jones, C.W.; Nimishakavi, L.; Sun, W.; Tedaldi, M.; Yacoot, A. Advances in engineering nanometrology at the National Physical Laboratory. Meas. Sci. Technol. 2012, 23, 074002.
24. Murakami, H.; Katsuki, H.; Sajima, T.; Suematsu, T. Study of a vibrating fiber probing system for 3-D micro-structures: Performance improvement. Meas. Sci. Technol. 2014, 25, 094010.
25. Murakami, H.; Katsuki, A.; Onikura, H.; Sajima, T.; Kawagoshi, N.; Kondo, E. Development of a System for Measuring Micro Hole Accuracy Using an Optical

Fiber Probe. J. Adv. Mech. Des. Syst. Manuf. 2010, 4, 995–1004.

26. Ito, S.; Kodama, I.; Gao, G. Development of a probing system for a micro-coordinate measuring machine by utilizing shear-force detection. Meas. Sci. Technol. 2014, 25, 064011.

27. Manning, B. The Use of Non-Contact Thin Gap Sensors in Controlling Coater Gap Uniformity; Capacitec, Inc.: Ayer, MA, USA, 2012.

28. Furukawa, M.; Gao, W.; Shimizu, H.; Kiyono, S.; Yasutake, M.; Takahashi, K. Slit width measurement of a long precision slot die. J. Jpn. Soc. Precis. Eng. 2003, 69, 1013–1017.

29. Claverley, J.D.; Leach, R.K. A review of the existing performance verification infrastructure for micro-CMMs. Precis. Eng. 2015, 39, 1–15.

30. Muralikrishnan, B.; Stone, J.A.; Stoup, J.R. Fiber deflection probe for small hole metrology. Precis. Eng. 2006, 30, 154–164.

31. Thermo Scientific. Available online: http://www.thermoscientific.com/en/product/ (accessed on 5 May 2016).

32. Ito, S.; Chen, Y.L.; Shimizu, Y.; Kikuchi, H.; Gao, W.; Takahashi, K.; Kanayama, T.; Arakawa, K.; Hayashi, A. Uncertainty analysis of slot die coater gap width measurement by using a shear mode micro-probing system. Precis. Eng. 2016, 43, 525–529.

33. Working Group 1 of the Joint Committee for Guides in Metrology (JCGM/WG1). JCGM 100. In Evaluation of Measurement Data—Guide to the Expression of Uncertainty in Measurement (GUM); Bureau International des Poids et Mesures: Paris, France, 2008.

CHAPTER 15

IDENTIFICATION OF A CRITICAL TIME WITH ACOUSTIC EMISSION MONITORING DURING STATIC FATIGUE TESTS ON CERAMIC MATRIX COMPOSITES: TOWARDS LIFETIME PREDICTION

Nathalie Godin, Pascal Reynaud, Mohamed R'Mili and Gilbert Fantozzi

INSA de Lyon, MATEIS (UMR CNRS 5510), 7 avenue Jean Capelle, 69621 Villeurbanne Cedex, France

ABSTRACT

Non-oxide fiber-reinforced ceramic-matrix composites are promising candidates for some aeronautic applications that require good thermomechanical behavior over long periods of time. This study focuses on the behavior of a SiC_f/[Si-B-C] composite with a self-healing matrix at intermediate temperature under air. Static fatigue experiments were performed below 600 °C and a lifetime diagram is presented. Damage is monitored both by strain measurement and acoustic emission during the static fatigue experiments. Two methods of real-time analysis of associated energy release have been developed. They allow for the identification of a characteristic time that was found to be close to 55% of the measured rupture time. This critical time reflects a critical local energy release assessed by the applicability of the Benioff law. This critical aspect is linked to a damage phase where slow crack growth in fibers is prevailing leading to ultimate fracture of the composite.

INTRODUCTION

Ceramic Matrix Composites (CMCs), and more particularly SiC/SiC composites are very attractive candidates for many high-temperature structural applications, because of their excellent creep resistance, high-temperature strength and light weight. Damage tolerance is achieved through the use of low shear strength fiber coatings that deflect cracks along the interfaces [1,2,3,4,5,6,7]. The multi-layered matrix was introduced in the new generations of SiC_f/[Si-B-C] composites in order to improve the lifetime under medium and high temperatures thanks to the formation of sealant glasses [8,9].

Future engine applications in civil aircrafts are foreseen for such composites. These applications require that the material can resist both severe conditions (temperatures up to 1200 °C under air) and very long static loadings at intermediate temperatures (400–700 °C during tens of thousands of hours). Various authors have studied the mechanical behavior of SiC_f/SiC composites and the damage mechanisms occurring at high temperatures [1,2,3,4,5,6,7]. Now more information is needed at intermediate temperatures: the oxidation kinetics of the different constituents are complex, and the effect of matrix sealing on the lifetime of the composite has to be examined. Expected lifetimes in use conditions are several thousands

Citation: Nathalie Godin, Pascal Reynaud, Mohamed R'Mili and Gilbert Fantozzi, Identification of a Critical Time with Acoustic Emission Monitoring during Static Fatigue Tests on Ceramic Matrix Composites: Towards Lifetime Prediction, doi:10.3390/app6020043

of hours, which can hardly be reached by laboratory tests for practical reasons. Therefore, a real-time prediction of the remaining lifetime during tests is necessary. It requires the monitoring of damage evolution for which AE measurement is a suitable technique. In fact, the AE technique consists in the recording and analysis of elastic waves created during material damage. It provides real-time data on initiation and evolution of damage in terms of location and mode.

Common analysis of AE is based on diagrams of cumulative hits or counts or histograms of amplitudes. Nevertheless, in the case of composite material, this approach is inadequate in order to identify the acoustic emission signature of several damage mechanisms. It was improved by grouping signals of similar shapes into clusters using classifier parameters [10]. Many works [11,12,13,14,15,16] have shown that AE techniques and multivariable classification techniques are the basis of pattern recognition tools. Kostopoulos [12], Godin [13,14], Moevus [15,16] have identified different classes of AE signals which were attributed to damage modes in oxide/oxide, glass/polyester and SiCf/[Si-B-C] composites. Moevus [16] analyzed the AE data using an unsupervised multi-variable clustering technique. This analysis showed that several types of signals may be distinguished, and that the obtained clusters were consistent with the expected damage modes in 3D SiC/SiC composites. In this way, a careful analysis of acoustic emission signals can lead to the discrimination of the different damage mechanisms occurring in a composite material. It is a possible solution for identification of damage during service with a view to component lifetime control.

A second analysis is based on a global AE analysis, on the investigation of liberated energy, with a view to identify a critical point. Many researchers investigated the elastic energy release during the failure process of materials [17,18,19,20,21,22,23,24]. They all observed that the energy release accelerated following a power law. These models are representative of an avalanche behavior very similar to that observed in seismicity.

This study focuses on static fatigue experiments realized at intermediate temperature under air on a SiC_f/[Si-B-C] composite with a self-healing matrix. A lifetime diagram is presented, and the evolution of several damage indicators is discussed. The objective of this approach is to propose a method based on acoustic energy in order to evaluate the remaining lifetime during long-term mechanical tests. This approach is based on the determination of energy released and identification of a critical point, in energy release during mechanical test. Therefore, two criteria have been defined which allow predicting the end of a static fatigue test knowing the AE activity emitted during the first half of the test. Therefore, beyond this characteristic point the criticality can be described by a power-law in order to evaluate time to failure. Moreover, a supervised classification method was used to differentiate the signals generated during fatigue tests performed on composites at intermediate temperature and build a specific library [16]. This library was used to identify the damage modes generated during fatigue tests performed at various temperatures, in order to establish a link between this critical time and the damage mechanisms.

EXPERIMENTAL PROCEDURE

Material

The material, manufactured by SAFRAN-HERAKLES Group (Bordeaux, France), had a multilayered [Si-B-C] reinforced matrix obtained by chemical vapor infiltration on SiC fibers and a PyC interface layer (Figure 1). This 2.5D woven composite is composed of Hi-Nicalon fibers, a pyro-carbon interphase layer and a self-healing matrix. This matrix has been processed by means of several chemical vapor infiltrations with various compositions, based on the ternary Si-B-C system. The external surface is protected by a seal-coat. The composite contains a volumic fraction of fibers equal to 35 vol.% and 20 vol.% of porosity. Dog-bone shaped specimens were used with the following dimensions: thickness 4 mm, width 16 mm, working length 40 mm with a constant cross section of 64 mm^2.

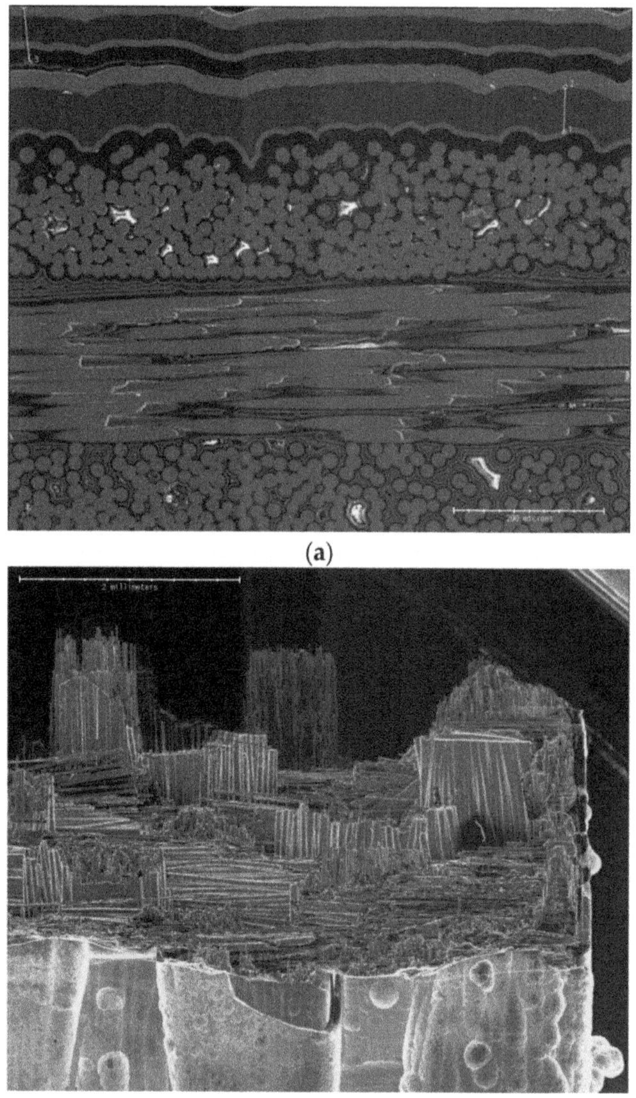

(a)

(b)

Figure 1. (**a**) SEM micrograph of the cross-section of SiC_f/[Si-B-C] composites; (**b**) Typical fracture surface of SiC_f/[Si-B-C] composite.

Static Fatigue Tests

Static fatigue tests were performed under uniaxial tensile loading with one direction of fibers parallel to the loading axis.

Tests were carried out at 450 °C, 500 °C and 560 °C under air, on a pneumatic testing machine. Before loading, specimens were heated up to the testing temperature at a rate of 50 °C/min and held during 30 min to get a uniform temperature distribution in the gauge section. Specimens were loaded at a constant rate of 600 N/min up to a constant stress chosen in the range of 44% to 95% of the rupture stress. Specimen elongation was measured using a high temperature extensometer.

Seldom unload-reload cycles were applied every twelve hours in order to measure the secant elastic modulus changes during static fatigue and to evaluate the damage parameter D (Figure 2). E_0 is the elastic modulus in the undamaged state.

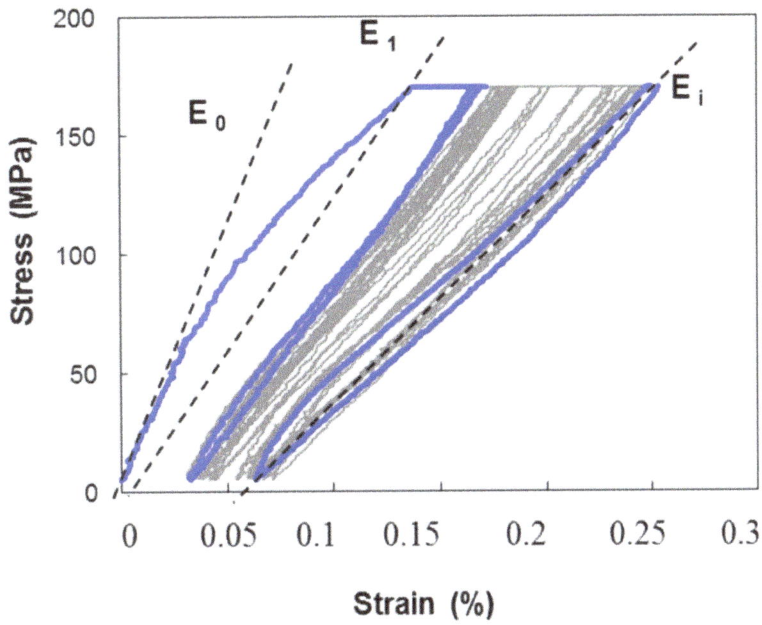

Figure 2: Schematic hysteresis loop and measure of the hysteresis loops modulus, E_0 is the initial Young's modulus, E_i is the secant elastic modulus for the unloading-reloading cycle *i*.

Acoustic Emission Monitoring

To detect acoustic emission, two MICRO-80 sensors (MISTRAS Group, Sucy en Brie, France) were put on the specimen inside the grips with vacuum grease as a coupling agent. Acquisition parameters were set as follows: preamplification 40 dB, threshold 32 dB, peak definition time 50 µs, hit definition time 100 µs, hit lockout time 1000 µs. AE signal descriptors such as amplitude, energy, duration, counts, average frequency, rise time and location as well as time, load and strain were recorded in real time by a MISTRAS 2001 data acquisition system.

For AE sources localization, AE wave velocity Ce_0 was determined before tests using a pencil lead break procedure on as received composites. This velocity was measured equal to 10,000 m/s. Since the elastic modulus decreases as damage occurs in the material, it is important to take into account the evolution of *Ce* during the mechanical test in order to better evaluate the location of the AE sources. As proposed by Morsher [25], the elastic secant modulus during unloading $E_i(\varepsilon)$ was measured during a cycled tensile test, where hysteresis loops were obtained at different strains. The velocity $Ce(\varepsilon)$ was then determined by using the Equation (1):

$$\frac{Ce(\varepsilon)}{Ce_0} = \sqrt{\frac{E_i(\varepsilon)}{E_0}} \quad (1)$$

where Ce_0 and E_0 are respectively the velocity and the elastic modulus in the undamaged state and $C_e(\varepsilon)$, $E_i(\varepsilon)$ respectively the velocity and the elastic modulus under a maximum strain ε. At the end of the tensile test, the velocity on SiC/SiC composite was found to be equal to 6480 m/s, instead of 10,000 m/s on the undamaged state. This decrease of wave velocity is thus not negligible. The location of sources has been calculated using the difference in times of arrival on each sensor. Only the signals coming from the working length of the specimens are then analyzed.

ACOUSTIC EMISSION ANALYSIS

Definition of the Acoustic Energy

Energy of an AE signal includes the energy released by the source at crack initiation and is affected by various parameters: wave propagation distance, damage attenuation, sensor/material surface coupling and sensor frequency response. Wave theory states that the energy of an acoustic wave decreases exponentially with the increase of propagation distance. Therefore, the following equation was proposed to describe the energy of recorded AE signals (for instance, at sensor 1) received from the source *n* [26,27]:

$$E_1(n) = E_s(n) \cdot A_1 \cdot e^{B(L+x(n))} \quad (2)$$

$E_s(n)$ is the energy released at source n in the form of elastic waves. Due to differences in coupling between sensors and material surface or in sensors frequency responses, for a source at equal distance, the sensors may record significantly different amounts of energy. Thus, A_1 is the proportion of source energy that is recorded by sensor 1. It is a constant characteristic of sensor. $L + x(n)$ is the propagation distance from source n to the sensor 1 (2L is the distance between two sensors). The attenuation coefficient B is linked to the propagation medium, which may change with damage evolution. Similarly, AE signal energy received at sensor 2 is expressed as:

$$E_2(n) = E_s(n) \cdot A_2 \cdot e^{B(L - x(n))} \tag{3}$$

The source energy is then defined as the square root of the product of the amounts of energy received at both sensors for each source:

$$E(n) = \sqrt{E_1(n) \cdot E_2(n)} \tag{4}$$

Attenuation Coefficient B

In order to evaluate energy attenuation [28], the ratio of AE signal energies recorded at both sensors is calculated for each source n. For an easier identification of attenuation coefficient B, $x(n)$ is defined as the natural logarithm of this ratio. From Equations (2) and (3), it comes:

$$X(n) = \log\frac{E_1(n)}{E_2(n)} = \log\frac{A_1}{A_2} - 2 \cdot B \cdot x(n) \tag{5}$$

The second term corresponds to the effect of propagation, related to source location $x(n)$ and to attenuation coefficient B. Both parameters can be estimated from $x(n)$ for various sources (1,2,..., n,...) since $x(n)$ is a linear function of x. The coefficient of attenuation B can be determined from the slope of the linear fit.

Because attenuation varies with damage, parameters estimation is done at successive time intervals. For a given time interval, values of $x(n)$ are analyzed in many space intervals in order to consider uncertainties in localization of AE sources. Estimation of attenuation constants was performed as follows. Attenuation is evaluated for each time interval with the median values of $x(n)$ in all space interval (width: 10 mm, overlapping: 5 mm). Each time interval is defined by 2000 AE successive sources. Each median value of $x(n)$ corresponds to a few hundred AE sources located in the same space-time interval. The linear approximation is repeated in consecutive time intervals. At every increment of 500 sources, a new time interval is considered. The overlapping is set in order to accurately monitor evolution of both attenuation parameters.

Coefficient of Emission R_{AE}

The coefficient of emission RAE is defined [29], as the increment of energy ΔE recorded during a time increment Δt, divided by the total energy emitted during the initial loading of the sample:

$$R_{AE}(t) = \frac{1}{E_{loading}} \frac{\Delta E}{\Delta t} \tag{6}$$

where $E_{loading}$ is the cumulative AE energy for all the signals recorded during the initial loading up to the nominal load of the test, ΔE is the cumulative AE energy for all signals recorded during the interval $[t; t + \Delta t]$.

Power Law

Benioff law [30], suggested initially for precursory phenomena of earthquakes, has been applied originally on composites [29]. Based on Equation (7), the increase of AE during fatigue is analyzed.

$$\Omega(t) = \sum_{i=1}^{N(t)} \sqrt{E_i} = \Omega_R + B\,(t_R - t)^{1-\gamma} \tag{7}$$

where E_i is the energy of the ith AE signal detected and $N(t)$ is the number of AE signals recorded and located along the gauge length until time t.

ΩR is the value of $\Omega(t) = \sum_{i=1}^{N(t)} \sqrt{E_i}$ when $t = t_R$, t_R is the failure time. $B = \frac{\phi}{1-\gamma}$ is negative, $1-\gamma$ is an exponent and φ is a constant.

The optimum circle method (OCM) [31], also developed in seismology, is used here to assess the relevance of the Benioff law. In addition, it determines which AE sources should be considered in order to achieve the best approximation. The OCM method was used to evaluate the time t_{start} when the AE energy is well simulated by the Benioff law. Two approximations are carried out on the energy release resulting from each time interval $[t_{start}; t_r]$: a power-law approximation using the Benioff law and a linear approximation ($\Omega(t) = \alpha t + \beta$) used as reference. The c-value is defined by the ratio of the root mean square error of the approximation by the Benioff law over that of the linear fit. When the c-value is lower than 1, there is a positive contribution of the Benioff law since the approximation error is lower than that of a linear fit. It is a relative validation of the relevance of the approximation by the Benioff law. Therefore, to ensure quality of the approximation, only$c$-values lower than 0.5 are considered to be relevant.

Identification of Damage Mechanisms with Supervised Clustering

In such CMCs, matrix cracking occurs in several modes of cracking [16]. First, cracks initiate in the external seal-coat and in macropores inside the composite, then propagate through the inter-yarn matrix. Afterwards cracks propagate inside the transverse yarns through fiber/matrix interfaces. Multiple matrix cracking finally occurs inside the axial yarns. These cracks are deviated by the fiber/matrix interphase layer, yielding to fiber debonding and overloading. Some fiber breaks occur under high stresses. They rapidly lead to unstable fracture of entire yarns and of the composite. Hence, the sources of AE are the various matrix cracks, fiber/matrix interfacial debonding, individual fiber fractures and yarn ruptures.

The supervised classification technique requires a data base of signals that have been labelled: the training set. This training set was created by merging data collected during tensile test [16]. As described in the previous paper [16], the analysis of AE signals, observation of microstructures and analysis of the mechanical behavior of the composite led to the identification of 4 types of AE signals and to the following labelling of classes:

Class A: Signals of cluster A are assigned to two damage mechanisms which are chronologically separated: seal coat cracking and tow breaks.

Class B: cluster B contains also signals from two damage mechanisms which are chronologically well separated: longitudinal matrix cracking and individual fiber breaks in the fracture zone just before failure.

Class C: Cluster C contains signals with relatively short duration, short rise time and low amplitude when compared to the others: transversal matrix cracking.

Class D: this cluster is the last one to be activated and it becomes more active as strain increases and the D-type signals have a longer rise time when compared to other signals: sliding at fiber/matrix interfaces, fiber/matrix debonding.

In order to establish the training set of labelled signals for the supervised analysis, the same amount of signals (500 signals) in each class (A, B, C and D) was used. This training set included all the damage modes that may operate in this composite.

The supervised method consists in comparing each detected signal to those of the library and to assign it to the class of the K nearest neighbors [32] in the library. First, it is necessary to determine the number K of nearest neighbors required to obtain the best classification. For this purpose, a self-validating procedure was used which consists in comparing each signal of the training set to all the others. The classification error rate was estimated for all the values of K using the "leave-one-out" method [33]. The optimal value of K corresponded to the lowest value of error, and in the present study $K = 11$, and error = 2.5%. Then, for each AE signal of a given test, the Euclidean distance to all signals of the training set was calculated and the signal was labelled according to the class the most frequent class among its 11 nearest neighbors. The supervised clustering procedure was carried out on the different fatigue static tests.

RESULTS AND DISCUSSION

Mechanical Analysis

The lifetime diagram, obtained at several temperatures and under various stresses, of the composite for static fatigue at intermediate temperature is plotted in Figure 3a. All the points are aligned in the logarithmic representation and follow the power-type law:

$$t \cdot \sigma^n = A \qquad (8)$$

where t is the lifetime, σ is the applied stress, A and n are constants depending on the material, the temperature and the environment. The stress exponent n is estimated at 3.2 ± 0.3. A comparison between our results on composite and on the fibers in literature [34] under static fatigue at 500 °C (Figure 3b) shows that at intermediate and high stresses the lifetime of the composite is longer than the lifetime of the fibers. This is characterized by two different values of n, n is equal to 3.2 for the composite and to 8.45 for the fibers [34].

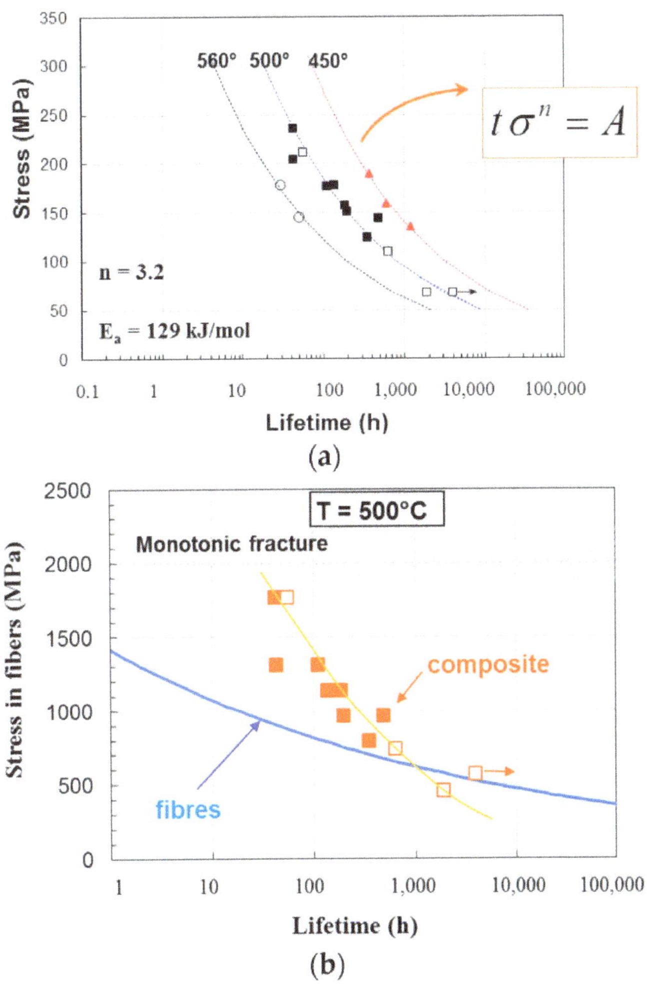

Figure 3: (**a**) Lifetimes obtained during static fatigue at 450 °C, 500 °C and 560 °C under air on SiC_f/[Si-B-C] composites for several applied stresses; (**b**) Lifetimes on the fibers bundles [34] and on SiC_f/[Si-B-C] composites in static fatigue at 500 °C under air.

For this range of applied stress, slow crack growth of the fibers needs the diffusion of oxygen but is controlled by the chemical reaction at the crack tip. As shown by Gauthier and Lamon [34], the activation energy 129 kJ/mol corresponds to the reactivity of carbon (present in the fibers) to oxygen. In the composite the diffusion of oxygen toward the fiber surfaces is slackened by the oxygen consumption due to reaction with self-healing matrices and with interphases. Under lower stresses, the lifetimes of the composite and of the fiber bundles are similar. For this low level of stresses reaction kinetics with fiber surfaces are slower than diffusion kinetics of oxygen into the matrix crack. Hence the subcritical crack growth kinetics in dry bundles and in the fibers of the composites are similar.

Stress as a function of strain is shown in Figure 4. The grey curve represents the monotonic tensile curve whereas the other curves correspond to the static fatigue curves for various ratios of stress and with some loading and unloading sequences. One may notice that strain increases significantly during the test, and that hysteresis secant loops evolve as well: the mean elastic modulus decreases whereas the loops width increases. The increase in the loops area is linked to a change in the interfacial shear stress, suggesting that some debonding occur during static fatigue. The evolution of the Young's modulus was evaluated during the tests by measuring the hysteresis loops modulus E_i. Then the damage parameter $D = 1 - E_i/E_0$ was calculated, E_0 being the initial Young's modulus of the composite. The evolution of D is plotted *versus* time inFigure 5, for several σ/σ_R ratios. The biggest increase of D occurs during the loading step and the first 24 h of static fatigue. Then D is observed to rise monotonically up to the failure of the specimen. The general trend is that the highest the σ/σ_Rratio, the highest the D parameter for similar durations of test. The final D values obtained at failure are not exactly the same: they vary in the range 0.6–0.8, and it seems that this value decreases when the σ/σ_R ratio decreases. However it seems difficult to define a failure criterion based on the damage parameter.

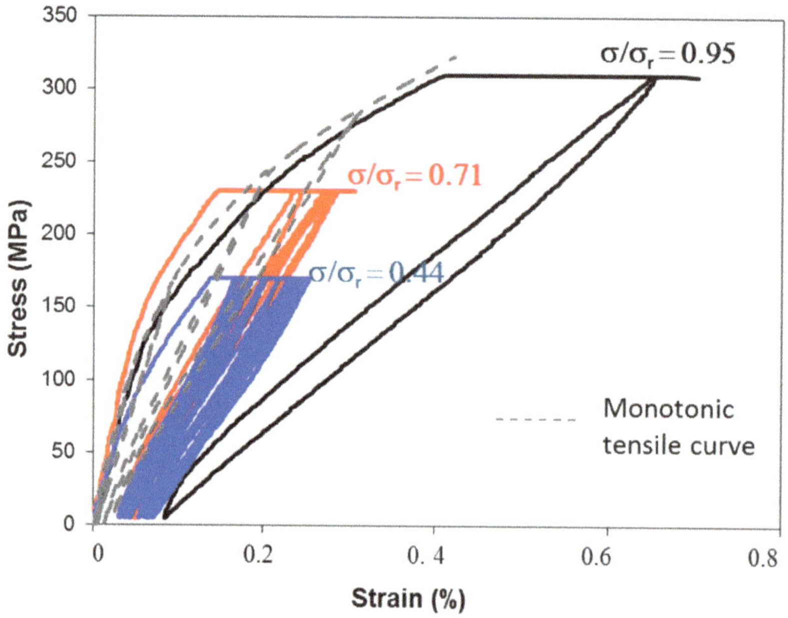

Figure 4: Stress *versus* strain during static fatigue experiments for different σ/σ_R ratios: 0.44, 0.71 and 0.95 at 500 °C on SiC_f/[Si-B-C] composites. The grey curve represents the monotonic tensile curve.

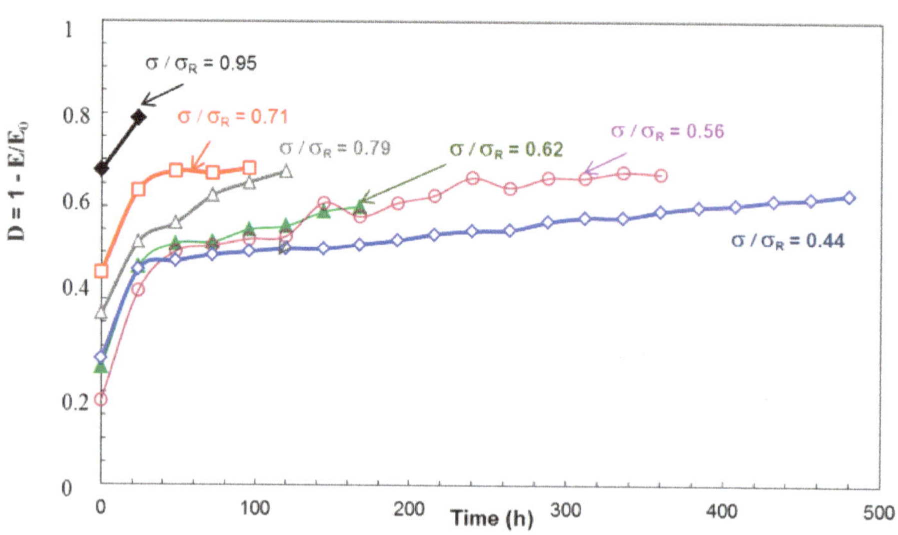

Figure 5: Evolution of the damage parameter D during static fatigue experiments for various σ/σ_R ratios 0.44, 0.56, 0.62, 0.71, 0.79 and 0.95 at 500 °C on SiC_f/[Si-B-C] composites.

The evolution of strain and acoustic emission events is shown in Figure 6 for a specimen loaded at $\sigma/\sigma_R = 0.44$. Under the steady loading the strain rate is important at the beginning then decreases before reaching a constant value. At this time, strain rises monotonically up to the final failure for the majority of the specimens (Figure 6b). The AE activity recorded during the constant load hold is also plotted in Figure 6b. The acoustic emission activity evolves with the same trend. Figure 6c shows the evolution of the cumulated acoustic energy. The main activity occurred at the beginning of the test and then reached a plateau. The renewal of AE activity (Figure 6c) just before failure seems to be an interesting way to anticipate the fracture but was not always noticeable on the AE energy *vs.* time plot.

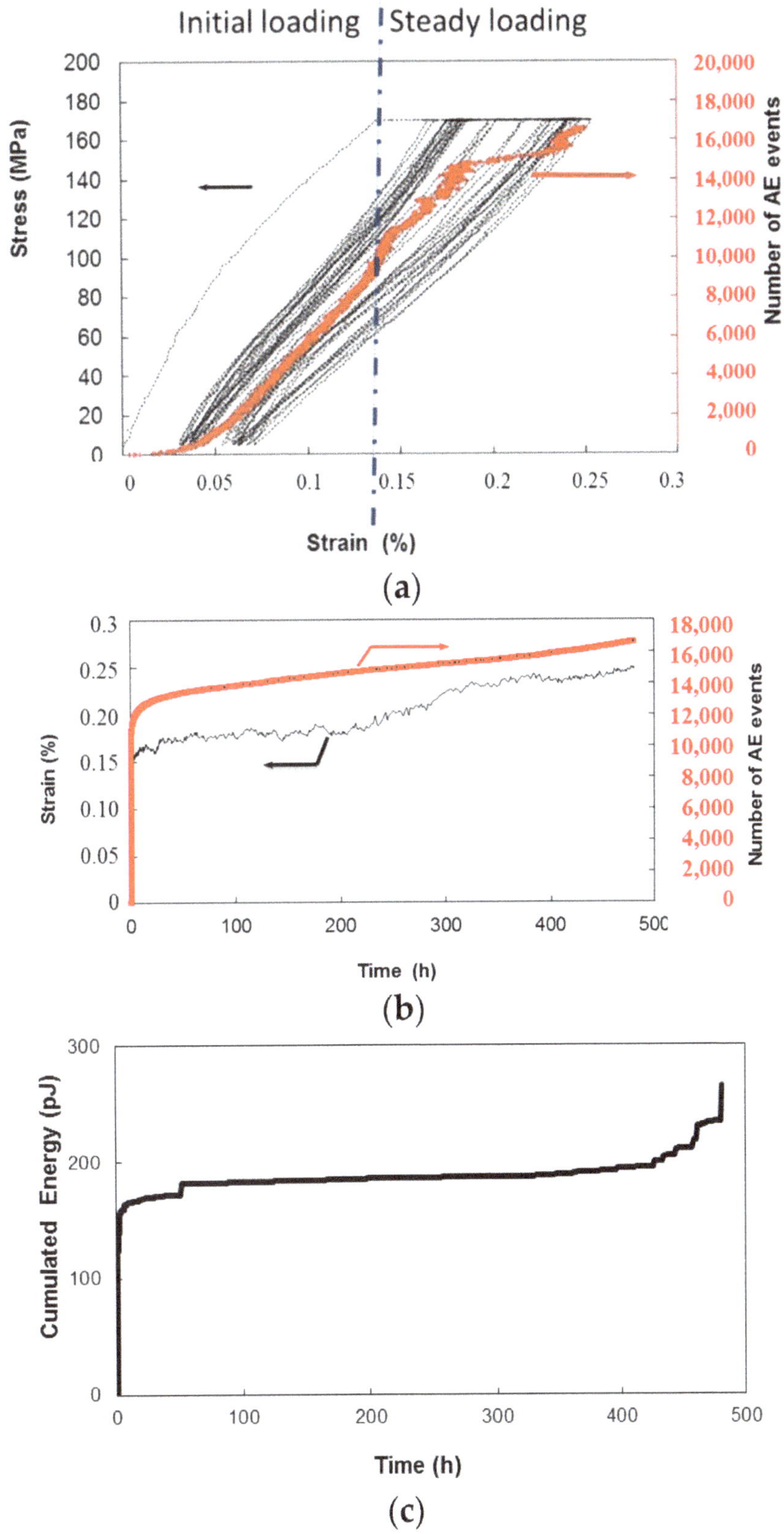

Figure 6: (**a**) Stress-strain curve and cumulated number of acoustic emission (AE) events recorded during initial loading up to the nominal load, and during the static loading at a constant load ($\sigma/\sigma_R = 0.44$ and $T = 500$ °C); (**b**) Strain and cumulated number of AE events *versus* time; (**c**) Cumulated acoustic energy *versus* time.

The main activity occurs at the beginning of the static fatigue. Then the AE rate decreases without stopping. During the initial loading up to the nominal load, the number of events increases with the applied stress, which is consistent with the increasing number of cracks in the matrix (the main source of acoustic emission) as the applied stress increases. Then, during the static loading, the number of events before failure is lower under high stress than under low stress (Figure 7): damage needs to progress more under low stress to cause the final failure of the composite, than under high stress where the matrix already contains a lot of cracks.

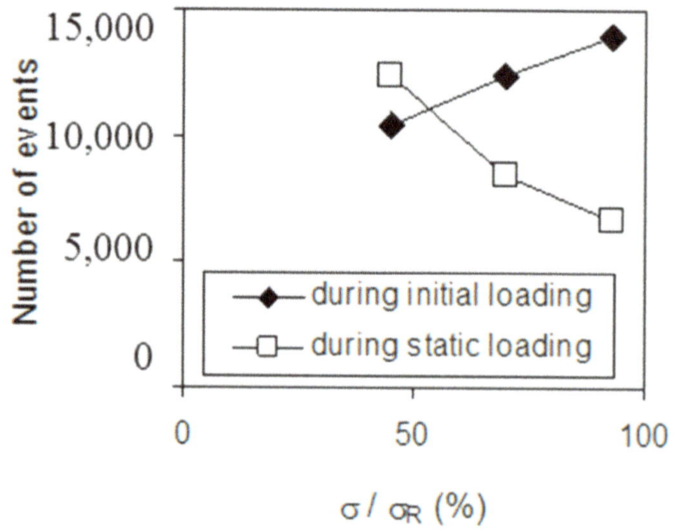

Figure 7: Number of AE events recorded during initial loading up to the nominal load and during the static loading at a constant load.

Identification of Critical Time

The evolution of RAE is plotted *versus* time in Figure 8 for three experiments. This coefficient is observed to decrease first, then to reach a minimum value for $t = t_m$, and then to increase up to the failure of the composite at time t_R. This ultimate increase in AE activity was not always visible on the curves of AE energy *versus* time, but it was revealed by this representation. The ratio t_m/t_R appears to be quite reproducible: $t_m/t_R = 0.57 \pm 0.06$ (Figure 9). Therefore, the RAE ratio may be used as a criterion for predicting the remaining lifetime of a specimen under static loading. During static fatigue test, attenuation coefficient *B* growths significantly during the first part of tests up to a plateau value near 50% of the rupture time (Figure 10). For each test, the slight increase of *B* observed during initial loading is very low compared to the one occurring during static fatigue. Therefore, the increase of attenuation coefficient *B* may be related to matrix crack opening and to the recession of interfaces.

The detection of the plateau during the typical evolution of attenuation coefficient B may be considered as an indicator for lifetime prediction. As the minimum of coefficient R_{AE}, the value plateau of *B* indicates the beginning of the critical damage phase and provides an estimation of the remaining lifetime.

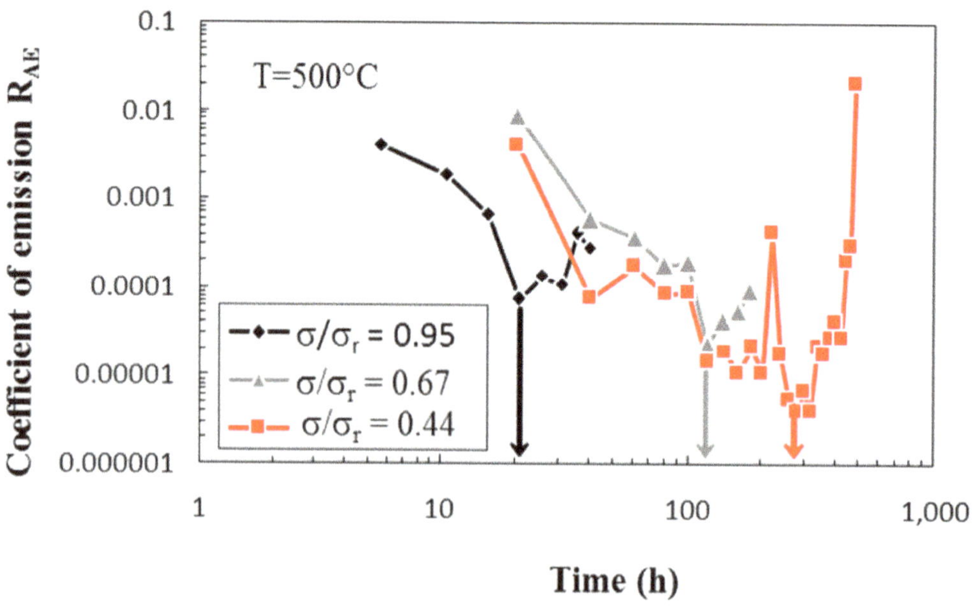

Figure 8: Evolution of the R_{AE} coefficient during the static load hold on SiC_f/[Si-B-C] composites.

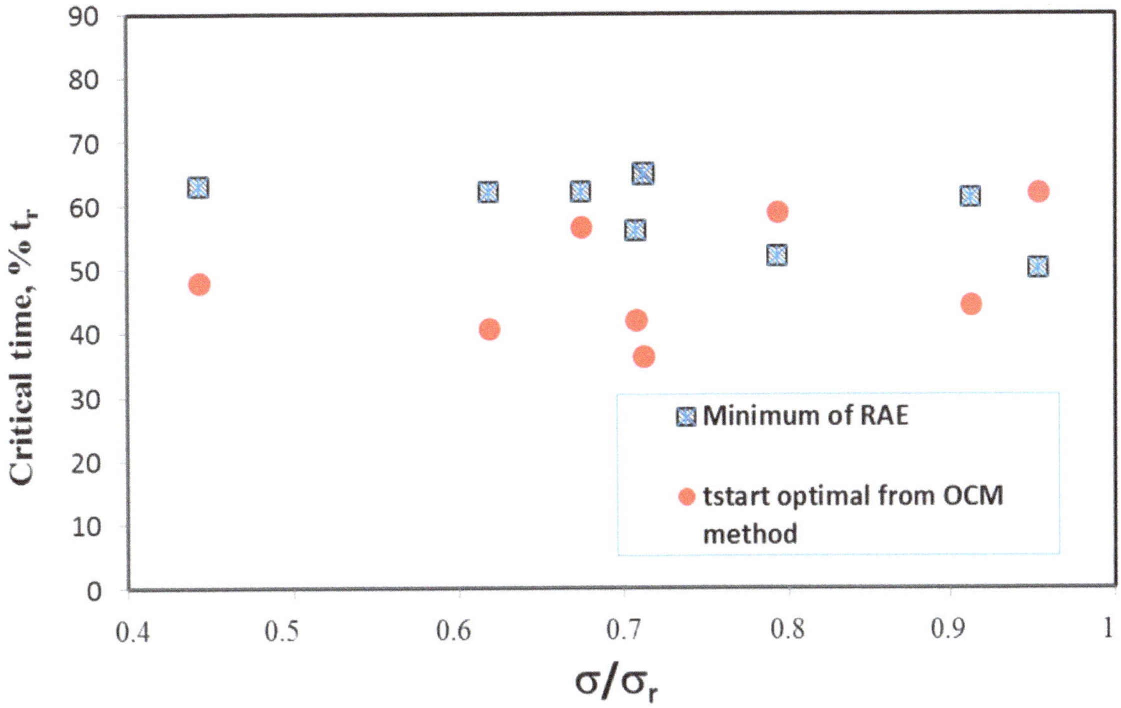

Figure 9: Critical times obtained with the minimum value of R_{AE} and the optimumt_{sart} for the Benioff law obtained with the OCM method.

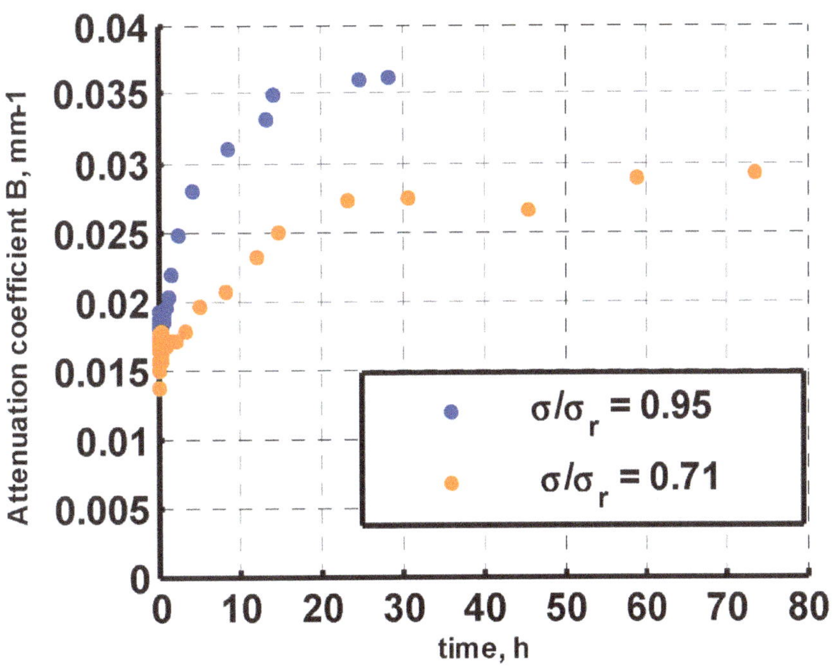

Figure 10: Evolution of the attenuation coefficient B during the static load hold ($\sigma/\sigma_R = 0.95$, $\sigma/\sigma_R = 0.71$ and $T = 500$ °C).

The critical damage phase prior to final rupture may be attributed to the avalanche fibers breaks, controlled by fibers oxidation and by recession of interfaces. In order to confirm this hypothesis, the coefficient R_{AE} is calculated for several damage mechanisms identified with clustering analysis of AE data. The Figure 11 show the activities of several damage mechanisms identified with the supervised analysis. During static fatigue, clusters B and D are more active. The evolution of the coefficients *B* obtained for the two classes A and B go through a minimum, contrary to those of classes C and D (Figure 12). For the B class (mainly fiber failure at the end of the tests), the minimum value of the coefficient *B* is observed around 65% of the lifetime. For the A class (mainly yarn fractures or collective fiber breaks at the end of the tests), the minimum is also observed at 65%. One may noticed that the minimum is observed only for clusters A and B corresponding to fibers breaks during the second part of the test. If the growth of the attenuation coefficient *B* is linked to matrix crack opening, coefficient B also allows considering the plateau observed on the evolution of attenuation coefficient *B* shows that matrix crack opening leads to an equilibrium state near 40%–50% of the rupture time. The significant increase of matrix crack opening pointed out before 50% of the rupture time is linked to carbon oxidation in the interphases provoking an increase in length of the debonded region of fibers in the vicinity of matrix cracks. Beyond 50% of the rupture time, the oxygen flux, determined by the degree of matrix crack opening, controls the rate of fibers break by slow crack growth. This critical time corresponds to the beginning of a second damage phase where slow crack growth in fibers is prevailing, leading to the ultimate fracture of the composite.

The minimum of coefficient RAE and the value plateau of B appear to be a promising tool to perform "short" static fatigue experiments which could be stopped once the minimum value of RAE is reached, rather than waiting the final failure of the specimen. In this way the test duration could be divided by 1.5 to 2.

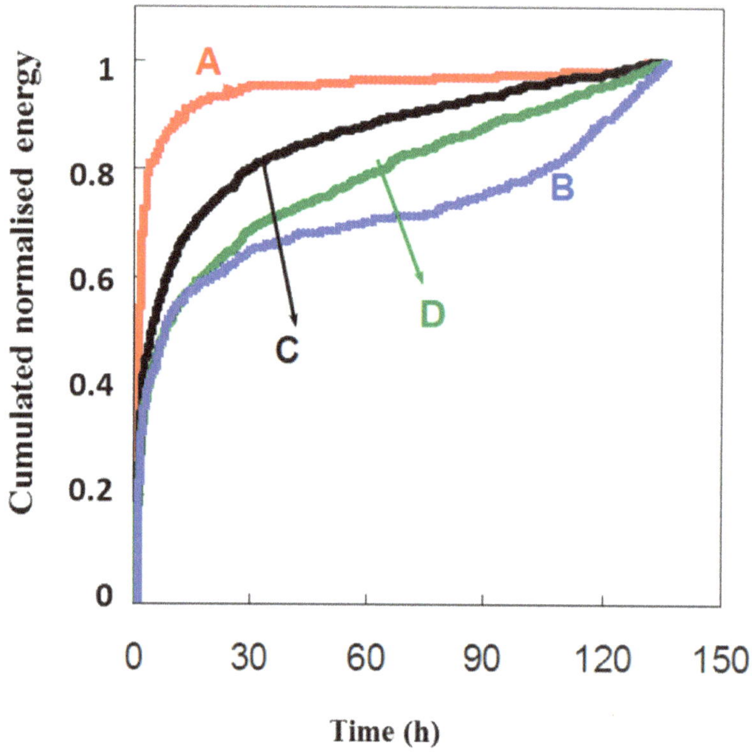

Figure 11: Activities of several clusters during static fatigue test at 500 °C and $\sigma/\sigma_R = 0.79$.

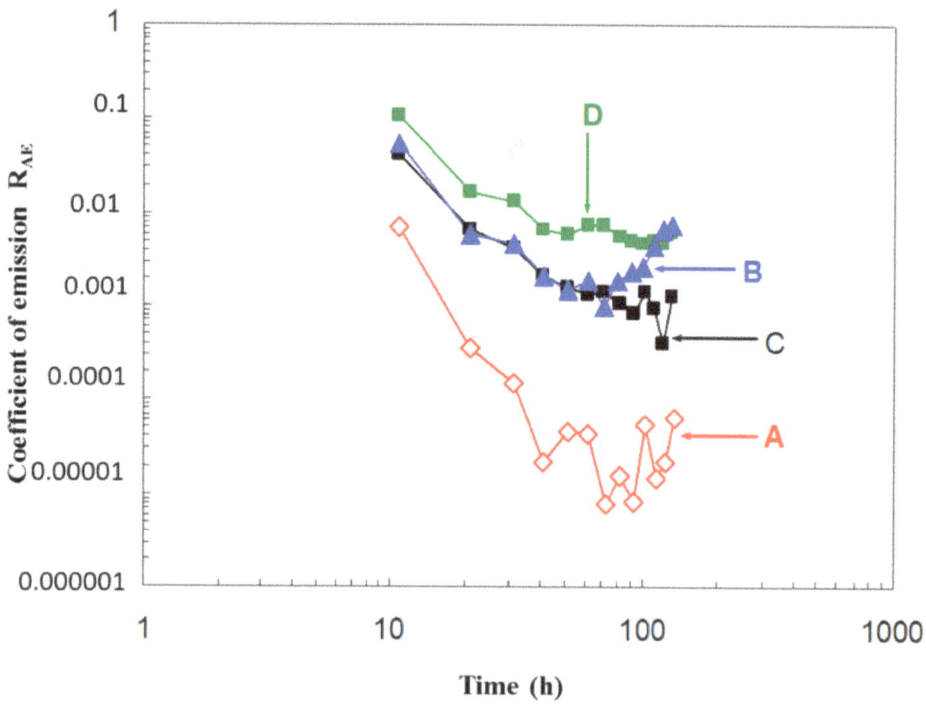

Figure 12: Evolution of the coefficient of emission for several clusters at 500 °C and $\sigma/\sigma_R = 0.79$.

Toward Lifetime Prediction

The Benioff law was applied to AE data recorded during the static fatigue tests in order to evaluate the lifetime of the studied material. This law is based on a power law description of the cumulative energy (Ω) released by the composite (Equation (2)) and characterized by 4 parameters (Ω_r, γ, φ and t_r).

Momon [11] studied the applicability on AE signals recorded for C/SiC and SiC/SiC composites. Data beyond the minimum of R_{AE} were in better agreement with the Benioff law than those before the minimum. This point was confirmed with the estimation of the *c*-value.

The Optimum Circle Method was applied to the fatigue tests. Optimum values (leading to the minimum *c*-value) of t_{start} are reported on the Figure 9, which means that the approximation by the Benioff law was relevant for the data collected behind this time. A value of t_{start} was taken every hour in time interval of 10% to 90% of rupture time t_r. The minimum *c*-value appears clearly around on average 40%–50% of rupture time. The energy release prior to rupture under static fatigue exhibits a critical evolution at 50% of rupture time regardless of the applied stress level. Thus, the Benioff law, initially used to study the activation of earthquakes, may also be applied to the damage of composites. However, this procedure requires preliminary tests until rupture to determine γ and φ for the studied material.

CONCLUSIONS

Static fatigue experiments were performed at intermediate temperatures under air to assess lifetimes of a SiC_f/[Si-B-C] composite. The lifetime as a function of applied stress follows a power-type law, which can be used to predict lifetimes. Additional information is obtained from strain measurement and AE monitoring during the tests. Two criterion based on the AE cumulative energy has been defined, which can be used to predict the final failure of a specimen if the AE activity during the first half of the test is known. This could be a way to shorten such static fatigue experiments, and to divide their duration by 1.5. The coefficient R_{AE} and the attenuation coefficient *B* confirm the existence of two distinct phases during damage of CMCs in static fatigue at intermediate temperatures. The first phase being mainly attributed to interfacial changes and the second one to the predominance of subcritical crack growth in fibers. Beyond this characteristic point, energy release may be modelled with the Benioff law in order to extrapolate AE activity and hence evaluate time to failure. The same analysis is in progress for the behavior during cyclic fatigue tests. Future works will focus on the use of the Benioff law as a predictive model.

ACKNOWLEDGMENT

The authors gratefully acknowledge Snecma Propulsion Solide, CNRS and DGA for supporting this work in the frame of the research program: "Modélisation, extrapolation, validation de la durée de vie des composites à matrice céramique".

AUTHOR CONTRIBUTIONS

All the authors have contributed to perform mechanical tests, to analyze data and to write this paper.

CONFLICTS OF INTEREST

The authors declare no conflict of interest.

REFERENCES

1. Fantozzi, G.; Reynaud, P.; Rouby, D. Thermomechanical behaviour of long fibres ceramic-ceramic composites. *Silic. Ind.* 2002, *66*, 109–119.
2. Reynaud, P.; Rouby, D.; Fantozzi, G. Effects of temperature and of oxidation on the interfacial shear stress between fibres and matrix in ceramic-matrix composites. *Acta Mater.* 1998, *46*, 2461–2469.
3. Naslain, R.; Christin, F. Sic-matrix composite materials for advanced jet engines. *MRS Bull.* 2003, *28*, 654–658.
4. Lamon, J. Micromechanics-based approach to the mechanical behavior of brittle matrix composites. *Compos. Sci. Technol.* 2002, *58*, 2259–2272.
5. Ohnabe, H.; Masaki, S.; Onozuka, M.; Miyahara, K.; Sasa, T. Potential applications of ceramic matrix composites to aero-engine components. *Compos. A* 1999, *30*, 489–496.
6. Lamon, J. CVI SiC/SiC composites. In *Handbook of Ceramics Composites*; Bansal, N.P., Ed.; Springer: Berlin, Germany, 2005; pp. 55–76.
7. Fantozzi, G.; Reynaud, P. Mechanical behaviour of SiC fiber reinforced ceramic matrix composites. In *Comprehensive Hard Materials*; Sarin, V.K., Ed.; Elsevier: Philadelphia, PA, USA, 2014; pp. 345–366.
8. Quemard, L.; Rebillat, F.; Guette, A.; Tawil, H.; Louchet-Pouillerie, C. Self-healing mechanisms of a SiC fiber reinforced multi-layered ceramic matrix composite in high pressure steam environments. *J. Eur. Ceram. Soc.* 2007, *27*, 2085–2094.
9. Forio, P.; Lamon, J. Fatigue behavior at high temperatures in air of a 2D SiC/SiBC composite with a self-sealing multilayered matrix. In *Ceram Transactions*; Bansal, N.P., Singh, J.P., Lin, H.T., Eds.; American Ceramic Society: Westerville, OH, USA, 2001; Volume 128, pp. 127–141.
10. Ono, K.; Huang, Q. Pattern Recognition Analysis of Acoustic Emission Signals. *NDT and E Int.* 1997, *30*, 109.
11. Yan, T.; Holford, K.; Carter, D.; Brandon, J. Classification of acoustic emission signatures using a self-organization neural network. *J. Acoust. Emiss.* 1999, *17*, 49–59.
12. Kostopoulos, V.; Loutas, T.H.; Kontsos, A.; Sotiriadis, G.; Pappas, Y.Z. On the identification of the failure mechanisms in oxide/oxide composites using acoustic emission. *NDT E Int.* 2003, *36*, 571–580.
13. Pappas, Y.Z.; Kontsos, A.; Loutas, T.H.; Kostopoulos, V. Failure mechanisms analysis of 2D carbon/carbon using acoustic emission monitoring. *NDT E Int.* 1998, *31*, 571–580.
14. Godin, N.; Huguet, S.; Gaertner, R. Integration of the Kohonen's self-organising map and *k*-means algorithm for the segmentation of the AE data collected during tensile tests on cross-ply composites. *NDT E Int.* 2005, *38*, 299–309.

15. Moevus, M.; Rouby, D.; Godin, N.; R'Mili, M.; Reynaud, P.; Fantozzi, G.; Fayolle, G. Analyse of damage mechanisms and associated acoustic emission in two SiC/[Si-B-C] composites exhibiting different tensile curves. Part I: Damage patterns and acoustic emission activity. *Compos. Sci. Technol.* 2008, *68*, 1250–1257.

16. Moevus, M.; Godin, N.; Rouby, D.; R'Mili, M.; Reynaud, P.; Fantozzi, G.; Farizy, G. Analyse of damage mechanisms and associated acoustic emission in two SiC/[Si-B-C] composites exhibiting different tensile curves. Part II: Unsupervised acoustic emission data clustering. *Compos. Sci. Technol.* 2008, *68*, 1258–1265.

17. Smith, R.L.; Phoenix, S.L. Asymptotic distributions for the failure of fibrous materials under series-parallel structure and equal load-sharing. *J. Appl. Mech.* 1981, *48*, 75–82.

18. Curtin, W.A. Theory of mechanical properties of ceramic-matrix composites. *J. Am. Ceram. Soc.* 1991, *74*, 2837–2845.

19. Newman, W.I.; Phoenix, S.L. Time dependent fiber-bundles with local load sharing. *Phys. Rev. E Stat. Nonlin. Soft Matter Phys.* 2001, *63*, 021507.

20. Turcotte, D.; Shcherbakov, R. Can damage mechanics explain temporal scaling laws in brittle fracture and seismicity?*Pure Appl. Geophys.* 2006, *163*, 1031–1045.

21. Guarino, A.; Garcimartin, A.; Ciliberto, S. An experimental test of the critical behaviour of fracture precursors. *Eur. Phys. J. B* 1998, *6*, 13–24.

22. Nechad, H.; Helmstetter, A.; El Guerjouma, R.; Sornette, D. Andrade and critical time-to-failure laws in fiber-matrix composites: Experiments and model. *J. Mech. Phys. Solids* 2005, *53*, 1099–1027.

23. Deschanel, S.; Vanel, L.; Vigier, G.; Godin, N.; Ciliberto, S. Experimental study of crackling noise: Conditions on power law scaling correlated to fracture precursors. *J. Stat. Mech. Theory Exp.* 2009.

24. Shaira, M.; Godin, N.; Guy, P.; Vanel, L.; Courbon, J. Evaluation of the strain-induced martensitic transformation by acoustic emission monitoring in 304L austenitic stainless steel: Identification of the AE signature of the martensitic transformation and power-law statistics. *Mater. Sci. Eng. A* 2008, *492*, 392–399.

25. Morscher, G.N. The velocity and attenuation of acoustic emission waves in SiC/SiC composites loaded in tension. *Compos. Sci. Technol.* 2002, *62*, 1171–1180.

26. Maillet, E.; Godin, N.; R'Mili, M.; Reynaud, P.; Lamon, J.; Fantozzi, G. Analysis of Acoustic Emission energy release during static fatigue tests at intermediate temperatures on Ceramic Matrix Composites: Towards rupture time prediction. *Compos. Sci. Technol.* 2012, *72*, 1001–1007.

27. Maillet, E.; Godin, N.; R'Mili, M.; Reynaud, P.; Lamon, J.; Fantozzi, G. Determination of acoustic emission sources energy and application towards lifetime prediction of ceramic matrix composites. In *Mechanical Properties and Performance of Engineering Ceramics and Composites VII*; Singh, D., Salem, J., Halbig, M., Mathur, S., Eds.; Wiley: Hoboken, NJ, USA, 2012; Volume 33, pp. 15–25.

28. Maillet, E.; Godin, N.; R'Mili, M.; Reynaud, P.; Fantozzi, G.; Lamon, J. Real-time evaluation of energy attenuation: A novel approach to acoustic emission analysis for damage monitoring of ceramic matrix composites. *J. Eur. Ceram. Soc.*2014, *34*, 1673–1679.

29. Momon, S.; Moevus, M.; Godin, N.; R'Mili, M.; Reynaud, P.; Fantozzi, G.; Fayolle, G. Acoustic emission and lifetime prediction during static fatigue tests on ceramic-matrix-composite at high temperature under air. *Compos. A Appl. Sci. Manuf.* 2010, *41*, 913–918.

30. Bufe, C.G.; Varnes, D.G. Predictive modeling of the seismic cycle of the Greater San Francisco Bay Region. *J. Geophys. Res.* 1993, *98*, 9871–9883.

31. Bowman, D.D.; Ouillon, G.; Sammis, C.G.; Sornette, A.; Sornette, D. An observational test of the critical earthquake concept. *J. Geophys. Res.* 1998, *103*, 24359–24372.

32. Likas, A.; Vlassis, N.; Verbeek, J. The global *k*-means clustering algorithm. *Pattern Recognit.* 2003, *36*, 451–461.

33. Lee, M.S.; Keerthi, S.; Jin Ong, C.; Decoste, D. An efficient method for computing leave one out error in support vector machines with gaussian kernels. *IEEE Trans. Neural Netw.* 2004, *15*, 750–757.

34. Gauthier, W.; Lamon, J. Delayed failure of Hi-Nicalon and Hi-Nicalon S multifilament tows and single filaments at intermediate temperatures. *J. Am. Ceram. Soc.* 2009, *92*, 702–709.

CHAPTER 16

AUTOCORRELATION AND FREQUENCY-RESOLVED OPTICAL GATING MEASUREMENTS BASED ON THE THIRD HARMONIC GENERATION IN A GASEOUS MEDIUM

Yoshinari Takao [1], Tomoko Imasaka [2], Yuichiro Kida [1] and Totaro Imasaka [1,3]

[1]Department of Applied Chemistry, Graduate School of Engineering, Kyushu University, 744, Motooka, Nishi-ku, Fukuoka 819-0395, Japan

[2]Laboratory of Chemistry, Graduate School of Design, Kyushu University, 4-9-1, Shiobaru, Minami-ku, Fukuoka 815-8540, Japan

[3]Division of Optoelectronics and Photonics, Center for Future Chemistry, Kyushu University, 744, Motooka, Nishi-ku, Fukuoka 819-0395, Japan

ABSTRACT

A gas was utilized in producing the third harmonic emission as a nonlinear optical medium for autocorrelation and frequency-resolved optical gating measurements to evaluate the pulse width and chirp of a Ti:sapphire laser. Due to a wide frequency domain available for a gas, this approach has potential for use in measuring the pulse width in the optical (ultraviolet/visible) region beyond one octave and thus for measuring an optical pulse width less than 1 fs.

INTRODUCTION

Many papers report attosecond pulse generation via high-order harmonic generation (HHG) for use in the studies of ultrafast phenomena related to electrons in an inner-shell orbital of an atom and molecule [1]. On the other hand, an ultrashort optical pulse approaching 1 fs, the spectrum of which is extended from the ultraviolet (UV) to the visible (VIS) region, would be useful for studies of ultrafast phenomena related to electrons in an outer-shell orbital, which are directly concerned with chemical bond/reaction and are more important in chemistry. For example, a molecular ion in mass spectrometry can be strongly enhanced by decreasing the optical pulse width in the femtosecond regime, which is then useful for more reliable identification of the explosive substances [2]. Several methods have been developed for generating an extremely-short optical pulse. For example, numerous emission lines have been generated in the entire VIS region, based on non-resonant four-wave mixing in a thin fused silica plate. This technique has been utilized to generate 2.2-fs optical pulses [3]. On the other hand, coherent supercontinua have been generated from the UV to near-infrared (NIR) by focusing the beam in a hollow-core fiber filled with neon gas, and a three-channel optical field synthesizer has been utilized to generate 2.1-fs pulse [1]. As recognized from the uncertainty principle, a wider spectral domain is essential for the generation of a 1-fs optical pulse. The generation of high-order Raman sidebands extending from the deep-UV (DUV) to the NIR region was first reported in a few decays ago [4]. To date, an extremely wide spectral region extending from 183 to 1203 nm has been covered, based on resonant vibrational four-wave Raman mixing in molecular hydrogen, suggesting that the generation of a 1-fs optical pulse by a phase control of the emission lines is possible [5].

Citation: Yoshinari Takao , Tomoko Imasaka , Yuichiro Kidaand Totaro Imasaka, Autocorrelation and Frequency-Resolved Optical Gating Measurements Based on the Third Harmonic Generation in a Gaseous Medium, doi:10.3390/app5020136

A variety of techniques, including autocorrelation (AC) and frequency-resolved optical gating (FROG), have been developed to measure optical pulse widths [6]. In these techniques, a nonlinear optical effect such as second harmonic generation (SHG), self-diffraction (SD), and others have been utilized. To measure a 1-fs optical pulse, it is necessary to use a nonlinear optical effect that is usable in a wide frequency domain in the UV-VIS region. A GaN diode with a nature of two-photon absorption in the VIS region (at round 400 nm) has been employed as a detector in a fringe-resolved autocorrelator (FRAC) [7]. In a previous study, we reported on the development of an FRAC, in which a mass spectrometer was employed as a two-photon-response device in the DUV region (at around 267 nm) [8]. However, the frequency domain of these techniques is limited to one octave [9]. As a result, it is difficult to measure optical pulse widths less than 1 fs in the UV-VIS region. In order to overcome this problem, the use of a nonlinear optical effect that can be used in the spectral region wider than one octave would be necessary. One of the approaches would be the use of a third harmonic generation (THG) as a nonlinear optical effect, since it has a frequency domain of twice one octave [10]. A surface-sensitive THG has been successfully used for the autocorrelation measurement [11,12]. Moreover, several papers have reported on the FROG system based on THG (THG-FROG), including the THG on the surface of a glass plate [13], in organic films [14], and in a glass coverslip used in multiphoton microscopy [15]. The use of a solid material is simple and easy-to-use. It is, however, difficult to transmit the THG beam generated in the vacuum-UV (VUV) region and to reduce dispersion to negligible levels, especially in the DUV region. On the other hand, a gas such as helium or argon is transparent, even in the VUV region, and the dispersion sufficiently small to be negligible. Thus, the THG in a gaseous medium would be useful for measuring a pulse width in a wide frequency domain. To our knowledge, a THG-based technique such as THG-FROG using a gaseous medium has not been reported to date.

In this study, we used argon or air as a nonlinear optical medium for THG to measure the pulse width of a fundamental beam of a Ti:sapphire laser emitting at 800 nm as a proof-of-principle experiment. Two types of AC systems, i.e., fringe-resolved AC (FRAC) and intensity-AC (IAC), were developed and were utilized for measuring the pulse width. In addition, a FROG system was developed that permits the pulse width and the chirp of the pulse to be evaluated more accurately. The results were compared with values obtained using FRAC and IAC.

EXPERIMENTAL SECTION

Fringe-Resolved Autocorrelation

Figure 1 shows a block diagram of the instrument used in this study. In measuring an AC trace, a fundamental beam of a Ti:sapphire laser (Elite, 800 nm, 35 fs, 1 kHz, 4 mJ, Coherent) was passed through a Mach-Zehnder interferometer, consisting of beam splitters made of BK7 plates with a thickness of 3 mm. Using a lens (BK7) with a focal length of 300 mm, the aligned beam was focused into argon gas contained in a cell (700 mm long, 10 mm i.d.) equipped with fused silica windows with a thickness of 0.5 mm. The third harmonic emission generated in the optical components such as the beam splitters was suppressed by passing the beam through a cover glass for microscopy and was confirmed to be negligible levels by evacuating the gas cell. The third harmonic emission generated in the argon gas was passed through an aqueous solution of $NiSO_4$ (500 g/L in water) to suppress the fundamental beam; a substrate of Si was used to remove the fundamental beam of a Ti:sapphire laser in the reported work [16]. The THG beam emitting at 267 nm was further isolated using a monochromator and was measured using a photomultiplier (R1332, no response at 800 nm, Hamamatsu Photonics) designed for measuring a large optical pulse. In the experiment, the signal intensity was measured at the level sufficiently lower than the level of signal saturation. The output signal was fed into a boxcar integrator, and the averaged signal was recorded using an oscilloscope. A function generator provided a signal for a piezoelectric transducer equipped with a translational stage with a retroreflector mounted on it. An autocorrelation trace was collected using the oscilloscope by recording the signal against the time delay between the pulses.

Frequency-Resolved Optical Gating

A FROG trace was measured using the instrument shown in Figure 2. The fundamental beam of the Ti:sapphire laser was separated into two parts using a pair of D-shape mirrors to remove the dispersion arising from the beam splitters. The beams were reflected by means of a pair of aluminum mirrors and were focused with an off-axis parabolic mirror into a nonlinear optical medium such as argon or air. The beam was collimated using a fused silica lens with a focal length of 10 cm and was introduced into a slit of a monochromator (CT-10, Jasco) using a rotating mirror. One of the THG beams, corresponding to $E_{sig}(t, \tau) = E(t)^2E(t - \tau)$, was measured as a signal. Electronics similar to those used in the FRAC experiment were employed for measuring the third harmonic emission. The spectrum of the THG beam was measured using the second-order diffraction of the grating to improve the spectral resolution of the monochromator. A FROG trace was measured by scanning the wavelength of the monochromator at different positions of the delay for one of the D-shape mirrors. The data were measured and analyzed using a program provided by Femtosoft Technologies.

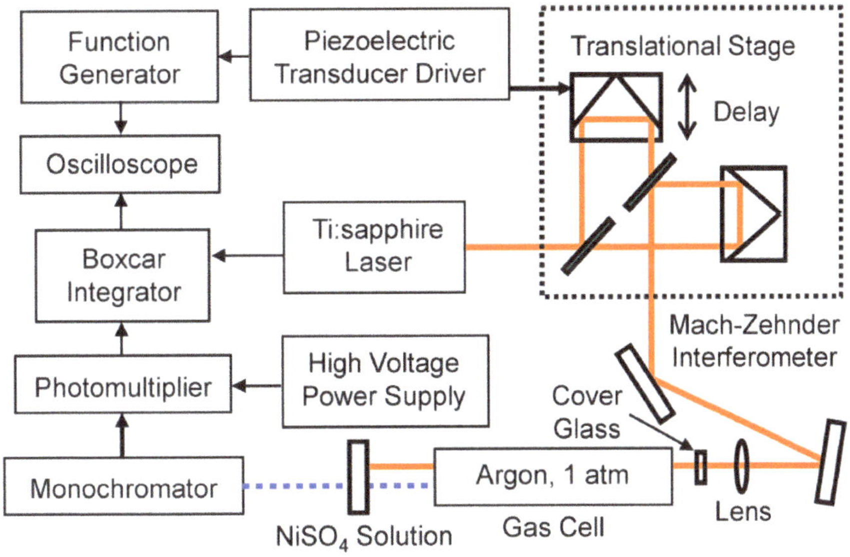

Figure 1: Experimental apparatus for third harmonic generation-fringe-resolved autocorrelator (THG-FRAC). The orange (solid) and blue (broken) lines show the fundamental and THG beams, respectively.

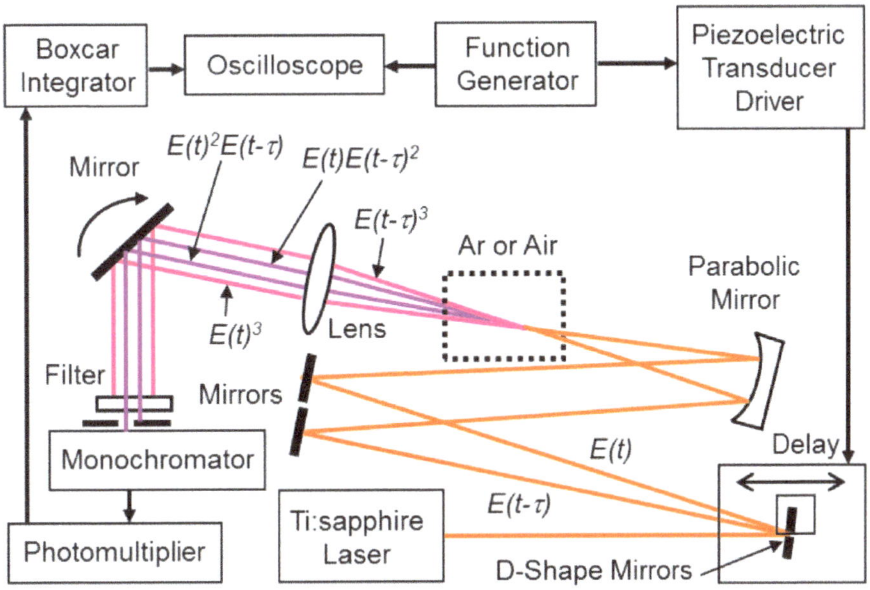

Figure 2: Experimental apparatus for THG-frequency-resolved optical gating (FROG).

Intensity Autocorrelation

An IAC trace was obtained using a non-collinear configuration developed for use in a FROG system shown in Figure 2. Thus, the instrument consists of only reflective optics except for a window of the gas cell, allowing a nearly dispersion-free experiment.

RESULTS AND DISCUSSION

Fringe-Resolved Autocorrelation

The spectral width measured for the fundamental beam of the Ti:sapphire laser was 27 nm (420 cm^{-1}). The Fourier-transform-limited (FTL) pulse width calculated by assuming a Gaussian temporal profile was 35 fs, identical to the value provided by the manufacturer of the laser. An FRAC trace is shown in Figure 3A. Suppression of the signal to the background level by destructive interference, in addition to a full modulation of the signal shown in the expanded view (Figure 3B), suggests the pump beams were superimposed in generating the THG beam. The ratio of the signal and the background was ca. 25, slightly smaller than the predicted value of 32 for an FTL pulse using the third-order nonlinear effect such as THG [10]. Figure 3C shows the FRAC trace calculated for FTL pulse. The observed trace is slightly wider than this trace, suggesting the chirp of the pulse. Figure 3D shows the FRAC trace calculated under the assumption that the FTL pulse (35 fs) is chirped to 41 fs with a group delay dispersion (GDD) of 500 fs^2, which is nearly identical to the value of GDD (ca.400 fs^2) roughly estimated from the thickness of the optical components in the beam path. The pulse width calculated from the IAC trace, which can be obtained by low-pass filtering the data shown in Figure 3A, was 41 fs, which is in good agreement with the above value. Another possible explanation for the discrepancy between the results of (A) and (C) would be the error in the measurement of the spectrum since a wider band width would be observed due to a finite resolution of the spectrometer, which provides a shorter transform-limited pulse width; see a lack of small wings at round ±50 fs in the observed data (A), which is in contrast to the calculated data (D), suggesting that the effect of chirp is minimal.

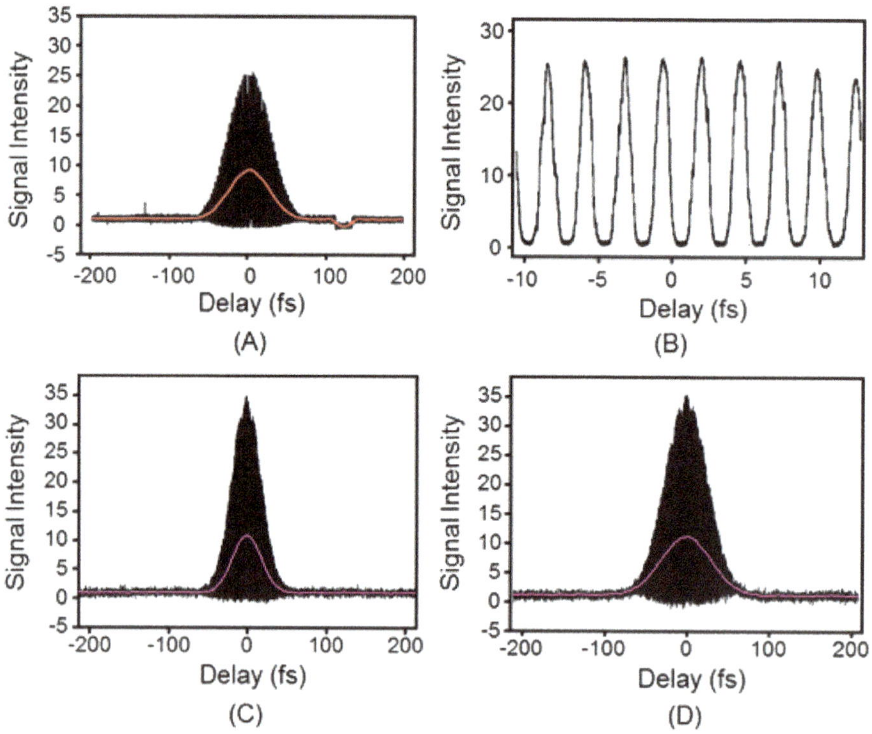

Figure 3: Autocorrelation trace for (**A**) observed data (**B**) expanded view (**C**) theoretically predicted trace for a Fourier-transform-limited (FTL) pulse (**D**) theoretically predicted trace for a chirped pulse (GDD = 500 fs^2). A random noise was calculated using a computer and was added to the calculated data for better visual comparison. In order to check the baseline level of the observed trace (A), the laser beam was interrupted during a period of 110–130 fs in the experiment, suggesting that the signal was suppressed to zero at the bottom of the trace.

Intensity Autocorrelation

The IAC trace observed in this study is shown in Figure 4. The full width at half maximum (FWHM) of the trace was 44 fs, suggesting a pulse width of 36 fs [10]. This value is slightly smaller than that obtained using an FRAC system, which can be attributed to the use of the reflective optics with no dispersion in the IAC system.

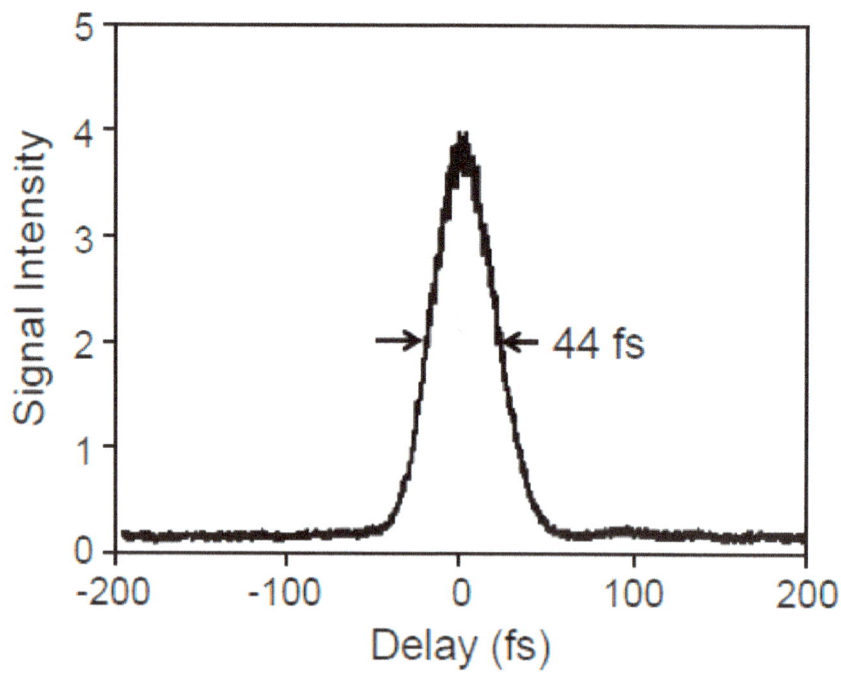

Figure 4: Intensity autocorrelation trace observed for the Ti:sapphire laser.

Frequency-Resolved Optical Gating

The chirp of the pulse can be more accurately evaluated using a FROG system. The efficiency of THG induced in air was similar to that obtained using argon. Because of this, air was used as a nonlinear optical medium in this study for the sake of simplicity. Before recording the FROG trace, the THG beam pattern was measured to properly extract the signal of $E_{sig}(t, \tau) = E(t)^2E(t - \tau)$ [6]. The THG beams corresponding to $E(t)^3$, $E(t)^2E(t - \tau)$, $E(t)E^2(t - \tau)$, and $E^3(t - \tau)$ could be clearly observed by rotating the angle of the mirror (see the arrow in Figure 2). Therefore, the intensity of the THG beam for $E_{sig}(t, \tau) = E(t)^2E(t - \tau)$ was monitored for measuring the FROG trace. The observed and retrieved data obtained for negatively-chirped, FTL, and positively-chirped pulses are shown in Figure 5. The pulse width measured for an FTL pulse was 32 fs, which appears to be similar to or slightly shorter than the values obtained using IAC. On the other hand, the chirped pulse provided a slightly longer pulse width (positive 42 fs, negative 44 fs) due to the GDD value being calculated to be 300 fs^2.

The minimal value of the pulse width measured using conventional SHG-FROG or SD-FROG is determined by the spectral region that is usable as a nonlinear optical medium: although a broad bandwidth spanning a multi-octave frequency domain can be covered using a thin optical crystal, the spectral region is practically limited to the VIS-NIR region [17]. On the other hand, SHG, which would restrict the spectral region of THG, does not occur in an isotropic gas, and a noble gas such as argon is transparent and has a small dispersion in the VUV to IR region. As a result, the present approach using a gas in conjunction with THG-FROG can be utilized to measure pulse widths in a wide frequency domain extending twice one octave. Because of this, this technique would be applied to the measurement of an ultrashort optical pulse less than 1 fs especially in the VUV-DUV region: it should be noted that a higher carrier frequency is desirable for generating a shorter optical pulse. However, in order to avoid the dispersion arising from the optics such as cell windows, it would be necessary to use a nozzle for introduction of a rare gas (not air) into a vacuum. Efficient generation of THG (even HHG) reported to date suggests sufficient sensitivity of this THG-based method using a gas for the measurement of an ultrashort pulse width.

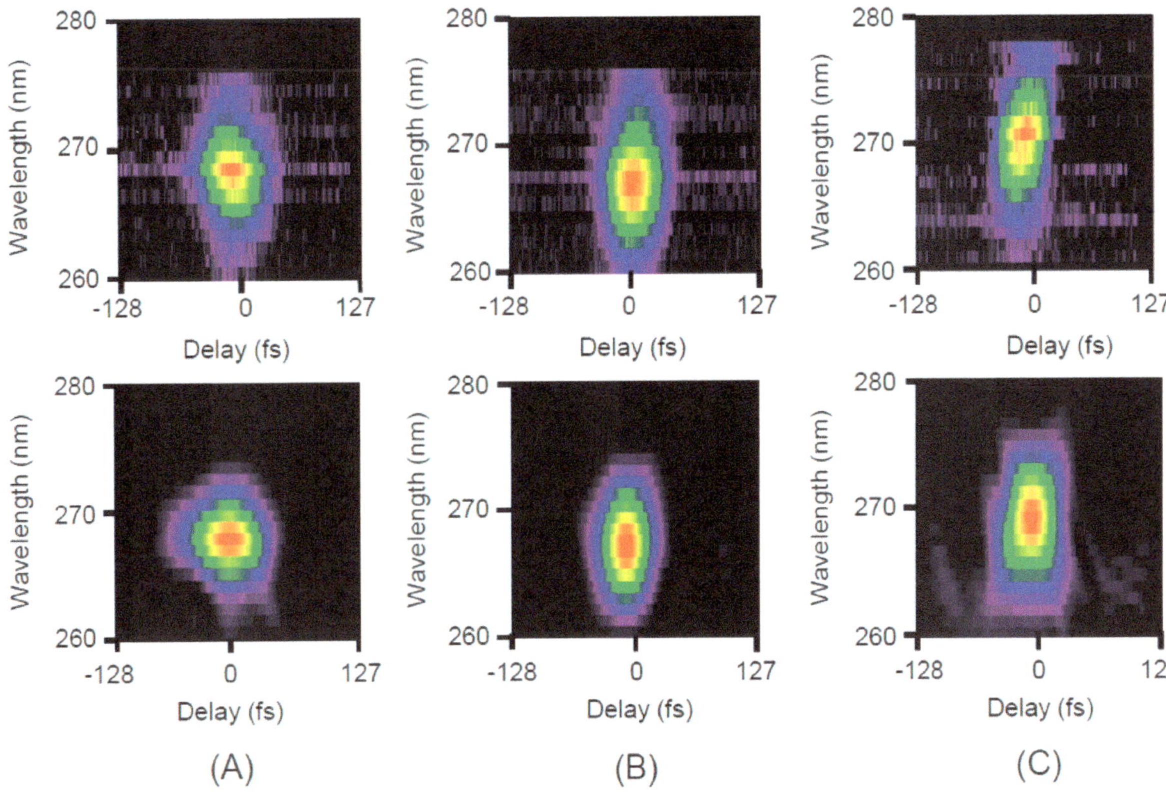

Figure 5: THG-FROG traces obtained for (**A**) negatively-chirped (**B**) FTL (**C**) positively-chirped pulses. The chirp of the pulse was adjusted by changing the position of the grating in the compressor of the Ti:sapphire laser to the value, at which the energy of the THG pulse decreased to a half of the value obtained using the FTL pulse. Above, original traces; below, retrieved data. FROG error: (**A**) 0.6% (**B**) 0.6% (**C**) 1%.

CONCLUSIONS

In this study, we report on the development of FRAC, IAC, and FROG systems based on THG using a gaseous medium. This approach can be used to measure an ultrashort optical pulse at any wavelength from the UV to the IR region although the emission generated by THG is located in the VUV-VIS region. Therefore, this technique can be applied to the lasers used in various areas of science and technology.

ACKNOWLEDGMENTS

This research was supported by a Grant-in-Aid for the Global COE program, "Science for Future Molecular Systems" from the Ministry of Education, Culture, Sports, Science and Technology of Japan and by Grants-in-Aid for Scientific Research from the Japan Society for the Promotion of Science (JSPS) KAKENHI Grant Number 23245017, 24510227, 26220806, and 15K13726. This study also received support from the Steel Industry Foundation for the Advancement of Environmental Protection Technology.

AUTHOR CONTRIBUTIONS

Performing the experiment: Yoshinari Takao, Numerical simulation: Tomoko Imasaka, Drafting of manuscript: Totaro Imasaka, Critical revision: Yuichiro Kida, Planning and supervision of the research: Totaro Imasaka.

CONFLICTS OF INTEREST

The authors declare no conflict of interest.

REFERENCES

1. Wirth, A.; Hassan, M.Th.; Grguraš, I.; Gagnon, J.; Moulet, A.; Luu, T.T.; Pabst, S.; Santra, R.; Alahmed, Z.A.; Azzeer, A.M.; et al. Synthesized light transients. Science 2011, 334, 195–200.
2. Hamachi, A.; Okuno, T.; Imasaka, T.; Kida, Y.; Imasaka, T. Resonant and nonresonant multiphoton ionization processes in the mass spectrometry of explosives. Anal. Chem. 2015, 87, 3027–3031.
3. Weigand, R.; Mendonça, J.T.; Crespo, H.M. Cascaded nondegenerate four-wave-mixing technique for high-power single-cycle pulse synthesis in the visible and ultraviolet ranges. Phys. Rev. A 2009, 79, 063838.
4. Imasaka, T.; Kawasaki, S.; Ishibashi, N. Generation of more than 40 laser emission lines from the ultraviolet to the visible regions by two-color stimulated Raman effect. App. Phys. B 1989, 49, 389–392.
5. Shitamichi, O.; Imasaka, T. High-order Raman sidebands generated from the near-infrared to ultraviolet region by four-wave Raman mixing of hydrogen using an ultrashort two-color pump beam. Opt. Express 2012, 20, 27959.
6. Trebino, R. Frequency-Resolved Optical Gating: The Measurement of Ultrashort Laser Pulses; Kluwer Academic Publishers: Boston, MA, USA, 2002.
7. Zürch, M.; Hoffmann, A.; Gräfe, M.; Landgraf, B.; Riediger, M.; Spielmann, Ch. Characterization of a broadband interferometric autocorrelator for visible light with ultrashort blue laser pulses. Opt. Commun. 2014, 321, 28–31.
8. Zaitsu, S.; Miyoshi, Y.; Kira, F.; Yamaguchi, S.; Uchimura, T.; Imasaka, T. Interferometric characterization of ultrashort deep ultraviolet pulses using a multiphoton ionization mass spectrometer. Opt. Lett. 2007, 32, 1716–1718.
9. Imasaka, T.; Imasaka, T. Searching for a molecule with a wide frequency domain for non-resonant two-photon ionization to measure the ultrashort optical pulse width. Opt. Commun. 2012, 285, 3514–3518.
10. Meshulach, D.; Barad, Y.; Silberberg, Y. Measurement of ultrashort optical pulses by third-harmonic generation. J. Opt. Soc. Am. B 1997, 14, 2122–2125.
11. Tsang, T.Y.F. Optical third-harmonic generation at interfaces. Phys. Rev. A 1995, 52, 4116–4125.
12. Squier, J.A.; Fittinghoff, D.N.; Barty, C.P.J.; Wilson, K.R.; Müller, M.; Brakenhoff, G.J. Characterization of femtosecond pulses focused with high numerical aperture optics using interferometric surface-third-harmonic generation. Opt. Commun. 1998, 147, 153–156.
13. Tsang, T.; Krumbügel, M.A.; Delong, K.W.; Fittinghoff, D.N.; Trebino, R. Frequency-resolved optical-gating measurements of ultrashort pulses using surface third-harmonic generation. Opt. Lett. 1996, 21, 1381–1383.
14. Ramos-Ortiz, G.; Cha, M.; Thayumanavan, S.; Mendez, J.; Marder, S.R.; Kippenlen, B. Ultrafast-pulse diagnostic using third-order frequency-resolved optical gating in organic films. Appl. Phys. Lett. 2004, 85, 3348–3350.
15. Chadwick, R.; Spahr, E.; Squier, J.A.; Durfee, C.G.; Walker, B.C.; Fittinghoff, D.N. Fringe-free, background-free, collinear third-harmonic generation frequency-resolved optical gating measurements for multiphoton microscopy.Opt. Lett. 2006, 31, 3366–3368.
16. Graf, U.; Fieβ, M.; Schultze, M.; Kienberger, R.; Krausz, F.; Goulielmakis, E. Intense few-cycle light pulses in the deep ultraviolet. Opt. Express 2008, 16, 18956.
17. Birkholz, S.; Steinmeyer, G.; Koke, S.; Gerth, D.; Bürger, S.; Hofmann, B. Phase retrieval via regularization in self-diffraction based spectral interferometry. 2014. arXiv:1412.2965.

CHAPTER 17

ADEQUACY OF TECHNICAL AND COMMERCIAL ALTERNATIVES APPLIED TO MACHINE TOOL VERIFICATION USING LASER TRACKER

Sergio Aguado [1], Jorge Santolaria [1], David Samper [1], Jesús Velázquez[1], Carlos Javierre [2] and Ángel Fernández [2]

[1]Design and Manufacturing Engineering Department, Universidad de Zaragoza, Zaragoza 50018, Spain

[2]Mechanical Engineering Department, Universidad de Zaragoza, Zaragoza 50018, Spain

ABSTRACT

Besides presenting a volumetric verification technique that allows characterization of the different geometric errors of a machine tool (MT) depending on its kinematic chain and geometry through a kinematic model, this paper investigates the influence of measurement tools and techniques available on the final accuracy of the MT. Volumetric verification based on a laser tracker (LT) relates the coordinates of the tool with the coordinates of the LT, including it into the kinematic model. Using a non-lineal optimization process, approximation functions that characterize the joint influence of MT geometric errors are obtained. However, measurement data will be affected by previous compensation of the MT, the accuracy of the measurement system, LT measurement technology, the type of retroreflector used, and techniques used to improve data accuracy, among other sources of errors. This paper studies the adequacy of different commercial alternatives such as: retroreflectors, LTs from different manufacturers, *etc.*, that can be applied in MT verification using a long-range MT. As the accuracy is strongly affected by the uncertainty of its angular encoders, the multilateration technique tries to improve data accuracy using only LT radial information. Nonetheless, a new bundle adjustment which uses radial and angular information is presented in current metrology software. This paper studies both techniques and analyzes their adequacy for MT verification too.

INTRODUCTION

The development and evolution of the technological field in the machine tool (MT) industry has provided continuous improvements in the characteristics and properties of MTs. So, customers are demanding improvements in the accuracy of the work performed as well as increased capacity and reliability of production systems.

However, many current production systems are obsolete and cannot achieve the required specifications. To avoid excessive errors and decay of the dimensional quality of parts, it is necessary to perform routine maintenance to approximately indicate the MT's errors and, when necessary, to realize a complete verification of and compensation for its errors. One of the principal ways to reduce machine errors is through processes involving positioning accuracy, because geometric errors result from structural elements that affect the repeatability and accuracy of the machine.

Citation: Sergio Aguado , Jorge Santolaria, David Samper, Jesús Velázquez, Carlos Javierre and Ángel Fernández , Adequacy of Technical and Commercial Alternatives Applied to Machine Tool Verification Using Laser Tracker, doi:10.3390/app6040100

Nowadays, there are two different methods of measurement: direct methods, which study the effect of individual errors, and indirect methods, which study the combined effect of errors. Historically, the accuracy of MTs have been guaranteed by using direct measurement methods, in which errors of each axis of movement are measured independently [1,2,3,4]. The main disadvantage of this type of measurement lies in the fact that the verification time required is much higher than that of indirect measurement methods [4]. Meanwhile, indirect methods reduce this time by providing global information about the influence of all geometric errors of the MT (volumetric error) by measuring a set of verification points in the MT workspace using multi-axis movement of the MT [5,6,7,8,9,10,11,12].

Techniques that use indirect measurement systems such as ball bars, laser tracers (LCs), laser trackers (LTs), and so on to characterize geometric errors using the kinematic model of the MT are called volumetric verification methods [5,6,7,8,13]. The verification of long-range MTs using interferometric measurement systems such as LTs and LCs is based on non-linear optimization techniques [6,7,8]. Researchers have studied the parameters of the regression function used, optimization methodology, distribution of measured points, techniques of multilateration, LT self-calibration to improve data accuracy, and so on [14,15,16,17,18,19].

This paper is focused on studying and comparing different commercial alternatives to deal with volumetric verification of long-range MTs. To do this, the same MT is verified using three different LTs from leading manufacturers using a traditional spherically mounted retroreflectors (SMR) or a new motorized retroreflector (active target). Moreover, the paper investigates trilateration and bundle adjustment techniques in verification as techniques for improving the accuracy of measurement with LTs and their influence on the verification of a real machine.

FEATURES OF THE MACHINE TOOL TAKEN INTO CONSIDERATION IN VOLUMETRIC VERIFICATION

The principal aim of volumetric verification is to mathematically characterize the combined effect of all geometric errors of the MT for further compensation; using the kinematic model of the MT, the difference between theoretical points introduced for numerical control (NC) and measurement of nominal points with an LT is found. Therefore, the kinematic model of the MT and its control software must be obtained and studied before carrying out verification.

Structural Configuration and Kinematic Model of the Machine Tool

An MT is a compound machine that carries out work and generates metal chips. MTs are classified as compound machines that transmit power (thermal, hydraulic, electrical) and machines that do not start making metal chips (cutting, stamping, compression).

Within these two groups, the combination of different structural elements such as guides, joints, or screws that define the kinematic structure of the MT shapes its kinematic chain. Assuming that moving cars of the kinematic chain are rigid bodies, transformation and rotation matrices can be implemented by using a mathematical model (kinematic model) based on the MT configuration [6,7,11,13]. Then, the position of the tool with regard to the part to be machined is obtained in relation to the programmed nominal position, the position of the tip of the tool with respect to the reference machine (offsets), and the geometric errors of the axes.

To obtain the combined effect of all geometric errors of the MT through its kinematic model, the measurement system (LT) must replace the part and the retroreflector must replace the tool (Figure 1). The kinematic model of the MT to be verified is represented by XFYZCB, where F determines the fixed part of the machine, the *X*-axis drives the worktable, *Y* and *Z* are combined to actuate the tool, and the *C*/*B* rotational axis allows head orientation, providing five degrees of freedom to the MT (Figure 1, Table 1).

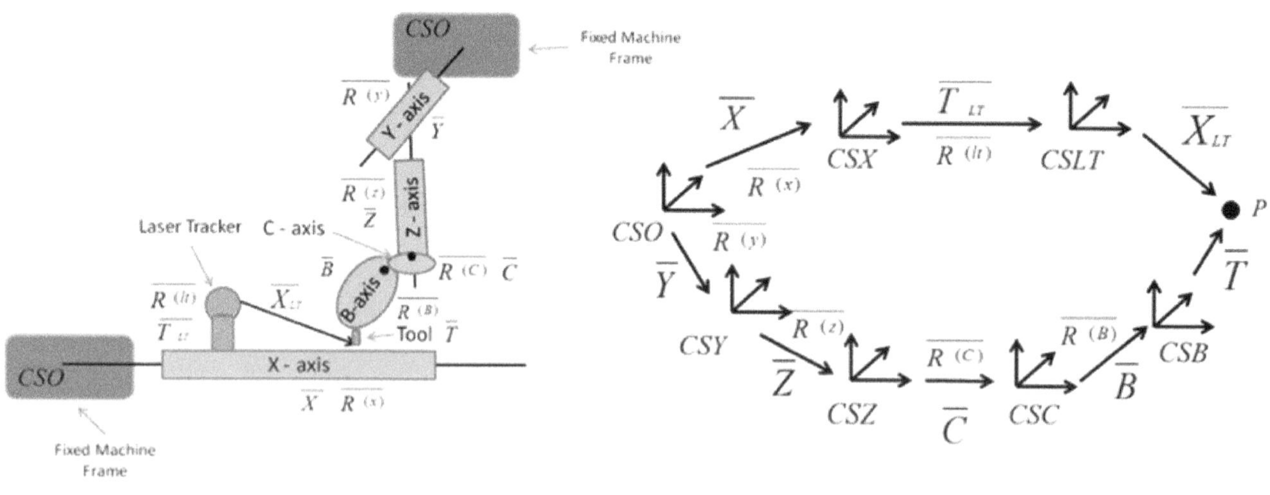

Figure 1: Kinematic chain and kinematic model of Zayer TB 5000 machine tool (MT).

Table 1: Nomenclature of kinematic chain of Zayer TB 5000 machine tool (MT).

Symbol	Definition	Symbol	Definition
CSO	Coordinate system of the machine tool (MT)	$\overline{X}$	Linear error vector in the X-axis
CSY	Coordinate system of the Y-axis	$\overline{Y}$	Linear error vector in the Y-axis
CSZ	Coordinate system of the Z-axis	$\overline{Z}$	Linear error vector in the Z-axis
CSC	Coordinate system of the C-axis	$\overline{C}$	Linear error vector in the C-axis
CSB	Coordinate system of the B-axis	$\overline{B}$	Linear error vector in the B-axis
CSX	Coordinate system of the X-axis	$\overline{T_{LT}}$	Translation vector CS0-CSLT
CSLT	Coordinate system of the laser tracker (LT)	$\overline{X_{LT}}$	Coordinates of the LT
$\overline{R(x)}$	Rotational error matrix of the X-axis	$\overline{T}$	Offset of the tool
$\overline{R(y)}$	Rotational error matrix of the Y-axis	$\overline{R(c)}$	Rotational error matrix of the C-axis
$\overline{R(z)}$	Rotational error matrix of the Z-axis	$\overline{R(b)}$	Rotational error matrix of the B-axis
P	Nominal MT coordinates	-	-

As a result of studies performed on this MT, its axes of rotation relative to *C* and *B* were locked. The final structural configuration of the MT to be verified is *XFYZ*, with six geometric errors per axis (position, horizontal and vertical straightness, yaw, pitch, and roll) and three squareness errors associated.

The equation of movement that relates the nominal coordinates of the MT with measured coordinates of the LT is presented in Equation (1).

$$\overline{X_{LT}} = \overline{R_{LT}^{-1}}\left(\overline{R_X^{-1}}\left(\overline{R_y}\left(\overline{R_z}\,\overline{T} + \overline{Z}\right) + \overline{Y} - \overline{Z}\right) - \overline{T_{LT}}\right) \quad (1)$$

where the components of the equations are constituted as follows: $\overline{T}$ is the offset of the tool.

$$\overline{T} = \begin{pmatrix} x_t \\ y_t \\ z_t \end{pmatrix} \quad (2)$$

$\overline{R_k}$ represents the rotational error matrix in the axis k of the tool with $k = x, y, z$.

$$\overline{R_{(k)}} = \begin{pmatrix} 1 & -\varepsilon_z(k) & \varepsilon_y(k) \\ \varepsilon_z(k) & 1 & -\varepsilon_x(k) \\ -\varepsilon_y(k) & \varepsilon_x(k) & 1 \end{pmatrix} \quad (3)$$

$\overline{X}$ represents the linear error vector in the x-axis of the milling machine.

$$\overline{X} = \begin{pmatrix} -x + \delta_x(x) \\ \delta_y(x) \\ \delta_z(x) \end{pmatrix} \quad (4)$$

$\overline{Y}$ represents the linear error vector in the y-axis of the milling machine.

$$\overline{Y} = \begin{pmatrix} \delta_x(y) - y\cdot\varepsilon_{xy} \\ y + \delta_y(y) \\ \delta_z(y) \end{pmatrix} \quad (5)$$

$\overline{Z}$ represents the linear error vector in the z-axis of the milling machine.

$$\overline{Z} = \begin{pmatrix} \delta_x(z) - z\cdot\varepsilon_{xz} \\ \delta_y(z) - z\cdot\varepsilon_{yz} \\ z + \delta_z(z) \end{pmatrix} \quad (6)$$

where εx(k), εy(k), and εz(k) are three rotation errors of an axis $k = x, y, z$; δk(k) is the position error of the axis $k = x, y, z$; δk(j) with $k \neq j$ is the straightness error in the k direction; and εxy, εxz, and εyz are squareness errors.

$\overline{T_{LT}}$ represents the translation vector between the coordinate systems of the machine, CSO, and laser tracker, CSLT.

$$\overline{T_{LT}} = \begin{pmatrix} oX_{LT} \\ oY_{LT} \\ oZ_{LT} \end{pmatrix} \quad (7)$$

$\overline{RLT}$ represents the rotation matrix θ between CSLT and CSO around a unitary vector $u = (u_x, u_y, u_z)$, where u2x+u2x+u2x=1.

$\overline{X_{LT}}$ represents the coordinates of a machine point measured with an LT.

$$\overline{X_{LT}} = \begin{pmatrix} X_{LT} \\ Y_{LT} \\ Z_{LT} \end{pmatrix} \quad (8)$$

Control System of the Machine Tool

Currently, new compensation systems have been developed by control system manufacturers, such as Siemens' Volumetric Compensation System (VCS), Heidenhain's Kinematic Comp, and Fanuc's 3D Compensation/3D Rotary Compensation, which use the approximation function of each geometric error of the MT to compensate for the influences of errors [20,21].

However, the most common method of compensating for geometric errors of an MT is based on the compensation table of the MT control software. Generally, not all geometric errors of the MT have their own compensation tables. In the same way, some MTs have their own control routines to compensate for errors during the machining process using a personal computer (PC) [22].

Otherwise there is another way to compensate for the influence of geometric error: by numerical control (NC) program reconstruction [23]. This method uses the kinematic model of the machine as a post-processor of the NC. Therefore, the MT accuracy is improved regardless of the control of the machine as this is not employed in the compensation. So, a new post-processor has been developed following this scheme:

Nominal coordinates of a point from the original NC are introduced into the kinematic model of the machine, where they are treated by approximation functions obtained in the verification process.

The kinematic model provides the coordinates of a new point associated with another error measured in the verification.

Through a process based on the Levenberg–Marquardt method, the software looks for the point coordinates where the influence of geometric errors is smallest. It takes into consideration the nominal coordinates and their nominal error.

Once all the nominal points have been compensated, the software (Figure 2) provides a new NC program with which the accuracy of the obtained approximation functions is validated and a more accurate part is made.

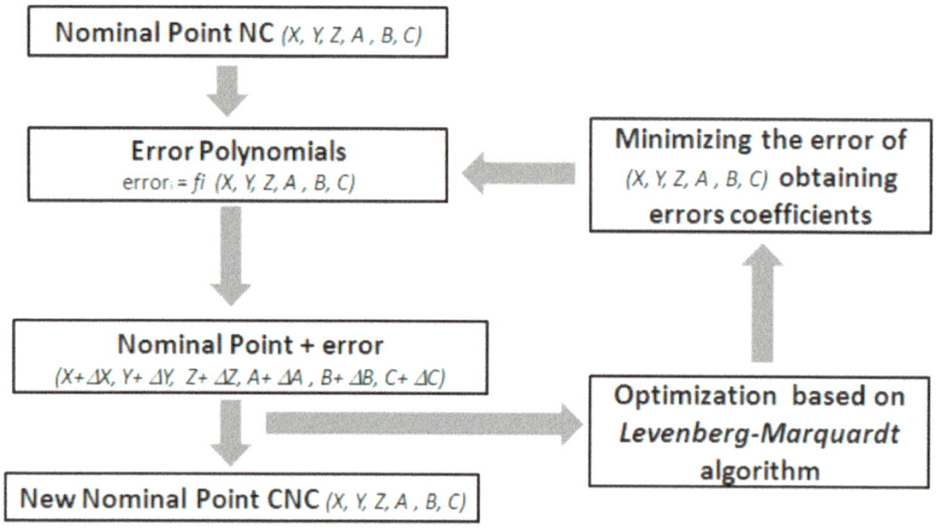

Figure 2: Working principle of post-processor software.

Due to its flexibility, this method was selected in this paper as the ideal one to perform the compensation for the geometric errors in this MT, allowing us to carry out all of the required tests.

LASER TRACKER AS A MEASUREMENT SYSTEM

An LT is a portable measurement system that provides the position of a point in spherical coordinates. This position is determined by comparing a measurement beam and a reference beam from the combination of a laser interferometer and the readings of the azimuth and polar coordinates of the angular encoders. Due to its versatility, LTs are used in many industrial applications such as in metrology and quality departments. Due to its long measurement range, accuracy, and versatility, the LT was selected as an adequate measurement system for volumetric verification, allowing the purchase of a dedicated system to be avoided and reducing costs.

The three biggest LTs manufacturers are Leica Geosystems, Faro Inc. (formerly SMX, Lake Mary, FL, USA), and API (Automated Precision Inc., Rockville, MD, USA). All of them have their own design and system characteristics. Also, the origins of their own coordinate systems are in different places (Figure 3).

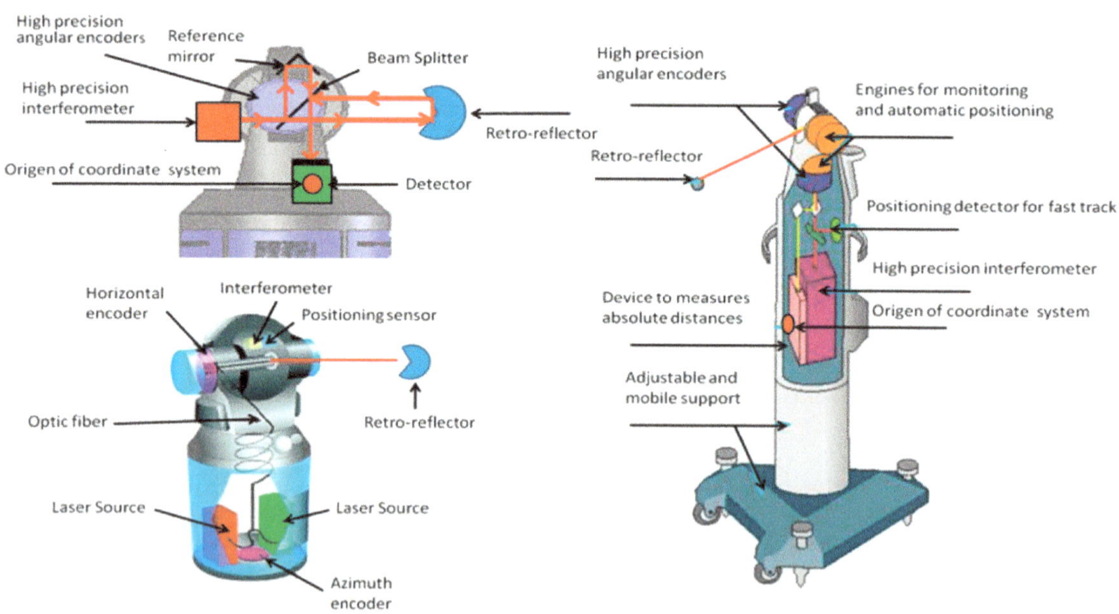

Figure 3: Laser trackers from different manufactures.

The main sources of errors in LT accuracy that affect verification are divided into errors due to environmental influences, assembly, and calibration errors, errors of LT components such as encoders or sensors, and errors due to the retroreflector and measurement technologies [24,25,26,27,28].

Laser Tracker Measurement Technologies

Currently, there are two different types of technologies used in LTs to determine the distance measurement: incremental distance measurements with an interferometer and a frequency-stabilized helium-neon laser (IFM), and absolute distance measurement (ADM) with an electronic control that analyzes the beam signal to determine its time of flight, providing the distance from the LT to the retroreflector. Meanwhile the measuring beam cannot be broken in IFM mode; if it is broken in ADM mode, the beam can be recovered and the measurement can continue. However, ADM is less accurate than IFM. The difference in accuracy between the two measurement techniques is shown in [24].

Retroreflector

A retroreflector is an item or surface that reflects light back towards the source regardless of the angle of incidence. This behavior is not equivalent to a mirror since it only reflects beams of light in the direction of incidence.

The most commonly used retroreflectors are the SMR and cat's eye. A traditional corner cube retroreflector SMR of 1.5″ has an incidence angle of ±30°, while a cat's eye has an approximate angle of ±90°. As in MT verification the LT replaces the part and the retroreflector is placed on the tool, the relative movement between LT and retroreflector causes its angle of incidence to exceed these values. As shown by [28], the incidence angle between beam and retroreflector affects the measurement coordinates and workspace to be verified.

Since 2010, a new motorized target retroreflector (Active Target) developed by API allows any API LT to automatically position itself, so it will never lose the beam. The Active Target reflector works in a similar way to any SMR. The difference is that the Active Target features motorization technology that automatically positions the reflector adjusting the beam so it is perpendicular to the reflector.

Measurement Uncertainty

As a result of the design, calibration, and accuracy of components used, the LT has a measurement uncertainty [24,25,26]. Therefore, manufacturers provide specifications of their LTs, and their accuracy is evaluated based on the standards ISO DIS 10360-10 and ASME B 89.4.19-2005. In order to reduce measurement uncertainty as a result of LT angular encoders, different multilateration techniques can be used, and the ones most commonly used in MT verification are studied in Section 4.

LASER MULTILATERATION IN MACHINE TOOL VERIFICATION

Of all the random sources of errors of an LT, the measurement noise that is formed by interferometer uncertainty and angular encoder uncertainty is the greatest contributor to the measurement uncertainty. Thus, angular uncertainty is removed using multilateration techniques; and point coordinates using only radial information of the points measured from at least three different positions (trilateration techniques) is obtained.

However, new metrology software includes angular information in the solution, allowing each individual measurement to be weighted based upon its type and accuracy; this is known as bundle or weighted adjustment trilateration [17,18].

Traditional Trilateration Technique

Trilateration aims to obtain a point with less involvement of the measurement uncertainty. Therefore, it uses only the radial component of each LT to determine points' coordinates. To do this, the new point is defined as the intersection of the three spheres that define the radial component of each LT, with (x_i, y_i, z_i) defining the center of a sphere LTi (i = 1, 2, 3).

$$R_i^2 = (x - x_i)^2 + (y - y_i)^2 + (z - z_i)^2 \quad \text{with } i = 0, \ldots, 3 \tag{9}$$

The equations system defined by LT coordinates can be solved analytically. However, the most useful technique, thanks to its accuracy, is trilateration using optimization [14,15]. This technique is based on an iterative process of parameter identification. The producer minimizes the difference between distances measured for each LT (Equation (10)).

$$Rest^2 = \sum_{j=1}^{n} \sum_{i=1}^{4} \left\{ \left[(x_{mj} - X_i)^2 + (y_{mj} - Y_i)^2 + (z_{mj} - Z_i)^2 \right]^{\frac{1}{2}} - (d_{ij}) \right\}^2 \tag{10}$$

The value *Rest* is the difference between the radial coordinate of point P_j measured with LTi (d_{ij}) and the distance between the coordinate system of point P_j (x_{mj}, y_{mj}, z_{mj}). The difference is calculated in the LT coordinates system, with (X_i, Y_i, Z_i) coordinates of the LT in the reference coordinate system (Figure 4). Parameters to be identified in the optimization process can be reduced if a new reference system of the LT is created as shown in Figure 4 [14,15].

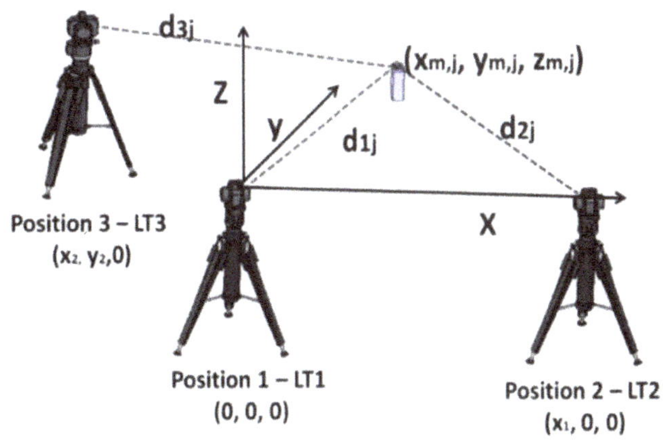

Figure 4: Trilateration and multilateration coordinate system.

Therefore, parameters to be identified are the LT positions X_1, X_2, and Y_2, and the multilateralized coordinates $x_{m,j}$, $y_{m,j}$, and $z_{m,j}$ with j = 1 n (n = number of points). These parameters are obtained by non-lineal optimization techniques based on the Levenber-Marquadt algorithm as shown in Figure 5.

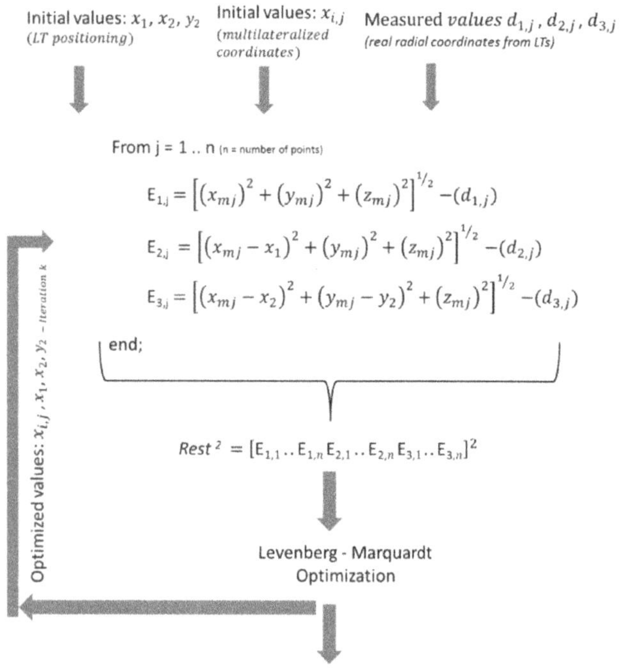

Figure 5: Scheme of multilateration optimization.

Weighted Trilateration

A weighted or bundle adjustment is a numerical adjustment that uses redundant measurements to obtain the best possible coordinates of a point. It simultaneously varies the positions of the measurement system and point until the sum of squares is minimized.

There are several studies of different bundle adjustment techniques [17,18]. However, their operating principle is the same: the importance of different parameters in the optimization process in relation to the uncertainty of the components of the LT is controlled. Generally, LT uncertainty is characterized in relation to standard deviations. Weighted trilateration gives greater weight to more accurate measurements than less accurate measurements (Equations (12)–(14)):

$$H \pm \Sigma h \tag{11}$$

$$V \pm \sigma V \tag{12}$$

$$R \pm \sigma R \tag{13}$$

where H is the horizontal angle, V is the vertical angle, and R is the radial coordinate of an LT.

The gradual increase in the use of this technique, as well as a greater demand for more accurate measurements, has led to the introduction of dynamic weights. For example, the weight associated with the angular measurement of an LT varies in relation to the range and angle measurement accuracy (Equations (15)–(17)) [28]:

$$H \pm \sigma H(\theta) \tag{14}$$

$$V \pm \sigma V(\varphi) \tag{15}$$

$$R \pm \sigma R(R) \tag{16}$$

WORKING PRINCIPLE OF VOLUMETRIC VERIFICATION

Volumetric verification aims to reduce the influence of geometric results based on volumetric error from LT measurement. This is obtained using the kinematic model of the machine including the measurement system on it (as shown in Section 2).

The relationship that provides this equation generates a non-linear equation system. Using the Levenberg–Marquardt algorithm in a non-lineal optimization process, the approximation functions of geometric errors are obtained reducing the volumetric error.

The working principle is shown in Figure 6. Nominal points introduced in the machine kinematic model are influenced by the geometric errors of the machine, modifying the point coordinates. As the error values are not known, regression functions provide the values of each error at each point to approximate the geometric errors of the kinematic model.

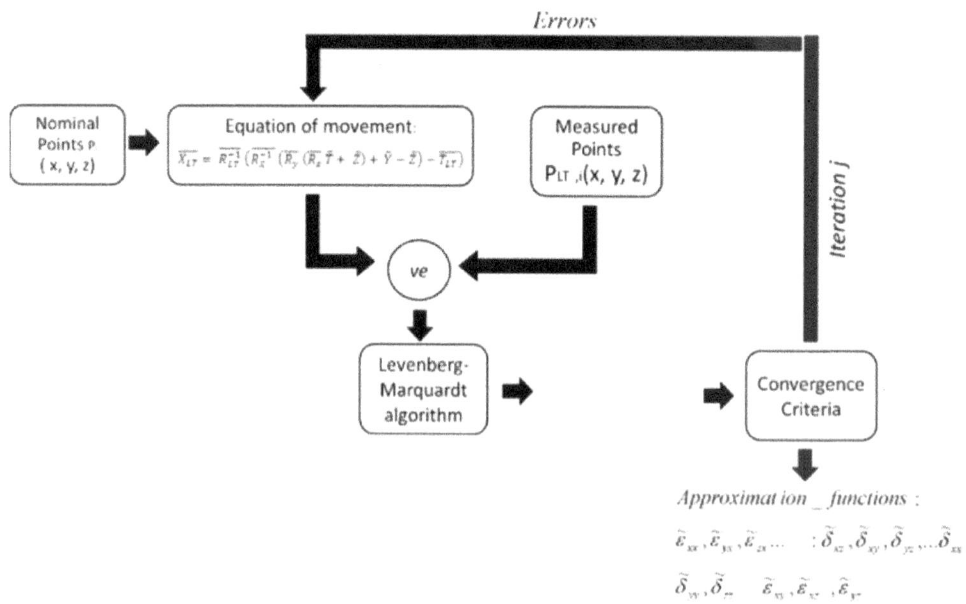

Figure 6: Procedure for the identification of errors.

To reduce the volumetric error of the machine, the values of the coefficients that define the regression functions change during the optimization process. When the verification process is complete, one of the convergence criteria is fulfilled. Then, the regression functions obtained define the approximation function of each geometric error.

WORKING PRINCIPLE OF VOLUMETRIC VERIFICATION

Besides studying the adequacy of volumetric verification to improve the accuracy of long-range MTs, tests were conducted with the aim of studying and comparing the equipment and techniques that currently exist. For this purpose, the linear axis of a Zayer TB5000 machine center was tested with all of them, providing relevant information to management and metrology departments as well as researchers interested in the use of LT and multilateration techniques.

Verification of Machine Tool Using API Motorized Retroreflector with and without Previous Compensation

The use of a motorized retroreflector as an active target allows us to increase the MT workspace to be verified. This is especially relevant when the part to be machined is above the axes of movement, because the LT must replace the part to show the same errors as the part (Figure 7), and the incidence angles of the SMR and cat's eye reflectors are limited. However, the motorized retroreflector can only be used by the API LT in ADM mode.

Figure 7: MT volumetric verification using Active Target.

Using an MT workspace with 0 mm ≤ X ≤ 5050 mm, 0 mm ≤ Y ≤ 2050 mm, 0 mm ≤ Z ≤ 750 mm, C = 0°, and B = 0°, the system can verify a workspace of 0 mm ≤ X ≤ 5000 mm, 0 mm ≤ Y ≤ 2000 mm, 0 mm ≤ Z ≤ 750 mm, C = 0°, and B = 0° using as intervals 500 mm to the X-axis, 500 mm to the Y-axis, and 250 mm to the Z-axis. It provides a verification mesh with 219 points (Figure 7).

Based on information of reference [27] the preheating of the MT is carried out using specific NC cycles. It is formed with the eight points that define the large workspace of the machine and its diagonals. This vacuum cycle prior to capturing points is repeated for 45 min until all MT components are hot.

Among the non-geometric errors of the MT such as repeatability, deformations, and so on, one of the most relevant is the backlash error of the axis of movement. It is not introduced in the kinematic model of the machine. However, to reduce its influence, verification points are measured twice; in the first measurement the effect of backlash has a certain direction while in the second one it has the opposite direction. In this case, the coordinates of the measured point are the average of the coordinates obtained in the backward and forward measurements. In the same way, temperature variations should be controlled during verification, preventing the occurrence of thermal gradients due to light sources, air currents, and so on. This source of error changes the machine structure and affects the measured points. Therefore, it is necessary to monitor the temperature variations during and between tests.

Firstly, the current status of the machine and the suitability and performance of its compensation were analyzed. For this purpose, the accuracy of the MT was obtained using verification points with and without compensation. The variations in the temperature of the environment during measurement and validation are presented in Figure 8. The variation is approximately 1 °C in both cases. Variation between measurements is also reduced because both points are measured twice. Consequently, temperature variations are not depreciable but assumable. However, temperature influence is deeply studied in order to reduce it [29,30]. This one is not considered in the verification process proposed but it is minimized.

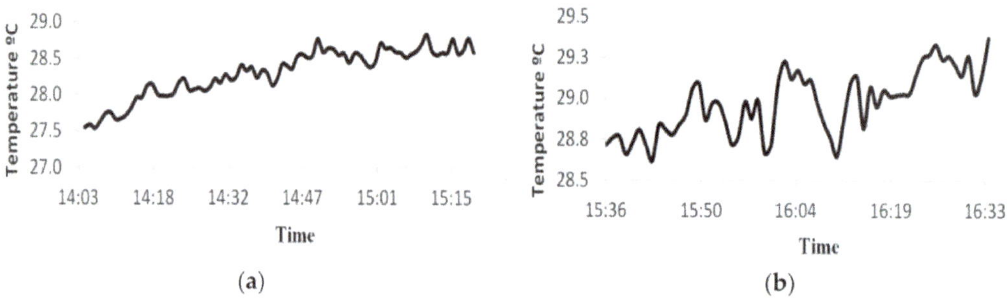

Figure 8" Evolution of environmental temperature: data acquisition (**a**) and validation results (**b**).

Calibration times are divided in three steps with three different times, preparation, acquisition, and identification:

- Preparation is the step which requires more time. It is approximately 2 h. It is because LT should be located on the correct place in order to maximize the workspace of the MT to verify. Moreover, the LT requires as little as 30 min of heating time before measure. Additionally, a numerical control program (NC) taking into consideration characteristic of the MT to verify should be created to automatize data capture.
- Acquisition or data captured time is presented in Figure 8, 1 h, it is time required to measure all verification points of the NC created previously. This information is used to obtain approximation functions of each MT geometric error during identification process.
- Identification was carried out approximately in 20 min using optimization process (Section 5). Once it is finished, MT compensation can be carried out. Therefore, the MT verification required approximately 3 h 30 min.

On analyzing the initial errors after thermal variations, the results displayed in Figure 9, which represent an error histogram of verification points, show that its current compensation reduces the maximum error from 246.3 to 197.7 µm, the average volumetric error from 112.9 to 83.1 µm, and the minimum error from 15.1 to 10.1 µm. Therefore, it reduces the average positioning error of the MT by 26.5%.

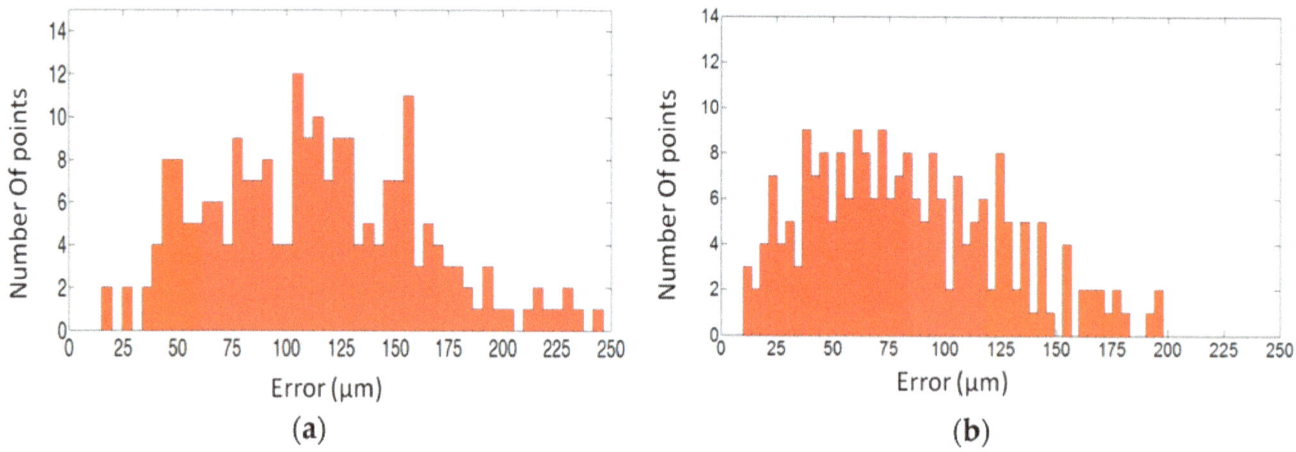

Figure 9: Error histograms before verification with MT compensation inactive (**a**) and active (**b**).

The current MT compensation method is based on compensation tables provided by manufacturers, which are not easy to access or use. Therefore, compensation was carried out using an NC post-processor. In the same way, as volumetric verification is a mathematical correction that cannot provide a complete physical meaning to the approximation function obtained, verification can be carried out when MT compensation software is activated (verification with compensation).

Table 2 shows the results of the adequacy of approximation with and without previous compensation using simple polynomials of degree three as regression functions.

Table 2: Verification results using Active Target.

Compensation Table	Error (µm)		
	Maximum	Average	Minimum
Compensation inactivated	246.3	112.9	15.1
Compensation activated	197.7	83.1	10.1
Verification without compensation	221.5	57.4	15.1
Verification with compensation	105.5	47.1	4.5

When compensation prior to verification is studied, it can be observed that the maximum error is reduced by 40 µm, while the average volumetric error is reduced by 30 µm. In the same way, if compensation is inactive in data acquisition (verification without compensation), the average error reduction is reduced to 57.4 µm. Therefore, volumetric verification improves the accuracy of the MT by 49%, whereas an improvement of 26% was obtained with the previous method. In the same way, if data with prior verification is used, the approximation functions obtained reduce the error to 58%. This is because the residual errors of previous compensation are the result of uncompensated geometric errors. Therefore, the MT has been compensated after volumetric verification, providing an increase of MT positioning accuracy.

The histograms in Figure 10 show that the global error of the machine has been reduced. However, to find out how errors are distributed in the MT workspace, color maps such as that in Figure 11 are studied. It shows that volumetric verification provides a homogeneous correction on the whole of the MT workspace, avoiding areas of failure at the ends of the axes of movement. Among the non-geometric errors of the MT such as repeatability, deformations, and so on, one of the most relevant is the backlash error of the axis of movement. It is not introduced in the kinematic model of the machine. However, to reduce its influence, verification points are measured twice; in the first measurement, the effect of backlash has a certain direction while, in the second one, it has the opposite direction. In this case, the coordinates of the measured point are the average of the coordinates obtained in the backward and forward measurements. In the same way, temperature variations should be controlled during verification, preventing the occurrence of thermal gradients due to light sources, air currents, and so on. This source of error changes the machine structure and affects the measured points. Therefore, it is necessary to monitor the temperature variations during and between tests.

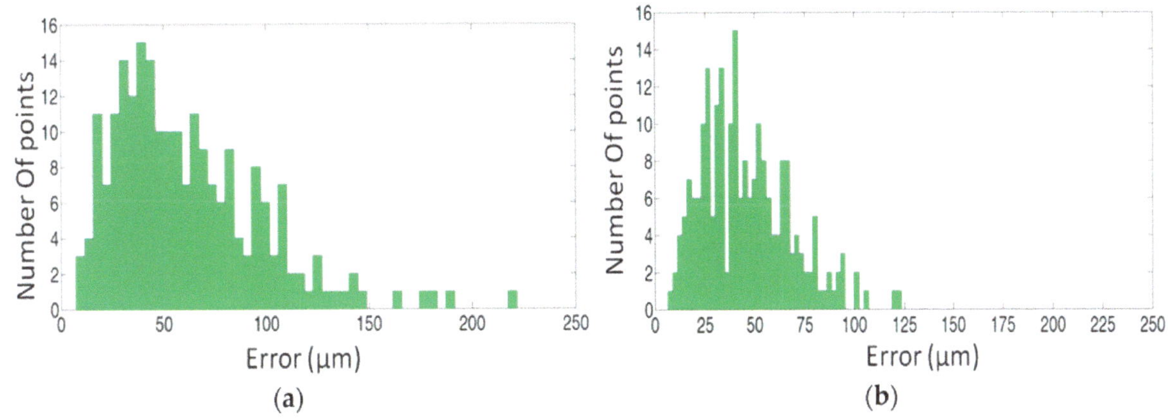

Figure 10: Error histograms after compensation with MT compensation inactive (**a**) and active (**b**).

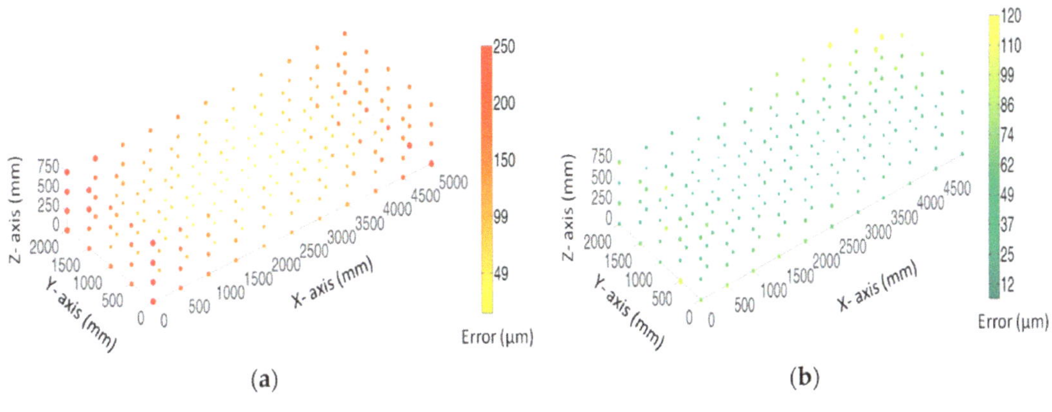

Figure 11: Color map of error distribution: before (**a**) and after compensation (**b**).

Influence of Laser Tracker (LTs)' Measurement Characteristics on Verification Results

As the active target cannot be used with LTs from different manufacturers, it is replaced by a traditional retroreflector, the Spherically Mounted Retroreflector, SMR-1.5″. The workspace of the MT to be verified is reduced because the angle of vision of the SMR is ±30°. Therefore, the characteristics of the new verification process are defined below:

MT workspace: XFYZ: 0 mm $\leq X \leq$ 5050 mm, 0 mm $\leq Y \leq$ 2050 mm, 0 mm $\leq Z \leq$ 750 mm, C = 0°, and B = 0°; MT workspace to be verified: 0 mm $\leq X \leq$ 2000 mm, 0 mm $\leq Y \leq$ 1000 mm, 0 mm $\leq Z \leq$ 450 mm, C = 0°, B = 0°, X-axis interval = 250 mm, Y-axis interval = 250 mm, and Z-axis interval = 150 mm.

To analyze the influence of the LT on verification, the same mesh of points was measured with three LTs from three different manufacturers with the measurement specifications presented in Table 3.

Table 3: Laser tracker specifications.

Laser Tracker	A	B	C
Measurement volume	-	-	-
Horizontal (°)	±235°	±320°	±360°
Vertical (°)	±45°	77°/−60°	77.9°/−52.1°
Range (m)	0–40 m	0–30 m	0–60 m
Measurement Technology	-	-	-
IFM	Si	Si	No
ADM	No	Si	Si
Accuracy	-	-	-
Angular	0.14 arcs	3.5 μm/m	20 μm + 5 μm/m
Absolute accuracy	±10 μm/m	±5 ppm	16 μm + 0.8 μm/m

Although the influence of geometric errors on verification points is the same for all LTs, the MT errors that each LT provides are different due to the different sources of error such as SMR errors, LT measurement characteristics, non-geometrical error, thermal variations, and so on. By measuring the same mesh of points simultaneously with all LTs, the influence of extra error in the measurement data is only the result of the LT characteristics and SMR errors.

Table 4 presents the results of the identification process using the three LTs independently. The initial errors that each LT provides are different. LT A provides an average initial error of 33 μm using IFM technology, which is smaller than the errors of 40 and 54 μm provided by LTs C and B using ADM. Thus IFM may introduce less error than ADM technology. Similarly, LT C introduces less error than LT B using the same technology. If the final average error is analyzed, all LTs provide similar results. Therefore, the approximation functions obtained with LT B compensate for non-geometric errors.

Table 4: Verification results in relation to laser tracker (LT) used.

Laser Tracker	A	B	C
Maximum initial error (μm)	66.5	102.9	80.9
Average initial error (μm)	33.5	54.1	40.1
Minimum initial error (μm)	8.0	6.9	8.9
Maximum final error (μm)	54.7	86.7	62.6
Average final error (μm)	22.3	28.9	25.5
Minimum final error (μm)	5.1	4.4	7.2

Influence of Improving Data Accuracy through Traditional or Weighted Trilateration

In MT verification, the spatial LT distribution is strongly affected by the MT configuration and visibility of the reflector.

To find out whether the new weighted trilateration is better than traditional trilateration reducing the increase of radial noise influences on final MT positioning errors, three LTs were positioned so as to create a spatial angle of approximately 36° (Figure 12), which is far from the ideal distribution of 90° [24].

Figure 12: MT verification through simultaneous measurement by trilateration.

In this test, all LTs simultaneously measured the same points, avoiding the influences of thermal variation and SMR error when both techniques were compared, because the input data were the same for both. Once the points had been measured and the verification points of each LT had been created, two meshes of points were created. A weighted trilateration mesh was created using Spatial Analyzer software (SA 2015.02.25, New River Kinematic, Williamsburg, VA, USA, 1994), while traditional trilateration was carried out using Matlab algorithms (2013a, MathWorks, Natick, MA, USA, 1984).

The results in Table 5 show that multilateration provides an average initial error similar to verification with only one LT when using both multilateration techniques. In the same way, the residual error obtained is similar. Consequently, bundle adjustment and trilateration have similar behavior in the verification of this MT.

Table 5: Verification results using trilateration with simultaneous measurement by three LTs.

Laser Tracker	A	B	C	Traditional Trilateration	Weighted Trilateration
Maximum initial error (μm)	66.5	102.9	80.9	75.0	80.9
Average initial error (μm)	33.5	54.1	40.1	38.2	40.0
Minimum initial error (μm)	8.0	6.9	8.9	7.9	8.8
Maximum final error (μm)	54.7	86.7	62.6	64.3	71.3
Average final error (μm)	22.3	28.9	25.5	21.7	25.7
Minimum final error (μm)	5.1	4.4	7.2	2.5	5.3

Although multilateration might improve the MT's accuracy, as shown by the synthetic results [14], poor relative positioning of LTs does not improve verification results using weighted trilateration. Therefore, the LT's positioning must be studied carefully to avoid unnecessary measurements, as it is the most influential factor.

CONCLUSIONS

Volumetric verification is a real solution in MT verification. Thanks to it, mathematical compensation can be used to improve MT accuracy, even if previous compensation is not eliminated. Previous compensation of the MT provides a volumetric error reduction of the MT of 26.7%. Meanwhile, volumetric verification reduces the volumetric error by between 49.7% and 58.7%.

The use of the Active Target as a retroreflector has led to an increase in the size of the MT workspace verified from 0.9 to 7.5 m^3. This retroreflector is especially useful when the LT is associated with an axis of movement and the available space limits the visibility of a traditional retroreflector. In relation to the LT measurement technique, both ADM and IFM provide similar results. However, the influence of measurement noise in volumetric error is smaller when using IFM mode. In the same way, the use of a multilateration technique that eliminates the influence of angular measurement noise might improve the verification process. However, the radial noise has been increased by the same order as the angular reduction due to the spatial angle between the LTs. This behavior could not be improved using weighted trilateration.

ACKNOWLEDGMENT

This work was supported by Spanish Government (Ministry of Economy and Competitiveness) through the research project METRAP—Development of new techniques of verification and compensation of manufacturing systems for traceable in-process metrology (DPI2013-46979-C2-1-P), Scientific and Technical Research of Excellence Development National Program, and by Aragon Government (Department of Industry and Innovation) through the Research Activity Grant for research groups recognized by the Aragon Government (2014-T62 Manufacturing Engineering and Advanced Metrology Group).

AUTHOR CONTRIBUTIONS

Sergio Aguado and Jorge Santolaria design the expreiments; David Samper, Jesús Velazquez, Sergio Aguado and Jorge Santolaria developed required software; Carlos Javierre, Ángel Fernández, Jorge Santolaria and Sergio Aguado performed the experiments; Jesús Velázquez, Sergio Aguado and David Samper analyzed the data; Sergio Aguado wrote the paper.

CONFLICTS OF INTEREST

The authors declare no conflict of interest.

REFERENCES

1. Chen, J.S.; Kou, T.W.; Chiou, S.H. Geometric error calibration of multi-axis machines using an auto-alignment laser interferometer. *Precis. Eng.* 1999, *23*, 243–252
2. Barakat, N.A.; Elbestawi, M.A.; Spence, A.D. Kinematic and geometric error compensation of a coordinate measuring machine. *Int. J. Mach. Tools Manuf.* 2000, *40*, 833–850.
3. Kunzmann, H.; Trapet, E.; Wäldele, F. A uniform concept for calibration, acceptance test, and periodic inspection of coordinate measuring machines using reference objects. *Ann. CIRP* 1990, *39*, 561–564.
4. Schwenke, H.; Knapp, W.; Haitjema, H.; Weckenmann, A.; Schmitt, R.; Delbressine, F. Geometric error measurement and compensation of machines—An update. *CIRP Ann. Manuf. Technol.* 2008, *57*, 660–675.
5. Zargarbashi, S.H.H.; Mayer, J.R.R. Assessment of machine tool trunnion axis motion error, using magnetic double ball bar. *Int. J. Mach. Tools Manuf.* 2006, *46*, 1823–1834.
6. Schwenke, H.; Franke, M.; Hannaford, J.; Kunzmann, H. Error mapping of CMMs and machine tools by a single tracking interferometer. *CIRP Ann. Manuf. Technol.* 2005, *54*, 475–478.
7. Aguado, S.; Samper, D.; Santolaria, J.; Aguilar, J.J. Identification strategy of error parameter in volumetric error compensation of machine tool based on laser tracker measurements. *Int. J. Mach. Tools Manuf.* 2012, *53*, 160–169.

8. Aguado, S.; Samper, D.; Santolaria, J.; Aguilar, J.J. Towards an effective identification strategy in volumetric error compensation of machine tools. *Meas. Sci. Technol.* 2012, *23*.
9. Uriarte, L.; Zatarain, M.; Axinte, D.; Yagüe-Fabra, J.; Ihlenfeldt, S.; Eguia, J.; Olarra, A. Machine tools for large parts.*CIRP Ann. Manuf. Technol.* 2013, *62*, 731–750.
10. Zhenya, H.; Jianzhong, F.; Liangchi, Z.; Xianhua, Y. A new error measurement method to identify all six error parameters of a rotational axis of a machine tool. *Int. J. Mach. Tools Manuf.* 2015, *88*, 1–8.
11. Velenosi, A.; Campatelli, G.; Scippa, A. Axis geometrical error analysis through a performace test to evaluate kinematic error in a five axis tilting-rotary table machine tool. *Precis. Eng.* 2015, *39*, 224–233.
12. Wenjie, T.; Weiguo, G.; Dawei, Z.; Tian, H. A general approach for error modelling of machine tools. *Int. J. Mach. Tools Manuf.* 2014, *79*, 17–23.
13. Slocum, A.H. *Precision Machine Design*; ISBN: 0-13-690918-3. Society of Manufacturing Engineers: Dearborn, MI, USA, 1992.
14. Aguado, S.; Santolaria, J.; Samper, D.; Aguilar, J.J. Influence of measurement noise and laser arrangement on measurement uncertainty of laser tracker multilateration in machine tool volumetric verification. *Precis. Eng.* 2013, *37*, 929–943.
15. Aguado, S.; Santolaria, J.; Samper, D.; Aguilar, J.J. Study of self-calibration and multilateration in machine tool volumetric verification for laser tracker error reductions. *J. Eng. Manuf.* 2013.
16. Linares, J.M.; Chaves-Jacob, J.; Schwenke, H.; Longstaff, A.; Fletcher, S.; Flore, J.; Uhlmann, E.; Wintering, J. Impact of measurement procedure when error mapping and compensationg a small CNC machine using a multilateration laser interferometer. *Precis. Eng.* 2014, *38*, 578–588.
17. Predmore, C.R. Bundle adjustment of multi-position measurement using the Mahalanobis distance. *Precis. Eng.* 2010,*34*, 113–123.
18. Meid, A.; Sandwith, S. Dynamic Weighting of Laser Tracker Measurement for Bundle Adjustment. Available online: http://www.leica-geosystems.fr/fr/bundle_sandwith.pdf (accessed on 5 April 2016).
19. Sandwith, S.; Predmore, R. Real Time 5-Microm Uncertainty with Laser Tracking Interferometer Systems using Weighted Trilateration. Available online: http://www.kinematics.com/images/TrilaterationPaperR1.pdf (accessed on 5 April 2016).
20. Kono, D.; Marsubara, A.; Yamaji, I.; Fujita, T. High-precision machining by measurement and compensation of motion error. *Int. J. Mach. Tools Manuf.* 2008, *48*, 1103–1110.
21. Khan, A.W.; Chen, W. A methodology for systematic geometric error compensation in five-axis machine tools. *Int. J. Adv. Manuf. Technol.* 2011, *53*, 615–628.
22. She, C.H.; Chang, C.C. Design of a generic five-axis postprocessor based on generalized kinematics models of machine tool. *Int. J. Mach. Tools Manuf.* 2007, *47*, 537–545.
23. Gui, G.; Lu, Y.; Li, J.; Gao, D.; Yao, Y. Geometric error compensation software system for CNC machine tool based on NC program reconstructing. *Int. J. Adv. Manuf. Technol.* 2012, *63*, 169–180.
24. Conte, J.; Majarena, A.C.; Aguado, S.; Acero, R.; Santolaria, J. Calibration strategies of laser trackers based on network measurements. *Int. J. Adv. Manuf. Technol.* 2016, *83*, 1161–1170.
25. Teoh, P.; Shirinzadeh, B.; Foong, C.; Alici, G. The measurement uncertainties in the laser interferometry based sensing and tracking technique. *Meas. Sci. Technol.* 2002, *32*, 135–150.
26. Zhuang, H.; Motaghedi, S.H.; Roth, Z.S.; Bai, Y. Calibration of Multi-Beam Laser Tracking Systems. *Robot. Comput. Integr. Manuf.* 2003, *19*, 301–314.
27. Aguado, S.; Samper, D.; Santolaria, J.; Aguilar, J.J. Protocol for machine tool volumetric verification using a commercial laser tracker. *Int. J. Adv. Manuf. Technol.* 2014, *75*, 425–444.
28. Cheu, X.; Zhang, G.X.; Zhao, S.Z.; Duan, F.J. Analysis and measurement fir the optical error of the cat's eye retro-reflector. *J. Phys. Conf. Ser.* 2006, *48*, 64–68.
29. Mian, N.S.; Fletcher, S.; Longstaff, A.P.; Myers, A. Efficient estimation by FEA of machine tool distortion due to environmental temperature perturbations. *Precis. Eng.* 2013, *37*, 372–379.
30. Gebhardt, M.; Mayr, J.; Furrer, N.; Widmer, T.; Weikert, S.; Knap, W. High precision grey-box model for compensation of thermal errors on five-axis machines. *CIRP Ann. Manuf. Technol.* 2014, *63*, 509–512.

CHAPTER 18

FUNDAMENTALS OF HIGHLY NON-DEGENERATE CASCADED FOUR-WAVE MIXING

Rosa Weigand [1] and Helder M. Crespo [2]

[1]Departamento de Óptica, Facultad de Ciencias Físicas, Universidad Complutense de Madrid, Avda. Complutense s/n, Madrid 28040, Spain

[2]Departamento de Física e Astronomia e IFIMUP-IN, Faculdade de Ciências, Universidade do Porto, R. do Campo Alegre 687, Porto 4169-007, Portugal

ABSTRACT

By crossing two intense ultrashort laser pulses with different colors in a transparent medium, like a simple piece of glass, a fan of multicolored broadband light pulses can be simultaneously generated. These newly generated pulses are emitted in several well-defined directions and can cover a broad spectral range, from the infrared to the ultraviolet and beyond. This beautiful phenomenon, first observed and described 15 years ago, is due to highly-nondegenerate cascaded four-wave mixing (cascaded FWM, or CFWM). Here, we present a review of our work on the generation and measurement of multicolored light pulses based on third-order nonlinearities in transparent solids, from the discovery and first demonstration of highly-nondegenerate CFWM, to the coherent synthesis of single-cycle pulses by superposition of the multicolored light pulses produced by CFWM. We will also present the development and main results of a dedicated 2.5-D nonlinear propagation model, *i.e.*, with propagation occurring along a two-dimensional plane while assuming cylindrically symmetric pump beam profiles, capable of adequately describing noncollinear FWM and CFWM processes. A new method for the generation of femtosecond pulses in the deep-ultraviolet (DUV) based on FWM and CFWM will also be described. These experimental and theoretical results show that highly-nondegenerate third-order nonlinear optical processes are formally well understood and provide broader bandwidths than other nonlinear optical processes for the generation of ultrashort light pulses with wavelengths extending from the near-infrared to the deep-ultraviolet, which have many applications in science and technology.

INTRODUCTION

Four-wave mixing (FWM) processes result from the interplay between four electromagnetic waves coupled through the optical Kerr nonlinearity of a medium, given by the third-order susceptibility $\chi^{(3)}$ [1]. Since all media, either isotropic or anisotropic (with any kind of crystal symmetry), have a non-zero third order susceptibility, FWM processes have a universal character. The four interacting waves can all have the same frequency, in which case the process is degenerate and the corresponding susceptibility is given by $\chi^{(3)}(\omega;\omega,\omega,-\omega)$. In this case, three fields with frequency ω are mixed in the medium to produce another field also with frequency ω. The energy conservation law that determines the frequency of the

Citation: Rosa Weigand and Helder M. Crespo, Fundamentals of Highly Non-Degenerate Cascaded Four-Wave Mixing, doi:10.3390/app5030485

newly generated field is given in this case by: $\omega = \omega + \omega - \omega \equiv 2\,\omega - \omega$. In practice, this process can be observed by crossing two intense pulses in a nonlinear medium at a small angle. For incident wavevectors $\boldsymbol{k}_0$ and $\boldsymbol{k}_1$, momentum conservation dictates that the newly generated beams will be emitted in the directions given by $\boldsymbol{k}_2 = 2\boldsymbol{k}_1 - \boldsymbol{k}_0$ and $\boldsymbol{k}'_2 = 2\boldsymbol{k}_0 - \boldsymbol{k}_1$, as shown in Figure 1. This process can be seen as the diffraction of the incident beams by the nonlinear index grating (laser induced grating) produced by the same beams [2]. In bulk dispersive media this process is intrinsically phase mismatched (see, e.g., [3]), with a mismatch given approximately by $\delta k = 2k(1 - \cos\theta) \simeq k\theta^2$, where $k = |\boldsymbol{k}_0| = |\boldsymbol{k}_1|$. This means that the efficiency of the process drops for increasing interaction (and hence emission) angles.

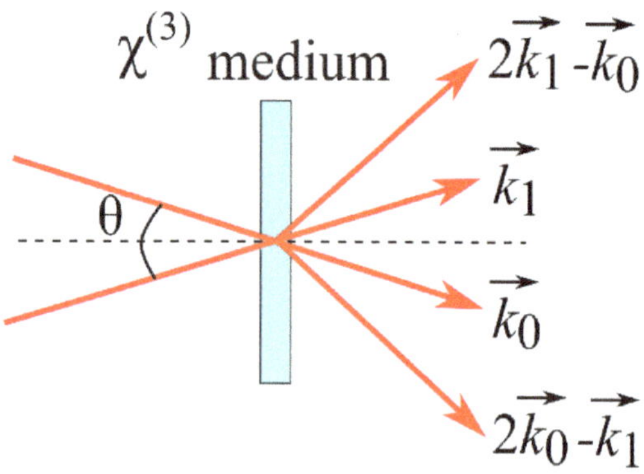

Figure 1: Basic geometry of nonlinear self-diffraction.

The waves can also have different frequencies, so a new field with frequency ω_2 can be generated departing from two fields with frequencies ω_0 and ω_1 ($\omega_1 > \omega_0$). In this case, the nonlinear polarization terms which describe diffraction from a moving laser-induced grating give rise to frequency upshifted and downshifted pulses with frequencies $\omega_2 = 2\omega_1 - \omega_0$ and $\omega'_2 = 2\omega_0 - \omega_1$, respectively, so the output consists of four beams with different colors emitted in different directions. Penzkofer and Lehmeier [4] analized theoretically the noncollinear phase-matched amplification of ultrashort light pulses via four-wave mixing in isotropic media, having derived explicit expressions for the phase-matching angles and gain.

If the fields are intense enough, the third order susceptibility $\chi^{(3)}$ can continue mixing the fields, as long as the energy and momentum conservation laws are fulfilled for the new processes. This way, new fields are generated via cascaded four-wave mixing (CFWM). For noncollinear interactions, this gives rise to additional beams that can be seen as higher orders of diffraction of the moving grating (note that the moving grating corresponds to the spatially and temporally varying nonlinear refractive index change inside the medium, since the medium itself is stationary in the laboratory frame). For the degenerate case, the larger the interaction angle, the smaller the observed number of diffracted orders (due to increasing phase-mismatch), but for the nondegenerate case we will see that the phase-mismatch in cascaded processes can be reduced by proper choice of the interaction angle between the two initial beams.

Collinear fifth-harmonic generation in a crystal as a result of cascaded processes of a cubic nonlinearity was reported in an early work [5]. Extensive work has been done in this field by mixing near-degenerate nanosecond laser pulses in optical fibers and obtaining several pairs of sidebands [6,7,8]. Cascaded processes can also occur with second order processes and have been observed in BBO crystals [9] or in an hybrid form with third order processes in picosecond optical parametric amplifiers [10].

Further progress was made when highly nondegenerate CFWM was demonstrated by mixing femtosecond visible laser pulses with different colors. The first demonstration was done in bulk isotropic media (thin glass slide) [11], and resulted in the generation of multiple broadband light pulses extending from the infrared to the ultraviolet starting from two femtosecond laser pulses in the visible range. Nonresonant nondegenerate CFWM in the femtosecond regime was subsequently observed and demonstrated in other nonlinear media and spectral regions, from semiconductors pumped with mid-IR pulses [12] to gases and plasmas pumped in the NIR to uv. In particular, Misoguti *et al.* [13] successfully extended CFWM to the vacuum-uv range (down to 160 nm) using a gas-filled hollow waveguide pumped with ω and 2ω pulses from a Ti:Sapphire laser. Raman-assisted noncollinear CFWM has also been obtained in Raman-active media, such as diamond, using dual color laser pulses and chirped broadband pulses (see, e.g., [14] and references therein).

This paper gives a review of the work done by the authors in this field and includes the experimental procedures and a complete theoretical model explaining not only the spectral characteristics of the generated beams, but also their intensity, spectral phases, emission angles and energies. Important examples of the application of the generated pulses namely for the synthesis of single cycle pulses, will be discussed, as well as the possibility of using CFWM for generating broadband light pulses in the deep ultraviolet spectral region.

BASIC EXPERIMENT AND INTERPRETATION

The first experiment on highly nondegenerate noncollinear CFWM was done by H. Crespo *et al.* at LOA (Laboratoire d'Optique Appliquée in Palaiseau, France) [11] and the experimental setup can be seen in Figure 2a. The laser source is a dye laser-amplifier system delivering an orange beam (frequency ω_0, $\lambda_0 = 618$ nm, 80 fs) and a green beam (frequency ω_1, $\lambda_1 = 561$ nm, 40 fs). Both beams are horizontally polarized and with near-Gaussian spatial profiles (approximately 5 mm diameter). The green beam was sent through a delay-line and both beams were coupled through a 50/50 beam splitter to impinge on a large section plane-convex lens (f = 30 cm) at an adjustable small angle. The lens focuses both beams in a thin slide of BK7 glass with 150 μm thickness at an external angle $\theta = 2.9°$. A thin glass slide was chosen to minimize self-phase modulation processes and it was also placed 1 cm before the focus of the lens to avoid laser-induced damage. At the plane of the slide both beams had 20 μJ energy, corresponding to intensities of 1.0 TW cm^{-2} for the orange and 2.1 TW cm^{-2} for the green beam. When both beams were temporally synchronized at the plane of the slide, a large set of multicolored beams (Figure 2b) was produced at both sides of the incident beams, consisting of two frequency-downshifted beams (we will call D_n the downconverted beam of order n) in the red and near infrared, as well as 11 frequency-upshifted beams (we will call U_n the upconverted beams of order n) extending into the blue and ultraviolet regions of the spectrum. The ultraviolet beams are seen as blue in the picture because the screen used was fluorescent, whereas the second downconverted beam could be observed with the help of a NIR viewer. The spectrum of the generated beams was registered by collecting the generated fan of beams with a large aperture aluminum-coated parabolic mirror M4 to minimize chromatic aberration and UV absorption and by sending the collimated beams to a spectrograph equipped with a CCD camera where the resulting spectrum is given in Figure 3. Two downconverted bands (D_1 and D_2) can be seen to the left of the pump beams, while only 5 upconverted bands (U_1 to U_5) are seen to the right, due to sensitivity limitations of the CCD in the blue and UV spectral ranges. All generated orders are broadband, with bandwidths of the order of those of the pump and signal beams (16 nm approximately) which are larger than the original laser bandwidths (5–6 mm) due to some self- and cross-phase modulation taking place within the slide. Energy measurements were done with a calibrated photodetector and the whole set of newly generated beams carried about 5%–10% of the total energy of the orange and green pulses.

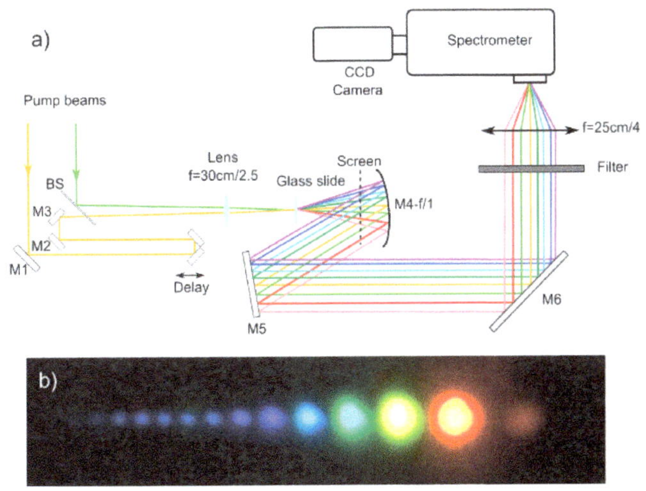

Figure 2: (**a**) Experimental arrangement: BS, beam splitter; M1–M3, silver mirrors; M4, parabolic mirror; M5, M6, aluminum mirrors. Spacing between colors not at scale; (**b**) Picture of output beams as taken on a white card. Adapted from [11].

The spectrum of Figure 3 can be explained, to a very good approximation, by assuming a cascaded FWM process in the central frequency approximation. The two pump pulses drive the $\chi^{(3)}$ medium at the modulation frequency $\omega_m = \omega_1 - \omega_0$, giving rise to multiple pulses with frequencies $\omega_n = \omega_0 + n\omega_m$, where n is the integer beam order and $n > 1$ ($n < 0$) denotes frequency upconverted (downconverted) pulses. This can also be written as $\omega_n = \omega_{n-1} + \omega_m$ or $\omega_n = n\omega_1 - (n - 1)\omega_0$. Note that even though the total number of photons involved in a given process is $2n$, this does not correspond to the order of the corresponding nonlinearities ($2n - 1$), but to that of an effective nonlinearity obtained via cascaded $\chi^{(3)}$, *i.e.*, third-order, processes. Table 1 shows, in an equivalent way, how the frequencies of the different orders are generated from the previous order plus or minus the modulation frequency.

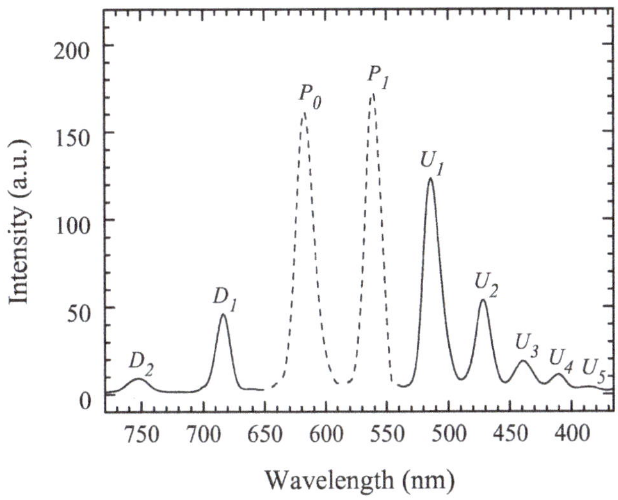

Figure 3: Spectra of the beams in the cascade: dashed curve, original spectra of the orange (P_0) and green (P_1) pump beams after crossing the glass slide but without temporal overlap (scaled for comparison); solid curve, spectra of the simultaneously generated beams. Adapted from [11].

Table 1: Approximate frequencies generated by the Cascaded Four-Wave Mixing process.

Order n	Frequency $\omega_n = \omega_0 + n\omega_m$	Equivalence
$-n$	$\omega_{-n} = \omega_0 - n\omega_m = (n+1)\omega_0 - n\omega_1$	$\omega_{-n} = \omega_{-n+1} - \omega_m$
.	.	.
.	.	.
.	.	.
-2	$\omega_{-2} = \omega_0 - 2\omega_m = 3\omega_0 - 2\omega_1$	$\omega_{-2} = \omega_{-1} - \omega_m$
-1	$\omega_{-1} = \omega_0 - \omega_m = 2\omega_0 - \omega_1$	$\omega_{-1} = \omega_0 - \omega_m$
0	$\omega_0 = \omega_0$	ω_0
1	$\omega_1 = \omega_0 + \omega_m = \omega_1$	ω_1
2	$\omega_2 = \omega_0 + 2\omega_m = 2\omega_1 - \omega_0$	$\omega_2 = \omega_1 + \omega_m$
3	$\omega_3 = \omega_0 + 3\omega_m = 3\omega_1 - 2\omega_0$	$\omega_3 = \omega_2 + \omega_m$
.	.	.
.	.	.
.	.	.
n	$\omega_n = \omega_0 + n\omega_m = n\omega_1 - (n-1)\omega_0$	$\omega_n = \omega_{n-1} + \omega_m$

An approximate description of the cascaded FWM process is completed by considering the phase-matching condition for the wave vectors of the upconverted and downconverted beams. Let us call $\boldsymbol{k}_0$ and $\boldsymbol{k}_1$ the wave vectors of the lower frequency (orange) and higher frequency (green) pump, whereas $\boldsymbol{k}_{-n}$ and $\boldsymbol{k}_n$ denote the wave vector of a downconverted beam or an upconverted beam of order n respectively. With $\boldsymbol{k}_m = \boldsymbol{k}_1 - \boldsymbol{k}_0$ the modulation (or grating) wave vector, the newly generated wave vectors are approximately given by $\boldsymbol{k}_n = \boldsymbol{k}_0 + n\boldsymbol{k}_m$, or equivalently by $\boldsymbol{k}_n = \boldsymbol{k}_{n-1} + \boldsymbol{k}_m$ or $\boldsymbol{k}_n = n\boldsymbol{k}_1 - (n-1)\boldsymbol{k}_0$. We will see that approximate phase matching can be achieved for a large number of frequency upconversion processes in a medium with normal dispersion, whereas frequency downconversion processes are intrinsically phase-mismatched, which explains the observed asymmetry between the two processes. For the same reason, the contribution of backwards processes such as $\boldsymbol{k}_n = \boldsymbol{k}_{n+1} - \boldsymbol{k}_m$ is small and therefore can be neglected.

Table 2 shows how the wave vectors of the different orders are generated from the previous one plus or minus the modulation wave vector.

Table 2: Approximate wave vectors generated by the Cascaded Four-Wave Mixing process.

Order n	Wave Vector $k_n = k_0 + nk_m$	Equivalence
$-n$	$k_{-n} = k_0 - nk_m = (n+1)k_0 - nk_1$	$k_{-n} = k_{-n+1} - k_m$
⋮	⋮	⋮
-2	$k_{-2} = k_0 - 2k_m = 3k_0 - 2k_1$	$k_{-2} = k_{-1} - k_m$
-1	$k_{-1} = k_0 - k_m = 2k_0 - k_1$	$k_{-1} = k_0 - k_m$
0	k_0	k_0
1	$k_1 = k_0 + k_m = k_1$	k_1
2	$k_2 = k_0 + 2k_m = 2k_1 - k_0$	$k_2 = k_1 + k_m$
3	$k_3 = k_0 + 3k_m = 3k_1 - 2k_0$	$k_3 = k_2 + k_m$
⋮	⋮	⋮
n	$k_n = nk_1 - (n-1)k_0$	$k_n = k_{n-1} + k_m$

Figure 4 shows the diagrams for the vector additions in the cascaded FWM processes for both the upconverted and downconverted beams, along with their exiting angles β_n.

Moreover it is possible to calculate the interaction angle θ_n that the main beams should have inside the medium for phase matching to occur for a particular process of order n, under the assumed approximation. Let us choose a coordinate frame in which $k_1 = k_1 \times (1,0)$. In this frame k0k0 would be given by $k_0 = k_0(\cos\theta_n, \sin\theta_n)$ and let us have in mind that any wave vector is related with the refractive index of the medium through $k_i = n_r(\omega_i)\omega_i/c$. Thus for the upconverted beams the equation $k_n = nk_1 - (n-1)k_0$ is written in this coordinate system as

$$k_n = (nk_1 - (n-1)k_0\cos\theta_n, (n-1)k_0\sin\theta_n) \tag{1}$$

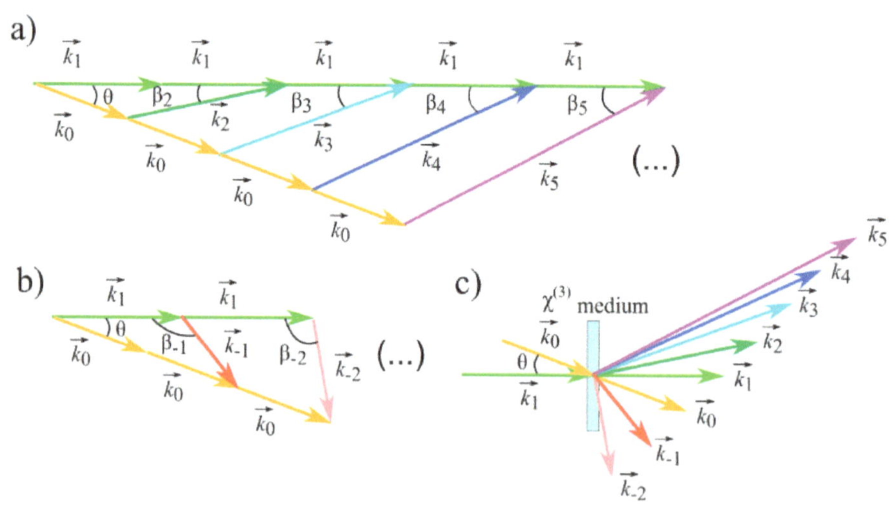

Figure 4: Wave vectors of the upconverted and downconverted beams generated by cascaded four-wave mixing (CFWM). (**a**) upconversion processes; (**b**) downconversion processes; (**c**) schematic of the wave vector distribution obtained by CFWM in real space.

Its square modulus is given by

$$k_n^2 = n^2k_1^2 - 2n(n-1)k_0k_1\cos\theta_n + (n-1)^2k_0^2\cos^2\theta_n + (n-1)^2k_0^2\sin^2\theta_n \tag{2}$$

and solving for $\cos(\theta_n)$ we obtain

$$\cos\theta_n = \frac{n^2{k_1}^2 + (n-1)^2{k_0}^2 - {k_n}^2}{2n(n-1)k_0k_1} \tag{3}$$

which is valid for the upshifts (n≥2) and downshifts (n≤−1). With reference to Figure 4a,b, the emission angles of the newly generated orders can be obtained by calculating the corresponding wave vectors (e.g., $k_2=2k_1-k_0$ for the first frequency-upshifted beam), with the ideal phase-matching angle given by Equation (3), and determining their angle with respect to the horizontal axis, which results in

$$\cos\beta_n = \frac{n^2{k_1}^2 - (n-1)^2{k_0}^2 - {k_n}^2}{2n(n-1)k_0k_1} \tag{4}$$

Hence it is possible to calculate the interaction angle between the beams inside the medium for the different orders to appear. Thus one can chose for example the phase matching angle which optimizes the first upconverted order. This order will be generated in optimum energy conditions and the remaining orders will be generated less efficiently, since perfect phase-matching is not set for them. Also, it is not possible to exactly phase-match frequency down conversion processes in media with normal dispersion, as Equation (4) does not have a real solution, which results in the observed asymmetry between frequency upconverted and downconverted beams. The fact that the interaction angle affects the generation of the cascade of colors was experimentally observed [11] and the fact that angle phase-matching was achieved thanks to the material dispersion was settled, as detailed below. Figure 5 shows the measured and calculated wave vectors generated in the first nondegenerate CFWM experiment, where the two-color pump beams cross in the medium at an external angle of 2.9°. The arrows denote the pump beam wave vectors; the ten consecutive upconverted beams appear above the horizontal axis, and the two downconverted beams are below this axis. The open squares falling outside the dotted box are the five upconverted beams for which only the wave vector direction was measured directly, and hence the amplitude was extrapolated from the phase-matching conditions. The filled circles are calculated with the phase-matching conditions, where k0k0 and k1k1 were obtained directly from spectral and angular measurements of the pump pulses. The open circles are the calculations for perfectly phase-matched beams and are shown for reference only, since a cascade of such ideal processes (which would required different angles for each process) cannot be obtained experimentally.

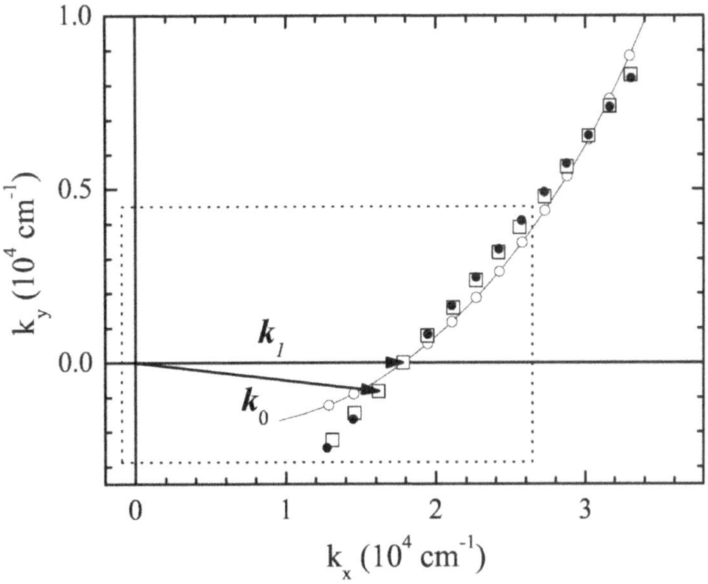

Figure 5: Representation of the wavevectors of all the beams in the pattern: open squares, measured points; filled circles, calculation with the approximate cascaded processes; open circles, calculated exactly phase-matched beams, with the interaction angle given by Equation (3). The solid curve is a guide. Adapted from [11].

Figure 6 shows the phase mismatch Δk_n for the optimized interaction angle of 2.9° as a function of beam order *n*, where *n*> 1 (*n* < 0) correspond to upconverted (downconverted) beams. The dashed curve is a calculation assuming collinear geometry (0°). As the interaction angle approaches 2.9°, the wave vector mismatch becomes clearly asymmetric, with the appearance of a region where the material dispersion is partially compensated for by the interaction angle. Since the self-phase modulated pump pulses have very large bandwidths, the geometry that results in phase-matching of the *n*th-order beam can give rise to multiple emission for all orders below *n*, provided that the maximum phase-mismatch for each order is less than approximately three times the bandwidth of the ultrafast pump pulses (~1350 cm^{-1}). We see that eleven upconverted beams and two downconverted beams all fall inside this region, and so the phase-matching conditions can be met in a quite relaxed way for several orders of the cascaded processes, within the frequency range allowed by the bandwidths, and in perfect agreement which the observed asymmetry between frequency upconversion and downconversion processes. Notice that, according to this criterion, a first frequency-upshifted beam should be generated even for a collinear interaction geometry (dashed curve of Figure 6). This is indeed the case, which has been confirmed by measuring the spectra of the resulting collinear and overlapping beams (with proper attenuation of the pump and signal pulses). Experimentally, the interaction angle is adjusted for maximum overall intensity and number of orders in the projected multicolored pattern. We expect the efficiency of a particular process to be determined by the magnitude of the nonlinear susceptibilities, the strength of the input fields, the phase-mismatch of any intermediate process, and ultimately absorption for any of the fields. Moreover, we see that the optimum yield does not necessarily occur under conditions that minimize the phase-mismatch for the final step in the process. Cascading will take place in the presence of some residual phase-mismatch in either the intermediate steps or the final step. The low degree of residual phase-mismatch also allows cascaded third-order processes to dominate over any high-order direct processes since cascaded processes will be favored if the phase-mismatch of the intermediate steps is sufficiently low. Storage of energy in the intermediate field then determines the conversion efficiency, similar to the case of an intermediate resonance. Although direct higher-order processes cannot be completely ruled out, they do not appear to take part in the observed phenomenon, as the detailed numerical simulations presented in the next section also demonstrate.

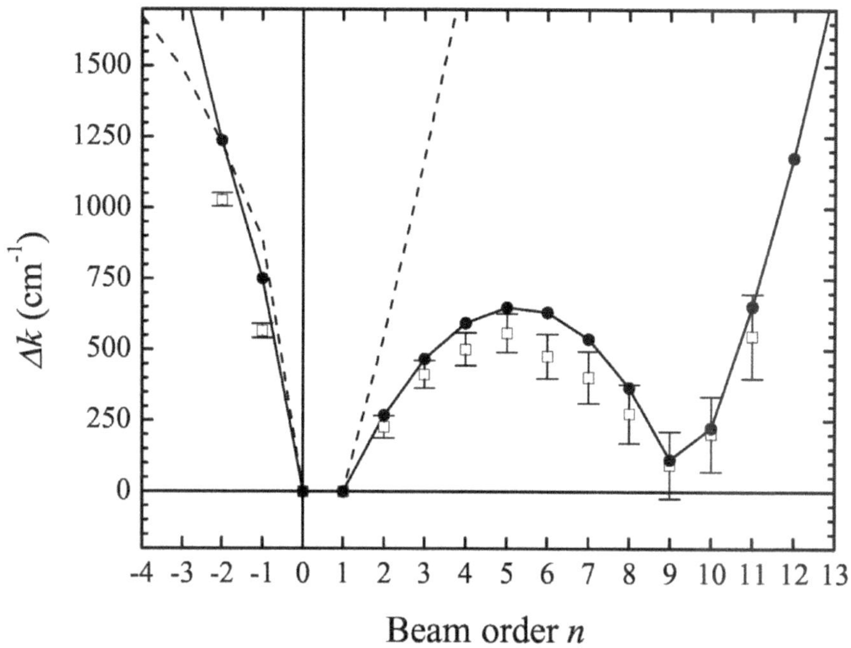

Figure 6: Wave-vector mismatch of the beams in the pattern: open squares, measured points; filled circles, calculation for the cascaded processes; dashed curve, calculation assuming collinear geometry (θ = 0°). Adapted from [11].

The experiment and the simple description based on momentum and energy conservation laws already show that the process is of universal character and will occur in any material with a third order susceptibility. Hence, all materials can show CFWM processes. Indeed CFWM processes was shortly thereafter shown to occur in gases [13] and later found to occur in other materials such as sapphire [15] and in crystals [16] or using two color beams from a fundamental laser and the output of a hollow fiber [17], which also demonstrated the intrinsic tunability of the process. Highly nondegenerate CFWM was also shown to occur in an optical parametric oscillator via cascaded $\chi^{(2)}$ processes giving rise to an effective third-order nonlinearity [18]. Recently, almost octave-spanning spectra were produced by CFWM of two synchronized ps lasers in optical microfibers [19]. Tunable CFWM has also been obtained by crossing two chirped pulses with the same central wavelength and variable delay in glass [20].

KNOWING THE FIELDS: THE COMPLETE THEORETICAL MODEL

Although the previous reasonings based on energy and momentum conservation allow us to understand the generation of multicolored light pulses departing from two-color pumps by cascaded FWM processes, enabling the estimation of the frequencies and emission angles of the generated beams, no predictions can be made for example regarding the energy carried by each color, nor the emission bandwidth and phase of the generated fields (including the time at which each beam is generated). Hence, a more complete model was required, which was developed based on the Slowly Varying Envelope Approximation (SVEA) for the propagation of the fields in 2.5 D (propagation along a plane while assuming cylindrical symmetry in each beam) that accurately reproduces the experimental results. The use of two spatial dimensions is important to account for angular phase-matching and actual beam overlapping region.

The wave equation for an electric field E(r,t) propagating in a nonlinear medium with a polarization comprising a linear and a nonlinear term as given by $P(r,t) = P_L(r,t) + P_{NL}(r,t)$, is

$$\nabla^2 E(r,t) - \frac{1}{c^2}\frac{\partial^2 E(r,t)}{\partial t^2} = \mu_0 \frac{\partial^2 P(r,t)}{\partial t^2} \tag{5}$$

with c the velocity of light and μ_0 the vacuum permeability, where we have written the electric field as a scalar quantity, which is correct for linearly polarized fields as is the case of our experiments. Considering that the electric field is well represented under the SVEA approximation, *i.e.*, $E(r,t)=1/2A(r,t)e^{i(k0z-w0t)}+c.c.$, changing the equation to the reference frame which moves with the group velocity of the pulse ($k_1=1/v_g$, where v_g is the group velocity of the pulse at frequency ω_0), separating the polarization into its linear and nonlinear parts and Fourier transforming to the frequency domain, the wave equation becomes

$$\frac{\partial^2 A(r,\omega-\omega_0)}{\partial z^2}+2i[k_0+k_1(\omega-\omega_0)]\frac{\partial A(r,\omega-\omega_0)}{\partial z}+ [k^2(\omega)-{k_0}^2-2k_0k_1(\omega-\omega_0)-{k_1}^2(\omega-\omega_0)^2+{\nabla_\perp}^2]A(r,\omega-\omega_0)= -2\mu_0\omega^2 P_{NL}(\omega)e^{-ik_0z} \quad (6)$$

This equation describes both linear and nonlinear propagation. It includes dispersion and diffraction, as well as a particular form of P_{NL} which includes the instantaneous non-resonant Kerr effect and self-steepening. Note that no expansion of the wave vector k was performed and therefore higher order terms of the dispersion are intrinsically included. To solve the equation, it was separated in its linear and nonlinear parts. For P_{NL}=0, the linear part of Equation (6) in the paraxial approximation ($\frac{\partial^2 A(r,\omega-\omega_0)}{\partial z^2}=0$) is given by

$$2i(k_0+k_1(\omega-\omega_0))\frac{\partial A(r,\omega-\omega_0)}{\partial z}+ [k^2(\omega)-{k_0}^2-2k_0k_1(\omega-\omega_0)-{k_1}^2(\omega-\omega_0)^2+{\nabla_\perp}^2]A(r,\omega-\omega_0)=0 \quad (7)$$

This linear part carries information about dispersion and diffraction and can be solved in the frequency domain by using a (2,2) Padé approximant for wide angle propagation.

The nonlinear part in the paraxial approximation is given by

$$2ik_0\frac{\partial A(r,\omega-\omega_0)}{\partial z}=-2\mu_0\omega^2 P_{NL}(\omega)e^{-ik_0z} \quad (8)$$

The nonlinear polarization P_{NL} can be written in terms of a slowly varying envelope p as $P_{NL}(r,t)=1/2p(r,t)e^{i(k_0z-\omega_0t)}+c.c.$, with $p(r,t)=3/8\epsilon_0\chi^{(3)}|A(r,t)|^2A(r,t)$ and is better described in the time domain. By doing the inverse Fourier transform, we get

$$\frac{\partial A(r,t)}{\partial z}=\frac{3\chi^{(3)}}{8ik_0c^2}\left[-{\omega_0}^2|A|^2A-2i\omega_0\frac{\partial(|A|^2A)}{\partial t}+\frac{\partial^2(|A|^2A)}{\partial t^2}\right] \quad (9)$$

where the first term on the right hand side accounts for self-phase modulation effects and the second for self-steepening. The accurate evaluation of the self-steepening term is essential, because the spectral span of the generated CFWM pulses is of the same order as the central frequencies of the pumps, ω_0 and ω_1. It should be pointed out that no additional delayed response terms were required in order to faithfully reproduce the experimental results. This equation can be solved by a second-order Runge-Kutta method, which proved sufficiently accurate.

Equations (8) and (9) are very powerful in the context of the present work and can exhibit a very rich behavior, since they constitute a much better approach to the process than the simple algebraic description of the previous section. Let us consider the geometry of Figure 7 with spatial gaussian profiles $A(r,t)$ or $A(r,\omega-\omega_0)$ and transform limited pump pulses, for the two beams used in the first experiment: an orange beam (ω_0, λ_0 = 618 nm, 80 fs) and a green beam (ω_1, λ_1 = 561 nm, 40 fs), interacting at an internal angle of 1.93° (corresponding to an external angle of 2.9°) in a BK7 glass slide with a relative delay of 0 fs between them.

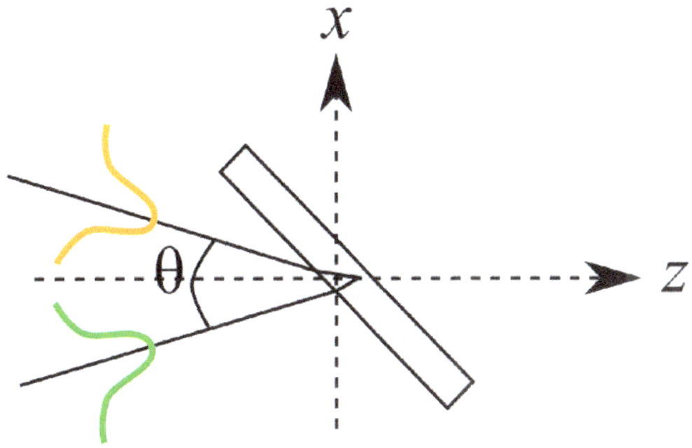

Figure 7: Interaction geometry.

The resulting simulated CFWM spectrum is shown in Figure 8 (black curve). We see that several upshifted orders are generated, which are broadband and have well-defined central wavelengths. Also, the total spectral intensity has an envelope which decays at around 400 nm and then increases again, in agreement with the experimental observations shown in the next section. If the interaction occurs at a slightly different angle, which is a feasible situation in an experimental setup, the efficiency and the generated central wavelengths change, as shown in Figure 8 (collinear pumps don't generate more than 1–2 CFWM orders since there is no phase-matching). For angles smaller than 1.93°, the generated orders don't reach so far into the UV, while for angles larger than 2° the overall efficiency drops rapidly (Figure 8a). For a fixed interaction angle, different time delays between the pump and the signal can be set, which also influences the central wavelength of the generated orders due to additional shifts caused by cross-phase modulation and the varying amount of temporal and spatial overlap between the two pump beams (Figure 8b). We see that for delays of ± 20 fs (typical reading precision of a manual delay line) around the optimum zero delay point, the overall efficiency drops by only a small amount (more noticeable for higher orders) since such delays are still well within the 40 and 80 fs FWHM pump pulse durations. For delays larger than ± 40 fs the efficiency starts to drop significantly, as illustrated in Figure 8b for the +40 fs case. The model also offers the possibility of spatially shifting the pumps at the entrance of the slide. Figure 8c shows the dependence of the generated field intensity for different spatial overlaps of the pump beams. For ± 50 µm shifts in the overlap of the pump beams (each 100 µm in diameter) at the entrance plane of the medium, the overall efficiency decreases roughly by a factor of 2 with respect to the perfect overlap (0 µm) case. This is again more noticeable for the higher orders in the ultraviolet and near-infrared regions.

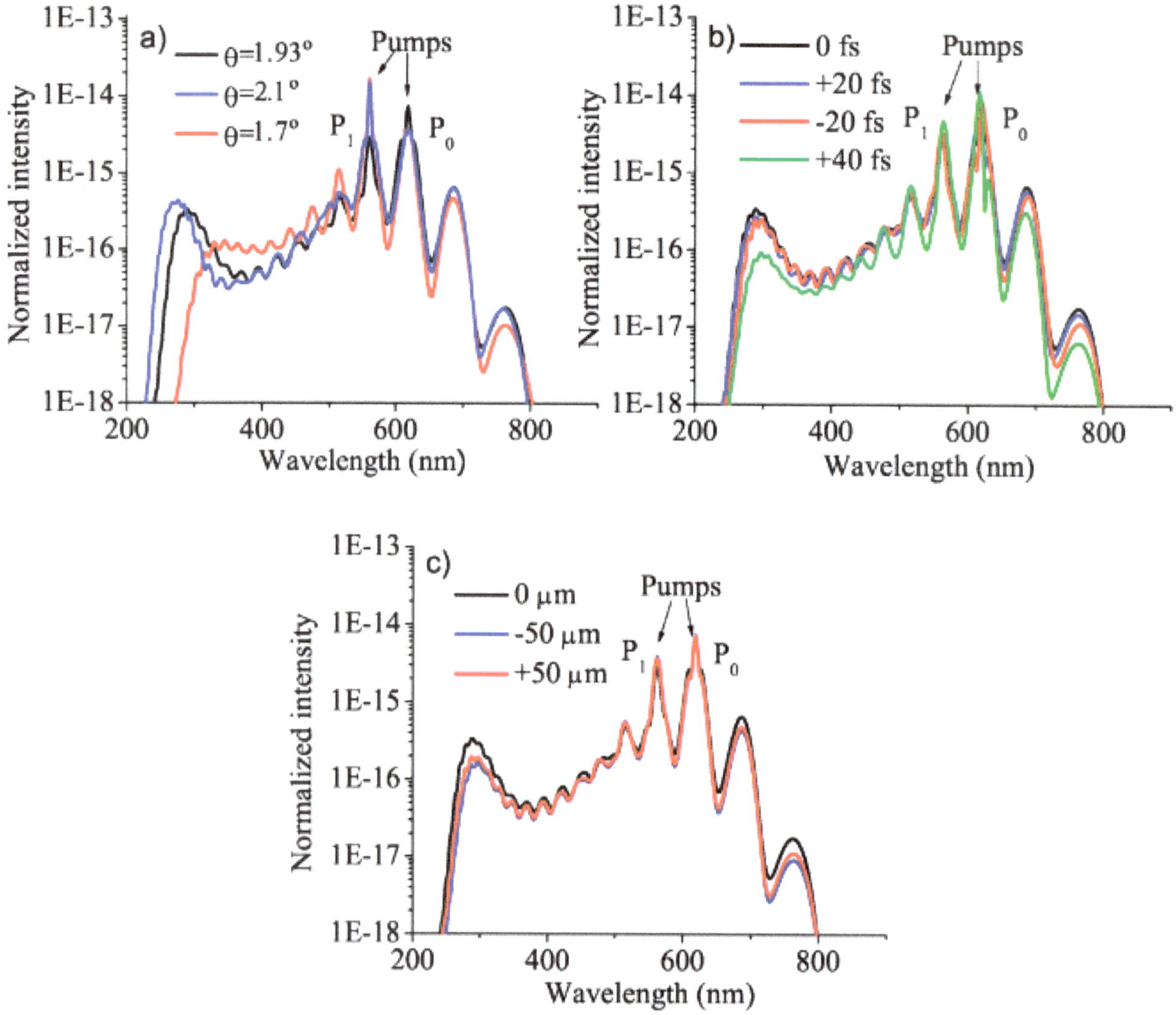

Figure 8: (**a**) Generated spectra for different interaction angles; (**b**) Generated spectra for 1.93° and different relative time delays between pumps (positive delay P_1 arrives before P_0, negative delay P_0 arrives before P_1); (**c**) Generated spectra for $\theta = 1.93°$, 0 fs relative time delay and different entrance *x*-positions at the slide.

Hence it is possible to tune the generated spectra by varying the interaction angle and/or the delay, in very good agreement with experimental observations. This means that the terms considered in the linear and nonlinear sub-equations are sufficient to faithfully reproduce the experimental results, as seen further below. The experiment was repeated in a 150 μm fused silica slide to obtain more upconverted orders in the UV and the internal interaction angle was 1.57°. In this situation, multiple CFWM orders were generated up to the 21st upshifted order at 209 nm (made visible with the help of a fluorescent card) and were registered with a broadband fiber-coupled spectrometer, but due to spectral limitations the measurement reached only down to 250 nm. The temporal characteristics of the beams were measured by polarization-gating FROG (PG-XFROG). More details about these experiments are given in the next section.

The simulations provide the time dependent field distribution at the end of the glass slide as a function of the transverse coordinate x and frequency ω. The corresponding total spectrum (obtained by integrating along the transverse direction) for an interaction angle of 1.57° is given in Figure 9 and clearly shows the asymmetric generation of up and downconverted orders. The efficiency of the generated frequency upshifted pulses follows the same pattern as the experiment, although it has a sharper cutoff after the 12th order that depends on the interaction angle and pulse energy. This discrepancy is possibly due to experimental intensity calibration errors, which are larger at the spectral edges.

A look at the fields inside the slide shows that all orders are practically generated within the first 60 μm of material. The spatial resolution of the model allows determining at which angle each of the orders is generated. This can be seen in the $\theta - \lambda$ plot in Figure 10.

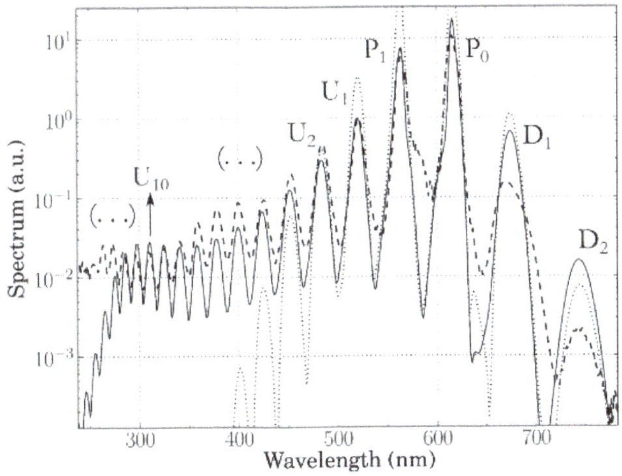

Figure 9. Spectrum of the multiple CFWM orders generated in fused silica spanning over 1.5 octaves (the two pump frequencies correspond to the two larger peaks). Dashed line: Experimental results (see [21]), Solid line: Numerical simulation, Dotted line: Numerical simulation for collinear pumps [22].

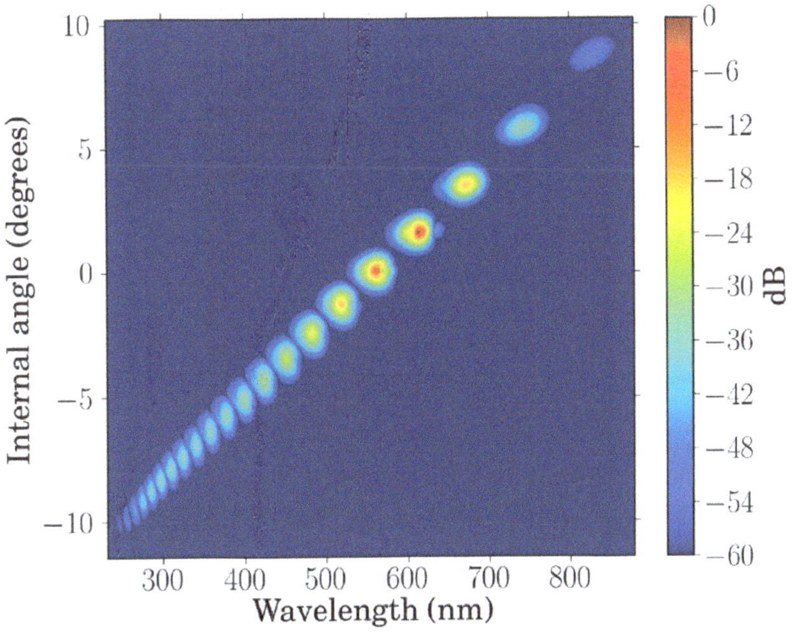

Figure 10: Simulated $\theta - \lambda$ spectrum of the CFWM beams, clipped at 10^{-6} of the maximum [22].

The temporal characteristics of the pulses can be also calculated, since the model gives us complete temporal and spectral information of the field. A given CFWM order can be spatially selected since they are emitted at different angles and we can simulate its interaction with another beam, as done in experiments for measuring the temporal duration. For instance, one can make a given generated order interact with the beam at ω_0 on a second thin slide of fused silica for different relative temporal delays and hence numerically obtain polarization-gating cross-correlated frequency resolved optical gating (PG-XFROG) traces. Figure 11 shows the measured and simulated PG-XFROG traces for the second frequency upshifted beam with central frequency ω_2 (top) and the first downshifted beam with central frequency ω_{-1} (bottom). I, the simulation results are in good agreement with the measurements regarding the broad bandwidth and short duration of the newly generated low-order pulses (which are shorter than the pumps). There are however differences in the central frequency and in the chirp of the pulses. The former can be due to several factors, from changes in the central frequency of the pumps that can occur between experiments, to differences in the interaction angle as well as in the relative pulse delay, all of which can result in frequency shifts, as described previously (see, e.g., Figure 8). Regarding the latter, the simulated pulses exhibit some positive chirp due to the nonlinear phase imposed by the pumps. This is clearer in the right plot of Figure 12, which shows the calculated PG-XFROG traces for the synthesized field of Figure 13. Another factor that directly contributes to the chirp of the newly generated pulses is the initial chirp of the pump pulses. In the simulations the pumps are assumed to be Fourier-limited, whereas in the experiments there can be slight deviations to this condition.

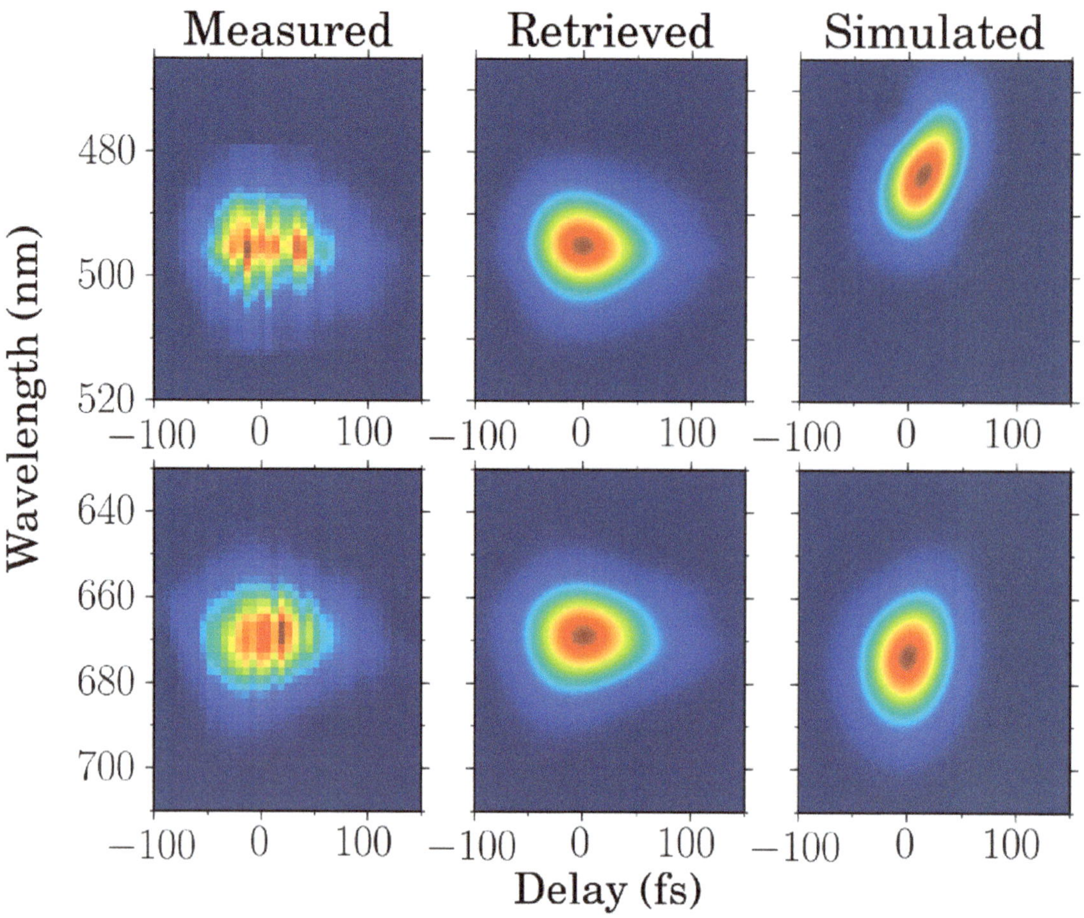

Figure 11: Experimental (measured and retrieved) and simulated polarization-gating FROG (PG-XFROG) traces of the second frequency-upshifted pulse (top row) and first downshifted pulse (bottom row). The retrieved pulse durations are 30.7 and 40.2 fs for the second upshift and first downshift respectively (FROG error was 0.014 and 0.008 for a 128 × 128 grid) [22].

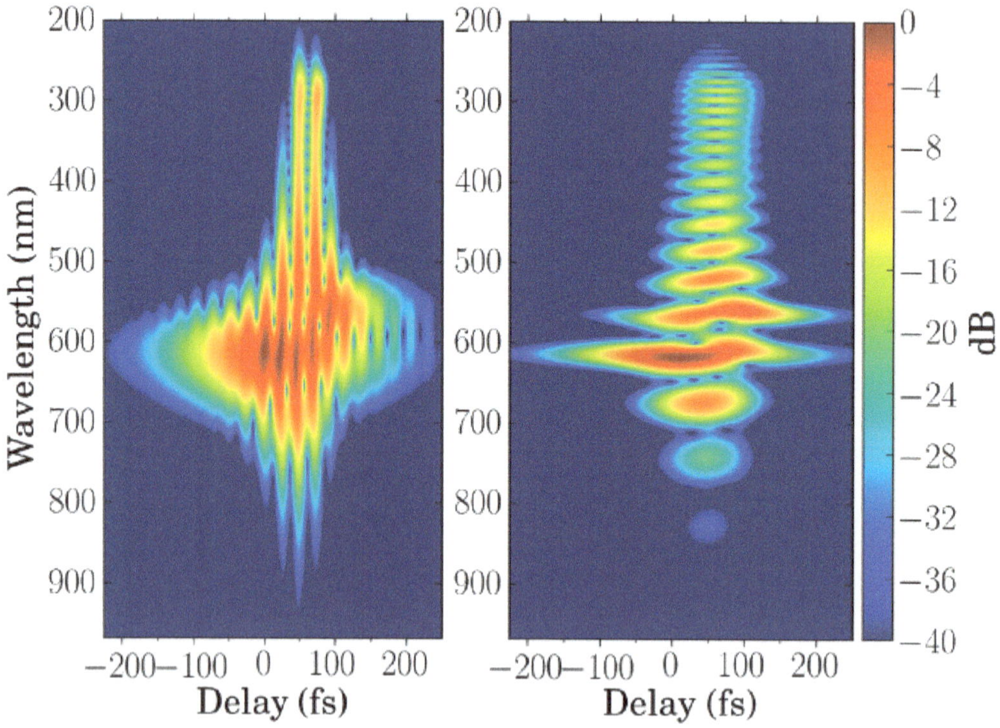

Figure 12: Simulated PG-XFROG, clipped at 10^{-4} of the maximum, for the recombined and focused CFWM field, assuming 20 fs (**left**) and 80 fs (**right**) gate pulses [22].

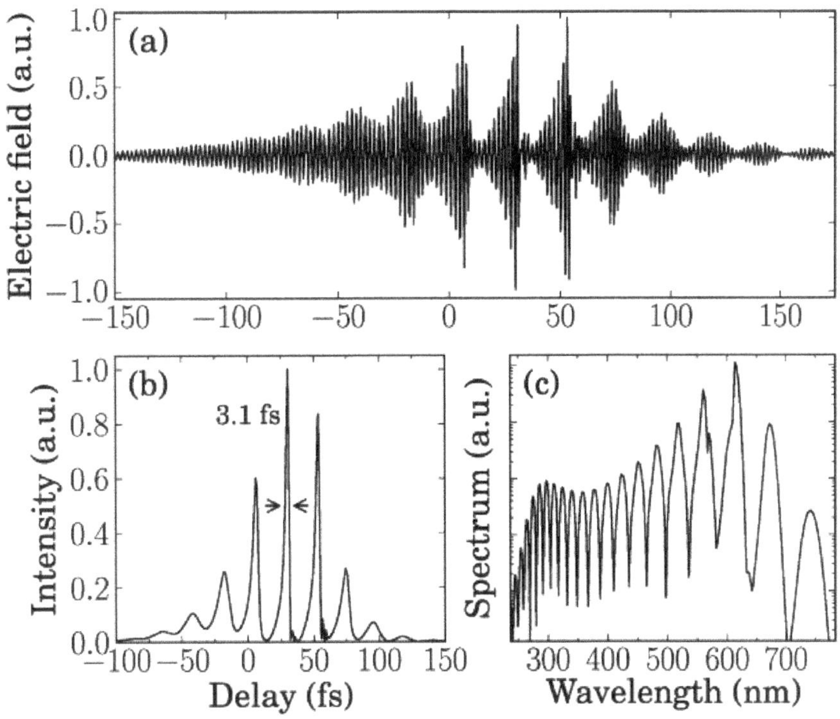

Figure 13: Recombined and focused CFWM pulses. (**a**) Electric field as a function of the delay (in fs); (**b**) Intensity; (**c**) Power spectrum (logarithmic scale) [22].

Other kinds of temporal processing are also possible. For example the total electric field can be calculated by synchronizing and spatially superimposing all the CFWM orders. Since after passing the slide the different orders have acquired a slightly different group delays, synchronization can be achieved by subtracting the linear phase. This is equivalent to focusing all the orders together in a point, which would be an experimental approach for having the total field in the same region. The total electric field obtained by this numerical procedure can be seen in Figure 12a, along with the field intensity in Figure 12b and its power spectrum in Figure 12c. We see that in the time domain we obtain a train of pulses separated by the pump beat period (23 fs) with the central peak having 3.1 fs in duration.

A PG-XFROG trace of this total field can also be simulated. The gate pulse can be of course arbitrarily chosen, but it looks reasonable and informative to use one of the pump pulses as gating pulse, as done in the experiments, namely the 80 fs pulse at ω_0. The PG-XFROG trace for the complete focused set of CFWM orders is shown in Figure 13 where the different sub-traces for the individual orders can be distinguished (on the right). For shorter gate pulses spectral resolution is lost but temporal resolution is improved revealing the fine temporal structure (pulse train) obtained in this case (on the left).

TOWARDS SINGLE-CYCLE PULSE SYNTHESIS

The described set of cascaded FWM orders will be a coherent broadband source provided the different orders can be spatially and temporally overlapped. Coherent spectra that are sufficiently broad can be used for generating few- and single-cycle pulses if properly compressed. A well-known technique for generating high energy few-cycle pulses is the temporal compression of the broadband supercontinuum produced by self-phase modulation (SPM) in e.g., gas-filled hollow core fibers [23] (discrete spectra, composed of evenly spaced near-monochromatic waves can also lead to the synthesis of a train of single-cycle optical pulses, as demonstrated using molecular modulation in a gas driven by two independent nanosecond lasers [24]). Moreover, the central frequency of the total CFWM spectrum and the separation between sidebands can be freely adjusted by tuning the pump frequencies, so in principle it should always be possible to obtain the commensurate sidebands required to synthesize a train of identical pulses. Two frequencies are commesurate if $m\omega_0 = n\omega_1$, with m, n integers. This also means that each frequency is an exact multiple of the frequency spacing $\omega_m = \omega_1 - \omega_0$.

Let us take a deeper insight into pulse synthesis by addition of a finite number of pulses with different central wavelengths. The pulses must have fixed relative phases in order to produce a stable coherent superposition of fields giving rise to an ultrashort pulse or to a train of ultrashort pulses. We also found this result in the previous section (Figure 12), where the coherent sum and focusing of the CFWM orders generated by numerically solving the equations of our model, resulted in a train of ultrashort laser pulses. For the CFWM pulses this can be understood using a simplified model based on the coupled amplitude equations between the pumps and the generated sidebands.

Using the wave equation in the slowly-varying envelope approximation, for two pump fields with complex amplitudes $\tilde{E}_0 = E_0 e^{i\phi_0}$ and $\tilde{E}_1 = E_1 e^{i\phi_1}$, frequencies ω_0 and ω_1, respectively, assuming a cubic nonlinearity, neglecting dispersion and pump depletion, and assuming perfect phase-matching and constant coupling coefficients, the solution for the complex amplitude $\tilde{E}_n$ of the field with frequency ω_n at the exit of a medium with length L can be written as [21,25]

$$\tilde{E}_n(s) = i^n e^{in\delta} \tilde{E}_0 J_n(s) - i^{n+1} e^{i(n-1)\delta} \tilde{E}_1 J_{n-1}(s) \quad (10)$$

where s=γL= is the nonlinear phase shift acquired in the medium, $\delta=\varphi_1-\varphi_0$ is the initial phase difference between pumps, J_n are the nth order Bessel functions of the first kind, and the nonlinear coefficient is given by $\gamma = 3\omega_0\chi^{(3)} E_0 E_1/(2n_0 c)$, with n_0 the linear refractive index at frequency ω_0 and c the speed of light in vacuum. Equation (1) predicts well-defined phase relationships between the several CFWM orders, thus establishing mutual coherence between the beams. The total field in the time domain is then

$$E(t) = \mathrm{Re}\left[\sum_n \tilde{E}_n(s) e^{-i(\omega_0 + n\omega_m)t}\right] \quad (11)$$

From Equation (11) and using the Jacobi-Anger identity $\sum_n J_n(z)\exp(in\theta)i^n = \exp(iz\cos\theta)$, we obtain

$$E(t) = Re\left[e^{i\phi_0}(E_0 e^{-i\omega_0 t} + E_1 e^{i\delta} e^{-i\omega_1 t}) \times e^{-is\cos(\omega_m t-\delta)}\right] \quad (12)$$

This result is similar to the frequency-modulation (FM) solution by Harris [24] for coherent addition of Raman sidebands in the approximation of negligible dispersion and limited modulation bandwidth (on its hand similar to the solution previously obtained by Lichtman *et al.* [6] for nondegenerate FWM of 100 ns pulses - for which dispersion was practically negligible - propagating in 400 m long optical fibers), with the important difference that in CFWM the synthesized pulse envelope strongly depends on the initial phase difference δ. In molecular modulation, the detuning between the driving frequency and the Raman transition dictates the preparation of a phased or an anti-phased molecular state, the latter resulting in a negative effective nonlinear coefficient γ for which the synthesized pulses could have an approximately negative chirp and hence could be further compressed by simple propagation in a normally dispersive medium [24]. In the present case of CFWM, the sign of γ is independent of the pump field (and usually positive for most optically transparent $\chi^{(3)}$ media), but the chirp of the synthesized pulses can nevertheless be controlled by adjusting δ alone. From Equation (12) we see that φ_0 only affects the carrier envelope phase (CEP) for the pulse train, defined with respect to the two-color beat envelope, whereas δ is associated with net temporal shifts and also determines the pulse chirp, as given by the oscillating nonlinear phase term. In particular, when δ = ±π/2, the synthesized pulses can have a negative chirp. In practice, a given phase difference can be introduced by slightly adjusting the delay between the two (phase-locked) pump pulses without significantly affecting their temporal overlap. Also, if the pump frequencies are commensurate, this will result in the generation of identical pulses that are CEP-stabilized within each train (although their CEP may change from shot-to-shot), since even though each CFWM step is not self-CEP-stabilized, the total recombined field will be. The corresponding pulse envelopes will also be identical from shot-to-shot, provided that the relative phase difference δ remains constant, which is a far less stringent requirement than the need of CEP stabilized pump pulses.

To demonstrate the feasibility of this approach for the generation of single cycle pulses, the basic set-up of Figure 2 was improved by using fused-silica instead of BK7 to generate more orders in the UV, and complemented with additional optics for pulse synthesis and measurement. This set-up is reproduced in Figure 14-I. As in the previous case, two-horizontally polarized visible femtosecond pulses from a dye laser-amplifier at 10 Hz were used as pumps: an orange pump with λ_0 = 615 nm, ~80 fs duration, 2 mJ energy and a green pump with λ_1 = 569 nm, ~60 fs duration and 200 μJ energy. Special care was taken in using Glan-Taylor polarizers (GTPs, 100,000:1 extinction ratio) to have perfectly polarized beams, which is important not only for the generation, but for the subsequent temporal measurements. In this laser system, the green pump beam is directly (optically) derived from the orange pump beam: a small portion of the orange beam is first used to generate supercontinuum in a cell filled with deuterated water, and then the green portion of this supercontinuum is amplified using green laser dyes. Even though supercontinuum generation and laser action are coherent processes, the relative phase between the two pulses is not actively locked and can fluctuate due to normal thermal and mechanical perturbations and drifts in the system. The pulses are commensurate to within 0.2% (well within their ~5 nm bandwidths), hence we can expect the resulting synthesized pulses to have the same CEP within the pulse train.

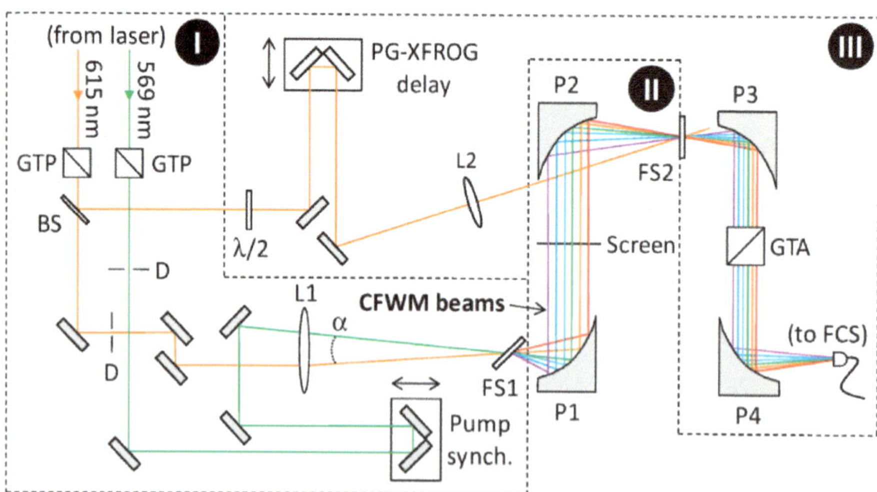

Figure 14: Experimental setup: (I) generation of CFWM pulses; (II) pulse recombination and synthesis; (III) pulse characterization by PG-XFROG (see text for details) [21].

The orange and green pulses are synchronized at the entrance plane of the fused silica slide FS1, and have energies of 32 and 38 μJ, respectively, near-transform-limited durations (measured using SHG-FROG), beam radii of approximately 80 and 100 μm, respectively, and similar intensities on the order of 2×10^{12} W/cm^2 as they cross on the slide at an angle of $\approx 3°$. Under these conditions, a fan of 20 upconverted CFWM orders was generated up to 209 nm, as shown in Figure 15a. The total measured energy in the cascaded beams (excluding the pumps, *i.e.*, orders 0 and 1) is > 6 μJ, so approximately 10% of the incident energy is transferred to the newly generated frequencies.

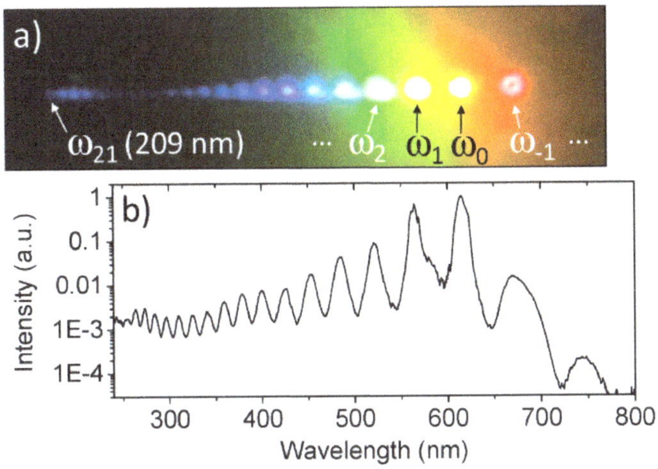

Figure 15: (**a**) Direct (no filtering) image of the fan of multicolored CFWM pulses as seen projected on a phosphor-coated paper screen (the dark arrows denote the pump beams); (**b**) Corresponding two-octave spectrum measured at the focal plane of mirror P2. [21].

To perform pulse synthesis all orders were collimated and recombined in a single 100 μm white light spot using two λ/8 Aluminum-coated off-axis parabolic mirrors P1 (f = 2.54 cm) and P2 (f = 5.08 cm) (Figure 14-II). A perforated screen was placed between these two mirrors to transmit only the central portions of the pumps and thus have a more balanced spectrum, but no additional amplitude filtering was performed. The total spectrum at the focal spot was measured with an intensity calibrated UV-NIR (200–1100 nm) spectrometer, but only 15 frequency upconverted orders could be detected due to the limited bandwidth of the mirrors, as shown in Figure 15b. All orders have been generated under simultaneous phase-matching conditions with a fast electronic nonlinearity, so they are phase-locked when exiting the first fused silica slide FS1. Additional propagation in air only adds negligible second-order dispersion and hence we expect a train of few-cycle pulses to be synthesized at the focus of P2.

Temporal characterization was done with polarization gating XFROG (PG-XFROG) using an 80 fs pulse as the gate (Figure 14-III). This gating pulse was obtained from the main orange beam with a 50/50 beamsplitter (BS) and its polarization was rotated 45° with a half-wave plate. This pulse induces birefringence in another fused silica plate (FS2) placed at the focal plane of mirror P2, where the pulse synthesis using all CFWM orders is taking place. A second Glan-Thompson polarizer was used as analyzer (GTA) crossed with the original GTPs, which required collimating the beams with a third parabolic mirror P3 (f = 2.54 cm). Finally the total spectrum of the gated pulses was measured by focusing the output of the GTA with mirror P4 (f = 5.08 cm) onto an optical fiber with a large 400-μm core coupled to a spectrometer (FCS). The PG-XFROG trace obtained by registering the spectrum of the gated pulses as a function of the delay of the gating pulse can be seen in Figure 16a, where up to 13 orders could be simultaneously gated, so they were synchronized, and also practically have no chirp. Since the gate pulse was long compared to the expected pulse train structure, information about the temporal structure of the traces is lost and retrieval of the whole field with this trace is not reliable (see previous section where a simulated PG-XFROG trace using a 20 fs gate pulse showed the fine temporal structure). However, unambiguous retrieval of the individual orders is possible. Figure 16b,c show the measured and retrieved PG-XFROG traces of the first upconverted CFWM pulse and its associated temporal shape with 30.6 fs in duration. Pulse compression and synthesis of the generated CFWM beams have also been studied and reported by the T. Kobayashi (see, e.g., [26]) and A. H. Kung [14] research groups.

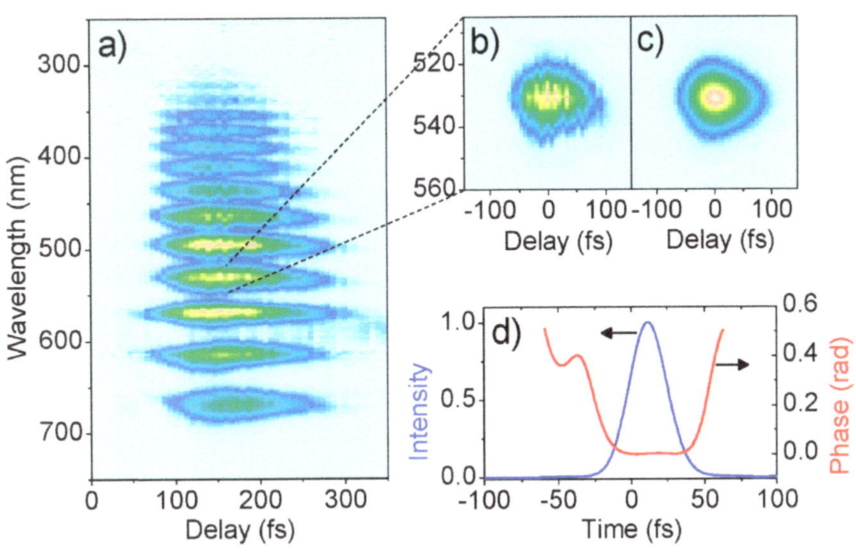

Figure 16: (**a**) Measured PG-XFROG trace of the synthesized field; (**b**) Measured and (**c**) retrieved PG-XFROG traces of the first frequency upconverted CFWM pulse, and (**d**) corresponding intensity and phase in the time domain. [21].

Adding the fields obtained for all measured orders while setting zero relative delay between each CFWM pulse, a synthesized field is obtained. This field has a main central transform-limited pulse with 2.2 fs duration and two smaller pulses, one at each side on the main pulse, separated 25 fs from the main pulse, which corresponds to the pump beat period. Figure 17 shows the synthesized field along with the pulse intensity and evidences the possibility of synthesizing single cycle pulses with this technique. The central pulse carries almost all the energy, which amounts to 5 μJ.

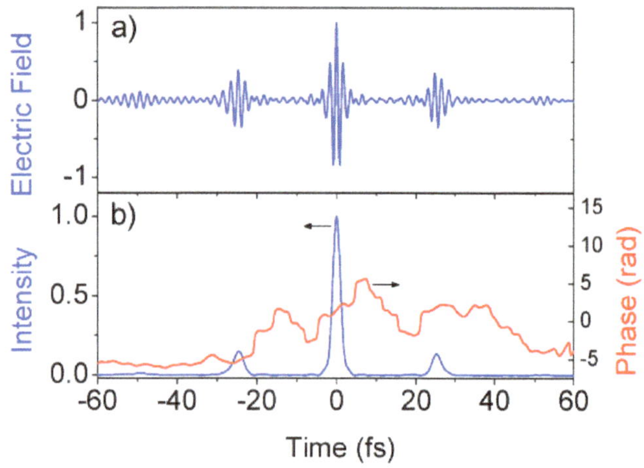

Figure 17: Total field (**a**) Normalized electric field and (**b**) intensity and phase of the pulses obtained by coherent addition of the retrieved electric fields of the 13 gated CFWM pulses. The main peak is a 1.3-cycle, transform-limited 2.2 fs pulse. [21].

GENERATION OF ULTRAVIOLET (UV), DEEP ULTRAVIOLET (DUV), VACUUM ULTRAVIOLET (VUV) AND HIGHER-ORDER HARMONICS BY CFWM FROM A STANDARD TITANIUM: SAPPHIRE LASER AMPLIFIER

Since the materials have usually resonances in the UV, the use of pumps in the visible part of the spectrum means working on the normal dispersion region. Angle phase-matching of upconverted orders in CFWM processes is possible thanks to dispersion. The shorter the wavelength of the resonance in the material the higher the number of upconverted orders that can be obtained, provided that the form of the dispersion curve is adequate. In the previous results we showed how cascaded upconverted orders comfortably reached the UV and DUV up to 209 nm departing from orange and green pump pulses from a dye laser-amplifier. A natural extension of the research in CFWM is trying to reach the UV, DUV, VUV or to generate high order harmonics with more common pump colors, available in many labs, such as the fundamental and the second-harmonic of a Titanium:Sapphire laser.

Ultrashort Pulses from the UV to the VUV by CWFM

Materials with large band gap such as alkali metal halide crystals are transparent in the VUV, so they are candidates for generating a cascade of upconverted orders via FWM up to the VUV, and we have tested this numerically [27]. The model developed in Section 3 assumes isotropic materials, and some alkali metal halides have cubic symmetry, so they are equivalent to an isotropic material and we used our model to test the feasibility of obtaining VUV pulses via CFWM in LiF using a pump beam at 400 nm and a signal beam at 800 nm. Equation (3) gives the internal crossing angle with would result in perfect phase-matching of the *n*th order, but for pump and signal beams as separated as a fundamental wavelength and its second harmonic, the internal angles for successive upconverted orders is too different for a cascade to be generated departing from only two beams, so the geometry of Figure 18 was devised, with a pump field at 400 nm and 3 signal beams at 800 nm, the latter satisfying the phase-matching condition for different consecutive orders.

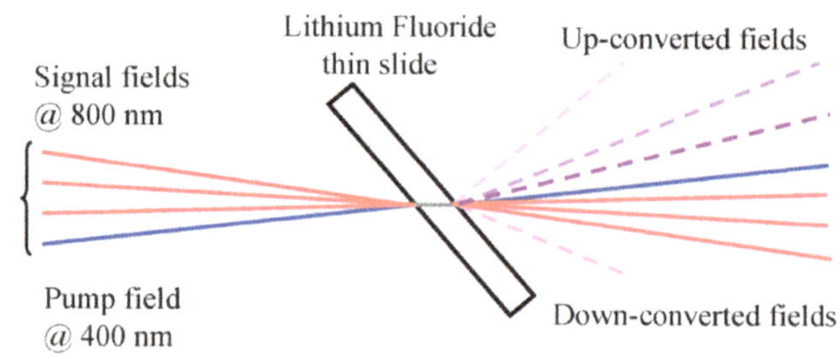

Figure 18: Scheme for multiple-beam CFWM with strong angular separation between pump and signal frequencies. [27].

We consider a pump beam at 400 nm and three signal beams at 800 nm, all transform-limited with 25 fs duration and focused to a 40 μm radius spot. All pump and signal beams have an irradiance of 5×10^{12} Wcm^{-2} (energy per pulse of ~ 6 μJ). The pump and the three signal beams form internal angles of 11.98, 15.65 and 20.06 degrees in a 300 μm thick LiF slide. The signal beams can be individually delayed to optimize the VUV generation. Figure 19 shows the CFWM spectrum generated for a time delay of 0 fs or 7 fs (for all signal beams) along with the spectrum generated by adding the individual spectra obtained by using the pump pulse and one of the signal beams at a time, also with a time delay of 7 fs.

In these conditions, we observe generation of CFWM beams up to the 6th harmonic of 800 nm, with good efficiency up to the 5th harmonic. Only a slight dependence on the time delay is seen and there is full evidence of the existence of CFWM processes, since the VUV spectrum obtained by addition of the individual spectra from the different pairs of pump and signal pulses is much weaker. As already seen in section 3 the generated orders are angularly separated and due to the fact that all three pump beams at 800 nm contribute to generate a specific order at a slightly different interaction angle and wavelength, because of the different phase-matching conditions, the generated orders have a complex spatial structure, although angular selection of a specific order can always be performed. Figure 20 shows the simulated $\theta - \lambda$ spectrum for 7 fs and 0 fs delay. A cleaner structure and also a better efficiency in the 4th and 5th harmonic is observed in the case of 7 fs delay compared to perfect synchronization (0 fs delay).

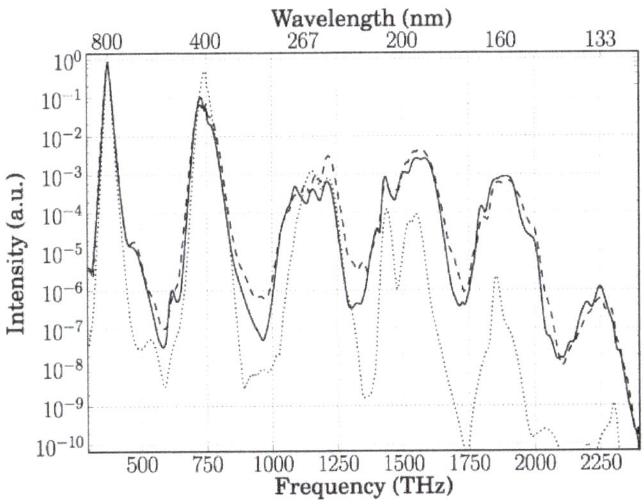

Figure 19: Simulated CFWM spectra in LiF. Solid (dashed) curve: spectrum generated by the interaction of all three signal pulses at 800 nm delayed by 7 fs (0 fs) with respect to the pump pulse at 400 nm – (see text); dotted curve: combined spectrum for three independent simulations, each optimizing an individual consecutive harmonic generated by each pair of noncollinear pump and 7 fs delayed signal pulses [27].

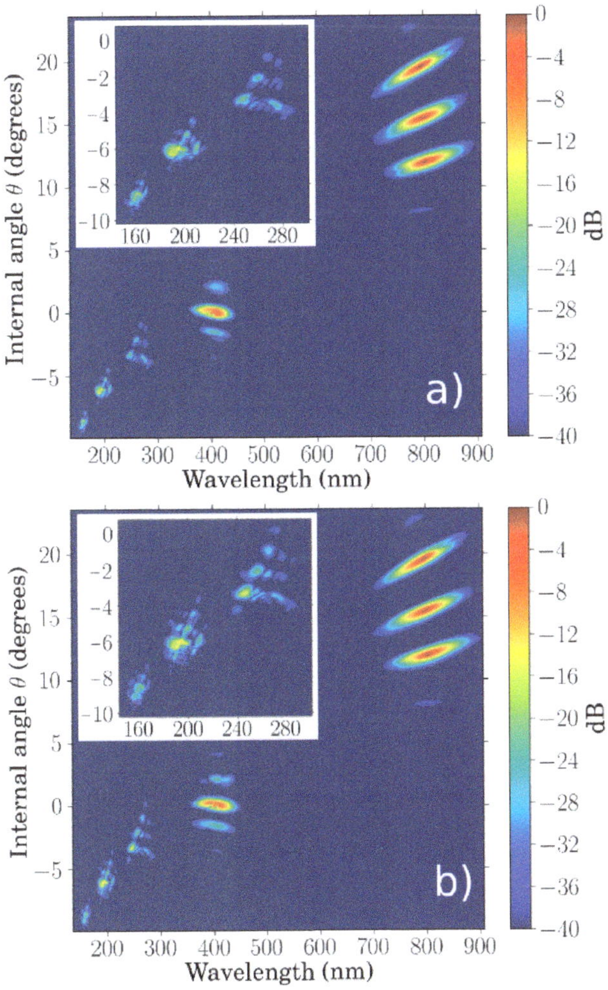

Figure 20: $\theta - \lambda$ spectrum for 7 fs (a) and 0 fs (b) relative delays between signal and pump pulses. Insets: magnified $\theta - \lambda$ spectra for the first three upconverted orders [27].

The temporal behavior is also as expected, with the generated orders having ultrashort durations of the order of the duration of the pump and signal pulses (31.9, 30.5, 30.1 and 20.4 fs for the third to the sixth harmonic respectively), due to the corresponding nonlinear spectral phases. This can be seen in Figure 21. The spectrum of all orders is broadband and assuming flat spectral phases the transform-limited durations are much shorter than those of the pump and signal pulses, namely 8.86, 8.1, 6.0 and 4.0 fs, respectively.

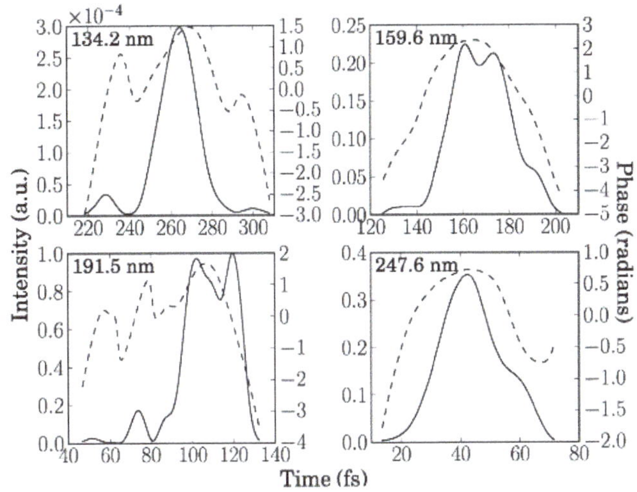

Figure 21: Temporal characteristics of the generated harmonics. Solid curve: intensity; dashed curve: phase [27].

DUV Ultrashort Pulse Generation by Highly Nondegenerate FWM

The observed fact that the generated orders are broadband deserves more attention, since the availability of sub-300 nm DUV ultrashort pulses with energies in the range of microjoules is important to study processes in photophysics, photochemistry and photobiology. Hence, although not a cascaded process, we also concentrated on studying the generation of the third harmonic of 800 nm via highly nondegenerate FWM [28]. Several techniques based on FWM in gaseous media have been employed by other groups, from third-order difference-frequency generation (DFG) of Ti:Sapphire laser pulses and their second-harmonic in hollow waveguides [3,29] and in filaments [30,31], to DFG of pre-chirped pulses in dual hollow-fiber systems [32]. Direct harmonic upconversion of few-cycle femtosecond pulses in gases [33,34] has also been performed. This last method has enabled the generation of the shortest (sub-3-fs) DUV pulses to date [35].

The setup used for the generation of broadband pulses at 266 nm by mixing two ultrashort pulses at 800 nm and 400 nm respectively in an isotropic solid is shown in Figure 22, along with the part for the temporal measurement. A Titanium: Sapphire amplifier provides the fundamental signal (or idler) pulses at 800 nm (ω) with 27 fs pulse duration, 1 mJ energy, horizontally polarized and at 1 kHz repetition rate. The pump pulses at 400 nm (2ω) are generated in a type-I 200 μm thick BBO crystal. To maximize the SHG, the chirp of the fundamental pulse was optimized and a refractive telescope reduced its beam size to 5 mm. A dichroic mirror separated the pump and signal beams and a half wave plate HWP was used to rotate the polarization plane of the signal beam by 90° so as to have the polarization parallel to the pump beam, which emerges perpendicularly polarized from the BBO crystal. Both pump and signal beams were focused with lenses to interact on a fused silica slide FS at an internal angle of 15.75° as given by Equation (3) (external angle θ = 23.2°) for the FWM process 3ω = 2 × (2ω) − ω. The slide was placed 2 cm before the focus of the lenses to avoid damage. Synchronization is achieved with a delay line in the signal branch of the setup and in the plane of the slide the signal and pump beams had 314 μJ and 191 μJ with estimated intensities of 2.6×10^{12} and 1.2×10^{12} W/cm^2, respectively. Under these conditions we obtained 5–6 μJ DUV pulses emitted at around 266 nm at an external angle of 7°.

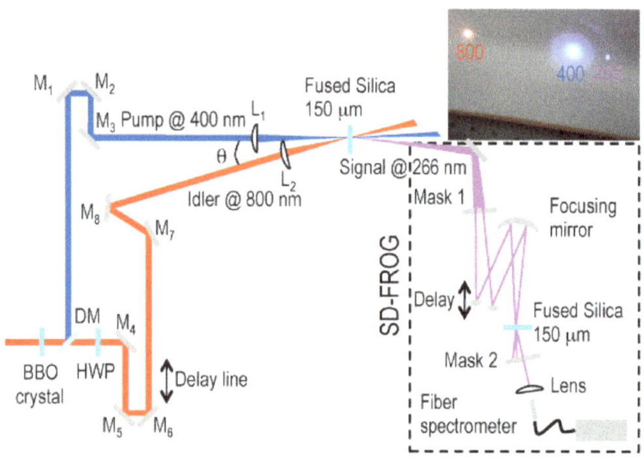

Figure 22: Experimental setup for ultrashort deep ultraviolet (DUV) pulse generation by highly nondegenerate four-wave mixing (FWM) and temporal characterization via SD-FROG. DM: Dichroic mirror, HWP: Half-wave plate; M1-M3: Aluminum mirrors; M4-M8: 45° dielectric mirrors. Inset: idler, pump and generated DUV beam, directly projected onto a white card [28].

Figure 23 shows the spectra registered with a fiber coupled spectrometer for different delays, defining the 0 fs delay for the situation in which higher energy pulses were obtained (6 μJ). Different broadband spectra were obtained for delays ranging from −99 fs to 99 fs and the dependence of the shape and central frequency can be explained by the cross-phase modulation between the pump and idler pulses. Integration of these spectra with reference to the energy measured for the one with 6 μJ (Figure 20, delay 0 fs) gives 1.5, 2.4, 3.85, 6.0, 4.75, 4.9 and 1.4 μJ for this series.

The spectra are broad enough (Figure 23 and Figure 24) to support ultrashort durations and the DUV (as well as the SHG pump pulses) were characterized by self-diffraction FROG (SD-FROG). The experimental details can be seen in Figure 22. Two holes with 1.6 mm diameter separated by 3.5 mm were drilled on a metallic mask and aligned so that both holes transmitted similar intensities from the central part of the DUV beam. A relative temporal delay was set between both pulses using d-shaped mirrors and a stepper motor stage (3.33 fs step). Both pulses were focused with an aluminum mirror (f = 200 mm) in a second fused silica slide were a self-diffraction beam was generated and later isolated to measure its spectrum in a fiber coupled spectrometer. Figure 25 shows the measured FROG traces and corresponding retrievals using standard Femtosoft FROG software, which gives 27.3 fs DUV pulses with a spectral width of 5.7 nm and a time-bandwidth-product of 0.67. The FROG measurement of the SHG pulses resulted in a duration of 48.6 fs.

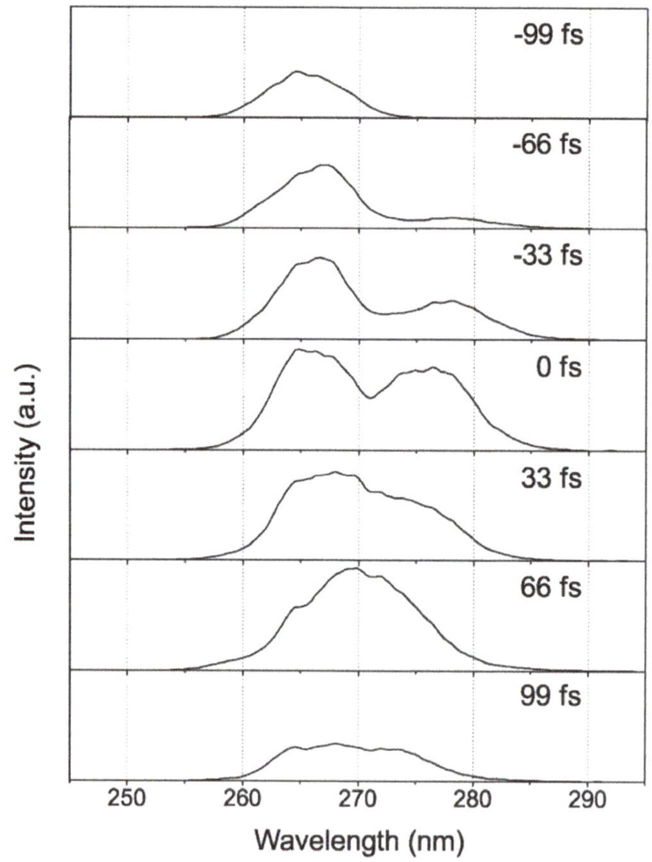

Figure 23: Experimental spectra of the generated DUV signal for different delays (separated by ~ 33 fs) between idler and pump pulses. Negative delay: pump arrives after idler, positive delay: pump arrives before idler [28].

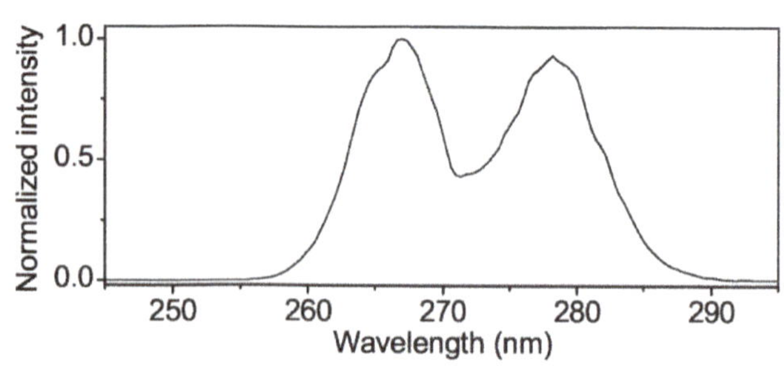

Figure 24: Broadest experimental spectrum obtained in our-set-up, capable of supporting sub-4-fs [28].

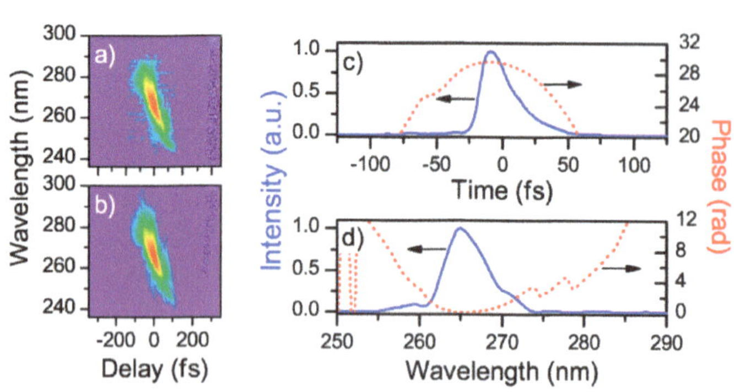

Figure 25: (**a**) Experimental and (**b**) retrieved SD-FROG traces of the DUV pulses (in log scale); (**c**) retrieved pulse in the time and (**d**) frequency domains [28].

The model described in Section 3 also predicts broadband generation of the third-harmonic of 800 nm at a central wavelength of 266 nm. Using the experimental parameters given in this section and for different internal angles θ = 15.0°, 15.7° (the angle of perfect phase-matching) and 16.4° different broadband spectra were obtained for different delays between pump and signal pulses. In Figure 26 (θ = 15.0°) we can see double band spectral structures which are enhanced in its blue portion for negative delays and in its redder portion for positive delays. The two situations are of course not symmetrical because of the different group velocities of the pulses, where the idler can catch up with the pump in situations of positive delay but will always come before the pump for negative delays. We also see that the different pulses have chirp and the more intense ones, generated at delays of −20 fs, 0 fs and 20 fs, have a clear positive chirp, mainly due to self- and cross-phase modulation. This positive chirp is in agreement with the experimental results of Figure 23, where the retrieved pulse shows also positive chirp. The calculated emission angle for the 266 nm pulse obtained for 0 fs relative delay between pump and idler was 7.3°, with an output energy efficiency of 3.6% with respect to the pump pulse and a pulse width of 19.7 fs (9 fs TL duration), all in reasonable agreement with the experimental results.

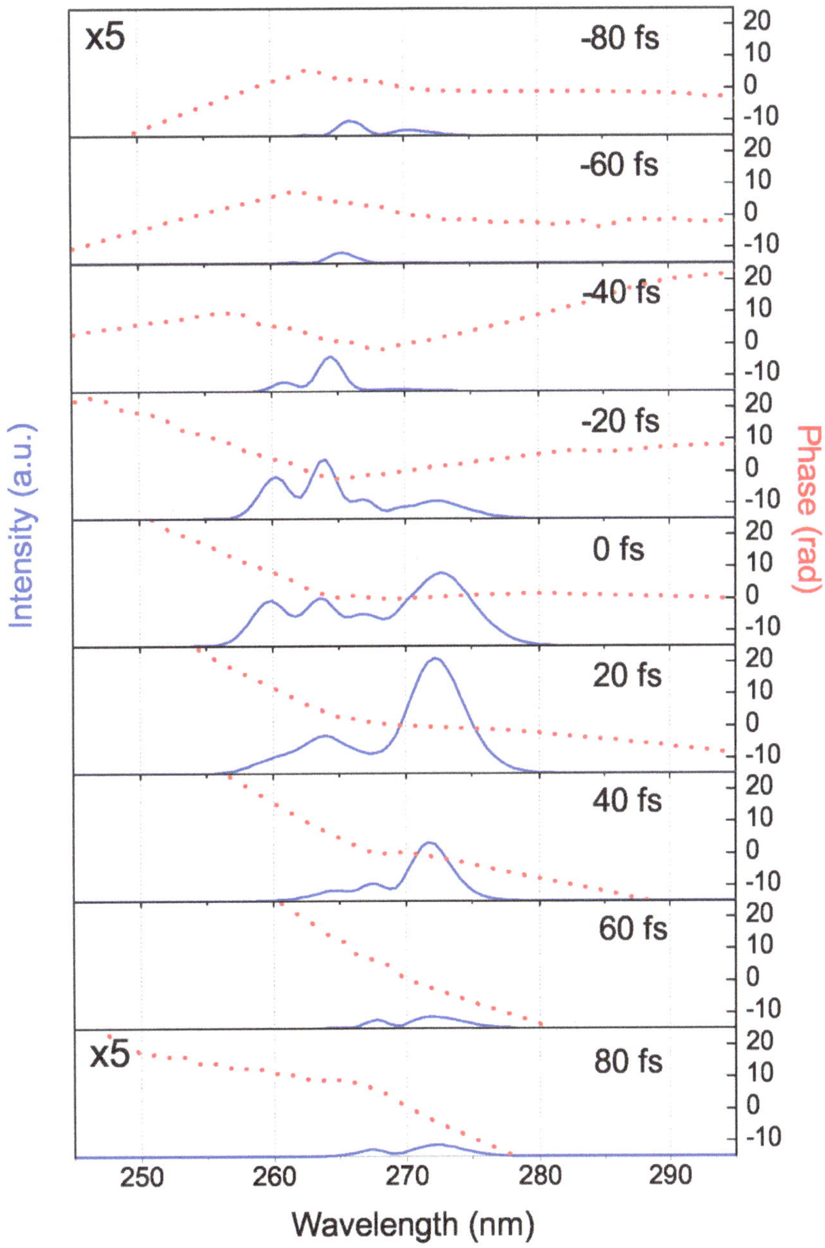

Figure 26: Simulated spectral intensity and phase of the generated DUV signal for θ =15.0° and different delays between idler and pump pulses (The intensities in the first and the last plots have been multiplied by a factor of 5). Negative delay: idler arrives before pump, positive delay: idler arrives after pump [28].

For other interaction angles (θ = 15.7° (Figure 27) and θ = 16.4° (Figure 28)) higher efficiencies are obtained (12.5% and 6.9% respectively) but with narrower bandwidths. The situation with θ = 15.0° seems to better represent the experimental one. We see that the change in interaction angle from 15 to 16.4 ° also produces a significant shift/tuning of the central frequency of the generated DUV pulses. This is due to the large angular sensitivity of the geometric phase matching condition for the highly nondegenerate FWM process involving ω and 2ω beams crossing at a very large angle.

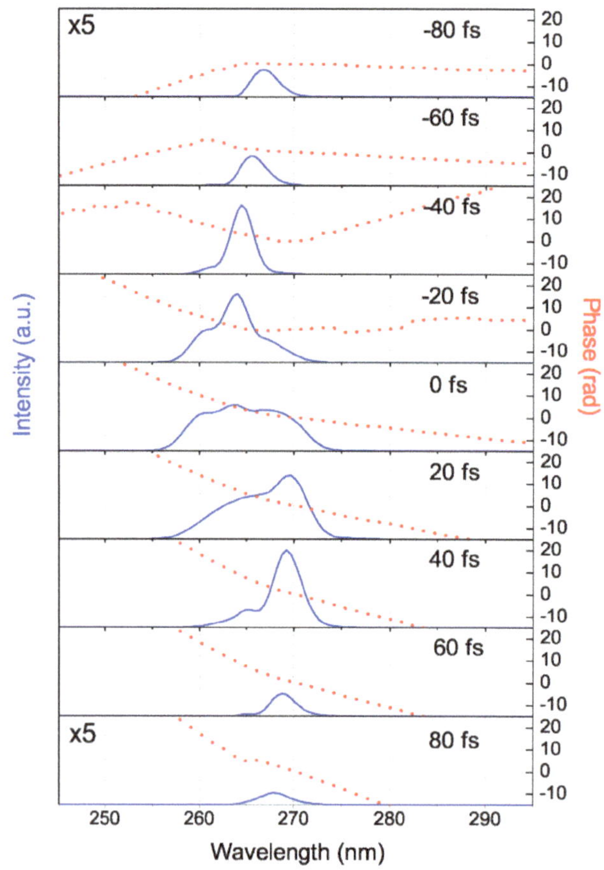

Figure 27: Simulated spectral intensity and phase of the generated DUV signal for θ = 15.7° and different delays between idler and pump pulses. (negative and positive delays as in Figure 23; note the scale factors for the first and last intensity plots) [28].

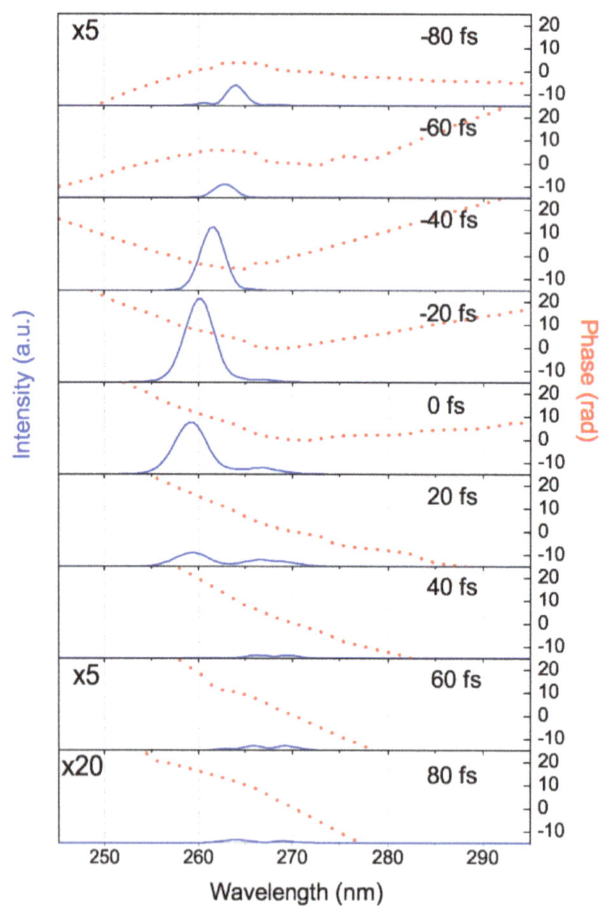

Figure 28: Simulated spectral intensity and phase of the generated DUV signal for θ = 16.4° and different delays between idler and pump pulses. (negative and positive delays as in Figure 23; note the scale factors for the first and last two intensity plots) [28].

Given the rather large angles used between the interacting beams, the generated pulse has angular chirp. Figure 29 shows the angle at which each frequency is generated for the 0 fs delay of Figure 26. We see that the spectrum spans from 255 nm to 280 nm only within 0.5°, which is small enough for many applications, namely those requiring focused pulses.

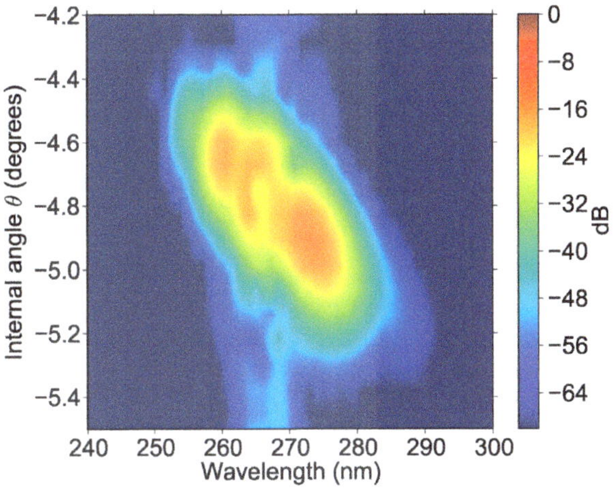

Figure 29: Angular chirp of the generated beam for the case of 0 fs delay in Figure 5. The intensity is given in dB with respect to the maximum intensity of the pump pulses) [28].

Many laboratories nowadays have Titanium:Sapphire amplifiers with pulses longer than those used here. Our model predicts that one can still obtain 42 fs DUV pulses when using 100 fs pump pulses interacting at 15°, with 0 fs relative delay and with the same intensities used in the experiment and the previous simulations, $I_\omega = 2.6 \times 10^{12}$ and $I_{2\omega} = 1.2 \times 10^{12}$ W/cm^{2}respectively. For $I_\omega = 2.0 \times 10^{12}$ and $I_{2\omega} = 2.0 \times 10^{12}$ W/cm^2 29.5 fs DUV pulses were obtained. These results show that it is possible to obtain sub-30 fs pulses starting from 100 fs pump and idler pulses, which might be of great interest in many labs.

Generation of Higher-Order Harmonics by CFWM

The generation of even higher-order harmonics via CFWM, by using a fundamental beam at 800 nm at its second harmonic at 400 nm, was suggested and studied numerically shortly after the first experimental demonstration of highly-nondegenerate CFWM by the same authors [25], where the simplified analytical model further developed in [21] was also presented. Unlike usual non-perturbative high-harmonic generation in gases, which involves ionization and recombination of an electron wave-packet in the presence of an intense laser field, the proposal in [25] was based on a simple perturbative model of CFWM. More recently, Bertrand *et al.* demonstrated that noncollinear high-harmonic generation of fundamental and second-harmonic ultrashort laser pulses can be fully understood in terms of perturbative nonlinear optical wave mixing [36].

CONCLUSIONS

In this paper we presented an overview of our work in ultrafast highly-nondegenerate cascaded four-wave mixing in bulk media, with emphasis on key experimental results as well as on a detailed theoretical model of the phenomenon that can be solved numerically. We described how to perform an experiment where, thanks to the instantaneous third-order nonlinearity of a thin isotropic transparent medium, two noncollinear femtosecond pump pulses with different colors can give rise to a fan of broadband multicolored beams (both frequency upconverted and downconverted) via cascaded four-wave mixing, with macroscopic efficiencies in the range of 5%–10%. The newly generated beams have broad bandwidths, extending from the infrared to the ultraviolet, with each beam corresponding to a particular order within the cascade. A simple algebraic model based on energy and momentum conservation laws enables calculating the optimum interaction angle for generating a particular order *n*, as well as estimating the frequencies of the newly generated beams and their corresponding emission angles.

Other important characteristics such as pulse energy, bandwidth, duration or generation efficiency have been calculated and compared with the experimental results by using a unique 2.5-D nonlinear propagation model developed in our group, which takes into consideration the effects of dispersion, diffraction, self-phase modulation and self-steepening in the SVEA approximation. With this model we have been able to faithfully reproduce all the experiments performed so far. We then showed how the fan of beams produced by cascaded four-wave mixing in thin low-dispersion media, which are mutually coherent and can cover two octaves in bandwidth, can be appropriately manipulated in experimental arrangements and thus be used to synthesize extremely short pulses with durations in the single-cycle regime, which was also corroborated by our theoretical model.

A main characteristic of the generated beams is their broadband spectrum, which can result in pulses shorter than the pump beams. This characteristic has been deeply studied for the efficient generation and measurement of broadband pulses in the deep-ultraviolet region by highly nondegenerate four-wave mixing of fundamental and second-harmonic pulses from a Titanium:Sapphire laser amplifier. Our theoretical model also anticipates the generation of multiple broadband pulses extending into the vacuum ultraviolet by cascaded four-wave mixing in fluorides, using a novel multiple pump beam configuration. All these results help to establish the capability and potential of cascaded four-wave mixing for producing new sources consisting of a set of mutually coherent broadband pulses with different colors, whose characteristics are very difficult to obtain otherwise, based on simple nonlinear media, and which can have numerous applications in ultrafast spectroscopy and other fields of science and technology.

ACKNOWLEDGMENTS

Helder M. Crespo gratefully acknowledges support from the Portuguese funding agency, Fundação para a Ciência e Tecnologia (FCT), through Project Grant PTDC/FIS/122511/2010 and Sabbatical Leave Grant SFRH/BSAB/105974/2015, co-funded by COMPETE and FEDER. Rosa Weigand acknowledges support from Grant GR3/14-910133 (Ayudas para los grupos de investigación UCM). The authors thank J. L. Silva for helpful comments.

AUTHOR CONTRIBUTIONS

Helder M. Crespo performed pioneering work on this subject and both authors further developed it. The authors contributed equally to this paper.

CONFLICTS OF INTEREST

The authors declare no conflict of interest.

REFERENCES

1. Shen, Y.R. *The Principles of Nonlinear Optics*; John Wiley & Sons: Hoboken, NJ, USA, 2003.
2. Eichler, H.J.; Günter, P.; Pohl, D.W. Diffraction and Four-Wave Mixing Theory. In *Laser-Induced Dynamic Gratings*; Tamir, T., Ed.; Springer: Berlin, Germany, 1986; pp. 94–121.
3. Durfee, C.G.; Backus, S.; Kapteyn, H.C.; Murnane, M.M. Intense 8-fs pulse generation in the deep ultraviolet. *Opt. Lett.*1999, *24*, 697–699.
4. Penzkofer, A.; Lehmeir, H.J. Theoretical investigation of noncollinear phase-matched parametric four-photon amplification of ultrashort light pulses in isotropic media. *Opt. Quantum. Electron.* 1993, *25*, 815–844.
5. Akmahnov, S.A.; Martynov, V.A.; Saltiel, S.M.; Tunkin, V.G. Observation of nonresonant six-photon processes in a calcite crystal. *JETP Lett.* 1975, *22*, 65–67.
6. Lichtman, E.; Friesem, A.A.; Waarts, R.G.; Yaffe, H.H. Exact solution of four-wave mixing of copropagating light beams in a Kerr medium. *J. Opt. Soc. Am. B* 1987, *4*, 1801–1805.
7. Hart, D.L.; Judy, A.; Brian Kennedy, T.A.; Roy, R.; Stoev, K. Conservation law for multiple four-wave-mixing processes in a nonlinear optical medium. *Phys. Rev. A* 1994, *50*, 1807–1813.
8. Hart, D.L.; Judy, A.F.; Roy, R.; Beletic, J.W. Dynamical evolution of multiple four-wave-mixing processes in an optical fiber. *Phys. Rev. E* 1998, *57*, 4757–4774.
9. Tan, H.; Banfi, G.P.; Tomaselli, A. Optical frequency mixing through cascaded second-order processes in β-barium borate. *Appl. Phys. Lett.* 1993, *63*, 2472–2474.
10. Varanavicius, A.; Dubietis, A.; Berzanskis, A.; Danielius, R.; Piskarskas, A. Near-degenerate cascaded four-wave mixing in an optical parametric amplifier. *Opt. Lett.* 1997, *22*, 1603–1605.
11. Crespo, H.; Mendonça, J.T.; Dos Santos, A. Cascaded highly nondegenerate four-wave-mixing phenomenon in transparent isotropic condensed media. *Opt. Lett.* 2000, *25*, 829–831.
12. Chin, A.H.; Calderón, O.G.; Kono, J. Extreme Midinfrared Nonlinear Optics in Semiconductor. *Phys. Rev. Lett.* 2001, *86*, 3292–3295.
13. Misoguti, L.; Backus, S.; Durfee, C.G.; Bartels, R.; Murnane, M.M.; Kapteyn, H.C. Generation of broadband VUV light using cascaded processes. *Phys. Rev. Lett.* 2001, *87*, 013601.
14. Lu, C.; Yang, L.; Zhi, M.; Sokolov, A.; Yang, S.; Hsu, C.; Kung, A. Generation of octave-spanning supercontinuum by Raman-assisted four-wave mixing in single-crystal diamond. *Opt. Express* 2014, *22*, 4075–4082.
15. Liu, J.; Kobayashi, T. Cascaded four-wave mixing and multicolored arrays generation in a sapphire plate by using two crossing beams of femtosecond laser. *Opt. Express* 2008, *16*, 22119–22125.
16. Liu, W.; Zhu, L.; Wang, L.; Fang, C. Cascaded four-wave mixing for broadband tunable laser sideband generation. *Opt. Lett.* 2013, *38*, 1772–1774.
17. Liu, J.; Kobayashi, T. Wavelength-tunable, multicolored femtosecond-laser pulse generation in fused-silica glass. *Opt. Lett.* 2009, *34*, 1066–1068.
18. Zhang, X.; Giessen, H. Four-wave mixing based on cascaded second-order nonlinear processes in a femtosecond optical parametric oscillator operating near degeneracy. *Appl. Phys. B* 2004, *79*, 441–447.
19. Li, Y.H.; Zhao, Y.Y.; Wang, L.J. Demonstration of almost octave-spanning cascaded four-wave mixing in optical microfibers. *Opt. Lett.* 2012, *37*, 3441–3443.
20. Zhang, H.; Zhou, Z.; Lin, A.; Cheng, J.; Liu, H.; Si, J.; Chen, F.; Hou, X. Controllable cascaded four-wave mixing by two chirped femtosecond laser pulses. *Appl. Phys. B* 2012, *108*, 487–491.
21. Weigand, R.; Mendonça, J.T.; Crespo, H.M. Cascaded nondegenerate four-wave-mixing technique for high-power single-cycle pulse synthesis in the visible and ultraviolet ranges. *Phys. Rev. A* 2009, *79*, 063838.
22. Silva, J.L.; Weigand, R.; Crespo, H.M. Octave-spanning spectra and pulse synthesis by nondegenerate cascaded four-wave mixing. *Opt. Lett.* 2009, *34*, 2489–2491.
23. Nisoli, M.; De Silvestri, S.; Svelto, O.; Szipöcs, R.; Ferencz, K.; Spielmann, Ch.; Sartania, S.; Krausz, F. Compression of high-energy laser pulses below 5 fs. *Opt. Lett.* 1997, *22*, 522–524.
24. Harris, S.E.; Sokolov, A.V. Subfemtosecond Pulse Generation by Molecular Modulation. *Phys. Rev. Lett.* 1998, *81*, 2894–2897.
25. Mendonça, J.T.; Crespo, H.; Guerreiro, A. A new method for high-harmonic generation by cascaded four-wave mixing. *Opt. Commun.* 2001, *188*, 383–388.
26. Kobayashi, T.; Liu, J.; Okamura, K. Applications of parametric processes to high-quality multicolour ultrashort pulses, pulse cleaning and CEP stable sub-3fs pulse. *J. Phys. B* 2012, *45*, 074005.
27. Silva, J.L.; Crespo, H.M.; Weigand, R. Generation of high-energy vacuum UV femtosecond pulses by multiple-beam cascaded four-wave mixing in a transparent solid. *Appl. Opt.* 2011, *50*, 1968–1973.
28. Weigand, R.; Crespo, H.M. Generation of high-energy broadband femtosecond deep-ultraviolet pulses by highly nondegenerate noncollinear four-wave mixing in a thin transparent solid. *Appl. Phys. B* 2013, *111*, 559–565.
29. Jailaubekov, E.A.; Bradforth, E.S. Tunable 30-femtosecond pulses across the deep ultraviolet. *Appl. Phys. Lett.* 2005, *87*, 021107.
30. Fuji, T.; Horio, T.; Suzuki, T. Generation of 12fs deep-ultraviolet pulses by four-wave mixing through filamentation in neon gas. *Opt. Lett.* 2007, *32*, 2481–2483.
31. Fuji, T.; Suzuki, T.; Serebryannikov, E.E.; Zheltikov, A. Experimental and theoretical investigation of a multicolor filament. *Phys. Rev. A* 2009, *80*, 063822.

32. Kida, Y.J.; Liu, J.; Teramoto, T.; Kobayashi, T. Sub-10 fs deep-ultraviolet pulses generated by chirped-pulse four-wave mixing. *Opt. Lett.* 2010, *35*, 1807–1809.

33. Backus, S.; Peatross, J.; Zeek, Z.; Rundquist, A.; Taft, G.; Murnane, M.M.; Kapteyn, H.C. 16-fs, 1-μJ ultraviolet pulses generated by third-harmonic conversion in air. *Opt. Lett.* 1996, *21*, 665–667.

34. Graf, U.; Fiess, M.; Schultze, M.; Kienberger, R.; Krausz, F.; Goulielmakis, E. Intense few-cycle light pulses in the deep ultraviolet. *Opt. Express* 2008, *16*, 18956–18963.

35. Reiter, F.; Graf, U.; Schultze, M.; Schweinberger, W.; Schröder, H.; Karpowicz, N.; Azzeer, A.; Kienberger, R.; Krausz, F.; Goulielmakis, E. Generation of sub-3 fs pulses in the deep ultraviolet. *Opt. Lett.* 2010, *35*, 2248–2250.

36. Bertrand, J.B.; Wörner, H.J.; Bandulet, H.C.; Bisson, É.; Spanner, M.; Kieffer, J.C.; Villeneuve, D.M.; Corkum, P.B. Ultrahigh-Order Wave Mixing in Noncollinear High Harmonic Generation. *Phys. Rev. Lett.* 2011, *106*, 023001.

INDEX

L

M

N

O

P

Q

R

S

T

U

V

W